Geographical Variation in Waders

Geographical Variation in Waders

Meinte Engelmoer and Cees S. Roselaar

KLUWER ACADEMIC PUBLISHERS
DORDRECHT / BOSTON / LONDON

A C.I.P. Catalogue record for this book is available from the Library of Congress

Additional material to this book can be downloaded from http://extras.springer.com.

ISBN 0-7923-5020-0

Published by Kluwer Academic Publishers,
P.O. Box 17, 3300 AA Dordrecht, The Netherlands.

Sold and distributed in North, Central and South America
by Kluwer Academic Publishers,
101 Philip Drive, Norwell, MA 02061, U.S.A.

In all other countries, sold and distributed
by Kluwer Academic Publishers,
P.O. Box 322, 3300 AH Dordrecht, The Netherlands.

Printed on acid-free paper

Printed in the Netherlands

Contents

Acknowledgements

This book is an accident. It would never have been written if Gerard Boere had not stimulated the production of a publication on the geographical breeding origin of waders in the Dutch part of the Wadden Sea. Such an article appeared to be impossible without a detailed analysis of the existing morphometrical variation in waders. Since then Ebel Nieboer, Rudi Drent, Gerard Boere and Jan Wattel supported us to write it.

We were very lucky to receive additional funding from the Prins Bernhard Fonds, the Koninklijke Nederlandse Academie voor Wetenschappen and the Dutch Ministry of Agriculture Nature and Fisheries. Without these grants it would have been impossible to reach our goal. Jo Rampen and with him the daily board of the Netherlands Ornithologists' Union has been very cooperative in organizing financial matters.

We would like to thank Anne-Marie Blomert, Gerard Boere, Ebel Nieboer, Frieda van Noordwijk (†), and Theunis Piersma for their companionship during travelling and for their assistance with measuring the skins. The Department of Behavioural Ecology of the University of Groningen provided accommodation and logistics for several years.

We are grateful for having had the opportunity to measure the skins in the following collections: Institute of Systematics and Population Biology (Zoological Museum, Amsterdam, The Netherlands), National Museum of Natural History (Leiden, The Netherlands), Frisian Nature Museum (Leeuwarden, The Netherlands), British Museum of Natural History (Tring, United Kingdom), University Zoological Museum (København, Denmark), Zoological Research Institute and Museum Alexander Koenig (Bonn, Germany), Zoological Museum of the Moscow University (Moscow, Russia), Zoological Institute of the Russian Academy of Sciences (St. Petersburg, Russia), Icelandic Museum of Natural History (Reykjavik, Iceland), Zoological Museum (Helsinki, Finland), Swedish Museum of Natural History (Stockholm, Sweden), Zoological Museum (Oslo, Norway), Royal Ontario Museum (Toronto, Canada), Canadian National Zoological Museum (Ottawa, Canada), American Museum of Natural History (New York, USA), United States National Museum of Natural History (Washington D.C., USA), Museum of Comparative Zoology (Cambridge, USA), Carnegie Museum of Natural History (Pittsburgh, USA), Museum of Vertebrate Zoology (Berkeley, USA), Burke Museum (Seattle, USA), Slater Museum of Natural History (Tacoma, USA). We have a good memory for the hospitality of and discussions with Allan J. Baker, Carla Cicero, Let Curvers, Carl Edelstam, Jon Fjeldså, W. Earl Godfrey, Mary LeCroy, Vladimir Loskot, Hans Meltofte, Henri Ouellet, Kenneth C. Parkes, Dennis Paulson, Raymond A. Paynter Jr., Ævar Petersen, Jay Scott, Sievert A. Rowher, Pavel Tomkovich, Risto Väisänen, Scott Wood, Chris Woods.

David Agro, J. Phillip Angle, Ernst Bauernfeind and Robert Prŷs-Jones provided us with details of the type specimens present in the collections of respectively Philadelphia, Washington, Vienna and Tring.

Support with the statistical analyses was obtained from Prof. Dr. Willem Schaafsma and Douwe van der Sluis. We grateful to Dick Visser for drawing all the maps and some figures. Timo Mank produced the front cover and the figure with axillary feathers of Bar-tailed Godwits. Pavel Tomkovich provided us with much additional information on the localities of nesting waders in Russia and/or Siberia. Elena Lebedeva helped us with translating Russian texts. Anne-Marie Blomert, Ebel Nieboer, Pavel Tomkovich, Ron Mes, Leo Zwarts, Joop Jukema and Gerard Boere commented on various parts of the manuscript. We gratefully acknowledge their critisism. Minke Kuipers typed the text of Annex 3: also thanks! Anne Morrissey corrected our english in the introductory chapters and tried to learn us the punctuation rules in english.

Finally and above all we thank our families and friends for their support in a great variety of ways. Their moral help has always been very important.

During this study training samples were constructed, which contained morphometric information of breeding populations. These samples were used for calculating the degree of resemblance of individual birds with various breeding populations. The calculations were based upon the computer package POSCON. Both the POSCON program and the training samples are included on diskette in this book. The methods and procedures of the POSCON program are presented in Annex 3.

Chapter 1

Introduction

Most Holarctic waders fly huge distances during their seasonal migrations. For instance, Alaskan Bar-tailed Godwits winter as far south as southern Australia. The southernmost wintering area of Red Knots breeding in N Canada is along the coasts of Patagonia in S America. Sanderling wintering in S Africa breed on the Siberian Taymyr Peninsula.

These migrations cause intermixing of different breeding populations during migration and wintering.The wintering populations of Turnstones on the Banc d'Arguin in Mauritania might serve as an example, since these wintering flocks include both N European and NE Canadian breeding birds. Such a process of intermixing is already long known and its occurrence obscures our knowledge of the migratory habits of distinct populations within the various wader species. This knowledge is essential for our understanding of wader biology and is necessary for the development of the proper conservation measures. An evident example is the estimation of population sizes, used to trace the key-sites for wintering waders. Population estimates are generally based upon the counting results of wintering populations, since waders nearly always winter in large concentrations in a limited number of areas, whilst they breed in low densities dispersed over huge areas. If accurate estimates of the population composition of wintering populations were available, we would have better estimates of total population sizes. It would then be much easier to identify the key-sites during wintering and migration.

Theoretically, migratory movements and wintering habits of waders can also be clarified with ringing information, but many waders breed and winter in areas without human settlements. Therefore, the migration patterns as obtained from ring recoveries are often biased: most recoveries originate from migration grounds with many human beings living around.

In the early 1980's we tried to find other ways to estimate the composition of wader populations during migration and wintering. The solution was sought in analysing morphometric information since there is considerable geographical variation in measurements within the various wader species. The research was restricted to the morphometric variation within breeding populations, since genetical information is only passed to a next generation on the breeding grounds and not on the wintering sites, even though different selectional forces in the various migration flyways and on the various wintering grounds have just as much their influence on the shape of waders. Moreover, there is a high degree of inheritance of body dimensions from parent to offspring (details further on in introduction) and waders are generally considered to be more site faithful to their breeding grounds as to their wintering area.

Once the attention was focused on the morphometrics of breeding populations, we went to the museum collections where many skinned birds from a great variety of breeding areas are stored. Since 1983 a database was filled containing the measurements of 4.950 birds. It was not until 1995, that this database was large enough to analyse geographical variation.

This book not only aims to describe the morphometrical variation in 15 wader species in relation to their systematics, but also provides tools to predict an individual's breeding origin. These predictions are generated with the 'POSCON' computer package and the corresponding samples of breeding populations collected during this study. The information presented in this publication concerns:

- A description of the geographical variation in morphometrics of Holarctic wader species.
- An evaluation of the potential quality of predicting the breeding origin of migrant and wintering waders as based on external measurements.
- Identification of those external measurements which increase the reliability of these predictions.

The description of the geographical variation within a species resulted in the construction of a typology per species and with this typology basically 4 different lines of analysis were followed (Fig. 1):

- so-called 'POSCON' samples were constructed in order to predict a bird's breeding origin or sex (explained in the section on the prediction of geographical breeding origin').
- the described geographical variation was taxonomically interpreted.
- the resulting pattern of phenotypic variation was compared with known patterns of mito-

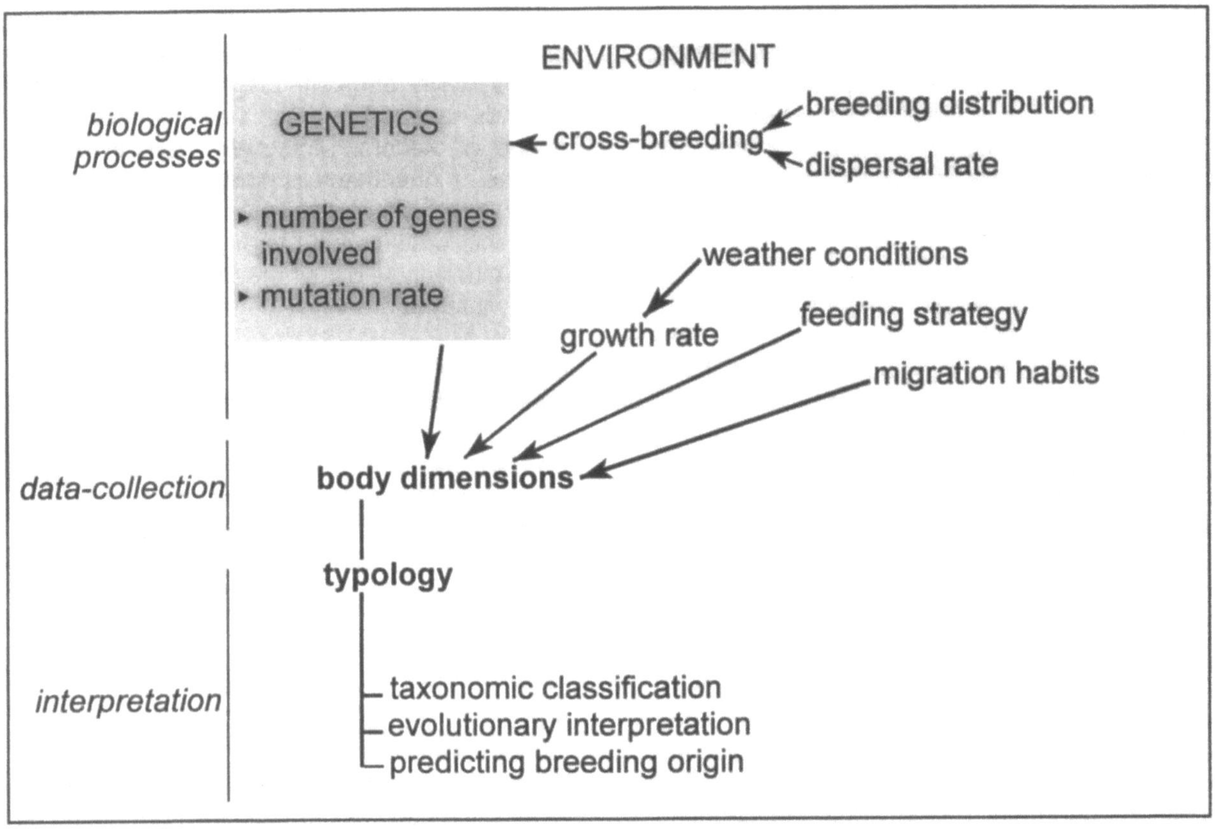

Fig. 1. *Schematic representation of the factors influencing body dimensions and the lines of analysis following the construction of a typology.*

chondrial DNA variation (Dunlin, Knot and Whimbrel only).

- the morphometric typologies -together with other information- were interpreted in evolutionary terms.

Some last remarks concern the special features which are associated with this sort of morphometric studies. Firstly (sub)arctic species breed in low densities in areas which were only visited by a limited number of ornithologists. This occasionally results in small sample sizes. Secondly most waders are long-distance migrants. Even when a bird is collected in (sub)arctic regions, it may be a migrant, a non-breeder, or a local breeding bird. Due to a shortage of specimens, former workers often needed to analyse the measurements of non-breeding birds. Thirdly the reliability of statistics used in older publications is doubtful because of the lack of standardization in the way measurements were taken and because of a lack of reproductibility studies. Even nowadays, methods of measurement are not always sufficiently standardized.

SKIN COLLECTIONS

The habit of collecting and preserving birds and eggs started in the first half of the 19th century (Fig. 2). The major purpose of these collection activities was to describe the richness of nature, but they were just as often a goal in themselves like present-day stamp-collection activities. Preserved birds made it easy to compare taxa from completely different regions. Comparison was virtually impossible with live birds. It was also well understood, that the best proof of a species to be claimed as new for science was its skin. In the second half of the 19th century much effort was made to find and collect the eggs (preferably with nest and bird) of arctic breeding species. Vaughan (1992) excellently described the reasons for collecting birds and eggs by collectors and ornithologists in the Arctic in times past. Most skin- and egg-collections were privately owned, and a vigorous exchange of items occurred between collectors. Many of these private collections were donated to public zoo-

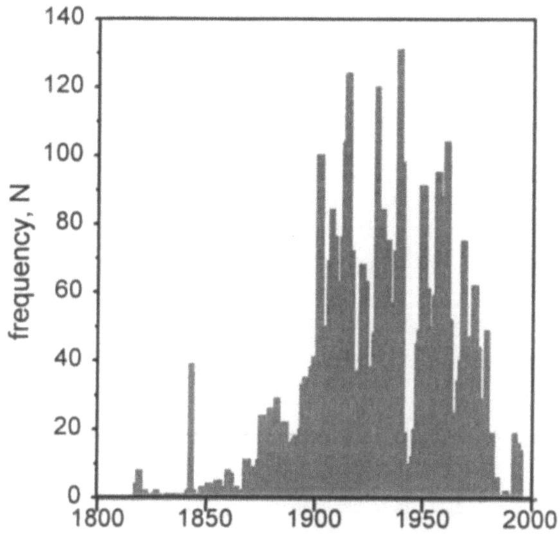

Fig. 2. *The number of skinned breeding birds collected since 1800 and processed for this study (n = 4.810). Another 136 skins with unknown year of collection were measured.*

logical collections in later years. At the turn of this century, when most species and their eggs were well-known, interest in bird collecting declined. However, with skins from more areas available to ornithologists, it appeared that species were not as uniform in appearance as previously thought, but that geographical variation occurred within many species. The introduction of the subspecies concept provided a framework for this variation, and led to many new expeditions by professional taxonomists (Kleinschmidt 1900, Ridgway 1919 and Hartert 1912-21). With the appearance of the 5th edition of the A. O. U. Check-list of N American Birds (American Ornithologists' Union 1957), of Vaurie's standard work on the geographical variation of Palearctic birds (Vaurie 1959, 1965) and of Mayr's work on systematics and geographic speciation (Mayr 1942, 1951, 1963) the variation of Holarctic birds was by most people considered to be fixed. The interest in bird taxonomy at the subspecies level declined. Nowadays, skins are often only made as a by-product during ecological studies and birds are usually trapped and measured alive. The information resulting from these collection and trapping activities has never been combined on a Holarctic scale. This is because of:

- the problems concerning data ownership,
- the former political circumstances, and
- the data from the literature has long been regarded as sufficiently adequate to make inferences about the resemblance between populations.

We collected most data from the collections of St. Petersburg, Ottawa, Copenhagen, New York and Tring (Fig. 3).

In the past collection activities were rather peaked, since the remote arctic regions were only visited by irregularly organized expeditions. Our data illustrates well the survey of arctic explorations presented by Vaughan (1992). The activities of A.F. Middendorff into Taymyr in 1843, the expeditions of Kleinschmidt to E Siberia in 1901, the Danish expeditions of A.L.V. Manniche to NE Greenland between 1906 and 1908, Sutton's activities in the Canadian Hudson Bay in the 1930's, and Portenko's research on Wrangel Island (E Siberia) in 1939 are clearly reflected in our database. Most waders have been collected during the first part of the 20th century (Fig. 2). The collection activities strongly declined during both World Wars. After the Second World War collecting activities in the Arctic never recovered to pre-war levels but instead became increasingly uncommon. A regional presentation of the data (Fig. 4) showed that:

- scientists had a strong interest in the Siberian Arctic long before biological expeditions were sent to Greenland or NE Canada;

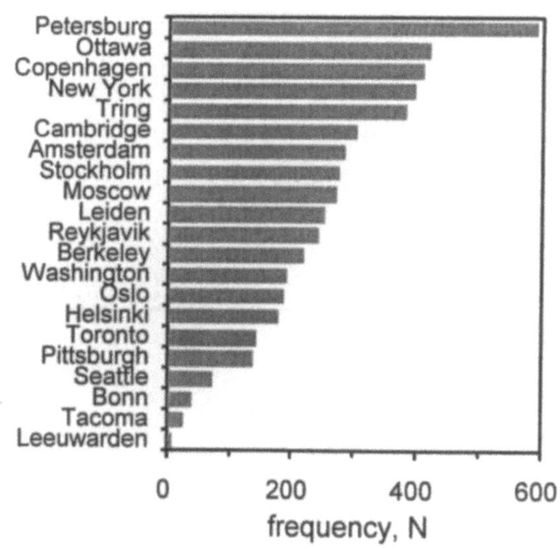

Fig. 3. *The number of breeding birds (n= 4.946) measured in the various museum collections.*

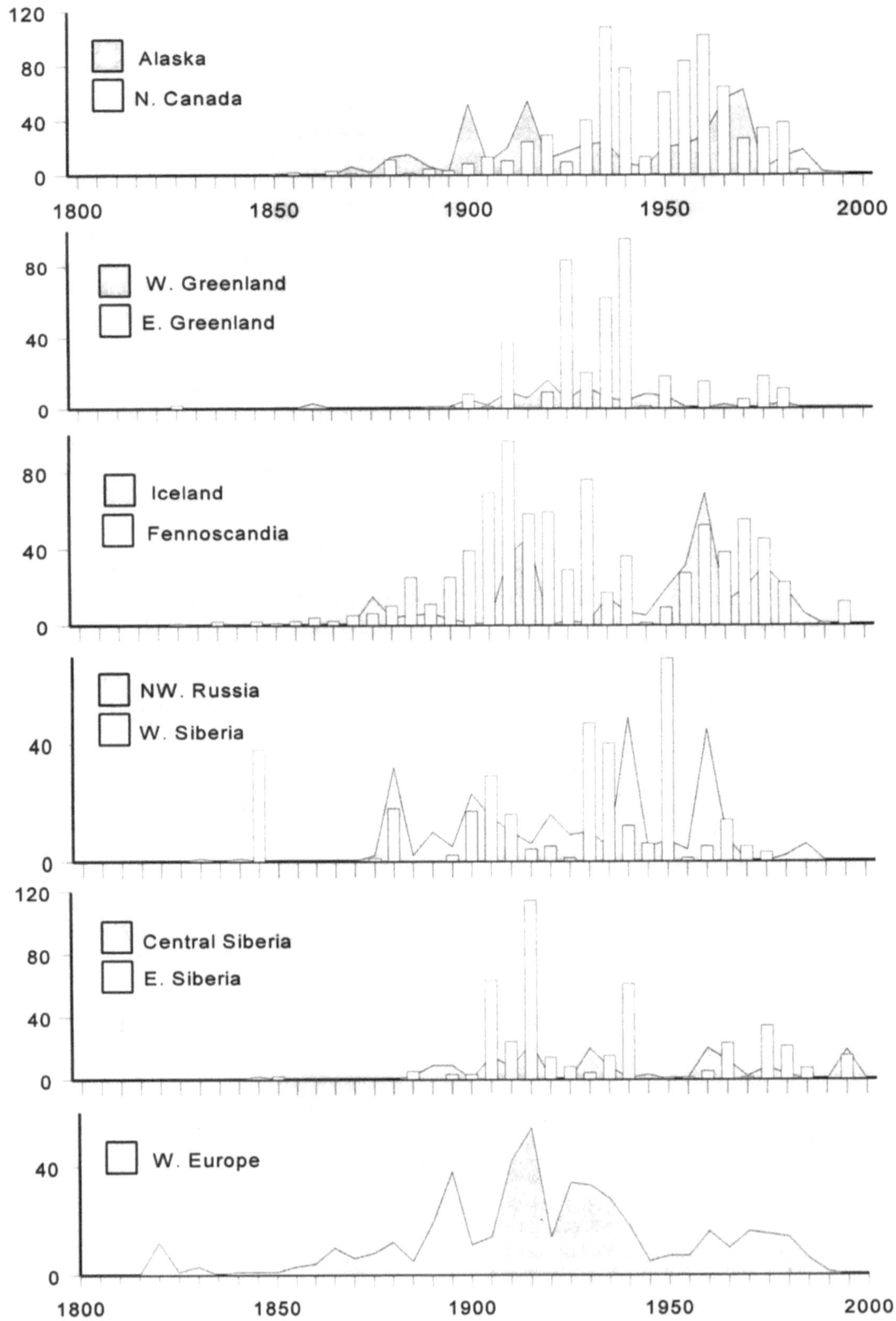

Fig. 4. *The number of skins used in this study and collected in various breeding regions since the start of the 19th century. These delineation of these regions is shown in Fig. 5.*

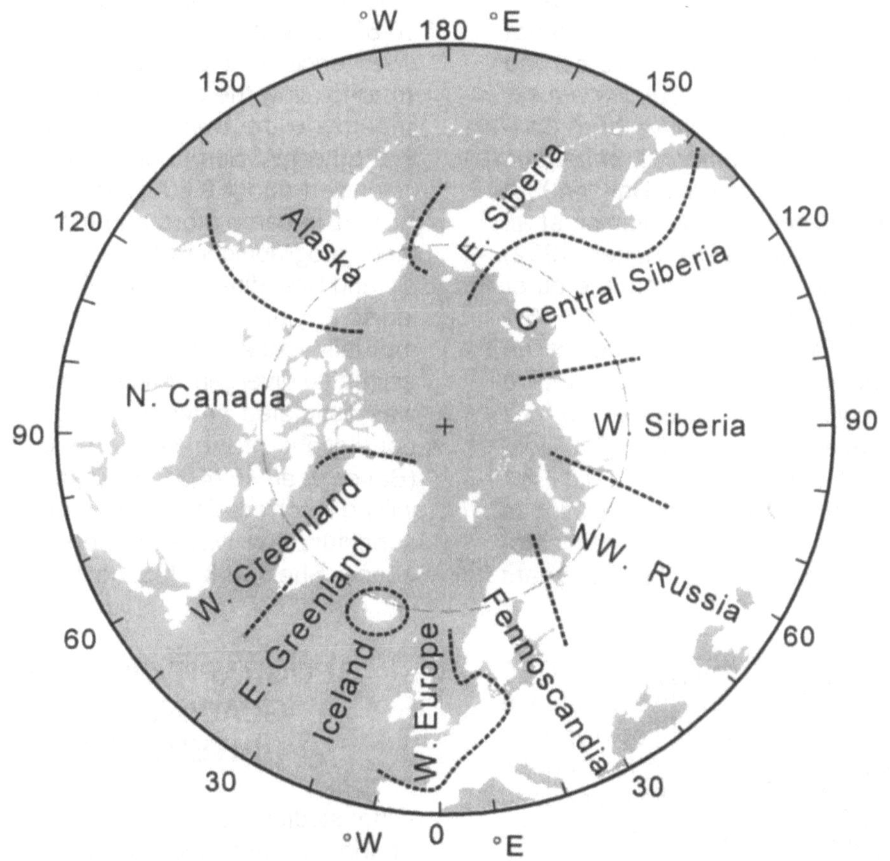

Fig. 5. *Delineation of the various breeding regions as presented in Fig. 4.*

- biological expeditions reached Alaska earlier than Arctic Canada;
- breeding grounds in W Greenland received earlier attention from ornithologists than those of N and E Greenland. Most skins from Greenland had been collected in the first half of the 20th century;
- the collection activities on Iceland, in Scandinavia, and in W European countries became structural after 1850, and reached high values prior to the First World War and during the period between both world wars;
- the Russian breeding grounds were thoroughly sampled after 1950, though collection activities had already been conducted in the early 19th century. Activities in E Siberia peaked after 1900.

THE PHYLOGEOGRAPHY OF WADERS

Archeopteryx must have lived in the Upper Jurassic about 140 million years ago (De Beer 1954 in Eldredge 1989). Since then many processes have shaped the dimensions and coloration of birds. Both genetic and environmental processes are intricately entangled and have lead to a huge variety of bird species. The Charadriiformes are amongst the eldest known orders (Carroll 1988, Olson 1988) with fossils found from the late Cretaceous (70 million years ago). Olson (1988) summarized the known age of the 'modern' families within the Charadriiformes:

- fossil remains of the Recurvirostridae date back to the middle Eocene (45 million years),
- the Scolopacidae to the late Eocene (40 million years),
- the Glareolidae to the early Miocene (20 million years), and
- the Haematopodidae to the early Pliocene (5 million years).

The evolutionary age of the Charadriidae is unclear, since Olson (1988) doubted whether the remains of the oldest fossils belonged to this group.

Estimating the evolutionary age of species or subspecies from fossil records is difficult, if not impossible. Recent DNA-studies were carried out in certain species: Dunlin *Calidris alpina*, Purple Sandpiper *Calidris maritima*, Knot *Calidris canutus*, Whimbrel *Numenius phaeopus* and Turnstone *Arenaria interpres* (Baker 1992, Baker *et al.* 1994, Wenink 1994, Wenink *et al.* 1994 & 1996, Zink *et al.* 1995). Baker (1992) showed the divergence of Knot from the other sandpiper species to be the oldest one: 4.6 - 6.1 million years ago. The evolutionary ages of the other *Calidris* species treated in this book were estimated as follows: Curlew Sandpiper 3.9 - 5.2 million years; Purple Sandpiper 2.5 - 3.3 million; Sanderling 0.9 - 1.2 million; and Dunlin 0.8 - 1.1 million. The most recent process of speciation in the genus *Calidris* has taken place in the Nearctic and/or E Palearctic about 5 to 7 million years ago.

Wenink *et al.* (1993, 1996) analysed the mtDNA patterns in Dunlin and were able to estimate the evolutionary age of the different haplotypes in this species. The same was done for Knot by Baker *et al.* (1994). However, caution is needed with these estimates, because the estimated divergence times were based upon the analyses of Shields & Wilson (1987) and Quinn (1992), having found that the mtDNA for control region I of geese diverged on average 20.8 % per million years. Sheldon & Bledsoe (1993) mentioned 2 important weaknesses of this estimate: (1) it is only based on a single regression estimate (8-9 % divergence per 4-500.000 years, thus leading to a range of 16-22.5 % per million years); and (2) there is no ground to expect different taxa of birds to share equal mtDNA evolutionary rate patterns. They mention several studies revealing substantial differences in sequence-level evolution of mtDNA between and within classes of vertebrates. Moreover, extrapolation on this time scale without due attention to the confidence limits of the underlying average estimates, is easily resulting in differences of 10.000's of years.

Calculation of the divergence times within Dunlin produced the following estimates (Wenink *et al.* 1996 with inclusion of the comments of Sheldon & Bledsoe (1993)): (1) the divergence of the Canadian breeding population 280-200.000 years ago (average 223.000), (2) the N European 145-105.000 years ago (average 117.000) and (3) the divergence between the Central Siberian, E Siberian, and Alaskan populations 90-65.000 years (average 75.000). More recent divergence was also observed, but was too small to allow the distinction of 6 or more major lineages in the Dunlin: the Icelandic as well as the other W-European breeding sites were colonized about 9.000 years B.P. when the N Atlantic became more and more ice free. Since then, geographical isolation between the Icelandic and Scandinavian breeding populations has led to 2 clusters of haplotypes, being almost mutually exclusive (Wenink *et al.* 1996). This estimated divergence time is about the same as was estimated between the different breeding populations of the Knot: 10-20.000 years ago (Baker *et al.* 1994). Of course these estimates will be incorrect if the rate of mtDNA nucleotide substitution in waders differs from the one of geese (Sheldon & Bledsoe 1993).

GEOGRAPHICAL VARIATION, GLACIATIONS, AND INTERSTADIALS

Older studies often used the selecting influence of the last glaciation in order to explain the present-day geographical variation (Johansen 1956, Moreau 1972, Ploeger 1968, Hale 1980, Hale 1984, Prater 1981, Greenwood 1986, Piersma 1994). This glaciation was considered to be the last major sieve on geographical variation. The aim of these studies was to locate so-called 'refugia': breeding areas where a population succeeded in producing offspring. These studies were based upon an extent of glaciation as presented by Flint (1947), Frenzel & Troll (1952), Ploeger (1968), Frenzel (1968), Denton & Hughes (1981). Recently Frenzel *et al.* (1992) made a major effort to combine all available knowledge in order to map the climatological conditions since the last interglacial (Fig. 6). But the discussion concerning the impact of the last glaciation on geographical variation became somewhat 'out of fashion', since so much argument had to remain unproven. Baker (1992), Baker *et al.* (1994), Wenink (1994), Wenink *et al.* (1993, 1994, 1996) also concluded that the estimates of divergence times continue to suggest an important impact by the quarternary glaciations on the patterns of geographical variation in waders as recognized nowadays: divergence nearly always coincided with the onset of a new period of glaciation. Whilst such a conclusion does not only concentrate on the selecting influence of the last glaciation, it also stresses the selecting impact of the different glaciations on the process of geographical variation. This is the process of isolation caused by *cold* circumstances.

There is also an interpretation based on the occurrence of *warm* climates: the selecting influence of warm interstadials. When comparing the estimates on divergence times in Dunlin and Knot from the mtDNA research with information from Paleoclimatical research (Shackleton 1994), divergence mostly coincided with prolonged warm periods (*contra* Wenink *et al.* 1993, 1996), Baker *et al.* (1994) - Fig. 7). Such an alternative interpretation is not new: Larson (1957) already mapped the position of tundra refuges during the last Interglacial (120.000 B.P.). This was done in order to argue, that (1) genetical variation must have been minimal by that time and thus (2) the degree of geographical variation increased during the following and most recent glaciation (Wisconsian/ Weichselien/ Würm glacial 15-75.000 years B.P.). The recent mtDNA results - mentioned above- clearly give evidence to a much older history of the subspecies known nowadays. Thus, we might forget about Larson's second point, but not the first one.

Berthold (1993, pers. comm.) argued in a discussion on the effects of global warming on the phenomenon of bird migration, that there will be no future for arctic breeding species migrating over long distances, since:

- most tundra will rapidly disappear;
- long-distance migrants are at a disadvantage with short-distance or partial migrants when competing for breeding sites, since long-distance migrants generally arrive later at the breeding grounds than short-distance migrants do.

Overleaf:

Fig. 6. *On the following 3 pages three stages in the glacial hitory are shown. On these maps important wader habitats are shown. The delineations of these habitats are interpretations of information presented by Larson (1957), Ploeger (1968), Times (1981), Ruddiman & Wright (1987), Van Rhijn (1991), Frenzel et al. (1992) and Piersma (1994).*
Three periods are compared: (1) 120.000 years B.P. (warmest period during the last Interglacial), (2) 20-18.000 years B.P. (maximum cooling during the last glaciation), and (3) present times.

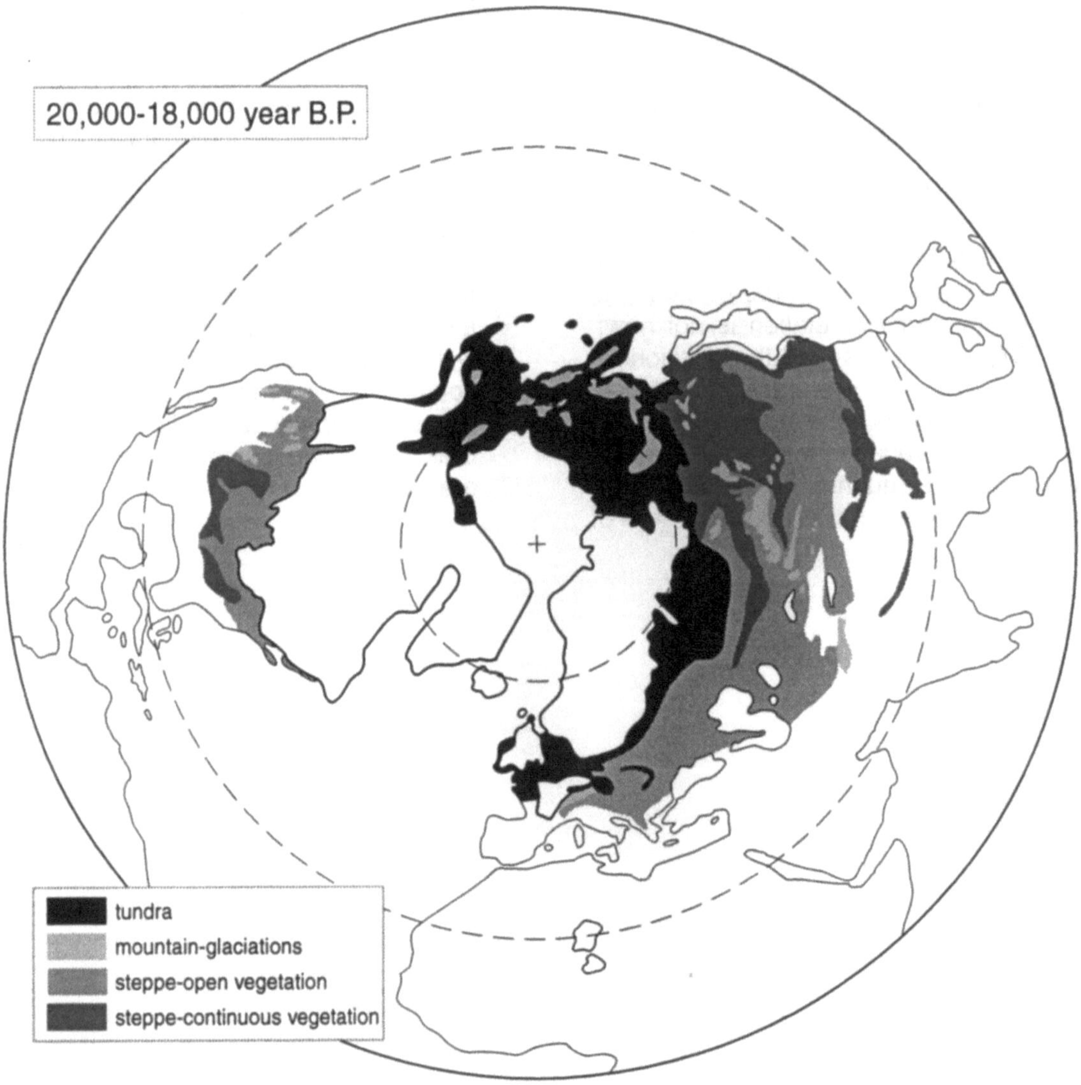
20,000-18,000 year B.P.
tundra
mountain-glaciations
steppe-open vegetation
steppe-continuous vegetation

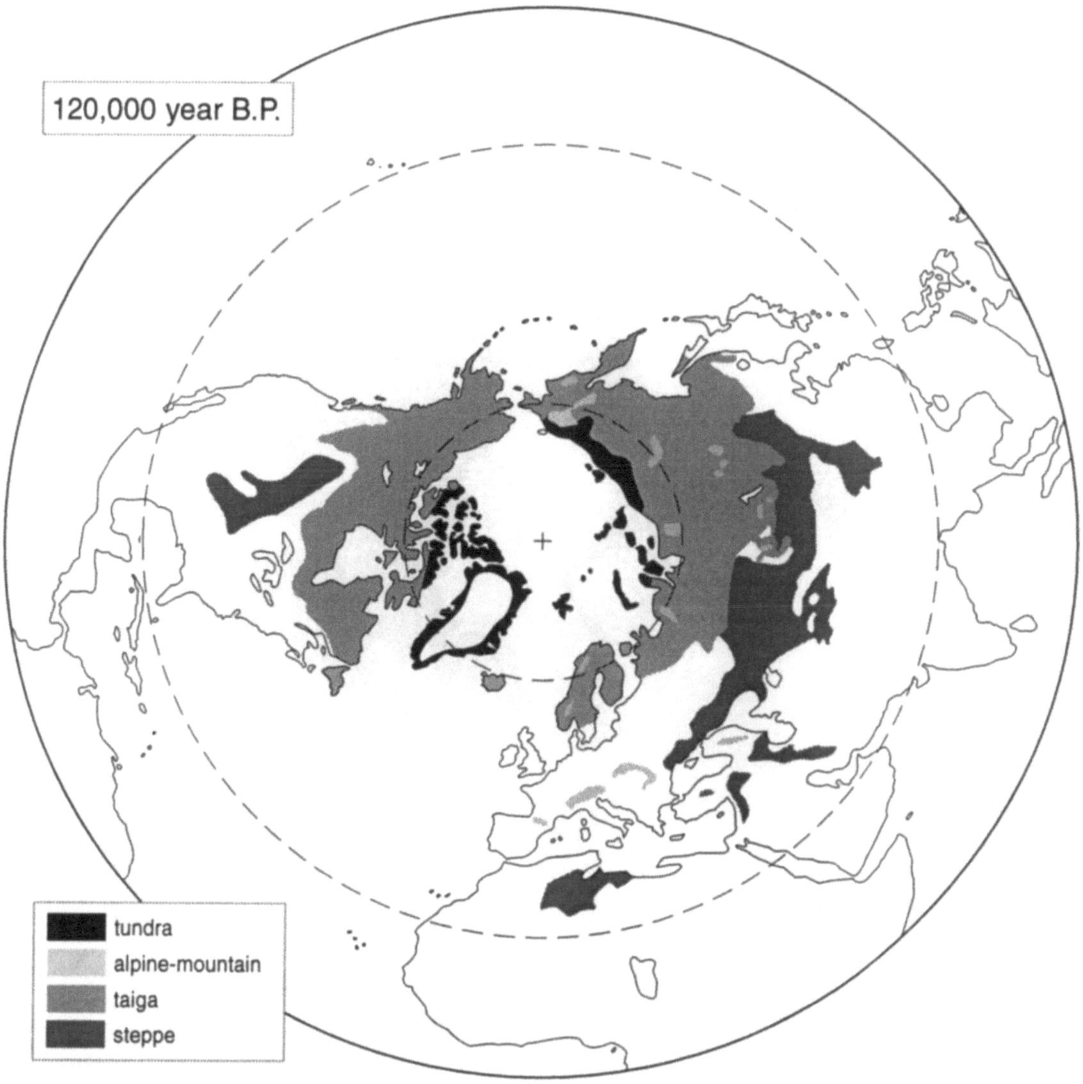
120,000 year B.P.
tundra
alpine-mountain
taiga
steppe

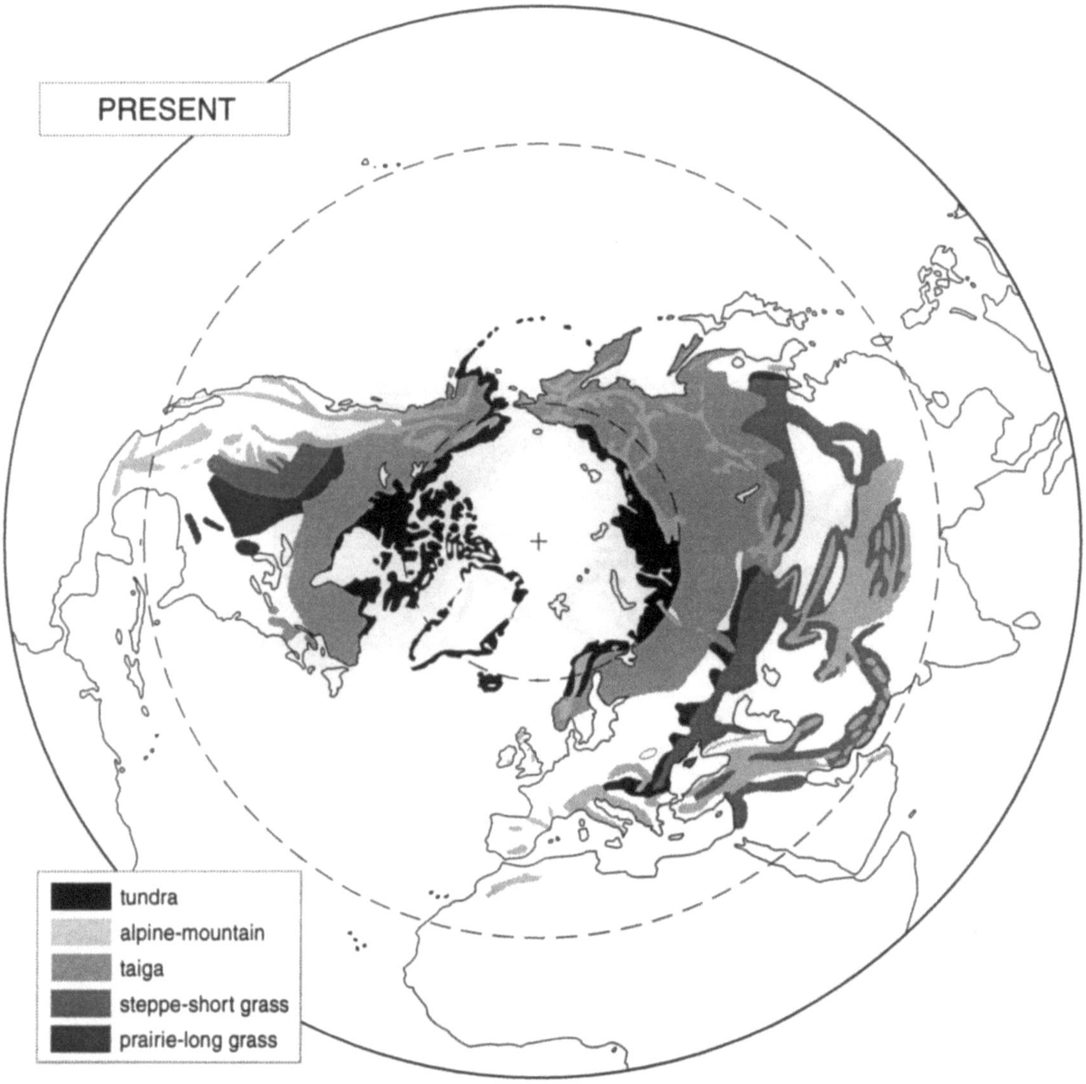
PRESENT
tundra
alpine-mountain
taiga
steppe-short grass
prairie-long grass

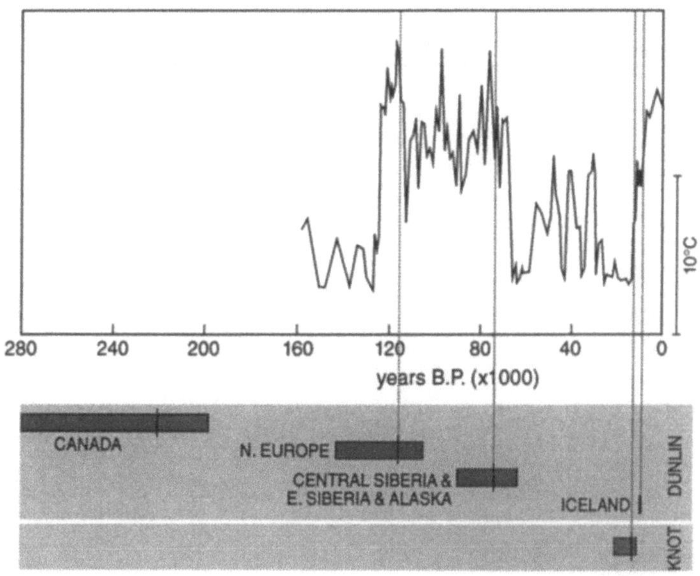

Fig. 7. *Reconstruction of the climate over the past 160.000 years (Shackleton 1994) in relation to the estimated age of different phylogeographic groups in Dunlin and Knot (Wenink et al. 1996, Baker et al. 1994).*

He calculated that it takes 10-30 generations for a middle-distance migrant species to become a short-distance migrant. Adaptation might cost a long-distance migrant about 50-100 generations which would probably be too long with respect to the process of global warming. If this holds, warm periods might well bottleneck the geographical variation in arctic long-distance migrants.

Therefore a straightforward answer concerning the effect of glaciations or interstadials on the geographical and genetical variation cannot be given and it might be better to search for those periods in the geological history having a minimum of breeding habitat available for a certain species. This period might then be regarded as the 'bottleneck' period for that species. The patchiness of this habitat might then reflect the largest degree of geographical isolation. Whatever it be that both the 'cold' and 'warm' interpretation have in common there were nearly no suitable breeding areas for arctic breeders in N Europe and W Siberia. These areas had to be mostly (re)colonized. Some other comments are made with respect to this topic:

- most of the attention of earlier biogeographic studies in waders was focused upon arctic breeding species. These are not the only ones studied here. The present-day breeding conditions of the studied species are presented in Table 1. If these conditions also reflect the historical ones, the delineation of these areas throughout the geological history ought to be known. For that purpose Fig. 6 was drawn and three periods were compared: (1) 120.000 years B.P. (the last interglacial), (2) 20-18.000 years B.P. (maximum cooling of the last glaciation), and (3) the present-day information on vegetation distribution.

- Environmental processes are not always straightforward: ice retreat is not always synonymous with a warmer climate (Williams

***Table 1**. Review of the nesting habitat of breeding waders (Cramp & Simmons 1983, Flint et al. 1984, Uspenski 1986, Godfrey 1986, Hayman et al. 1986, Rogacheva 1992).*

Species	Principally (C)oastal or (I)nland	Polar zone	Tundra	Forest tundra	Taiga	Mixed forest	Temperate marsh and bogs	Temperate saltmarsh and meadows
Ringed Plover	C	+	+	+	+	+	-	+
Semipalmated Plover	I/C	-	+	+	+	-	-	-
Eurasian Golden Plover	I	-	+	+	+	-	+	(-)
Grey Plover	I	+	+	+	-	-	-	-
Red Knot	I	+	+	-	-	-	-	-
Sanderling	I	+	+	-	-	-	-	-
Curlew Sandpiper	I	-	+	-	-	-	-	-
Purple Sandpiper	C/I	+	+	-	-	-	-	-
Rock Sandpiper	C/I	-	+	+	+	-	-	-
Dunlin	I	-	+	+	+	+	+	+
Black-tailed Godwit	I	-	-	-	+	+	+	+
Bar-tailed Godwit	I	-	+	+	-	-	-	-
Whimbrel	I	-	-	-	+	-	+	-
Eurasian Curlew	I	-	-	-	+	+	+	(+)
Spotted Redshank	I	-	-	+	+	-	-	-
Redshank	I	-	-	-	+	+	+	+
Greenshank	I	-	-	-	+	-	+	-
Ruddy Turnstone	C	+	+	+	+	-	-	(+)

& Bradley 1985, after Dyke & Morris 1990). Moreover, a process of glaciation not only involved a huge ice sheet possibly leading to more geographical isolation on the breeding grounds, it also involved extremely low sea levels (and thus different wintering grounds than nowadays), more desertification around the equator, shorter migration routes, and more windy circumstances during migration (Allen *et al.* 1994). It can easily be imagined with our present-day knowledge of the mechanisms regulating population size in waders, that these populations will be affected in various other ways than only on the breeding grounds (e.g., Ens *et al.* 1994).

- It is not clear how a 'long-term effective population size of ♀♀' (Baker *et al.* 1994, Wenink *et al.* 1996) can be interpreted in terms of real population sizes. But if it means that an effective population size of 500 ♀♀ in the Knot once corresponded with a reproductive population of 1.000 adults, then there is a real problem. This because the known breeding densities of Knots vary nowadays between 0.1 - 1.3 pair/km^2 (Meltofte 1985, Ebbinge & Boere 1991), and thus the total area of 'arctic refuge' for these 1.000 Knots might have ranged between 800 - 10.000 km^2. This is only 0.04 - 0.5 % the size of Greenland or 2 - 28 % of The Netherlands. Even the search for the present-day breeding areas of Slender-billed Curlew *Numenius tenuirostris* in Russia, with an estimated population size of 100 - 400 birds (Gretton 1991), is already on going for about 10 years without any success.

THE INHERITANCE OF MORPHOMETRIC CHARACTERS

Life-history characters are inherited in different ways (Fig. 1). When a trait is ruled by 1 or 2 genes in the cellular DNA Mendelian genetics can explain observed differences between groups (e.g., Buckley 1987). However, when more genes are involved, a quantitative approach to genetics is needed for explaining differences found (e.g., Boag & Van Noordwijk 1987). Wild populations are assemblages of different genotypes and this polymorphism affects different processes and characters like migratory patterns, wing lengths, and egg forms. Environmental selection on a polymorphic population results in (1) different fitnesses per genotype and trait, and (2) a huge variety of phenotypes. When taking external measure ments, we are measuring phenotypes: the length of an individual's bill is the product of many (both environmental and genetic) factors. Therefore, in the first part of this section some more attention is paid to the heritability of body dimensions.

The site fidelity of waders to their original breeding grounds is also important when searching for breeding populations with different measurements. If this site fidelity is low, it will be rather difficult to find differences in measurements as far as these have a genetic background. Therefore, literature was scanned on data concerning site fidelity in waders. The results are presented in the second part of this section.

Since morphometric differences between populations and individuals are phenotypic differences, and DNA studies deal with genetic differences, it is quite likely that morphometric studies can result in different conclusions than DNA-studies. Some comparative remarks are presented in the third part of this section.

The heritability of body dimensions

When analysing morphometrics it is important to know whether the measured characters are strongly inherited or whether the environment is strongly influencing the morphological characters of individuals and whole populations. For instance, Fjeldså (1983) and Hulscher (1985) showed the occurrence of processes of relatively rapid change in some morphological characters because of ecological processes. If such processes of character displacement regularly occur, it will be rather useless to characterize populations in terms of their morphometrics. In quantitative genetics the degree of inheritance is expressed in terms of the heritability value (Hailman 1986, Boag & Van Noordwijk 1987). Additionally, repeatability values are calculated in order to *'assess the magnitude of environmental variation resulting from measurement error or other temporary changes in character value between measurements periods'* (cited from Boag & Van Noordwijk 1987). In general, repeatability and heritability values are positively correlated (Fig. 8) apart from some outliers. Heritabilities seem to depend upon the measured repeatabilities. When repeatabilities are high, the heritability

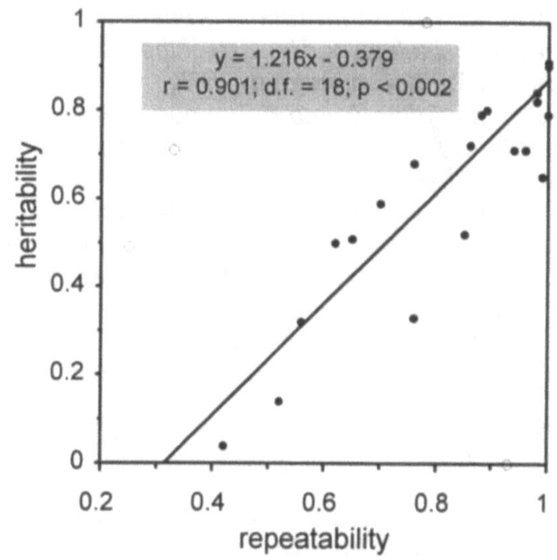

Fig. 8. *Correlation between heritability and repeatability-values (based upon data presented by Boag & Van Noordwijk 1987). Three of the 4 outliers (open circles) concern a study on Coal Tits (Van Noordwijk & Klerks 1983 after Boag & Van Noordwijk 1987).*

values of morphological characters in passerines are above 60 % which indicates a high degree of inheritance from parent to offspring (also Berthold & Querner 1982). There is no information on the inheritance of morphological characters in waders, but there are no specific reasons to assume different values.

Site fidelity in breeding waders

There is an important difference between natal philopatry and breeding site fidelity. Natal philopatry concerns the return rate of adults to the area where they were raised. Breeding site fidelity concerns the return rate of breeding birds to the breeding area of the previous year. An overview of published estimates of natal philopatry and breeding site fidelity (as represented by return rates) is presented in Table 2.

Some remarks are necessary:

- The return rates presented in this table include both site fidelity and survival. It would be better to separate both, but (1) such estimates are scarce and (2) survival estimates strongly depend upon many different assumptions and calculation techniques. Here it was our aim to stay as close as possi-

Table 2. *The return rate of waders to the breeding grounds. Natal philopatry is distinguished from breeding site fidelity. The size of the study area is given in km^2. The nature of the pair bond is presented under (B): M = monogamy; SPg. = serial polygamy; SPa. = serial polyandry; Pg. = polygamy; Pa. = polyandry (after Cramp & Simmons 1983, Hoyo et al. 1996). The parental care is presented under (C) and indicates whether one or both parents raise their chicks. 'St. p.' means the number of breeding seasons being studied. Additional remarks:*

1 *Principally percentage of fledged young, but if the percentage is placed between brackets it indicates the percentage of total number of chicks ringed;*
2 *serial polyandry is sometimes recorded;*
3 *range concerns the difference between unsuccessful and successful breeding birds.*
4 *range concerns the estimates under resp. coastal and inland conditions.*
5 *Parish et al. (1996) report the regular occurrence of polygamy in the Northern Lapwing.*

Species	Natal philopatry		Breeding site fidelity				Size area	Breeding system		St. p.	Source
	%	*n*[1]	♂♂	♀♀	both	*n*		B	C		
Eurasian Oystercatcher	19.7	157			91.3	1982	90	M	2	14	Schnakenwinkel 1970
Eurasian Oystercatcher			90.2	90.2	90.2	1487	106	M	2	14	Safriel *et al.* 1984
Eurasian Oystercatcher					90.0	1407	13	M	2	6	Hulscher 1989
Eurasian Oystercatcher					95.0		13	M	2	10	Ens *et al.* 1995
Eurasian Oystercatcher				90.9-96.1[4]				M	2		Goss-Custard *et al.* 1996
Black-winged Stilt					<30.0			M	2		Sordahl 1984
Pied Avocet	7.3	(149)						M	2	3	Cadbury & Olney 1978
Pied Avocet	3.4	88						M	2	23	Van Impe 1991
Pied Avocet	0.5	(220)			71.4	224	45	M	2	5	own data
American Avocet					29.0	21		M	2		Sordahl 1984
American Avocet	12.1	281	54.0	54.0	53.6	478	1250	M	2	3	Robinson & Oring 1997
Eurasian Golden Plover	59.1	44	78.0	77.0	77.7	112	300-1000	M	2	6	Parr 1980
Grey Plover					35.0	13	1400	M	2	2	Underhill *et al.* 1993
Ringed Plover	57.1	42	100.0	79.0	90.0	40	3600	M	2	4	Pienkowski 1984
Ringed Plover	0.0	(41)			71.4	42		M	2		Bub 1962
Ringed Plover	4.4	(68)			51.9	54		M	2		Laven 1940
Ringed Plover			86.0	57.0	71.1	199	1.6-4.8	M	2	4	Jackson 1994
Little Ringed Plover	1.5	(67)			37.9	29		M	2	3	Bub 1958
Killdeer	0.0	48	56.2	20.0	38.1	31	120	M	2	4	Lenington & Mace 1975
Piping Plover	12.2	90	75.0	56.0	67.7	57	15-30	M	2	6	Haig & Oring 1988a & b
Piping Plover	42.9	28			68.9	16		M	2	3	MacIvor *et al.* 1987
Piping Plover	2.9	35			56.3	16		M	2	11	Pike 1985
Piping Plover	21.0	70			68.0	47		M	2	3	Wiens 1986
Piping Plover	4.8	979			24.6	1173		M	2	20	Wilcox 1959
Piping Plover	6.0	123			55.2	111		M	2	2	Mayer & Ryan 1986
Piping Plover	2.1	48			36.8	19		M	2	2	Cairns 1982
Piping Plover	8.3	12			42.0	14		M	2	2	Whyte 1985
Kentish Plover	5.6	1220			59.3	410	0.7	SPg	1-2	9	Rittinghaus 1956
Kentish Plover			77.3	56.6	66.0	97	var.	SPg	1-2	5	Warriner *et al.* 1986
White-fr. Sandplover			100.0	100.0	100.0	18		M		4	Summers & Hockey 1980
Mountain Plover	0.9	229			62.5	8	1600	SPa	1	4	Graul 1975
Northern Lapwing	48.5	35	95.0	95.0	95.0	183	8.8	M/Pg[5]	2	3	Thompson *et al.* 1994
Spur-winged Plover	20.0	70	100.0	100.0	100.0	119		M	2	5	Barlow 1972

Species	Natal philopatry %	n[1]	Breeding site fidelity ♂♂	♀♀	both	n	Size area (km²)	Breeding system B	C	St. p.	Source
Spur-winged Plover	85.7	14						M	2		Thomas 1969
Sanderling					11-22	8	1400	M[2]	1-2	2	Underhill *et al.* 1993
Semipalmated S.piper	7.5	(375)	47.7	44.1	46.0	522	250	M	2	5	Gratto *et al.* 1985, Gratto 1988
Western Sandpiper	3.0	(185)	57.6	48.8	53.2	104	15	M	2	3	Holmes 1971
Little Stint					1-2	19	1400	Pg	1	2	Underhill *et al.* 1993
Temminck's Stint			93.0	46.0		63	20	SPg	1	9	Hildén 1975, 1978, 1979
Temminck's Stint	>6.0	(134)	43.8	57.6	53.1	49	1.5	SPg	1	3	Breiehagen 1989
Least Sandpiper			65.0	38.0		50		M	1		Miller 1977
Baird's Sandpiper					0.0	6		M	2	2	Norton in Soikkeli 1970a
Dunlin ssp. *pacifica*	5.9	(34)						M[2]	2	2	Holmes 1966
Dunlin ssp. *pacifica*					69.2	13		M[2]	2	2	Soikkeli 1970a
Dunlin ssp. *schinzii*	23.8	(453)	76.8	71.9	74.1	323	60	M[2]	2	8	Soikkeli 1970b
Dunlin ssp. *schinzii*	2.8	(448)	56.4	55.8	56.1	223		M[2]	2		Heldt 1966
Dunlin ssp. *schinzii*	20.1	(144)	89.4	76.8	83.8	396	60	M[2]	2	6	Jönsson 1991
Dunlin ssp. *schinzii*			90.0	65.0	77.5	250	1.6-4.8	M[2]	2	4	Jackson 1994
Curlew Sandpiper					2-5	26	1400	M	1(F)	2	Underhill *et al.* 1993
Stilt Sandpiper			56.0	48.0	44.0	50	4500	M	2	4	Jehl 1973
Black-tailed Godwit	0.0	243			75.0	< 50	450	M	2	5	Jonas 1979
Black-tailed Godwit	1.3	1231			83.3	222	52	M	2	5	Groen 1993
Whimbrel					26-99	51	500-650	M	2	4	Skeel 1983
Whimbrel			78.0	68.0	73.0	110		M	2		Grant 1991
Long-billed Curlew	4.1	(270)	71.4	69.0	69.8	43	1600	M	2	5	Redmond & Jenni 1982 & 1986
Redshank	7.0	186	42.0	34.7	38.3	527	200	M	2	3	Thompson & Hale 1989
Redshank			75.4	69.4	75.0	?	15	M	2	6	Zhmud 1992
Redshank			84.0	67.0	76.0	237	1.6-4.8	M	2	4	Jackson 1994
Greenshank	2.8	71					1500	M	2	7	Thompson *et al.* 1988
Willet					73.1	93	105	M	2	3	Howe 1982
Common Sandpiper	16.7	102	64.1	57.6	61.0	318	300	M	2	13	Yalden 1986, Holland & Yalden 1991
Spotted Sandpiper	16.7	263	29-63[3]	26-63[3]	47.9	152	10	Pa	1	8	Reed & Oring 1993; Alberico *et al.* 1992
Ruddy Turnstone					77.0	17	1400	M	2	2	Underhill *et al.* 1993
Wilson's Phalarope	4.8	314	19.0	5.0	14.6	276	140	Pa	1	6	Colwell & Oring 1989
Grey Phalarope	1.1	93	7.8	0.0	5.8	120	105	Pa	1	3	Schamel & Tracy 1977

ble to the observed field estimates. There is some information on the yearly mortality rates in waders: (1) overwinter mortality of adult ♂♂ Dunlin is estimated at 11 % and of first winter birds 44 % (Jönsson 1988); (2) average winter mortality of adult Oystercatchers 2-10 % (Hulscher 1989, Ens *et al.* 1995, Kersten 1997); (3) Evans (1991) reported mortality rates of adults of 5 wader species between 15 and 24 %. These last estimates were based upon studies in NE England with a relatively high winter mortality and might thus not be representative for a species in general. In all 3 studies 100 % site fidelity was assumed and this might not be correct. In that case survival would be higher than estimated and low return rates would then reflect lower breeding site fidelity.

- Estimates of natal philopatry include the mortality during fledging. It is then important to know whether estimates concern all ringed young or only the number of fledged young. Therefore, a distinction was made between both (Table 2).

- Calculated return rates are time- and scale related. For instance, a high winter mortality affects short time-series more than long ones. Also, a study area of 4.000 ha will result in higher return rates than an area of 4 ha.

Some conclusions were drawn from the combined data in Table 2:

- estimates of breeding site fidelity vary widely within families;
- there are important geographical differences within species as is shown in the Piping Plover *Charadrius melodus* (Haig & Oring 1988b) and in Dunlin;
- ♂♂ nearly always have higher return rates than ♀♀;
- estimates of natal philopatry are positively correlated with estimates of breeding site fidelity (Fig. 9), even though there is a large variation in the upper values.

The third conclusion is important in relation to gene flow within species and especially in relation to the inheritance of mtDNA-patterns from mother to offspring. Gene flow is also increasing with a low degree of natal philopatry No information was collected on dispersal rates, since these will inevitably be largely underestimated: a project on breeding birds concentrates on a certain area and nearly all dispersed birds from that area will remain largely undetected. However, there are some data on long-distance dispersal in waders. Reed & Oring (1993) collected the following maxima: 280 km in Dunlin (Soikkeli 1970a), 595 km in Piping Plover (Haig & Oring 1988c), and 150 km in Spotted Sandpiper (Reed & Oring 1993). Some additions include Oystercatcher 270 km, Ringed Plover 365 km, Redshank 8 km (all database EURING and Dutch Ringing Centre prior to 1982), and Avocet 900 km (Dietrich & Hötker pers. comm.).

Such a gene flow can strongly affect estimates of effective-population size (Payne 1990). Moreover, Grant & Grant (1992) recently showed that 10-20 % hybridization between bird species is quite common in nature. This rate of hybridization in birds is rather high because of a relatively slow rate of protein evolution (Prager *et al.* 1974, Prager & Wilson 1975). Berthold (1993) argued that such hybridization percentages are enough to bring new information in the genome of the pure species, since back-crossing of a hybrid with an individual of one of the parent populations can lead to fertile descendants. On the population level, cross-breeding between different breeding populations is likely to occur more frequently. For instance Wenink *et al.* (1996) found 10-20 % mixture between N European and Siberian Dunlin on the breeding grounds, and on the Yamal Peninsula it was up to 50 %.

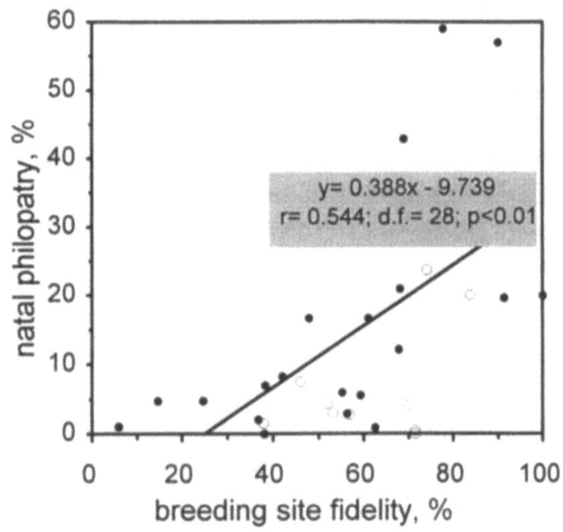

Fig. 9. *Correlation between natal philopatry and breeding site fidelity in waders as derived from the collected data in Table 6. Closed circles concern estimates of natal philopatry based on all ringed pulli, and open circles the ones based on the number of fledged young.*

Morphometrics and DNA research

Environmental selection acts upon the inheritance of cellular DNA but not of mitochondrial DNA. The reproduction of a 'foreign' ♀ in a population will remain visible in the mtDNA of future generations. The body dimensions are ruled by many different genes and their inheritance is coded in the cellular DNA. Cross-breeding is known to lead to body dimensions which are intermediate between the parents (e.g., Berthold & Querner 1982). And thus theoretically inherited DNA-haplotypes have just as little in common with wing lengths as wing lengths have with Mendelian inherited plumage patterns. Wilson & Brown (1953) already mentioned the 4 characteristics of geographical variation in external characters contributing to the difficulties in delimitation:

- most often characters show independent trends in geographical variation;
- the independent reoccurrence of phenotypically indistinguishable populations in widely separated areas - the polytopic subspecies;
- the occurrence of microgeographic races within formally recognized subspecies;
- the arbitrariness of the degree of distinction selected as justifying subspecific separation of slightly differentiated local populations.

In fact, with the introduction of DNA techniques other sets of independently varying characters are introduced as would be the case with e.g., categorical plumage variables.

GEOGRAPHICAL VARIATION AND TAXONOMY

Throughout this text we tried to stay as close as possible to the mensural data in a multivariate environment. The geographical variation within species was analysed by sampling *breeding populations* throughout the breeding range:

- these *samples* were tested with MANOVA;
- identical samples were grouped in so-called POSCON groups;
- predictions concerning breeding origin and sex were generated;
- the qualities of these predictions were tested.

Group distinction depends upon the (number of) measurements taken and the sampled number of birds. A different set of measurements of the same birds inevitably results in different dendrograms and different distances between samples. Likewise, significant differences in measurements between populations are strongly influenced by the sample size. What follows is the interpretation along a gradient from 'nearly no distinction' to 'totally different': local subpopulation (deme), subspecies, species, sibling species or superspecies. Of these, especially the concept of the subspecies has been much debated (e.g., Mayr 1963 and 1982, Amadon & Short 1992). We approved the arguments in favour of the concept, even though it is only a point along the scale of distinction. Our main argument in favour of the distinction of subspecies is a practical one: it is needed to facilitate the communication between ornithologists. Within species, subspecies can nearly never be separated totally from each other. If such was possible, these 'subspecies' ought to be considered as species. Subspecies are not uniform, individuals belonging to such subspecies are characterized by their degree of resemblance with the generalized subspecies type. The distinction of subspecies thus strongly depends upon the measured variables. Amadon & Short (1992) proposed to identify subspecies as '*those spatial subpopulations which are morphologically (phenotypically, and proven or presumed genetically) separable at some reasonable level e.g., 90 %*'. In our text, the normal statistical rules were adopted: a spatial subpopulation needed to differ from the other populations at the 5 % level or more when using MANOVA. This was done since the quality of *prediction* was sometimes too bad when using less restrictive criteria.

PHYLOGENETICS

In this study cluster analysis is used to show the (lack of) differentiation between different populations within a species. Dendrograms resulting from these cluster analyses are not the same as the ones resulting from phylogenetic analysis. Both phylogenetic and such 'UPGMA' trees have in common that they are interpretations of reality. In phylogenetics mostly categorical nonparametric characters are used and a distinction is made between 'ancestral' and 'descendant' characters (Brooks & McLennan 1991). In this study, it was not possible to use such analyses since:

- our variables mostly were normally distributed, parametric values;
- it is rather impossible to find good arguments for the thesis, that e.g. long wing lengths are more 'ancestral' than short ones.

Moreover, the major aim of this study was to search for methods to predict the spatial distribution of wader populations throughout the annual cycle, while using morphometric data.

And for that purpose normally distributed para metric data fit better than categorical non-parametric ones, since the first group of variables make probability calculations more robust than the second one.

Nevertheless, Voous (1992) ventilated a general objection against phylogenetic trees which just as much applies to UPGMA-trees: the probability that a derived tree reflects reality is quite low. This is partly because of the subjective choices needed during the process of analysis, and partly because of the problem that extinct types are nearly always unknown (also Grant 1992). This last objection is far from hypothetical in arctic breeding waders, where a temporary lack of suitable breeding areas -either during glaciations or warm periods- seems to have strongly reduced the genetical variation (Baker 1992, Baker *et al.* 1994). A hypothetical example is drawn in Fig. 10. Here any tree which was

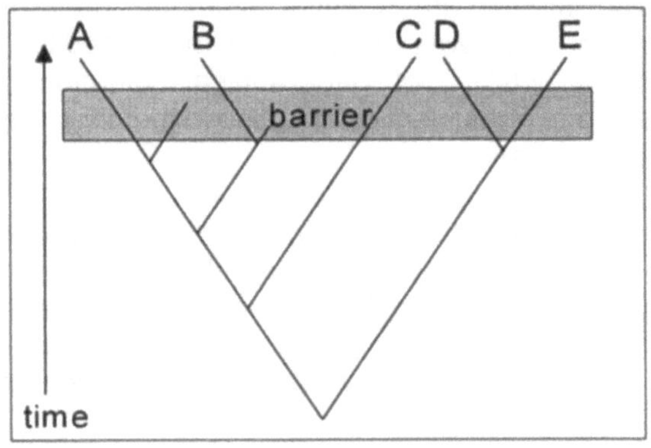

Fig. 10. *Hypothetical dendrogram or phylogenetic tree, showing several phenotypes with various 'evolutionary fates'.*

based upon present-day material, would not reflect reality. And thus the presented dendrograms must be regarded as working tools needed to provide insight in spatial variation.

Chapter 2

Material and methods

THE SELECTED SPECIES

The most frequently measured species were Dunlin , Ringed- & Semipalmated Plover, Turnstone, Redshank and Purple Sandpiper (Fig. 11). This frequency distribution was achieved due to:

- our selection on species occurring in the E Atlantic flyway, or
- their close relationship with one of these species, and
- their relative abundance in the museum collections.

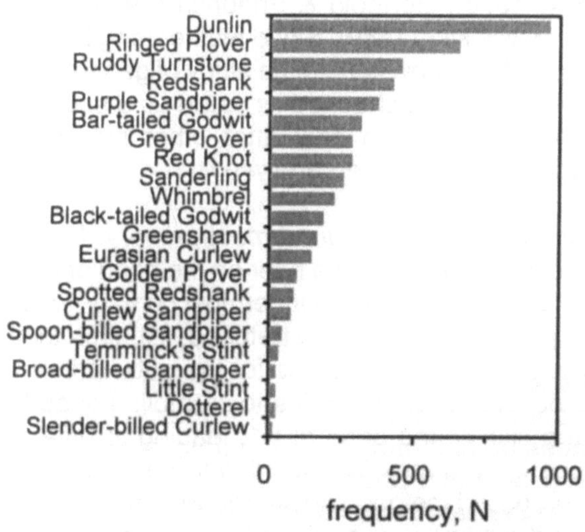

Fig. 11. *The number of skins per species examined during this study. Skins of Semipalmated Plovers are shared under 'Ringed Plover'.*

SELECTING BREEDING BIRDS

In order to avoid the inclusion of (non-breeding) migrants in a local breeding population, the following working method was employed:

- A check was made as to whether the locality on the label of the skin was within the breeding range of the species and whether the bird could be breeding at the moment of collection. If so, the bird was regarded as a 'probable' breeder. If the check failed, then the obtained measurements were not included in the data base on breeding waders.
- Birds were checked on having brood patches. In nearly all species we studied, both ♂♂ and ♀♀ developed brood patches. This check was not performed on the birds examined in the collections of Copenhagen and Tring. Birds collected with brood patches at the right time on the right spot, were considered as 'certain' breeders. Of course, the presence of brood patches are not an infallible proof for breeding on the site of collection: birds may have left their breeding haunts and have covered large distances without having had time to develop new feathers on their brood patches.
- Every note on the label of a skin indicating the presence of a nest, eggs, or chicks was considered sufficient proof of local breeding. However, such notes occurred surprisingly seldom (about 2 % of the labels checked).
- It was possible to incorporate a statistical check afterwards since POSCON groups were constructed for prediction purposes. It could then be tested whether birds were really resembling their own group. When a bird had an extremely low (< 5 %) probability to resemble its own group (in mathematical statistics one uses the term 'typicality probability') and a correspondingly high (> 95 %) probability to resemble another, the bird was omitted from further analysis. Data from very few skins were removed for this reason, since we considered the possibility rather small that birds (especially adults) from one breeding population occur on the breeding grounds of a morphometrically distinct population, especially since (1) most of the involved species are long-distance migrants undertaking long non-stop flights, and (2) only measurements of adults were collected.

The effects of a-typicality on sample means and standard deviations decrease with increasing sample sizes.

THE MEASURED VARIABLES

Measurements are taken in a huge variety: during the 1993 Annual Meeting of the International Wader Study Group up to 40 different measurements were practised by wader ringers. The selection of measurements for this study was based upon 4 criteria. The order of these criteria also reflects our priorities during data collection:

- measurements had to be commonly used by other researchers;
- measurements should not damage (part of) the skins;
- field- and museum data should result in the same values (systematic differences were allowed; see also section on shrinkage);
- newly developed ones needed to be useful for field studies.

The following variables were selected:

The maximum length of the wing. The wing was measured in mm from the bend of the wing to the tip of the longest primary, with the hand part of the wing maximally straightened and flattened against a ruler (maximum chord, Svensson 1992).

Culmen length. The exposed culmen, measured from the feather edge at the base of the culmen to the bill tip, taken with calipers to the nearest 0.1 mm (Fig. 12).

Tarsus length. Measured from the middle point of the joint between tibia and tarsus at the rear of the leg to the middle point of the joint between tarsus and middle toe in front of the leg (Fig. 12), in accordance with the method described by Cramp & Simmons (1977), and thus not conforming with the method used by Prater *et al.* (1977). Taken with calipers to the nearest 0.1 mm.

Feather length of the first secondary. Measured in mm by placing a ruler on the bend of the wing and on the tip of the outermost secondary with this secondary maximally stretched (Fig. 12).

Tail length. The maximum length from the implantation of the feathers to the tip of the longest tail feathers, measured with a pair of dividers to the nearest mm (Fig. 12).

Length of the middle toe. Exclusively measured in birds in which the middle toes were fully stretched after skinning, and measured with calipers to the nearest 0.1 mm (Fig. 12). We included the toe nail, in accordance with Cramp & Simmons (1977), whereas the data presented in Prater *et al.* (1977) had been collected without the nail.

Culmen height. The height of both mandibles taken at the gonydeal angle of the lower mandible, measured with calipers to the nearest 0.1 mm (Fig. 12). Used only in Grey Plover and Ringed Plover.

Feather lengths of the primaries 7, 8, 9, and 10. The primaries are numbered in the standard way with P1 being the innermost and P10 being the outermost primary. The lengths were measured in mm by using a pair of dividers which were placed at the implantation at the base of the feather and on its tip, with the primaries maximally stretched (Fig. 12). These measurements correspond with the ones proposed by Stravinsky (1974, 1976), Berthold & Friedrich (1979) and Jenni & Winkler (1989). Since feather length is a comparatively new item in the study of geographical variation, we checked its value as a source of extra information only after other morphometrical data had been analysed.

Feather length of the innermost primary. Measured in mm by placing a ruler on the bend of the wing and on the tip of the innermost primary, with this primary maximally stretched. At first the same method was used as for the outer primaries but problems arose due to crumbling of part of the skin at the base of P1, leading to differences up to 3 mm. We decided to deviate from the proposed method and started with measuring P1 in the same way as S1 (Fig. 12) considering that:

- the tissues at the base of P1 might have shrunken due to the drying process in the museum. Comparing data from museum skins with data from living birds might cause problems due to this phenomenon, just as is the case with the maximum wing length from museum specimens.
- it was not at all obvious as to whether the use of individual feather lengths in a wing formula was just as useful for waders as it is for passerines.

Whiteness score of the 7th primary. The length of the white vane along the leading edge of the 4th outermost primary (P7) of the Dunlin (Greenwood 1986). Greenwood based the definition of this character on the results of Horvath & Keve (1956), using a 1 to 13 scale for quantifying the amount of white. As the Dunlin was not the only species we dealt with, a scale with larger intervals was constructed, based on the estimated

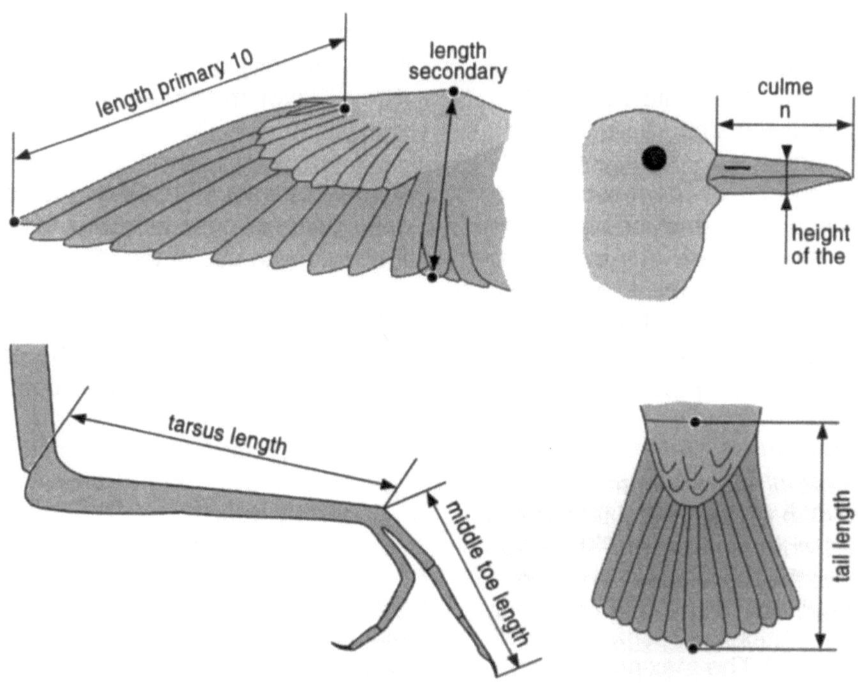

Fig. 12. *Illustrating some of the measuring techniques.*

length of the white vane in relation to the total length of the feather (Fig. 12). The meaning of the codes is:

'1' no white vane,
'2' a trace of white,
'3' more than a trace, but still less than ¼ of the total length is white,
'4' between ¼ and ½ white,
'5' between ½ and ¾ white,
'6' ¾ to almost completely white,
'7' white along the full length of the feather.

Miltenburg (1990) found that wear is strongly influencing the length of the white vane on P7, which makes P7 a useless character to score for birds with old primaries. After we had finished the data collection, correspondence between E. Nieboer and Greenwood lead to the conclusion that Greenwood's 4th outermost primary (P7) had in fact been the 4th innermost primary (P4) (Greenwood 1986; Nieboer pers. comm.). This seems more reasonable since the white vane of P4 is wearing less than that of P7.

Relative broadness of the white zone on the outer vane of the 4th primary. The outer vane is variable coloured white and brown. The maximum broadness of the white zone is related to the maximum broadness of the outer vane. The percentage amount of white is estimated.

Width of the breast band. The maximum width of the blackish breast band measured on the middle of the breast to the nearest mm. Used only in Ringed and Semipalmated Plover.

Last primary with white along its outer edge. The inner primaries of Ringed and Semipalmated Plover have partly white outer edges which are lacking in the outer primaries. The number of the last one having white on the outer edge was scored.

Length of the dusky border between the white patch on primary 5 and the tip of P5. Ringed and Semipalmated Plover have a white patch along the feather shaft of the 5th primary. The distance between the tip of the feather and the distal end of the white patch was measured with calipers to the nearest 0.1 mm.

Height of the white patch above the bill. The maximum height of this white patch of Ringed and Semipalmated Plovers was measured with calipers in mm.

White on P9. The length in mm of the white streak on the 9th primary of Grey Plover measured with calipers along the shaft.

White on P4. The distance from the feather tip to the distal end of the white patch on P4 of Grey Plover measured along the outer vane (not along the shaft) with calipers in mm.

White T1. Width of the first white band (starting from the top) on the innermost tail feather of Grey Plover measured along the inner side in mm.

Upper-tail covert. Based on the colour pattern of the upper-tail coverts of Bar-tailed Godwits and Curlews with use of the following scores:

'1' completely white,
'2' a little amount of brown barring,
'3' barred with equal amounts of white and brown,
'4' heavily barred brown,
'5' very heavily barred brown or almost completely dark,
'6' completely dark.

Axillary score. Nieboer *et al.* (1985) presented 7 different types of brown barring on the white axillaries of the Bar-tailed Godwit. These types are drawn in Fig. 13. Their description is as follows:

'A' feather totally white, occasionally some brown at the base;
'B' Brown spots or bars on only 1 side of the shaft;
'C' brown bars on 1 side and at most half of these at the other side;
'D' brown-barred both sides with the white parts being clearly wider than the brown bars;
'E' brown-barred on both sides, but the white parts are not anymore wider than the brown bars;
'F' brown-barred on both sides with the brown bars on one side fused into longitudinal bands;
'G' bars of both sides fused into wide longitudinal bands.

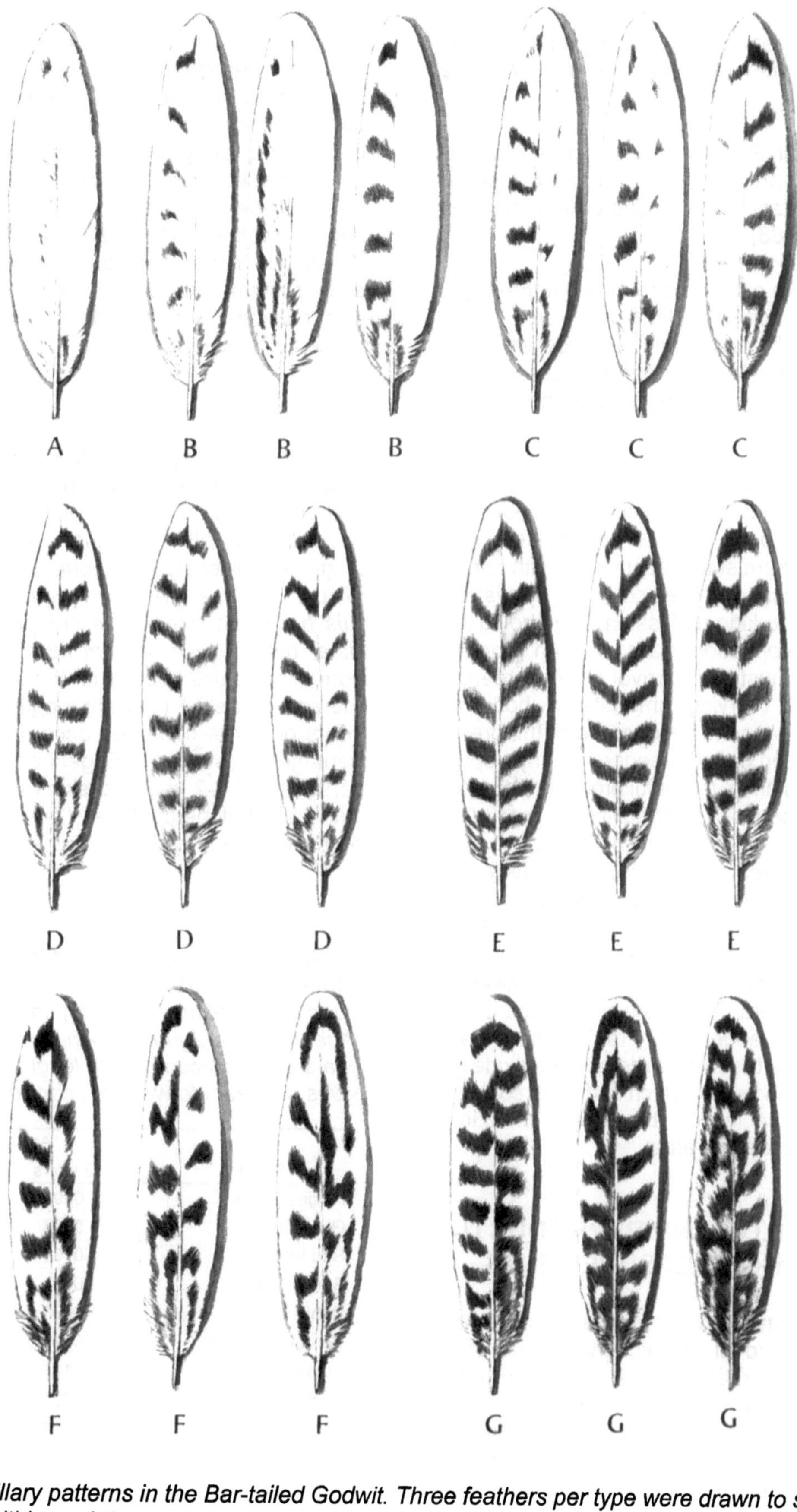

Fig. 13. *Axillary patterns in the Bar-tailed Godwit. Three feathers per type were drawn to show the variability within each type.*

The axillary patterns of Curlew and Whimbrel show the same variation and were scored in a similar way.

Primary moult. Primary moult was described by scoring each primary between 1 and 5 depending on their state of moult (Snow 1970 and Ginn & Melville 1983).

Body mass (in grams) as taken from the labels of the skins, if present. Since very little information is available concerning the body mass of breeding waders, all reported masses on the labels of the skins were noticed. Some remarks are needed:

- *The accuracy.* Although there are many uncertainties concerning the reliability of the label body masses, we are inclined to rely on them since most skins (with known body mass) had been collected by systematically working professional ornithologists.
- The *interpretation.* Mass variation is probably high just after arrival on the breeding grounds (varying amounts of body reserves) or in the second part of the breeding season (failed breeders, nearly departing birds and successful breeders occurring simultaneously). Variability is less in the first part of the breeding cycle, at least when sexes are treated separately: ♀♀ have a different mass cycle than ♂♂ in that period because of egg production. Birds collected during the breeding season can be migrants, non-breeders or (un)successful breeders. Therefore mass data of our 'certain breeders' only refer to data gathered from breeding birds with a fully developed brood patch which had been collected during the regular breeding period in a known breeding area.

RELIABILITY AND REPRODUCTIBILITY OF MEASUREMENTS

In general 2 types of errors are distinguished: systematic and stochastic ones. Systematic errors are due to e.g., wing shrinkage or differences between countries (workers having developed different standards). Systematic errors are especially disastrous when different sources of data are compared e.g., measurements from live birds and those of skins. Stochastic errors are due to intrinsic sources of variability becoming manifest if measurements are repeated by the same observer. The accidental exchange of labels of museum specimen, the making of typing or writing errors by the researcher etc., are also labelled as stochastic errors. Nisbet *et al.* (1970), Atkinson *et al.* (1981), and many others drew attention to inter-observer variation in measurements. Here we will discuss some systematic errors affecting the comparability of museum- and field measurements and will present correction values.

Wing length.

Post-mortem wing shrinkage is a well known process (Engelmoer *et al.* 1983 and authors cited in this paper). Shrinkage seems to continue several years after skinning. The data on wing lengths in this study were corrected for shrinkage (Table 3). *Thus the wing lengths presented in this book can directly be compared with field data, but not with original museum data!* The correction values depend on the species and on the age of a skin: (1) wing lengths of skins > 2 years old were corrected according to the percentages presented in Engelmoer *et al.* (1983), (2) those of 1-2 year old skins were corrected with half that percentage, (3) wing lengths of recently skinned birds were not corrected. When

Table 3. *Correction values used for wing length shrinkage in museum skins. These are based upon data presented by Engelmoer et al. 1983. 'Age' refers to age of a skin and stands for the years after collection. Wing lengths measured on skins < 1 year old were not corrected.*

	Age of skin	
Species	≤ 2 years	> 2 years
Ringed Plover	1.0096	1.0194
Grey Plover	1.0106	1.0215
Red Knot	1.0101	1.0204
Sanderling	1.0081	1.0163
Dunlin	1.0091	1.0183
Bar-tailed Godwit	1.0096	1.0194
Eurasian Curlew	1.0147	1.0299
Spotted Redshank	1.0121	1.0249
Redshank	1.0106	1.0215
Greenshank	1.0086	1.0173
Turnstone	1.0076	1.0152

no correction percentage was known for older skins of bird species not mentioned in Table 3, the shrinkage was assessed according to the following formula (Engelmoer *et al.* 1983):

$$\text{shrinkage} = 0.006 \times \text{wing length} + 0.976$$

The correction percentages did not exceed 3 % of the wing lengths measured.

Culmen length.

The comparability of culmen lengths of skins with those of live-trapped birds still provokes much discussion. In an earlier study we did not find any systematic post-mortem changes in culmen lengths (Engelmoer *et al.* 1983). Later Nicoll *et al.* (1988) proved the existence of a process of post-mortem shrinkage in culmen lengths of Purple Sandpipers, when comparing data of live birds with data collected at the moment of skinning as well as after skinning. The average culmen shrinkage for Purple Sandpipers was estimated to be 0.43 mm (= 1.4 %). We therefore started an additional analysis in order to look for systematic post-mortem changes in culmen length. Five data sets were analysed, including about 25.000 waders and 12 species (Table 4). These data sets included:

- Museum skins from those breeding grounds which probably belonged to the E Atlantic flyway.
- Skins of those migrants and/or wintering birds were selected which had been collected in W European wetlands. About half of all these birds had been collected in the Netherlands.
- Live-trapped waders in the Dutch part of the Wadden Sea. Details in Nieboer (1972), Boere (1976) and Goede *et al.* (1990).
- Waders trapped in Tunisia by expeditions of the Working Group International Wader and Waterfowl Research (WIWO). Details in Van Dijk *et al.* (1986) and Spiekman *et al.* (1993).
- Waders trapped on the wintering grounds in W Africa by WIWO teams during the 1980's. Details in Ens *et al.* (1988), Ens *et al.* (1990) and Altenburg *et al.* (1992).

In order to make a comparison of culmen lengths useful, wing lengths of the museum specimens were corrected for shrinkage as was described in the previous section. Then attention was focussed upon the relation between wing- and culmen length. It was hypothesized that there were no systematic differences between the 5 data sets concerning the relationship between wing- and culmen length.

Wing- and culmen lengths were obviously correlated in all 12 species, certainly when enough data were included. In theory, 120 correlations can be calculated when 2 age classes, 12 species and 5 data sources are available, but in practice, 103 remained because of paucity of

Table 4. *The sample sizes per species and data source as used in multiple regression. Age class 'I' stands for immature and 'A' for adult.*

		Breeding		Non-breeding								
		Museum		Museum		Wadden Sea		Tunisia		W Africa		
Species	Age	I	A	I	A	I	A	I	A	I	A	*n*
Ringed Plover		-	482	41	158	83	71	1	1	28	78	943
Grey Plover		1	261	55	55	172	441	-	36	13	76	1110
Knot		1	205	58	242	565	1003	1	7	117	552	2751
Sanderling		-	228	44	48	9	213	-	1	32	49	624
Curlew Sandpiper		-	68	31	25	104	141	5	17	40	203	634
Dunlin		7	646	304	460	2554	6036	136	629	340	522	11634
Bar-tailed Godwit		-	254	14	110	89	1128	-	-	30	215	1840
Curlew		5	95	52	98	109	289	2	5	-	13	668
Spotted Redshank		-	69	45	21	12	13	-	-	-	-	160
Redshank		4	326	20	40	700	1402	8	62	18	76	2656
Greenshank		4	145	24	24	33	211	1	-	-	1	443
Turnstone		-	337	23	66	174	670	-	27	129	423	1849
	n	22	3116	711	1347	4604	11618	154	785	747	2208	25312

data in the other 17 cases. Samples of more than 50 birds were available in 58 analyses, and 78 % of these correlated significantly (45 cases, with 5 at the 1 % level, and 40 at the 0.1 % level). Thus a relationship between wing- and culmen length is evident.

We then argued that if post-mortem changes in culmen length of skins occur, the wing-culmen relationships in the data sets of skins would systematically differ from the ones of live-trapped birds. A multiple regression analysis was performed with the culmen length as dependent and with wing length, age and data source as the explanatory ones. Ages were grouped in 2 categories: (1) immatures being in their first year of life were scored '1', and (2) adults being more than 1 year old were scored '2'.

The data sources were arranged in 4 designs according to the questions to be answered. Firstly it was questioned as to whether it was necessary to distinguish between the different data sources. If so, it was important to know whether their influence exceeded the effects caused by wing length and age. This is design I in Table 5. It is evident from this table that significant amounts of variation are explained by all 3 factors, with the least explained in Redshank and the most in Bar-tailed Godwit. Evidently the data source explained relatively much of the variation in nearly all species. However, the chosen design could not clarify whether the above shown source effect must be attributed to a systematic difference between museum- and field studies or whether it might be caused by seasonal differences (both museum sources contrast breeding and non-breeding situations) or area-specific differences (the field sources contrast W Europe to W and N Africa). Therefore 3 other designs were introduced:

- In design II a multiple regression analysis calculated the source effect when only both museum-studies were included.
- In design III the same was done with the field studies only.
- In design IV both museum studies were contrasted against the 3 field studies.

***Table 5**. Multiple regression of culmen lengths of waders depending upon 3 factors: wing length, age and data-source. The prime interest was focused upon the effect of the data source under 4 different analyses. In analysis I all 5 available sets of data were included separately; analysis II was limited to the 2 museum samples; analysis III was limited to the 3 field samples; in analysis IV the museum samples were contrasted against the field samples. The multiple regression analyses were performed by forcing all independents into the equation, followed by backward elimination of those variables which did not contribute significantly (P > 0.100). Presented are the T-values per analysis for the data source variable. The last 3 columns show (1) the order of importance of wing length (W), data source (D), and age class (A) for analysis IV; (2) the explained variance in % for analysis IV; and (3) the explained variance for analysis I.*

Species	I	II	III	IV	Order	R^2 IV	R^2 I
Ringed Plover	12.391***	11.314***	2.365*	8.131***	W>D	15.23	22.01
Grey Plover	12.229***	0.175	10.174***	8.320***	A>D>W	15.13	20.56
Knot	21.839***	8.362***	26.307***	3.408***	W>D	2.33	16.42
Sanderling	5.306***	1.063	-1.121	6.400***	W>D>A	28.9	27.49
Curlew Sandpiper	5.017***	1.439	-0.084	5.415***	W>D	11.88	11.32
Dunlin	8.976***	3.821***	-1.063	11.264***	W>D>A	37.68	37.43
Bar-tailed Godwit	5.073***	1.402	3.725***	3.236**	W>A>D	52.7	53.06
Curlew	6.626***	1.687	4.925***	4.662***	W>D	44.16	45.91
Spotted Redshank	-0.956	1.487	-	-1.892	W	15.57	16.07
Redshank	10.462***	0.67	6.839***	8.330***	D>W>A	4.96	6.34
Greenshank	3.189**	3.458***	-1.588	2.212*	W>D	8.64	9.72
Turnstone	23.470***	5.642***	6.529***	24.723***	D>W>A	32.44	30.74

Table 6. *Results of multiple regression analysis in 12 wader species concerning the dependency of culmen length on wing length, data source (museum data scored '1', field data '2'), and age ('1' immature and '2' adult). The regression coefficients of the significantly contributing factors and the constants are presented. The proportional shrinkage and the mean culmen lengths are presented in the last 2 columns.*

Species	Regression coefficient			Constant	% shrinkage	Mean culmen length
	Data source	Wing length	Age			
Ringed Plover	0.592	0.075	-	3.217	4.25	13.94
Grey Plover	0.832	0.042	1.064	17.797	2.8	29.67
Knot	0.37	0.052	-	24.27	1.09	33.87
Sanderling	0.646	0.227	-0.528	-3.799	2.6	24.81
Curlew Sandpiper	1.338	0.224	-	6.083	3.5	38.18
Dunlin	0.712	0.437	0.307	-21.517	2.19	32.59
Bar-tailed Godwit	1.492	0.816	-3.418	-89.667	1.69	88.25
Curlew	5.167	0.884	-	-147.706	3.86	133.82
Spotted Redshank	-	0.258	-	13.651	2.34	57.5
Redshank	0.984	0.038	0.335	33.219	2.34	41.98
Greenshank	0.564	0.137	-	26.948	1.04	54.4
Turnstone	1.453	0.079	-0.156	8.14	6.39	22.73

The data source is important in the museum comparison (design II) for 5 of the 12 species: Ringed Plover, Knot, Dunlin, Greenshank and Turnstone. The same applies to half of the number of species in the field design (design III): Ringed Plover, Grey Plover, Knot, Curlew, Redshank and Turnstone. The probable reason for this difference lies beyond the scope of this chapter. Here it is important to notice that not all the differentiation between data sources can be attributed to the contrast between museum and field studies: the relationship between wing- and culmen length appears to differ in nearly all species when the museum- and field studies are contrasted (design IV - Table 5) except in Spotted Redshank. The lack of contrast in the latter is apparently due to paucity of data. In most cases the data source is the second factor of importance next to the influence of wing length, but before the age effect (Table 5 - last column).

It is concluded from this analysis that culmen length decreases after death, just as is the case with wing length (regression coefficients in Table 6 - data source-column). Since skinned birds were scored a '1' and live-trapped birds were scored a '2', the regression coefficients are also a measure for the absolute estimated difference between both 'treatments'. The estimate varies between 0.37 mm for Knot and 5.17 mm for Curlew. This estimated shrinkage is not related to the size of the culmen ($R^2 = 0.013$; $n = 11$; $p > 0.20$) as is the case with post-mortem shrinkage in wings. A (probably logarithmic) relationship relies too heavily on the estimated shrinkage of Curlew bills. Separate inclusion of the data of both sexes in Bar-tailed Godwit and Curlew did not improve the outcome, since then any degree of relatedness depends solely upon the shrinkage estimate for Curlew ♀♀.

The average shortening of culmen length is 0.90 mm ± 0.40 SD when the Curlew estimate is excluded and 1.28 mm ± 1.34 SD when it is included. Assuming that the culmen-shrinkage percentages are not species specific, one might use one of both mean estimates. In Purple Sandpiper, the culmen shrinkage as estimated by Nicoll *et al.* (1988) amounts to about half the value we reached, viz. 0.43 mm against 0.90 mm, but these values were reached after a period of 6 months while our estimates rely on much older skins. The process of culmen shrinkage might also take some years as is the case for wing lengths (see above). Another explanation might be that shrinkage is species-specific, with the culmen of the Purple Sandpipers shrinking about as little as the one of skinned Knots. This is an argument in favour of using species-specific shrinkage figures. In this study we preferred to use species-specific values on culmen shrinkage. The shrinkage percentages are presented in Table 6. *Thus the culmen lengths used and presented in this study can directly be compared with field data, but not with original museum data!*

Other measurements.

There is no information on post-mortem changes of other measurements of museum skins. Shrinkage possibly plays a role in measuring the first primary- and first secondary lengths. *The lengths of the first primary and the first secondary presented in this study can probably not be compared directly with data obtained from live-trapped birds.*

STATISTICAL ANALYSIS OF THE DATA

The calculations were performed while using basically the SPSS programs versions 5.0 and later (Norušis 1992a, b and c) and the POSCON program (Van der Sluis *et al.* 1991). Generally there were 3 steps of analysis. It started with recognizing the various breeding populations. Most of the efforts were spent to find the proper delineations. Secondly, the measurements of various samples were compared and used to construct a typology. Thirdly, these calculations resulted in the construction of training samples, and the degree of resemblance of individuals with these training samples was calculated. These calculations allowed us to evaluate the quality of the predictions as well as the character of contact zones. These basic procedures are treated in some more detail in the following sections.

THE DISTINCTION OF BREEDING POPULATIONS

The breeding ranges of wader species can be distributed continuously like Redshank, or discontinuously like Red Knot. Discontinuous patterns are easier to analyse than continuous ones since morphometrically distinct populations might well coincide with a pattern of discontinuity. The distinction of (sub)populations within a species with a continuous breeding distribution is more difficult. Since the collection of data had not been random throughout the breeding range, it was natural to distinguish restricted populations accordingly. In order to reach subdivision in the few species with a more randomized data set, 2 additional criteria were used: (1) the degree of heterogeneity within the data set and (2) the possible existence of a plumage character which is characteristic for subdivision.

Several statistical procedures were tried to select and characterize the different samples. The distinction of initial samples was nearly always a cumbersome process involving previous knowledge, plumage characters, distribution patterns, and a detailed analysis of the heterogeneity within the preliminary data sets. Since nearly all studied species display considerable sexual dimorphism, it was essential to construct sex-specific samples.

Special attention is given to those situations where the breeding distribution of different taxa contact. Using POSCON procedures (explained in one of the next sections) it was possible to visualize the composition of such zones. This is of great help in understanding the degree of intermixing and the width of the zone of overlap.

CONSTRUCTION OF A TYPOLOGY

Some multivariate statistical procedures were used to analyse the degree of differentiation between samples:

- *Cluster analysis* was used to cluster the different samples by using the mean measurements per sample. Dendrograms visualize the distance between these samples. Grouping was done by calculating the squared Euclidean distances between samples with a hierarchical technique, based upon the average linkage between samples (Norušis 1992b, Johnson & Wichern 1992). All the dendrograms presented in this book are in UPGMA trees. Cluster analysis was often used in an exploratory way. It cannot be used for classification or prediction purposes, i.e. assigning individuals from unknown breeding origin to any of the a-priori distinguished samples. See also the chapter on 'geographical variation and taxonomy'.

Cluster analysis does not include density estimates around sample means which makes it in our situation less useful than MANOVA-analysis.

- *Multivariate analysis of variance.* Differentiation between samples was tested by using MANOVA (Norušis 1992c). Samples were pairwise contrasted with different sets of measurements and were constructed according to differences between breeding populations and sexes. When it appeared that the distinction between sexes did not play a role of importance in explaining the estimated variation, both sexes were combined. The same procedure was followed when the distinction between geographical samples appeared to be insignificant. The samples were contrasted pairwise in an orthogonal design. Multivariate differences were tested while using Pillai's trace test. The sums of squares were partitioned according to a regression approach. Pillai's criterion was transformed into a statistic with an approximate *F*-distribution.

THE DIFFERENCE BETWEEN 'TYPES' AND BREEDING POPULATIONS

A type is derived from a typology and cannot be regarded as a synonym for a breeding population, since a type is artificially constructed and is an abstraction from reality. A type represents the 'averaged' characters of a population. Also, nearly all breeding populations in the wild are mixtures of different genotypes which implies that polymorphism is the rule and not the exception. This is certainly true for inheritance patterns of characters being ruled by several genes (as is the case with morphometric data). A clear present-day example can be borrowed from the mtDNA studies in Dunlin (Wenink *et al.* 1996). Five haplotypes were distinguished but nearly all breeding populations in Fennoscandia, N Europe, and Siberia had a mixed composition of haplotypes. Even though the haplotypes can clearly be distinguished, it is not the case with the breeding populations.

It is one task to *distinguish* different types, but it is another (and more difficult) to know whether an individual or a non-breeding population *belongs* to any of these types. Firstly, there will always be individuals within a breeding population differing from the generalized type. This polymorphism will result in some birds having higher resemblances to other than the own populations. Secondly, even when 2 populations are called to be statistically different at the 5 % level, there will inevitably be overlap in measurements between both. The distinction at the 5 % level only means that the null hypothesis which states that both samples are drawn from the same breeding population, can be rejected with 95 % certainty. It does *not* mean that 95 % of the birds will be predicted to resemble its own group! Without overlap both groups would be easily recognized as separate species. This phenomenon of overlap in measurements complicates predictions on breeding origin whether it be individuals or populations. How smaller the differences between breeding populations, how worse the predictions will be with respect to these populations.

PREDICTING GEOGRAPHICAL BREEDING ORIGINS

Discriminant analysis is the proper tool to predict a bird's breeding origin. Generally, linear discriminant analysis (LDA) is used for this goal. However, this technique could not be applied to our analyses (Engelmoer 1993, Engelmoer 1995) since:

- The within-group covariance matrices were not equal for the involved samples in about 30 % of the analyses. This mostly happened when 2 or more samples had to be combined because of a too small degree of differentiation between both.
- Measured variables -mostly plumage scores- did not have a normal distribution.
- Measurements were often correlated and regularly strongly correlated.
- LDA requires the estimation of *prior probabilities* (more or less the same as prior knowledge) before calculating *posterior probabilities.* When no estimates of these prior probabilities can be given, they are automatically generated by equalling over the samples e.g., 0.5 in a 2-group sample and 0.125 in an 8-group sample. These automatically generated estimates were often evidently wrong, although no better ones could be given.

Generally, posterior probabilities estimate the probability that a bird *belongs to* a set of samples. We could not go further than to calculate the *degree of resemblance* of a bird with these samples. Once proper estimates can be made of the prior probabilities, it is possible to estimate posterior probabilities.

- The interpretation of posterior probabilities in LDA favoured the selection of a-typical birds. This is illustrated with an example (Fig. 14). The obtained probabilities were clearly not the ones needed: if a bird was rather a-typical for a certain sample, it might have a high probability of belonging to that sample.

In order to solve these problems, the following requirements were formulated:

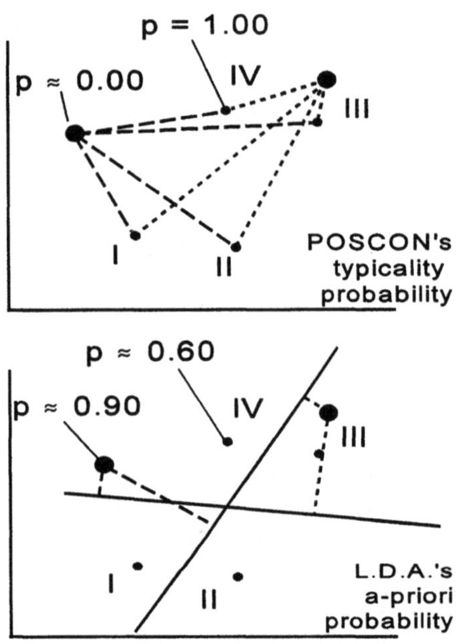

Fig. 14. *Schematical example of discriminant analysis with resp. POSCON and linear discriminant analysis as based on 2 variables and 4 groups (sexes and/or populations). The 4 stippled areas represent 1 S.D. around the population mean. The distances to the population means (upper figure - small circles) or the discriminant lines (lower figure - drawn lines) are presented with broken lines. The large black circles concern the measurements of individual birds. The p-values represent (roughly) the posterior probabilities.*

- the calculated probabilities had to be intuitively acceptable;
- the calculations should not depend on the assumption of homogeneity of covariance matrices;
- no estimates of prior probability ought to be necessary;
- the selected method had to deal with both categorical and metric variables;
- relatedness of measurements should not be a problem;
- it ought to be possible to incorporate secondary information like a bird's sex.

The computer package 'POSCON' (Ambergen 1989, Van der Sluis *et al.* 1991, Schaafsma *et al.* in prep.) clearly fitted these prerequisites. Only a short description of the procedures and calculations in POSCON are given here (details in Annex III):

(1) Calculation of *k* estimates of the (squared) *Mahalonobis distance*:

$$\Delta_x^2=(x-\mu_h)^T\Sigma_h^{-1}(x-\mu_h)$$

between the vector of scores *x* of an individual and the vector of expectations μ_h of population *h*. In practice μ_h and Σ_h have to be estimated from the sample from population *h*. Estimates of μ^h can be improved by pooling samples. Those of Σ_h can be improved as well without pooling samples but by making homogeneity assumptions e.g., of the form $\Sigma_3 = \Sigma_5 = \Sigma_8$.

(2) *Hotelling tests* are performed for testing the null hypothesis that an individual is from the same population as population *h*. The Hotelling T^2 statistic is defined (on the assumption of homogeneity) as

$$n_h(n-k-p+1)n^{-1}(n_h+1)^{-1}p^{-1}\hat{\Delta}_x^2$$

where $\hat{\Delta}_x^2$ is the appropriate estimate of Δ_x^2. It has an *F*-distribution with P and *(n-k-p+1)* degrees of freedom; *n* denoting the total sample size; *k* the number of groups; P the number of variables involved and n_h the number of cases in population *h*.

(3) *Typicality probabilities* are calculated as follows: if Δ_x^2 stands for the squared Mahalonobis distance between individual *i* and population *h*, then the typicality probability of this individual with respect to population *h* is defined as the probability that an individual randomly chosen from population *h* has a larger squared Mahalonobis distance than that observed in the individual of interest. Since the squared Mahalonobis distance between a randomly chosen individual and the sample mean follows a χ_p^2 distribution with P degrees of freedom, the typicality probability is equal to:

$$P(\chi_p^2 \geq \Delta_x^2)$$

The typicality probability of an individual for a certain sample (here called 'POSCON group') becomes smaller when the measurements of this individual become less typical for that POSCON group. If the individual is statistically different from the POSCON group at the 5 % level, the typicality probability will be ≤ 0.05. In practice Δ_x^2 must be estimated from the data (Van Vark & Schaafsma 1992). Confidence intervals can be constructed for the *k* typicality probabilities of an individual, under the assumption of normality (formulae in Ambergen 1989, p. 104).

THE ACCURACY OF PREDICTIONS

The quality of predictions can be evaluated by calculating their accuracy. Accuracy is defined here as the percentage occurrence, showing that known breeding birds are predicted to resemble their own POSCON group. When the resemblance between an individual and *k* different POSCON groups is calculated, this will result in *k* typicality probabilities varying between 0 and 1. If the breeding origin of an individual is known, the highest typicality probability out of these *k* probabilities would be expected to be the one of the own sample. In practice individuals of unknown breeding origin will be classified to the group with the highest calculated typicality probability. In general, accuracy concerns the quality of distinction between sample groups: a low degree of accuracy is associated with a rather bad distinction between groups.

THE EXCLUSIVITY OF PREDICTIONS

If it is not enough that the typicality probability for the own sample is the highest one but that it is also necessary that the remaining typicality probabilities (*k-1*) are below a certain level (e.g., 0.05), it is necessary to define the degree of exclusivity. This exclusivity is defined as the percentage occurrence, showing that birds are predicted to resemble just one sample group in between all other ones. This definition elaborates on the one used to define the accuracy. In general, a low degree of exclusivity is associated with a low quality of the predictions on the individual's level. If the aim of study is the prediction of an individual's breeding origin it is necessary to have a high degree of exclusivity. However, when the aim is to estimate the population composition within a staging area a high degree of accuracy is enough.

VARIABLE SELECTION FOR FUTURE USE

Selecting measurements is somewhat complicated, especially since most measurements are considerably correlated. Otherwise we would have used properly standardized discriminant coefficients under the linear discriminant-analysis assumptions in order to assess the relevance of the various measurements. Since this was impossible (see above), simple univariate *F*-statistics (ANOVA) were computed per measurement for testing homogeneity. These computations resulted in some ordering of the importance of the individual measurements. The estimated order of importance of the individual variables is presented in the tables of every species' chapter presenting the overall MANOVA-results.

Chapter 3

Introduction to the species' accounts

GENERAL INFORMATION

The morphometrical information is presented per species or per species complex. The sequence of chapters follows Voous' (1973) species order. The spelling of the American names of species follows Hayman *et al.* (1986). Each chapter starts with describing the breeding distribution and the present-day knowledge on morphometrical variation and systematics. Data analysis starts with a section on morphometrical differentiation. The importance of the various measurements is analysed. Since often only wing- and culmen lengths are measured in studies with live-trapped birds, the morphometric variation with these 2 variables is investigated. It is also analysed whether the scoring of different plumage characters improved the descriptions on geographical variation. The following section concerns the prediction of breeding origin as based on the described geographical variation. The quality of prediction is evaluated by estimating the degrees of accuracy and exclusivity (see Introduction under Statistical analysis of the data). The outcome of these predictions is used to analyse the character of contact zones between different populations. The last headings under the results' section concern body mass and the occurrence of primary moult on the breeding grounds. If necessary, another taxonomical division within the species(-complexes) is proposed in the discussion. Various related topics are also discussed.

THE BREEDING RANGE

In general, present-day ornithology has a good knowledge of the breeding range of waders in the Holarctic. Several distribution maps have already been published. However, when comparing different maps of the same species, important differences can regularly be observed even between maps published in the last 20 years. This is rather troublesome when one wants to know whether a skin concerns a breeding bird or not. Another complication can be that the published breeding ranges do not correspond anymore with the present-day ones. For example,

Explanation of the species' maps with the breeding distribution

The first maps in every chapter concern the breeding distribution of the species being analysed. These maps are composed out of 5 layers of information:

1. The continents have a light grey colour (10 % grey). Water bodies are white.
2. *Breeding ranges* are presented with a *20 % grey* shade. When breeding was confirmed by recent Atlaswork a *40 % grey* shade was used. *Question marks within broken lines* concern possible breeding ranges. Since the delineation of these breeding ranges was regularly based on an extrapolation from few data, a third layer was introduced:
3. Squares symbolize breeding sites as reported from literature: *dark grey squares* (40 % grey shade) were used when evidence of breeding was presented; *open squares* were used when evidence was doubtful.
4. The sites from where museum specimen were collected are represented with circles. *Black circles* concern collected birds with brood patches or birds which were reported with nests and/or young. Such was not evident from our database in relation to the *open circles*: these concerned skins without brood patches or skins for which it was not noticed.
5. Initial breeding samples were delineated for the analysis of geographic variation. These delineations are presented as *black drawn lines with corresponding black circles with a white number*. These numbers correspond with the sample numbers in the first column of the tables with the mean measurements.

the small breeding population of Spoon-billed Sandpiper near Kolyuchin Bay on Chukotsk was wiped out by explorers (Vaughan 1992). Other examples concern the retraction of the breeding range of the Whimbrel on the steppe areas between the Ural and the Volga (review in Glutz *et al.* 1977) and of that of the Dunlin subsp. *schinzii* in W Europe (reviews in Glutz *et al.* 1975 and in Cramp & Simmons 1983).

The delineation of breeding ranges is based upon literature. All references which were used to compile these maps are presented in Chapter 19, where they are marked. The maps were compiled by CSR and thoroughly commented by P.S. Tomkovich.

GEOGRAPHICAL VARIATION

The species' chapters nearly completely focus upon the geographical distribution during the breeding season. Mean measurements per breeding area and per sex are presented. These samples are used in cluster- and MANOVA analysis, where the measurements of the samples are compared. Various sets of measurements are used, since inclusion of additional measurements might lead to improved discrimination between samples. In general, the inclusion of more measurements in an analysis resulted in smaller sample sizes, since it was not always possible to quantify every variable in every bird. Most of our data gathering was strongly time- and budget limited. Therefore it was nearly always necessary to strike a balance between quick working (many birds - few measurements) or aiming at completeness (fewer birds - more measurements per bird). Mostly 3 MANOVA analyses per species were performed:

- all available measurements;
- the most promising measurements (depending, for each species separately, on the results of the first analysis), or wing-, culmen- and tarsus lengths;
- only wing- and culmen lengths.

The last type of analysis was added because many trapping studies only deal with these 2 measurements. Samples were contrasted in an orthogonal design of MANOVA. The resulting tables clearly reveal the differences between samples. Indistinguishable samples were combined when groups were composed for prediction purposes: the so-called 'POSCON groups'. Typicality probabilities were calculated and could be used to discuss the group membership of particular birds. Typicality probabilities were also used for:

- showing distribution patterns over the breeding range;
- searching for wrongly classified individuals; and
- analysing the quality of prediction.

In the discussion, the observed morphometrical variation is related to the taxonomy of the species. Moreover, the breeding range of populations is discussed as well as the type of contact between populations.

WING STRUCTURE

The feather lengths of the primaries 1, 7, 8, 9 and 10, and the length of the 1st secondary were related to the maximum wing lengths. These *relative feather lengths* were expressed in terms of percentages of the maximum wing length. This makes it possible in future studies to predict the maximum wing length from data of feather length when maximum wing length cannot be measured directly (primary moult, broken outer primaries). The relative feather lengths are presented in Annex II.

Chapter 4

Ringed Plover - *Charadrius hiaticula* and Semipalmated Plover - *Charadrius semipalmatus*

Fig. 15. *Breeding distribution of the Ringed Plover. For explanation of the used symbols see Chapter 3: 'Introduction to the species' accounts'.*

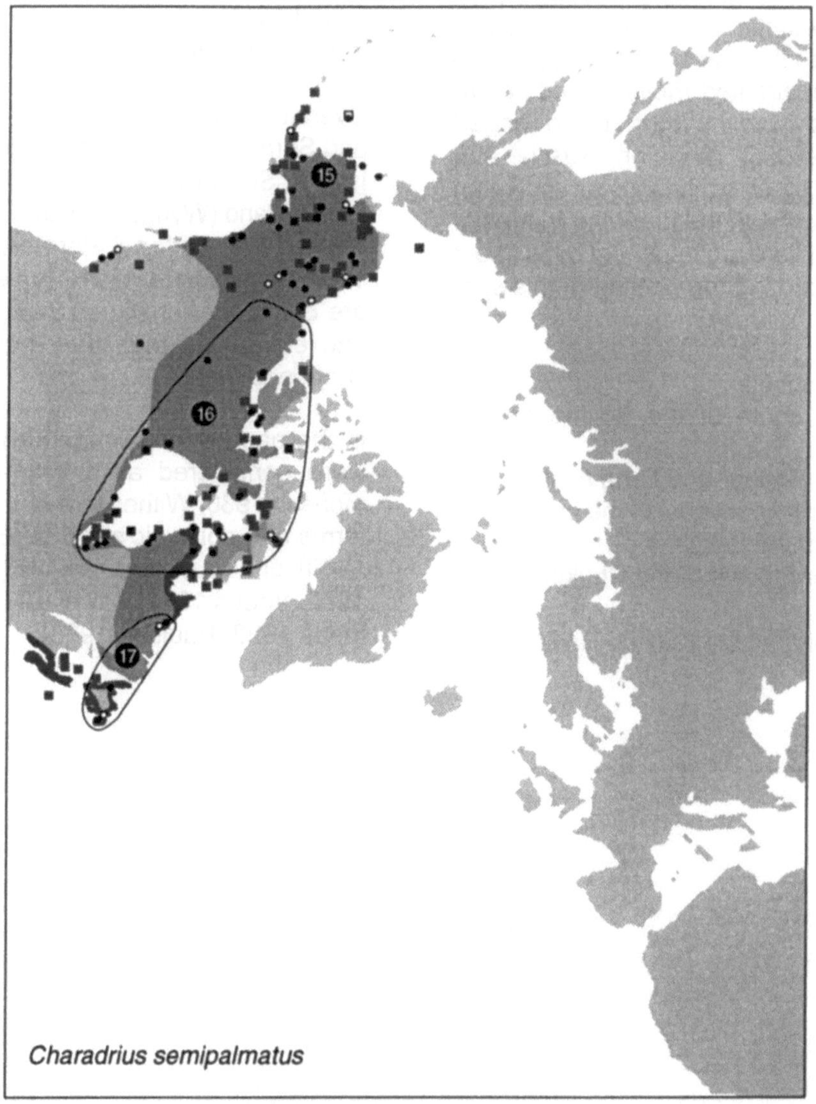

Fig. 16. *Breeding distribution of the Semipalmated Plover. For explanation of the used symbols see Chapter 3: 'Introduction to the species' accounts'.*

INTRODUCTION

Since Ringed Plover and Semipalmated Plover have a close phenotypical resemblance, the analysis of both is combined in a single chapter. The Ringed Plover has a primarily Palearctic breeding distribution under temperate, boreal, and arctic conditions, and the Semipalmated Plover is its Nearctic counterpart, breeding under more or less the same conditions (figs. 15 & 16). The Semipalmated Plover breeds more inland and not as far to the north as the Ringed Plover. The breeding range in Russia and Siberia might not be as continuous as shown on the published maps with the breeding ranges of Ringed Plovers. There are important gaps in our knowledge, concerning the nesting of this species on:

- the Kola Peninsula,
- around and south of the Bolshezemelsk tundra,
- between the Anabar and Lena river,
- along the north coast east of the Lena River as far east as the Indigirka,
- along the north coast of the Chukotsk Peninsula,
- the flood plains of the Kolyma River,
- the southernmost extent of the breedingarea in E Siberia.

The same applies to the Semipalmated Plover's breeding distribution in SW Alaska, the Yukon territory in Alaska, the MacKenzie district in Canada, Victoria Island, inland Quebec, and Labrador. But these anomalies in the Semipalmated Plover's breeding range might also stem from a too limited scanning of literature.

Overlap between the breeding ranges of Ringed and Semipalmated Plover is limited to Wrangel Island, St. Lawrence Island (Bering Sea), and Baffin Island (Wynne-Edwards 1952, Bock 1959, Smith 1969, Sutton & Parmelee 1955, Godfrey 1986, Stishov *et al.* 1991). But either one or both are only occasionally on Wrangel Island or St. Lawrence, and thus they have only a slight chance to meet.

In the past, Semipalmated and Ringed Plover were considered as a single species (Salomonsen 1930, Witherby *et al.* 1940) as allopatric forms of a polytypic species (Bock 1959, Smith 1969), or as 2 distinct species (Wynne-Edwards 1952, Voous 1960, Taylor 1979, Cramp & Simmons 1983, Paulson 1993).

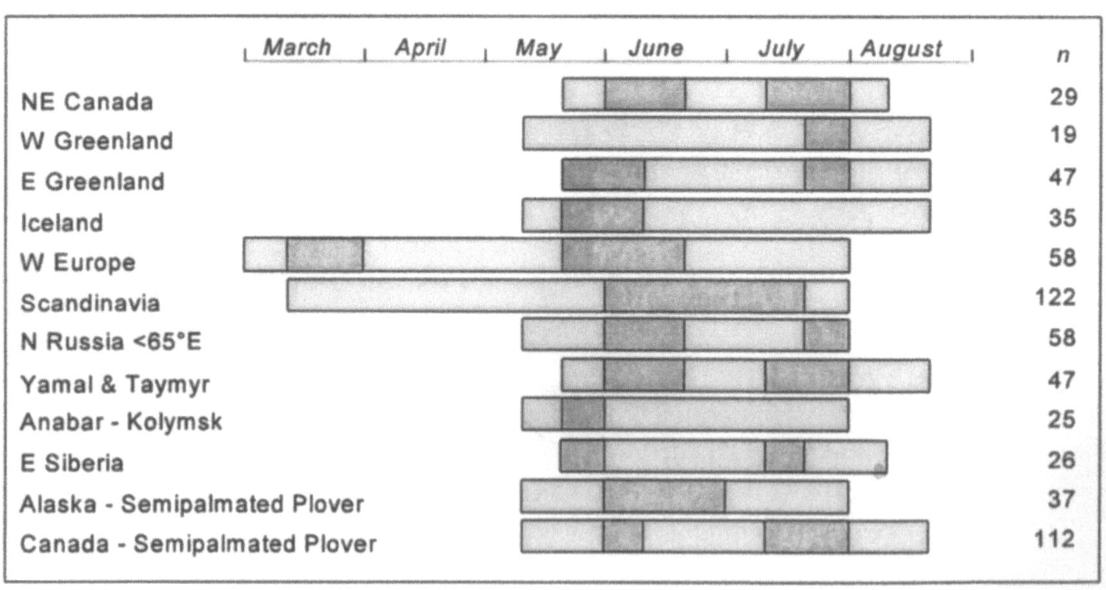

Fig. 17. *Collection dates of Ringed- and Semipalmated Plovers. Most of the birds had been collected during the periods marked by the heavily shaded bars. The sample sizes are presented on the right side.*

In the past 7 subspecies have been distinguished within the Ringed Plover (reviewed in Glutz *et al.* 1975). As most of the variation appeared to be clinal, subsequent research by various authors gradually reduced the number of subspecies to two, each subjected to some geographical variation, differing in measurements and plumage characters:

- *hiaticula*, breeding from NE Canada east to W Europe, Central Europe, and S Scandinavia.
- *tundrae*, breeding from Lappland east to the Chukotsk Peninsula (E Siberia).

Taylor (1979) and Cramp & Simmons (1983), favouring a wide subspecies concept, supported the recognition of the 2 subspecies mentioned above, after finding differences in morphometrics, moult cycle, and coloration of the upperparts. Kozlova (1962) and Dement'ev *et al.* (1969) did not recognize any subspecies at all. Hale (1980) recognized 4 subspecies, based upon the existence of different colour forms: (1) *hiaticula* - breeding around the North Sea, (2) *psammodroma* - breeding on Iceland and Greenland, (3) *tundrae* - breeding from N Scandinavia east to Central Siberia, and (4) *kolymensis* - breeding in E Siberia. Intergradation zones between *hiaticula* and *tundrae* were confined to Scandinavia (in agreement with Väisänen 1969, Taylor 1979, Cramp & Simmons 1983) and to Siberia south of Taymyr.

Taylor in Prater *et al.* (1977) and Cramp & Simmons (1983) considered the Semipalmated Plover to be monotypic, although differentiation in wing lengths between the populations of Alaska and E Canada was acknowledged.

MORPHOMETRICAL DIFFERENTIATION

In total, 574 Ringed and Semipalmated Plovers from the entire breeding range were examined. Initially, 17 breeding locations were distinguished (figs. 15 & 16). Most populations had been sampled throughout the breeding period. Southern breeding birds arrived by mid-March and arctic breeders by late May or early June (Fig. 17). Most of the arctic breeders had been collected soon after arrival. There were predominantly ♂♂ in the samples (64.5 %).

Nearly all initial breeding locations were represented with good samples, but few birds from Jan Mayen, Svalbard and Novaya Zemlya were examined. In the statistical analyses, the Ringed Plovers from Jan Mayen were considered as Greenland breeders, the ones from Svalbard as N European ones, and the ones of Novaya Zemlya as Central Siberian breeding birds (tables 7 - 11 and figs. 18 - 21). In order to reduce the number of samples, the ones from high arctic Canada, Greenland, and Iceland were combined. The differences between them were too small to allow separation as can be seen from the dendrogram in Fig. 21, resulting from cluster analysis.

The samples from the Semipalmated Plover in Alaska and Canada were also combined. Here the Canadian ♀♀ differed slightly from the Alaskan and Canadian ♂♂, but too little to validate recognition.

Semipalmated Plovers mostly had smaller measurements than Ringed Plovers (Table 7 and figs. 18-20), except for middle toe length, culmen height, and some plumage characters. The differences in wing- and culmen length, the width of the breast band and the length of the dusky border on primary 5 were marked. Both species showed distinct sexual and geographical variation (Table 10) with sexual differences being smaller than geographical differences. Nearly always, ♂♂ were larger than ♀♀, but surprisingly not in the W European Ringed Plover population (Table 9). Here ♂♂ were on average smaller than ♀♀. Sexual size dimorphism was largest in the Central and E Siberian populations of the Ringed Plover. This outcome is in contrast with earlier findings as summarized in e.g. Cramp & Simmons (1983), where sex differences were insignificant.

Most measurements were closely correlated (Table 10). Culmen heights were not related to any of the other measurements.

Table 7. *Mean measurements of Ringed and Semipalmated Plovers from the breeding grounds. The mean values with standard deviations and sample sizes are given. [1] Breeding areas in northern USA are also included.*

Locality	Wing length			Culmen length			Tarsus length			Secondary length			Tail length			Middle toe length			Culmen height		
	x	*SD*	*n*	*x*	*SD*	*n*	*x*	*SD*	*n*	*x*	*SD*	*n*	*x*	*SD*	*n*	*x*	*SD*	*n*	*x*	*SD*	*n*
Ringed Plover - ♂♂																					
NE Canada	133.6	4.0	19	13.8	0.8	19	24.8	1.0	19	57.2	2.5	19	54.5	2.6	19	22.8	1.7	3	3.5	0.2	18
W Greenland	135.8	3.6	13	14.2	1.1	13	25.3	0.7	13	56.9	1.5	9	55.8	3.2	9	21.3	1.7	3	3.8	0.2	9
E Greenland	134.4	2.9	29	14.6	0.6	28	25.7	0.7	29	55.6	0.9	5	55.2	2.6	5	20.7	0.9	4	3.9	0.2	5
Jan Mayen	134.0	2.0	4	15.0	0.3	4	25.1	0.3	4	-	-	0	-	-	0	-	-	0	-	-	0
Iceland	133.1	3.3	25	14.2	0.8	25	25.0	1.3	25	55.8	1.4	20	55.3	3.2	20	21.6	0.8	19	3.7	0.2	20
British isles	138.0	3.8	19	15.4	0.7	19	26.1	0.9	19	60.8	3.3	5	58.0	3.7	5	22.5	1.0	5	3.9	0.3	5
Netherlands	134.8	3.1	16	15.1	0.6	16	25.9	0.5	16	58.3	2.4	10	57.0	2.7	10	22.9	1.1	8	3.8	0.3	10
Germany	137.0	4.2	2	15.8	0.1	2	27.0	0.4	2	57.5	0.5	2	56.0	1.0	2	22.4	-	1	3.8	0.6	2
Denmark	138.0	-	1	14.8	-	1	25.9	-	1	59.0	-	1	56.0	-	1	22.0	-	1	3.6	-	1
S Scandinavia	132.6	4.7	21	14.1	0.6	21	26.2	1.0	21	56.8	1.8	20	55.3	2.9	18	22.5	0.9	19	3.7	0.3	19
N Scandinavia	130.7	3.6	60	14.0	0.7	60	25.8	1.1	60	54.6	1.9	34	53.3	2.3	20	22.4	1.0	31	3.6	0.2	32
Svalbard	133.3	3.0	6	13.9	1.1	6	25.7	0.7	6	55.3	1.2	3	54.0	3.0	3	21.7	1.0	3	3.6	0.1	3
N Russia	131.3	3.2	39	14.2	0.8	39	25.2	1.3	37	55.0	1.7	20	60.5	4.4	20	22.3	1.0	17	3.5	0.3	20
Novaya Zemlya	132.0	3.5	3	13.9	1.4	3	25.3	0.8	3	55.7	2.1	3	58.3	2.3	3	21.9	0.7	3	3.7	0.3	3
Yamal & Taymyr	131.1	2.9	30	13.9	0.7	30	25.3	1.0	30	56.2	1.3	25	55.5	2.1	25	22.3	0.8	21	3.5	0.2	25
Central Siberia	131.4	2.9	14	13.9	0.6	14	25.1	1.0	14	56.4	1.6	14	58.7	3.0	14	21.8	0.7	14	3.5	0.3	13
New Sib. Islands	132.0	2.8	2	13.6	1.1	2	25.7	0.5	2	55.5	0.7	2	53.5	5.0	2	21.8	0.0	2	3.3	0.1	2
E Siberia	132.4	2.7	17	14.2	0.5	17	25.9	0.9	17	55.5	1.3	17	56.3	5.4	17	22.4	1.3	13	3.7	0.2	15
Ringed Plover - ♀♀																					
NE Canada	132.6	2.9	10	14.1	0.8	10	24.7	0.9	10	56.5	1.7	10	55.3	2.1	10	22.3	0.8	5	3.6	0.2	10
W Greenland	133.7	3.4	6	14.4	0.8	6	24.8	0.6	6	55.7	0.6	3	55.7	0.6	3	21.3	-	1	3.6	0.3	3
E Greenland	134.9	4.2	18	14.3	0.8	18	25.2	0.7	18	55.6	1.8	5	56.4	3.9	5	21.5	0.5	3	4.0	0.2	5
Jan Mayen	132.8	3.4	4	14.5	0.1	4	25.2	0.7	4	-	-	0	-	-	0	-	-	0	-	-	0
Iceland	133.7	2.8	10	14.1	0.5	10	24.8	1.0	10	56.8	1.0	6	55.2	1.9	6	21.1	0.7	6	3.7	0.2	6
British isles	140.0	2.5	7	15.5	1.3	7	25.7	1.0	7	59.0	-	1	52.0	-	1	22.6	-	1	4.0	-	1
Netherlands	140.3	5.2	12	15.5	0.9	12	26.2	1.0	12	59.8	2.4	5	59.6	3.5	5	22.2	1.5	5	3.9	0.3	5
W Germany	138.0	-	1	15.7	-	1	26.0	-	1	58.0	-	1	53.0	-	1	22.0	-	1	3.6	-	1
S Scandinavia	132.2	4.9	14	14.3	0.9	14	25.9	1.2	14	56.5	1.9	13	55.3	3.2	11	22.3	1.0	11	3.6	0.3	13
N Scandinavia	131.2	3.4	27	13.9	0.7	27	25.3	0.9	27	55.3	2.1	15	55.7	3.4	9	22.1	0.5	15	3.6	0.2	12
Svalbard	132.0	1.2	4	13.6	0.6	4	25.4	0.6	4	55.5	2.1	2	55.5	0.7	2	20.9	0.5	2	3.5	0.1	2
N Russia	131.9	3.7	19	14.0	1.0	19	25.2	1.0	16	54.5	1.1	10	63.6	5.3	9	21.8	1.2	10	3.4	0.3	9
Novaya Zemlya	129.0	1.0	3	12.8	0.5	3	24.1	0.4	3	56.0	1.7	3	57.0	1.0	3	21.2	0.8	3	3.8	0.3	3
Yamal & Taymyr	132.4	3.6	17	13.8	0.6	17	24.9	1.3	17	56.6	1.3	16	55.5	2.9	15	21.7	0.8	15	3.4	0.2	16
Central Siberia	133.3	3.1	8	13.7	0.6	8	24.3	1.5	8	57.5	1.2	8	58.8	5.0	8	21.3	1.0	8	3.5	0.2	8
New Sib. Islands	133.0	-	1	12.5	-	1	23.1	-	1	57.0	-	1	59.0	-	1	20.4	-	1	3.2	-	1
E Siberia	133.2	4.2	9	13.6	0.8	9	25.0	0.7	9	56.0	1.3	9	58.4	5.3	9	21.8	1.1	8	3.4	0.2	9
Semipalmated Plover - ♂♂																					
Alaska	124.5	2.5	21	12.6	0.8	21	23.7	1.0	21	54.5	1.3	21	52.3	1.9	21	21.8	0.8	4	3.7	0.2	19
N Canada	124.9	3.2	56	12.8	0.8	56	23.8	1.1	56	54.2	1.6	56	51.9	2.5	56	22.2	0.9	23	3.8	0.2	53
S Canada[1]	124.5	2.8	14	12.7	0.5	14	24.2	0.9	14	53.9	1.7	14	51.7	3.5	14	22.0	0.9	12	3.7	0.2	13
Semipalmated Plover - ♀♀																					
Alaska	124.9	2.9	16	12.5	0.6	16	23.7	1.1	16	53.9	1.3	16	51.6	2.9	16	21.4	0.8	2	3.6	0.2	15
N Canada	126.7	3.8	35	12.4	0.5	35	23.5	1.0	35	54.8	2.2	35	51.8	2.0	35	21.4	0.6	14	3.7	0.2	35
S Canada[1]	124.9	2.9	7	12.7	0.7	7	23.8	0.9	7	53.6	1.5	7	52.0	1.6	7	21.2	1.0	6	3.6	0.3	7

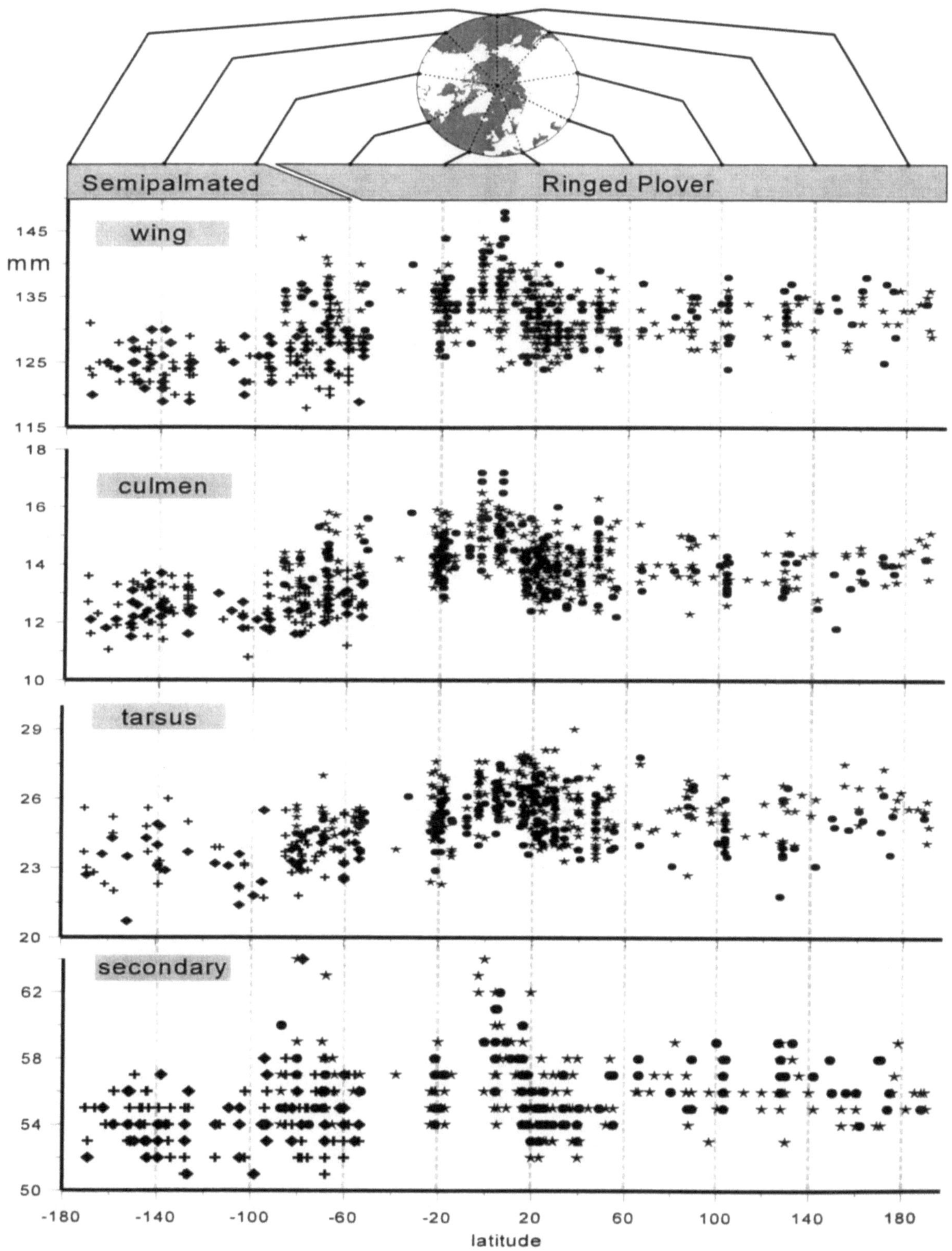

Fig. 18. *Measurements of Ringed- and Semipalmated Plovers from the breeding grounds ranging from Alaska on the left side of the plots to E Siberia on the right side. The following measurements are presented from top to bottom: wing length, culmen length, tarsus length and secondary length (described in chapter 2). Four groups are presented: ● Ringed Plover ♂♂, ★ Ringed Plover ♀♀, + Semipalmated Plover ♂♂, ◆ Semipalmated Plover ♀♀.*

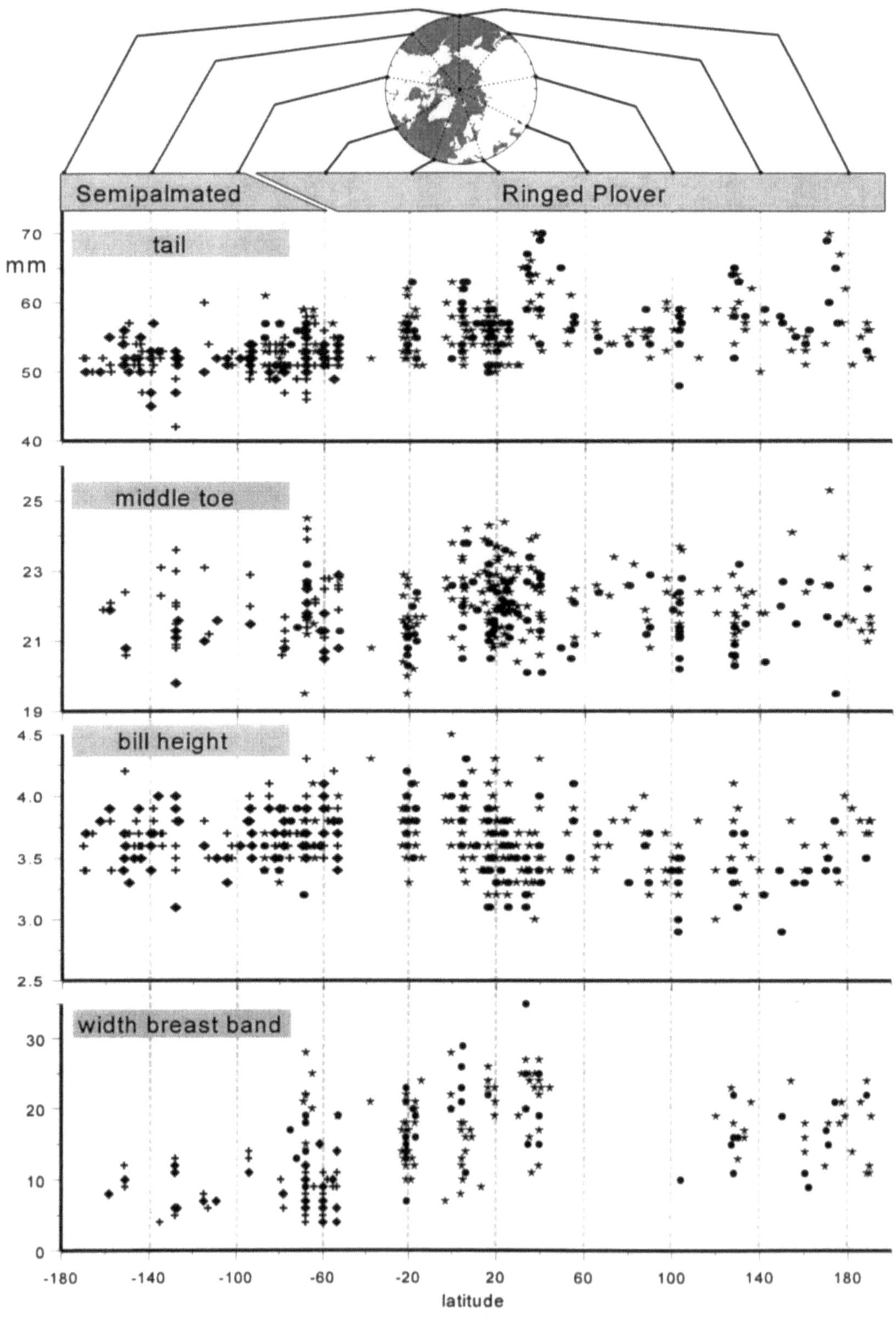

Fig. 19. *Measurements of Ringed- and Semipalmated Plovers from the breeding grounds ranging from Alaska on the left side of the plots to E Siberia on the right side. This following measurements are shown: tail length, middle toe length, culmen height, width of the breast band (see chapter 2 for detailed description). Four symbols are used: ● Ringed Plover ♂♂, ★ Ringed Plover ♀♀, + Semipalmated Plover ♂♂, ◆ Semipalmated Plover ♀♀.*

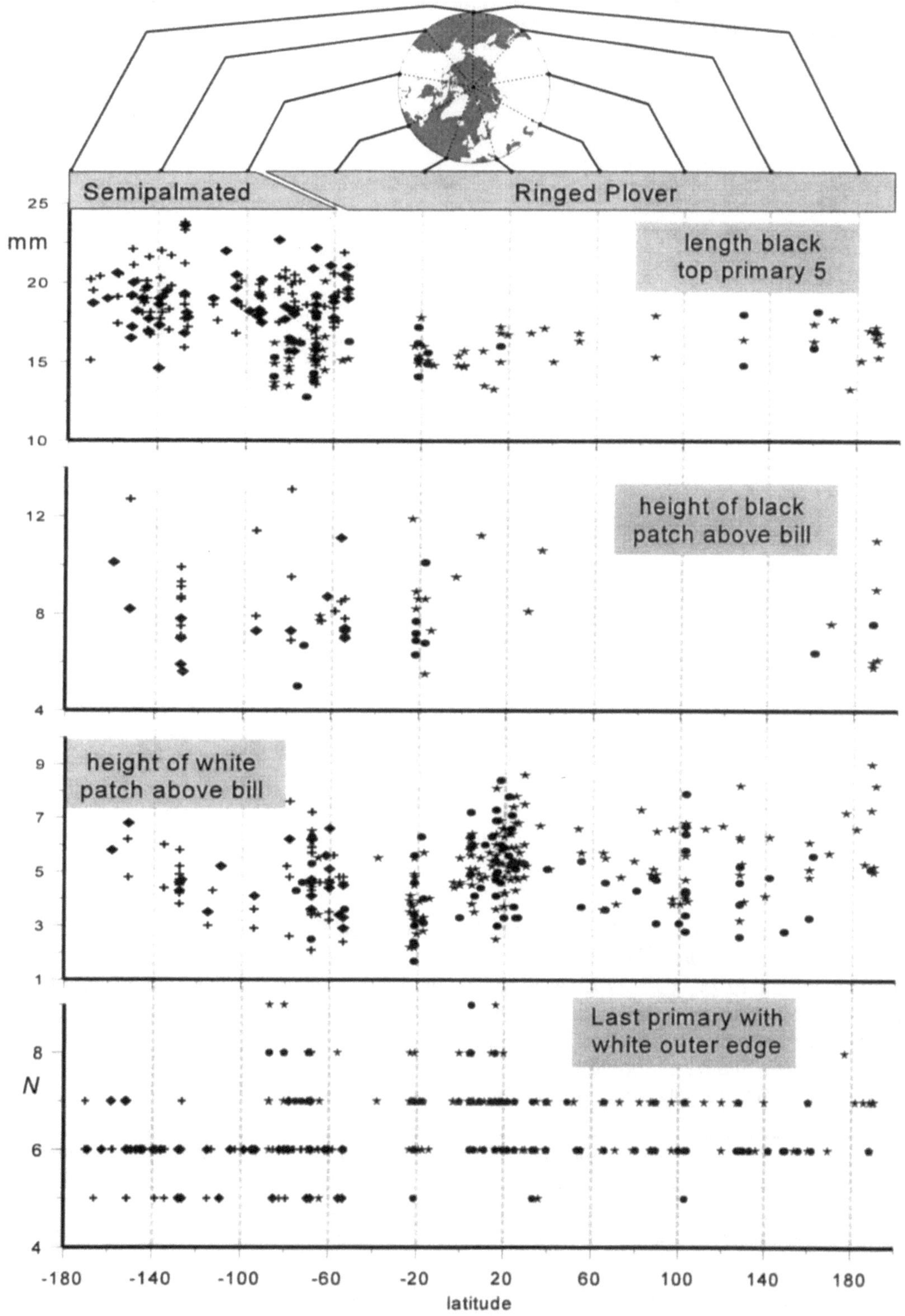

Fig. 20. *Measurements of Ringed- and Semipalmated Plovers from the breeding grounds ranging from Alaska on the left side of the plots to E Siberia on the right side. This figure includes the following measurements: length of the black top on P5, height of the black front above the bill, height of the white front above the bill, the last primary with along its outer edge (described in chapter 2). Four symbols are used: '●' Ringed Plover ♂♂, '★' Ringed Plover ♀♀, '+' Semipalmated Plover ♂♂, '◆' Semipalmated Plover ♀♀.*

***Table 8.** Overall results of MANOVA for all breeding populations of Ringed and Semipalmated Plover in combination with univariate ANOVA's in which the different measurements were analysed separately. * 0.01 < P < 0.05; ** 0.001 < P < 0.01; *** P < 0.001.*

Variables	Analysis I **wing, culmen, tarsus, secondary, tail, middle toe, culmen height**			Analysis II **wing, culmen, tarsus**			Analysis III **wing, culmen**		
n	257			593			598		
Multivariate analysis of variance									
Effects	Interaction	Area	Sex	Interaction	Area	Sex	Interaction	Area	Sex
Pillai's trace	0.143	1.181	0.108	0.039	0.759	0.047	0.033	0.686	0.024
F-value	1.020	10.730***	4.139***	1.547	39.383***	9.590***	1.973*	61.252***	7.058***
Hypoth. & error *df*	35/ 1215	35/ 1215	7/ 239	15/ 1743	15/ 1743	3/ 579	10/ 1172	10/ 1172	2/ 585
Univariate F tests									
df	5/245	5/245	1/245	5/581	5/581	1/581	5/586	5/586	1/586
wing length	2.341*	42.641***	1.700	2.785*	152.254***	9.399**	2.925*	152.381***	9.061**
culmen length	0.832	37.774***	3.470	1.354	133.731***	3.105	1.250	133.143***	2.844
tarsus length	0.427	13.540***	8.080**	0.953	58.503***	11.449***			
secondary length	0.527	37.142***	1.855						
tail length	0.760	18.371***	4.131*						
middle toe length	0.927	2.334*	9.847**						
culmen height	1.210	15.003***	4.559*						

***Table 9.** Sexual size dimorphism of Ringed and Semipalmated Plover. The mean percentage difference between the measurements of both sexes was calculated for every single variable and breeding population, according to: $(x_{♀}-x_{♂})/x_{♂}*100$. A negative value thus implies smaller measurements for ♀♀. The calculations were based upon the population means. The obtained values were averaged per variable and summed per breeding population. Some additional notes:*

[1] N Europe includes Svalbard and Novaya Zemlya.

[2] The data refer to Alaska, low arctic and S Canada, and the north part of the USA.

Locality	Wing	Culmen	Tarsus	Secondary	Tail	Middle toe	Culmen height	Σ
Ringed Plover								
Canada/ Greenland/Iceland	-0.1	-0.2	-1.1	-0.2	0.3	0.0	0.0	-1.3
W Europe	2.6	1.7	0.1	0.5	0.6	-2.0	1.9	5.4
N Europe[1]	0.0	-1.7	-1.3	0.1	2.5	-2.2	-1.7	-4.3
Central Siberia	1.0	-1.9	-2.7	1.2	0.2	-2.8	-3.9	-8.9
E Siberia	0.4	-2.9	-3.4	0.2	3.1	-2.7	-4.3	-9.6
Semipalmated Plover[2]	0.9	-2.0	-1.2	0.3	-0.4	-3.4	-1.5	-7.3
x	0.8	-1.2	-1.6	0.4	1.1	-2.2	-1.6	
SD	1.0	1.7	1.3	0.5	1.4	1.2	2.3	

Table 10. *Pearson's product-moment correlation between measurements of Ringed and Semipalmated Plovers. Both sexes are separately analysed. Two-tailed probabilities: * 0.01 < P < 0.05; ** 0.001 < P < 0.01; *** P < 0.001.*

	Wing	Culmen	Tarsus	Secondary	Tail	Middle toe	Culmen height	
Wing		0.686***	0.581***	0.604***	0.494***	0.320**	0.013	
Culmen	0.605***		0.663***	0.384**	0.350**	0.331**	0.036	
Tarsus	0.530***	0.529***		0.297**	0.294**	0.556***	0.008	
Secondary	0.609***	0.430***	0.283**		0.404***	0.155	0.050	♀♀
Tail	0.447***	0.391**	0.388**	0.378**		0.064	-0.171	
Middle toe	0.140	0.162	0.358**	0.105	0.229*		0.103	
Culmen height	-0.014	0.086	-0.049	0.104	-0.127	-0.044		
				♂♂				

Separation of both sexes results in a total of 12 samples used in MANOVA analysis (Table 11). The following conclusions were drawn from the analysis:

- Both sexes of Semipalmated Plover clearly differed from both sexes of all Ringed Plover samples.
- W European Ringed Plovers clearly differed from every other sample.
- The Canada / Greenland / Iceland samples differed from nearly all other samples, except for the E Siberian ones. They differed less from the Central Siberian population than from the W European one.
- N European Ringed Plovers differed from all other populations except for the E Siberian population.
- Central Siberian Ringed Plovers differed from all other populations except for the E Siberian ♀♀.

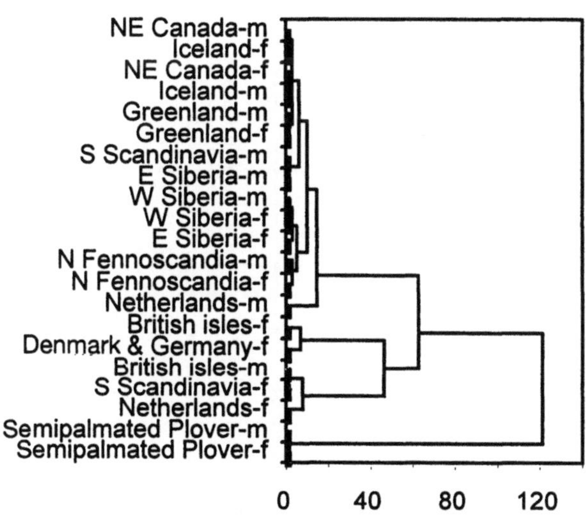

Fig. 21. *Results of cluster analysis in the Ringed and Semipalmated Plover as based on the mean measurements per breeding population and sex. Squared Euclidean cluster distances were calculated, with use of average linkage between groups. Explanation of the abbreviations: 'm' male; 'f' female.*

At least four geographical groups could thus be distinguished in the Ringed Plover: (1) NE Canada, Greenland, and Iceland; (2) N Scandinavia and N Russia east to 65°E; (3) Central Siberia between 65° and 150°E; (4) W Europe. The E Siberian population, breeding east of 150°E, had rather averaged characters, and was thus difficult to distinguish from other Palearctic populations. However, it was neither fully identical to any of these. Therefore, it was distinguished as a separate group.

Middle toe length, tarsus length, culmen height and tail length were important for the distinction between sexes (Table 8). Wing-, secondary-, culmen length, tail length, culmen height, tarsus length, and middle toe length (in decreasing order of importance) were important for the distinction between regions.

Both wing- and culmen lengths were important for the analysis of sexual and geographical variation (Table 8). More or less the same results were obtained as with all 7 variables included (Table 11). However, the N European and Siberian samples could nearly not be distinguished. There was more overlap between the

Table 11. *Results of multivariate analysis of variance (MANOVA) of morphometric differentiation in Ringed and Semipalmated Plover. Pillai's trace statistics are presented (Norušis 1992c). In these series of pairwise comparisons 12 groups (both sexes, 6 populations) were compared pairwise on the basis of 7 measurements in the upper part (wing-, culmen-, tarsus-, secondary-, tail- and middle toe length and culmen height) and 2 measurements in the lower part (wing- and culmen length). There were 7 and 239 df in the '7 variable' situation and 2 and 565 df in the second group of analyses. 'Can./Green./Ice.' refers to the breeding birds of high arctic Canada, Greenland and Iceland. Significance levels: *0.01 < P < 0.05; **0.001 < P < 0.01; ***P < 0.001.*

Locality	n	Can./ Green./ Ice. ♀	W Europe ♀	N Europe ♀ < 65°E	Siberia ♀ 65-160°E	Siberia ♀ > 160°E	Semipal. Plover ♀	Can./ Green./ Ice. ♂	W Europe ♂	N Europe ♂ < 65°E	Siberia ♂ 65-160°E	Siberia ♂ > 160°E
Can./Green./Ice. ♀	15											
W Europe ♀	7	0.120***										
N Europe ♀ < 65°E	18	0.125***	0.250***									
Siberia ♀ 65-160°E	25	0.122***	0.191***	0.111***								
Siberia ♀ > 160°E	6	0.053	0.111***	0.026	0.038							
Semipal. Plover ♀	22	0.216***	0.365***	0.223***	0.306***	0.190***						
Can./ Green./ Ice. ♂	29	**0.019**	0.123***	0.170***	0.174***	0.057*	0.316***					
W Europe ♂	14	0.129***	**0.030**	0.299***	0.236***	0.125***	0.422***	0.175***				
N Europe ♂ < 65°E	38	0.092**	0.225***	**0.054**	0.169***	0.036	0.295***	0.136***	0.274***			
Siberia ♂ 65-160°E	37	0.079**	0.200***	0.094**	**0.078****	0.043	0.275***	0.148***	0.217***	0.071*		
Siberia ♂ > 160°E	11	0.036	0.113***	0.094**	0.128***	**0.038**	0.243***	0.032	0.122***	0.034	0.067*	
Semipal. Plover ♂	35	0.292***	0.420***	0.309***	0.437***	0.238***	**0.074****	0.430***	0.483***	0.378***	0.375***	0.283***

Locality	n	Can./ Green./ Ice. ♀	W Europe ♀	N Europe ♀ < 65°E	Siberia ♀ 65-160°E	Siberia ♀ > 160°E	Semipal. Plover ♀	Can./ Green./ Ice. ♂	W Europe ♂	N Europe ♂ < 65°E	Siberia ♂ 65-160°E	Siberia ♂ > 160°E
Can./Green./Ice. ♀	48											
W Europe ♀	20	0.115***										
N Europe ♀ < 65°E	51	0.036***	0.207***									
Siberia ♀ 65-160°E	28	0.021***	0.160***	0.008								
Siberia ♀ >160°E	7	0.002	0.059***	0.006	0.002							
Semipal. Plover ♀	58	0.300***	0.426***	0.179***	0.150***	0.076***						
Can./Green./Ice. ♂	89	**0.001**	0.121***	0.056***	0.031***	0.003	0.379***					
W Europe ♂	36	0.070***	**0.024****	0.166***	0.125***	0.033***	0.437***	0.076***				
N Europe ♂ < 65°E	107	0.037***	0.214***	**0.008**	0.023***	0.007	0.275***	0.065***	0.170***			
Siberia ♂ 65-160°E	48	0.031***	0.197***	0.001	**0.012***	0.006	0.188***	0.049***	0.152***	0.003		
Siberia ♂ >160°E	15	0.001	0.086***	0.010	0.010	**0.002**	0.146***	0.002	0.046***	0.007	0.008	
Semipal. Plover ♂	91	0.350***	0.457***	0.217	0.187***	0.086***	**0.017****	0.450***	0.476***	0.331	0.220***	0.161***

samples in comparison with analyses, based on 7 measurements.

Additional measuring of some plumage characters might provide extra clues for separating populations. Four characters were scored: (1) the maximum height of the breast band, (2) the length of the dusky border on the tip of primary 5, (3) the width of the white front above the bill and (4) the number of the last primary with white along its outer edge. The last variable was the only one not having a normal distribution. Since these variables had low priority during measuring few data were gathered, but some conclusions can be drawn:

- Feather-field characters were more difficult to measure than colour patterns on individual feathers. This difficulty was tried to overcome by measuring the maximum size of the feather field characters.
- Semipalmated Plovers had narrow breast bands, whilst the ones of Ringed Plovers in W and N Europe were broad (Fig. 19). Since the N European Ringed Plovers had on average broader breast bands than the Central and E Siberian birds, this character may form an additional character for distinction.
- The size of the dusky border at the tip of primary 5 was larger in Semipalmated Plover and smaller in the western Ringed Plover samples. It provides a good additional character for differentiation between the Semipalmated and Ringed Plover.
- The white patch above the bill was widest in the N European and Siberian samples, though depended on sex in the Palearctic: wide in ♂♂, narrower in ♀♀.
- The Semipalmated Plover had less primaries with white along the outer edge than the Ringed Plover. Within the Ringed Plover samples, a shift from west (light) to east (dark) could be recognized (Fig. 20).

A MANOVA analysis with the 4 plumage characters (n = 98) showed all 4 to differ between areas while the third and fourth were also important for sex distinction. In a MANOVA with exclusively these 2 characters (n = 205), the size of the white patch did no longer contribute significantly to explain the variation between both sexes, unless the analysis was restricted to the N European and Siberian samples. Plumage characters did not provide better clues than measurements to separate N European and Siberian samples. This failure might be due to paucity of data (n = 22).

PREDICTION OF BREEDING ORIGIN

The MANOVA- and cluster analyses justified the distinction of 6 breeding locations (Table 9): (1) Canada, Greenland & Iceland, (2) W Europe, (3) N Europe west of 65°E, (4) Siberia between 65 and 160°E, (5) Siberia east of 160°E and (6) the breeding range of Semipalmated Plover. Both sexes were distinguished separately, thus resulting in the distinction of 12 POSCON groups. Two different sets of POSCON groups were constructed. In the first set, 6 variables were included (detailed in heading Table 12) and in the second only wing- and culmen lengths. All available measurements of any bird were used to calculate the typicality probability, that a bird resembled one of the constructed groups. In doing so, both the accuracy and the exclusivity of prediction is estimated (see Introduction under 'Statistical analysis of the data').

Accuracy

Semipalmated Plovers standed out from Ringed Plovers even with the exclusive use of wing- and culmen lengths. The quality of predictions was better when more measurements and less POSCON groups were included. If only Semipalmated and Ringed Plover were separated, the error rate was estimated to amount 4.6 - 8.8 % depending on the measurements included (Table 12). When 6 POSCON groups were distinguished and the sex of a bird was known, the error rate increased to 37.0 - 51.1 %. The most important sources of error could be eliminated by combining the N European and Siberian birds. As several Ringed Plovers from Canada, Greenland, and Iceland were mistaken for N European and Siberian birds, the measurements of both populations should be combined when only wing- and culmen lengths are measured (Table 12).

Table 12. *Accuracy of prediction in Ringed and Semipalmated Plover (sex* ***known*** *a priori). These predictions were based upon morphometric analysis and included (A) wing-, culmen-, tarsus-, secondary- and tail length and culmen height; (B) wing- and culmen length. The columns show the predicted proportions in relation to the total number per breeding area (last column).*

Locality	Predicted: Canada, Greenland, Iceland	W Europe	N Europe	Central Siberia	E Siberia	Semipalmated Plover	*n*
			(A) All measurements				
Ringed Plover							
Can./ Gr./ Iceland	44.0	10.7	17.3	20.0	5.3	2.7	75
W Europe	11.1	86.1	-	-	2.8	-	36
N Europe	1.8	-	75.0	10.7	7.1	5.4	56
Central Siberia	12.3	3.1	38.5	44.6	1.5	-	65
E Siberia	18.2	-	27.3	9.1	45.4	-	22
Semipalmated Plover	0.7	0.7	16.9	1.4	0.7	79.6	142
			(B) Wing- and culmen length				
Ringed Plover							
Can./ Gr./ Iceland	23.5	24.3	20.6	11.0	19.1	1.5	136
W Europe	10.7	76.8	5.4	1.8	5.3	-	56
N Europe	1.3	11.8	36.8	13.2	18.1	18.8	144
Central Siberia	12.2	2.7	32.4	32.4	12.2	8.1	74
E Siberia	34.8	4.4	13.0	13.0	34.8	-	23
Semipalmated Plover	-	-	6.2	1.4	3.4	89.0	145

Exclusivity

The degree of exclusivity was rather high (max. 22 %) for a morphometric analysis. This was mainly due to the good quality of separation between Semipalmated and Ringed Plover. Generally, predictions based on more measurements result in a slight increase of exclusivity. When 6 instead of 2 variables were used, exclusivity increased from 12.9 to 15.3 %, if sexes were unknown, and from 18.4 to 22 %, if sexes were known. These estimates were based on prediction with 3 POSCON groups. If the sample from Canada / Greenland / Iceland was separated from the N European & Siberian ones, the degree of exclusivity increased with 6-10 %. This slight improvement is due to the high degree of relatedness between measurements, which leads to a rather high degree of correlation (both positive and negative) between the typicality probabilities. And although the relatedness between typicality probabilities is decreasing with the inclusion of more measurements, it is not enough to result in a higher degree of exclusivity. We conclude that it is rather impossible to reach a satisfying degree of exclusivity, unless one accepts no more than 3 POSCON groups and collects 6 measurements. Then on average, 1 out of 5 birds can be assigned to a specific POSCON group.

We conclude from the predictive analyses with wing- and culmen length that it is only possible to separate between: (1) W European Ringed Plovers, (2) N Palearctic Ringed Plovers and (3) Semipalmated Plovers.

CONTACT ZONES

In the introduction it was already mentioned that overlap in breeding range between Semipalmated and Ringed Plover occurred on Baffin Island in Canada and on St. Lawrence and Wrangel Island in the Bering Sea region. Within the Ringed Plover, an intergradation zone exists in Fennoscandia. These situations are analysed below with POSCON-analysis.

Semipalmated and Ringed Plover. E Siberian Ringed Plovers were contrasted with Semipalmated Plover west of the Mackenzie River, and Ringed Plovers from Canada, Greenland, and Iceland were contrasted with Semipalmated Plovers east of this river. There was overlap in breeding range at St. Lawrence Island and at Baffin Island (Fig. 22). One of the 3 birds from Lawrence Island was most likely an E Siberian Ringed Plover, and both others were Semipalmated Plovers. The birds from S Baffin Island had highest resemblance with Semipalmated Plovers, but birds from the W-, E- and NE shore were a mixture of both species. Thus the zone of overlap is larger than previously thought. Occurrence of hybridization between both taxa on a large scale was not supported by our data: the percentage uncertainly classified birds (43 %, n = 28) on Baffin Island, due to having intermediate measurements, was comparable with the number estimated from the Hudson Bay (40.7 %, n = 27), and was lower than in eastern Quebec (72.7 %, n = 11). Both areas lie outside the overlap range.

Fennoscandian Ringed Plover. In Scandinavia, W European *hiaticula* and N European *tundrae* form a broad zone of intergradation. The breeding birds from the Kola Peninsula strongly resembled *tundrae* (Fig. 23). The degree of resemblance gradually decreased towards the south-west, and more abruptly in a westerly direction: many N Scandinavian Ringed Plover still had a fairly good chance of belonging to *hiaticula*. Ringed Plovers resembling birds from W Europe were clearly present in S Norway, on the Åland Islands, and near Trondheim along the west coast of Norway. Surprisingly, none of the S Swedish Ringed Plovers exclusively resembled the W European sample. One breeding ♀ from N Scandinavia was strongly and

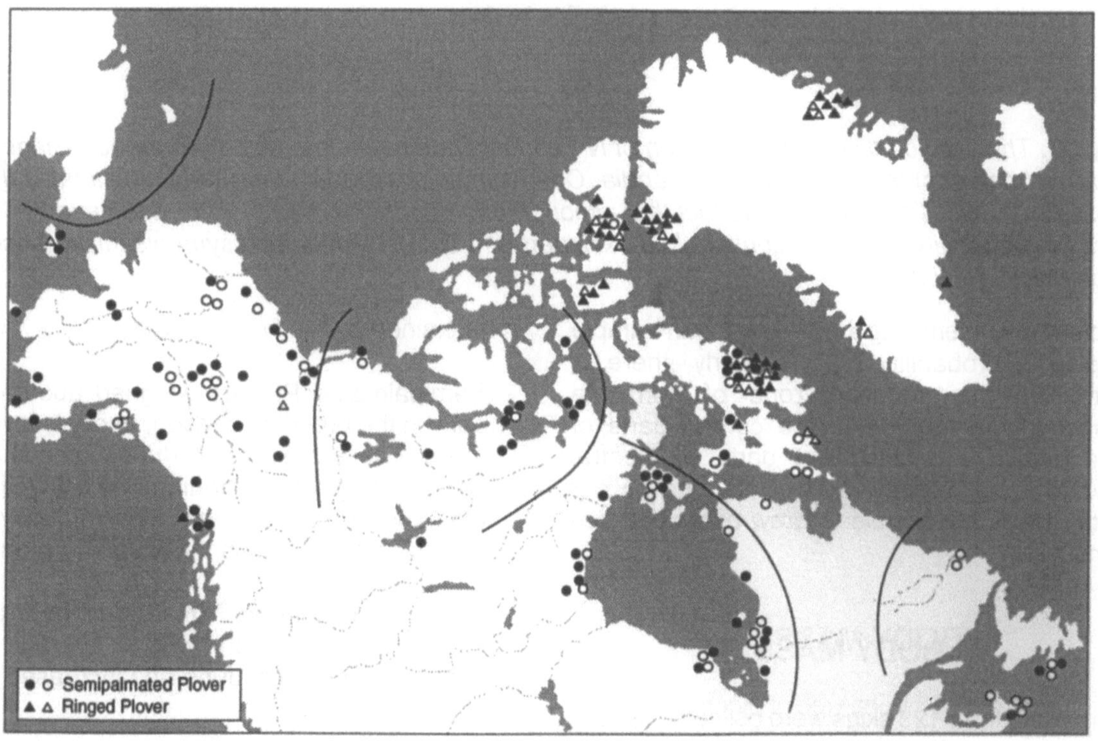

Fig. 22. *The predicted spatial distribution of Semipalmated- (circles) and Ringed Plover (triangles)in the Nearctic. Open symbols represent less certain estimates than closed symbols. This is due to the factthat although one population was predicted to have the highest probability, the other could not be excluded (> 0.05%). With closed symbols this opposite probalitity is < 0.05 %.*

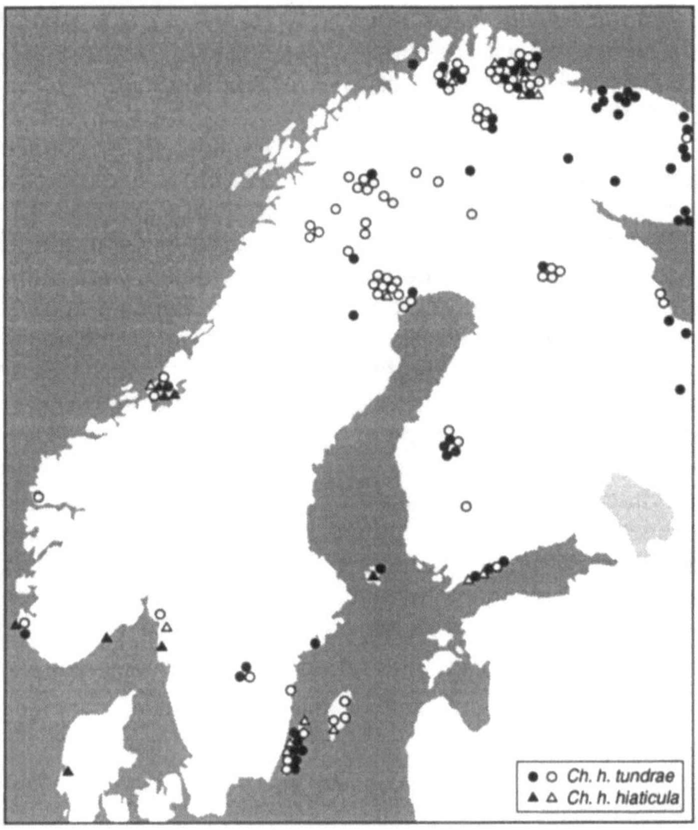

Fig. 23. *The predicted spatial distribution of N Fennoscandian (circles) and W European Ringed Plovers (triangles) throughout Fennoscandia. Open symbols represent less certain estimates than closed symbols. This is due to the fact that although one population was predicted to have the highest probability, the other could not be excluded (> 0.05%). With closed symbols this opposite probability is < 0.05 %.*

exclusively resembling the W European sample (typicality probability 0.87). Clearly, there is extensive intergradation zone of nominate *hiaticula* and *tundrae* in the central parts of Scandinavia. As important parts of Central Sweden and Norway are not covered by our data, it was impossible to draw exact delineations.

BODY MASS

Mass data from 123 skins were collected. Nearly all concerned Ringed Plovers, breeding in N Europe and Siberia, or Semipalmated Plover. Both the breeding location and the collection date were important for explaining the observed variation in body mass (Table 13), but the sex of a bird was not. Almost 40 % of the variation was explained by the first 2 factors.

Semipalmated Plovers weighed about 43 gm early in the breeding season, and this level was more or less maintained during breeding (Fig. 24). Pre-migratory fattening in the Semipalmated Plover was not evident, since no heavy birds had been collected on the breeding grounds.

The mass of Ringed Plovers from NE Canada, Greenland, and Iceland was on average 50 gm in the second half of June and increased during July.

This increase suggested pre-migratory fattening. Nearly all N European and Siberian Ringed Plovers had low but increasing masses during May and the first half June. A slight decline in body mass was observed after mid-June.

***Table 13**. The results of ANOVA in order to assess differences in body mass and primary moult between breeding populations of Ringed and Semipalmated Plovers. The breeding location and sex of a bird were distinguished as independent factors. Six populations were distinguished, according to the distinctions in Table 11. The day number was used as a covariate in 2 analyses.*

	Co-variate	Main effects		Inter-action	% ex-plained	*n*
Variable	Date	Breeding location	Sex			
df		4	1	4		
Body mass		15.916***	1.182	0.590	38.7***	113
Body mass	9.231**	13.270***	0.316	0.536	39.3***	109
Primary moult		2.404*	2.812	1.893	4.8	10
Primary moult	0.162	3.267*	3.154	2.188	5.1	10

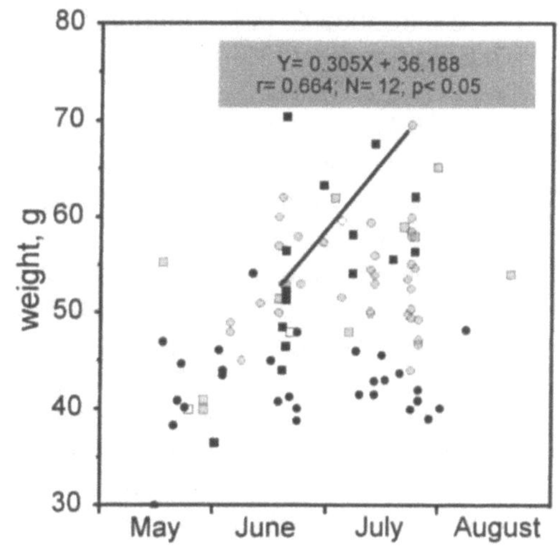

Fig. 24. *Body mass of Ringed- and Semipalmated Plovers during breeding. Explanation of the used symbols: ● Semipalmated Plover; ■ Ringed Plover - NE Canada, Greenland & Iceland; ○ Ringed Plover - N Europe; □ Ringed Plover - Siberia.*

PRIMARY MOULT

The occurrence of primary moult on the breeding grounds differed between breeding areas (Table 13). Of the 10 birds in active wing moult, 6 were Ringed Plovers breeding in NE Canada, Greenland, and Iceland. No W European breeders and nearly no birds from N Europe or Siberia showed active primary moult. Two Semipalmated Plovers were in advanced moult: 1 from Alaska with moult score 38 on May 20 and another from Baffin Island with score 46 on June 12.

DISCUSSION

The taxonomy of Ringed and Semipalmated Plover

Wynne-Edwards (1952) used 2 arguments to separate Semipalmated and Ringed Plover as 2 distinct species: (1) a lack of gene flow and (2) a difference in the amount of webbing between the toes. However, Smith (1969) noticed regular interbreeding between both taxa on NE and E Baffin Island. Chicks from mixed pairs were either *semipalmatus*- or *hiaticula* type as judged from the webbing of the toes and the dorsal coloration. Smith argued that webbing and coloration were inherited characteristics, indicating polymorphism rather than introgression, with the *semipalmatus* type being dominant over the *hiaticula* type. The presence of mixed broods could not be explained by the occurrence of adoption during the first days after hatching, since most of the chicks were marked within one day after hatching. However, the frequency of occurrence of 'egg dumping' in this area is not known. Paulson (1993) noticed several plumage and morphometric differences between both, with the Ringed Plover having: (1) a distinctly wider black breast band, being mostly wider than the wide area in front of it; (2) a more extensive white supercilium; (3) more extensive white on the forehead; (4) somewhat more black on the head; (5) a more conspicuous wing stripe; (6) almost no indication of naked skin around the eye; (7) a relatively longer bill with straight culmen; and (8) less-webbed toes. Our data on the size of the breast band, the white on the forehead, and the bill length are in line Paulson's (1993) conclusions. We can add, that Ringed Plovers also have (9) longer wings; (10) longer tarsi; (11) longer secondaries; (12) longer tails; (13) heavier weights; (14) smaller dusky borders at the tip of primary 5; and (15) more primaries with white along the outer edge.

The degree of morphometrical differentiation between both taxa allows good possibilities to predict the occurrence of any of both. We argued, that if the prediction results should reveal the existence of intermediate probabilities of belonging to both taxa, this would be an indication of a lack of reproductive isolation, which is an argument for considering both to be 1 species. If the predicted probabilities would not reveal any intermediacy, this is not necessarily an argument in favour of reproductive isolation. According to our data, 43 % of the birds from Baffin Island may belong to either taxon. This percentage was comparable with that of the Hudson Bay-sample and markedly lower than in E Quebec, both areas outside the overlap zone. This is an argument against intermediacy. Thus overlap in breeding range does not lead to introgression, either because of effective reproductive isolation (behavioural characters - pro Wynne-Edwards 1952) or to dominance of one type over the other (pro Smith 1969).

Whether the morphometric differentiation between both plover taxa is large enough to consider them as 2 separate allospecies (together forming a superspecies), rather than 2 distinct subspecies belonging to a single species, must be interpreted in relation to comparable situations in other species. For instance, Purple- and Rock Sandpiper are considered as 2 allospecies in 1 superspecies. Their measurements differ less than those of both plover taxa, but they can better be recognized from their plumage characters. Moreover, their breeding ranges are isolated from each other. Ringed and Semipalmated Plover are morphometrically well distinguished, but less so on their plumage while their breeding ranges border each other or do show a slight overlap. Wanting additional research, we propose Ringed and Semipalmated Plover to be considered as 2 allospecies belonging to 1 superspecies *Charadrius [hiaticula]*: *Charadrius [hiaticula] hiaticula* and *Charadrius [hiaticula] semipalmatus*.

Within *Charadrius [h.] hiaticula* 3 distinct breeding populations can be recognized. Our morphometrical data did not favour the uniting of the breeding populations in W Europe with the ones from NE Canada, Greenland, Jan Mayen, and Iceland in a single subspecies as was done in the past. The distinguished subspecies are:

- *C. [h.] h. hiaticula* is breeding in W Europe. The birds of this population are characterized by its large body dimensions: wing-, culmen-, tarsus-, secondary length and culmen height. *C. [h.] h. hiaticula* shows some geographical variation: their large size becomes gradually smaller towards the NE.
- *C. [h.] h. psammodroma* is breeding in NE Canada, on Greenland, Jan Mayen, and Iceland. The average values of wing length, culmen length and culmen height are intermediate between nominate *hiaticula* and *C. [h.] h. tundrae*. The length of the first secondaries is intermediate between the N European and W European Ringed Plover population, but is comparable with the Central Siberian ones. They have relatively short tails and small middle toe lengths. Also, the white patch above the bill is small in size. This population is distinct enough to be recognized as a separate subspecies, and we agree with Hale (1980) to re-introduce *C. [h.] h. psammodroma* of Salomonsen (1930).
- *C. [h.] h. tundrae* is breeding in N Europe and Siberia. The birds of this population are characterized by their short wing and culmen lengths and small culmen heights. They have intermediate middle toe lengths and their tail lengths are comparable with the ones of nominate *hiaticula*. The N European breeding birds have the shortest first secondaries of the Ringed Plovers measured, but the Siberian ones are somewhat longer. Their tarsus lengths are generally also short, although the E Siberian birds form an exception. *C. [h.] h. tundrae* shows geographical variation.

This classification implies that *C. [h.] h. kolymensis*, breeding in E Siberia (Hale 1980) is not regarded as a separate subspecies.

Shifts in breeding range in Fennoscandia

Väisänen (1969) pointed out, that *tundrae* is gradually replaced by nominate *hiaticula* in Fennoscandia. He based his conclusions on changes in egg volume of eggs of *hiaticula* over the past 100 years. A similar change is observed in Redshank *Tringa totanus* from Scandinavia (Hale 1980). Väisänen argued that such a change is due to competition between medium and small sized individuals, in which the medium sized birds benefit by starting earlier with breeding. This conclusion is doubtfull when one realizes that (1) Väisänen used wing length as a parameter to express size; in our study, wing length proved to be a poor character for estimating the structural size of a bird, and without other measurements it is impossible to decide whether Väisänens intermediate birds were small-sized *hiaticula* or large-sized *tundrae*; (2) several other factors than structural size influence the egg volume e.g., condition of the female, age of the female, and the feeding situation prior to egg-laying, and these factors need to be incorporated when using the egg volume as a predictor for the mean size of a population. Väisänen clearly showed that the breeding populations from the N Baltic and N Fennoscandia started to breed 5-10 days earlier over the last 100 years and that the mean egg volumes had increased. If *tundrae* is gradually replaced by *hiaticula* in Fennoscandia over the last 100 years, one should expect to find a greater number of *hiaticula* to be collected on the northern breeding grounds in more recent years. This expectation is supported by our results: 4 of 25 skins, collected north of 63°N since 1958, were mostly resembling nominate *hiaticula*, whilst none of 33 birds collected before 1958 resembled *hiaticula* ($Chi^2 = 4.762$; $df = 1$; $P < 0.05$). These data are thus a support of Väisänen's replacement theory.

SUMMARY

Semipalmated Plovers show little morphometrical variation over the breeding range, but Ringed Plover shows much. We propose to consider both as allospecies of a single superspecies *Charadius [hiaticula]*. The observed differences in morphometrics in the Ringed Plover resulted in the recognition of 3 subspecies: (1) *C. [h.] h. psammodroma* - NE Canada, Greenland, and Iceland; (2) *C. [h.] h. hiaticula* - W Europe; (3) *C. [h.] h. tundrae* - N Europe and Siberia. *C. h. kolymensis* is not considered to be a valid subspecies. Predictions of breeding origin were based upon the construction of 3-6 POSCON groups per sex, depending on the number of measurements analysed. All 7 measurements included were necessary for prediction purposes. Plumage characters might provide additional information on geographical differentiation. The zone of overlap between Semipalmated and Ringed Plover on Baffin Island is larger than previously thought. Ringed Plovers similar to those of W Europe occur as far N as Trondheim in W Norway. Our data support Väisänen's (1969) replacement theory.

Chapter 5

Eurasian Golden Plover

-

Pluvialis apricaria

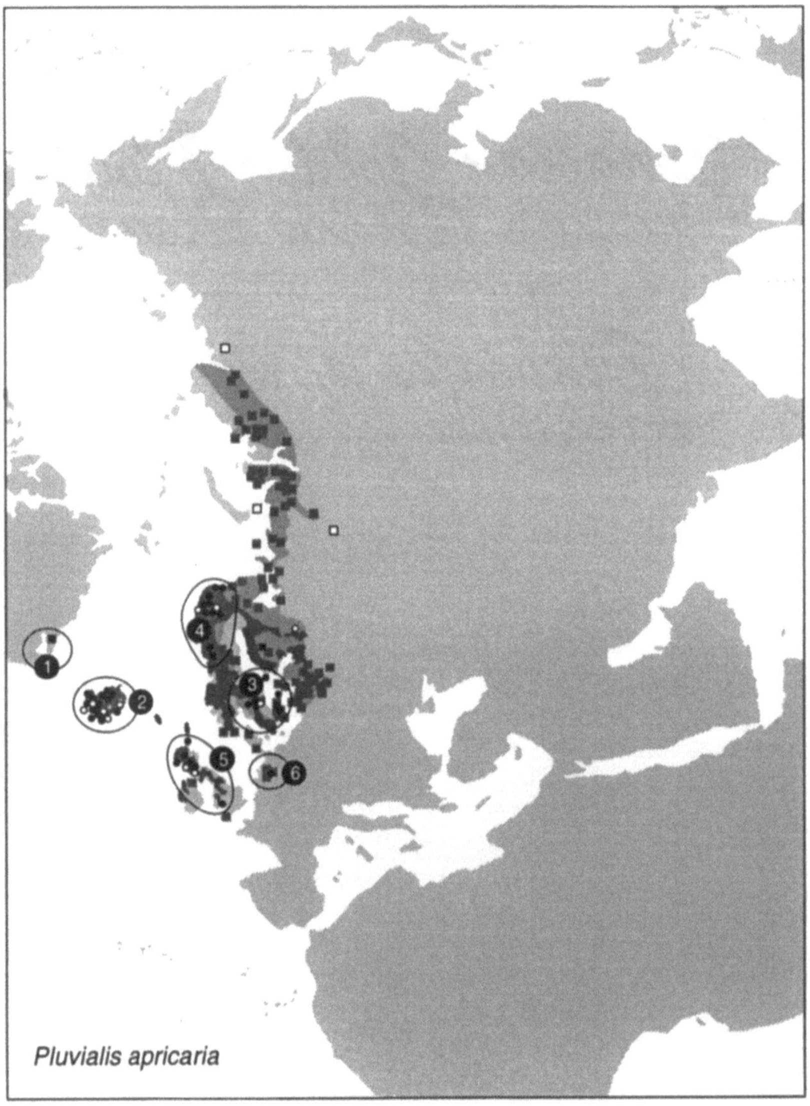

Fig. 25. *Breeding distribution of the Golden Plover. For explanation of the used symbols see Chapter 3: 'Introduction to the species' accounts'.*

INTRODUCTION

The Eurasian Golden Plover breeds from E Greenland to the Khatanga River in Siberia (De Korte 1975, Cramp & Simmons 1983, Tomkovich 1992b - Fig. 25). The W and Central Siberian breeding grounds are shared with the Pacific Golden Plover *Pluvialis fulva*, in E Siberia the Pacific Golden Plover is the only species, and across the N American breeding grounds the Pacific Golden Plover is replaced by the American Golden Plover *Pluvialis dominica* (Knox 1987, Connors 1983, Connors *et al.* 1993).

Fabricius & Hald-Mortensen (1969) and Glutz *et al.* (1975) summarized the geographical variation in the Eurasian Golden Plover. Most authors distinguished 2 different races:

- *P. a. apricaria* breeding on the British isles, in the Netherlands, Germany, Denmark, S Scandinavia and the S Baltic.
- *P. a. altifrons* breeding in E Greenland, Iceland, the Faeroes, N Scandinavia, NW Russia and W Siberia.

The trenchant characters between both races formerly consisted solely in the retention of the number of winter feathers in the breeding plumage. Without other criteria in existence, Glutz *et al.* (1975) supported the conclusion of Wynne-Edwards (1957) and Kozlova (1962) stating that both taxa cannot be considered as valid subspecies. An argument in favour of the recognition of 2 subspecies was the disjunct occurrence of 2 breeding populations, but Byrkjedal (1978) showed that SW Norwegian breeders had mixed characters, even though this population was earlier regarded as the only pure *apricaria* type of Norway. In more recent literature, some authors considered both populations to be colour forms (Parr 1980, Hayman *et al.* 1986), others preferred them to be subspecies (Byrkjedal 1978, Hale 1980, Jukema 1989). Cramp & Simmons (1983) explained the difference in plumage to be due to a difference in time of breeding moult available in spring: since (1) all populations start at about the same time, and (2) southern breeders stop earlier, they will have a less advanced pre-breeding moult. Moreover, it is not clear from Linnaeus' type locality, what population is involved. With Hartert's (1920) restriction of the type locality to Öland (Sweden), *apricaria* is in use for the southern population and *altifrons* for the northern. But some authors (e.g., Hale 1980) use *apricaria* for northern birds and *oreophilos* for southern ones. Here we follow Hartert's taxonomic treatment (Hartert 1912-21). Roselaar in Cramp & Simmons (1983) and Jukema (1989) showed that not only different proportions of winter feathers were retained, but that also breast feathers with different patterns were moulted (in different proportions) during the prenuptial moult.

In this study, virtually no attention was given to plumage differences, our prime attention being the multivariate analysis of measurements. Nevertheless, we checked the occurrence of the so-called 'yellow-striped' breast feathers in some of the breeders examined. Jukema & Piersma (1987) drew attention to these yellow-striped breast feathers, which were moulted during breeding.

MORPHOMETRICAL DIFFERENTIATION

In total, 97 specimens were examined from Iceland, Scandinavia and the British isles (Table 14). Icelandic and British breeders were present on the breeding grounds from the second half of March, S Scandinavian Golden Plovers by mid-

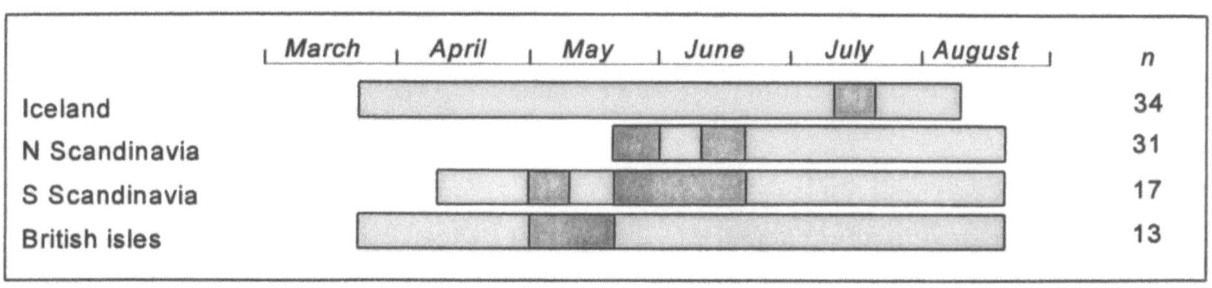

Fig. 26. *Collection dates of Golden Plovers. Most birds had been collected during the periods marked by the darkly shaded bars. The sample sizes are presented on the right side.*

Table 14. *Mean measurements of Golden Plovers from the breeding grounds. The mean values with standard deviations and sample sizes are given.*

Locality	Wing length			Culmen length			Tarsus length			Secondary length			Tail length			Middle toe length			Culmen height		
	x	*SD*	*n*	*x*	*SD*	*n*	*x*	*SD*	*n*	*x*	*SD*	*n*	*x*	*SD*	*n*	*x*	*SD*	*n*	*x*	*SD*	*n*
♂♂																					
2 Iceland	190.7	5.6	24	22.9	1.0	24	40.7	1.7	24	79.2	1.4	12	67.8	2.0	12	33.4	1.5	8	4.6	0.2	8
3 S Scandinavia	189.6	5.1	7	23.4	1.3	7	42.6	1.4	7	78.7	2.1	7	66.3	4.0	7	34.7	1.2	7	4.4	0.2	6
4 N Scandinavia	191.3	3.3	24	22.7	1.0	24	41.0	1.4	24	79.3	1.6	24	68.2	3.6	24	33.3	1.2	24	4.5	0.2	22
5 British isles	189.0	1.7	8	23.0	1.3	8	40.2	1.4	8	80.3	1.5	8	68.3	3.9	8	33.9	1.5	8	4.8	0.1	2
Pacific - China	181.0	-	1	24.6	-	1	43.5	-	1	72.0	-	1	60.0	-	1	32.0	-	1	5.0	-	1
♀♀																					
1 Greenland	188.0	4.3	2	23.0	1.1	2	40.0	1.9	2	78.6	5.6	2	63.2	1.4	2	33.5	1.5	2	-	-	0
2 Iceland	196.4	3.9	10	22.5	1.1	10	40.2	2.3	10	79.0	1.4	2	70.0	1.2	5	33.0	1.2	4	4.9	0.1	2
3 S Scandinavia	193.0	5.1	10	23.1	0.5	10	41.1	1.8	10	80.1	1.6	10	66.9	2.8	10	33.6	0.9	10	4.6	0.2	6
4 N Scandinavia	192.3	3.9	7	23.6	0.8	7	40.5	2.0	7	80.4	2.9	7	68.4	4.3	7	32.6	1.1	7	4.5	0.2	6
5 British isles	196.3	5.0	5	23.1	0.9	5	41.5	1.7	5	82.0	1.0	5	70.8	2.8	5	33.1	0.8	5	4.5	0.1	2
6 Netherlands	190.0	-	1	23.7	-	1	39.3	-	1	77.0	-	1	65.0	-	1	32.5	-	1	4.5	-	1

Table 15. *Overall results of MANOVA for 4 breeding populations of the Golden Plover in combination with univariate ANOVA's in which the different measurements were analysed separately. Significance levels: *0.01 < P < 0.05; **0.001 < P < 0.01; ***P < 0.001.*

Variables	Analysis I			Analysis II			Analysis III		
	wing, culmen, tarsus, secondary, tail, middle toe, culmen height			**wing, culmen, tarsus**			**wing, culmen**		
n	55			97			97		
Multivariate analysis of variance									
Effects	Interaction	Area	Sex	Interaction	Area	Sex	Interaction	Area	Sex
Pillai's trace	0.471	0.710	0.283	0.163	0.153	0.178	0.121	0.091	0.160
F-value	1.144	1.906*	2.317*	1.705	1.597	6.271***	1.905	1.418	8.366
Hypoth. & error *df*	21/ 129	21/ 129	7/ 41	9/ 267	9/ 267	3/ 87	6/ 178	6/ 178	2/ 88
Univariate F tests									
df	3/ 47	3/ 47	1/ 47	3/ 89	3/ 89	1/ 89	3/ 89	3/ 89	1/ 89
wing length	1.030	0.541	8.918**	2.012	1.143	16.135***	2.012	1.143	16.135***
culmen length	1.349	0.996	0.583	0.909	0.855	0.289	0.909	0.855	0.289
tarsus length	1.596	3.067*	0.013	1.673	2.019	0.446			
secondary length	0.189	1.602	0.936						
tail length	0.127	0.797	0.041						
middle toe length	1.258	2.817*	7.087*						
culmen height	1.032	3.007*	0.098						

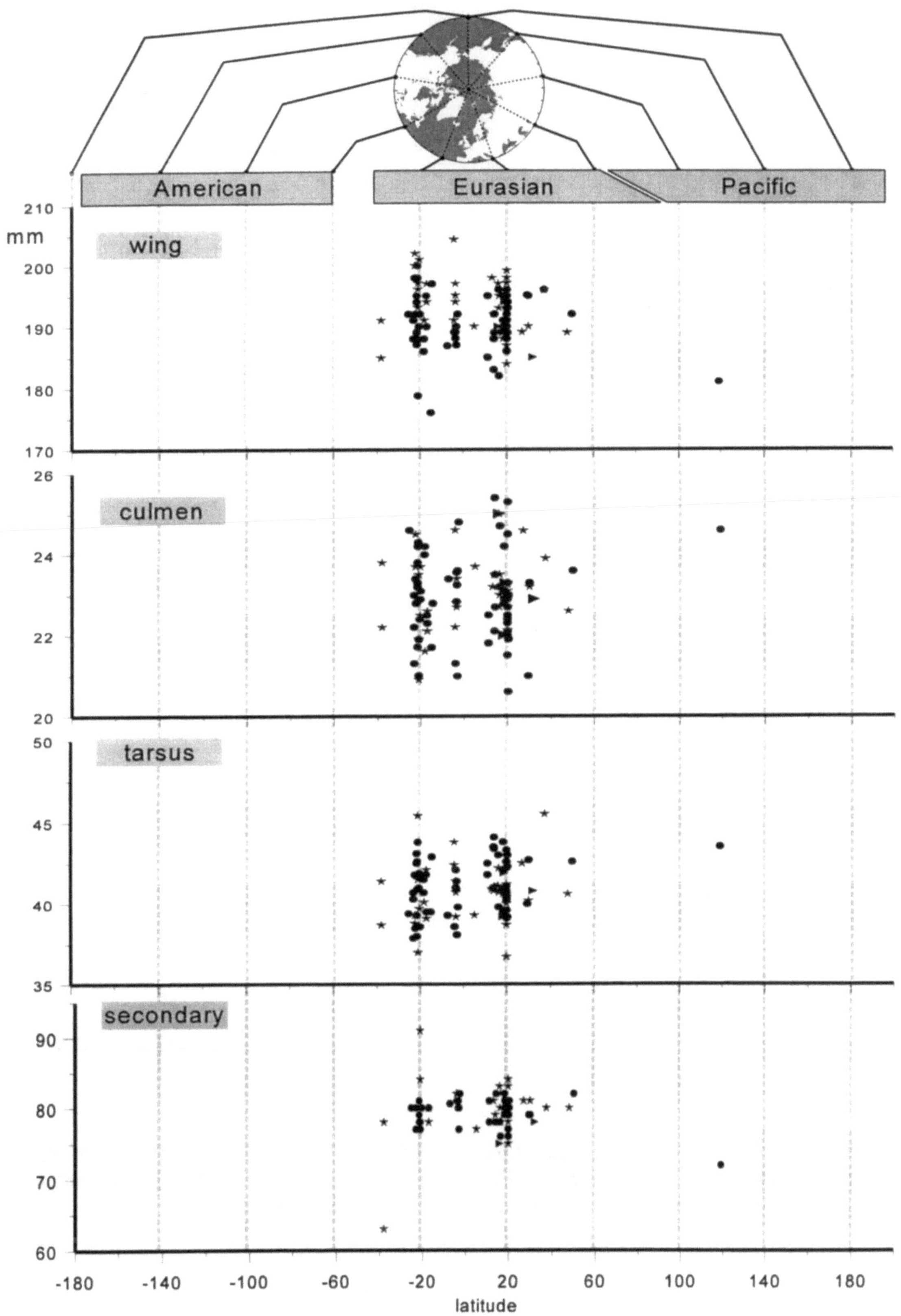

Fig. 27. *Measurements of Golden Plovers from the breeding grounds. The following measurements are presented from top to bottom: wing length, culmen length, tarsus length and secondary length (description in chapter 2). Explanation of the used symbols: ● ♂♂, ★ ♀♀, ▸ sex unknown.*

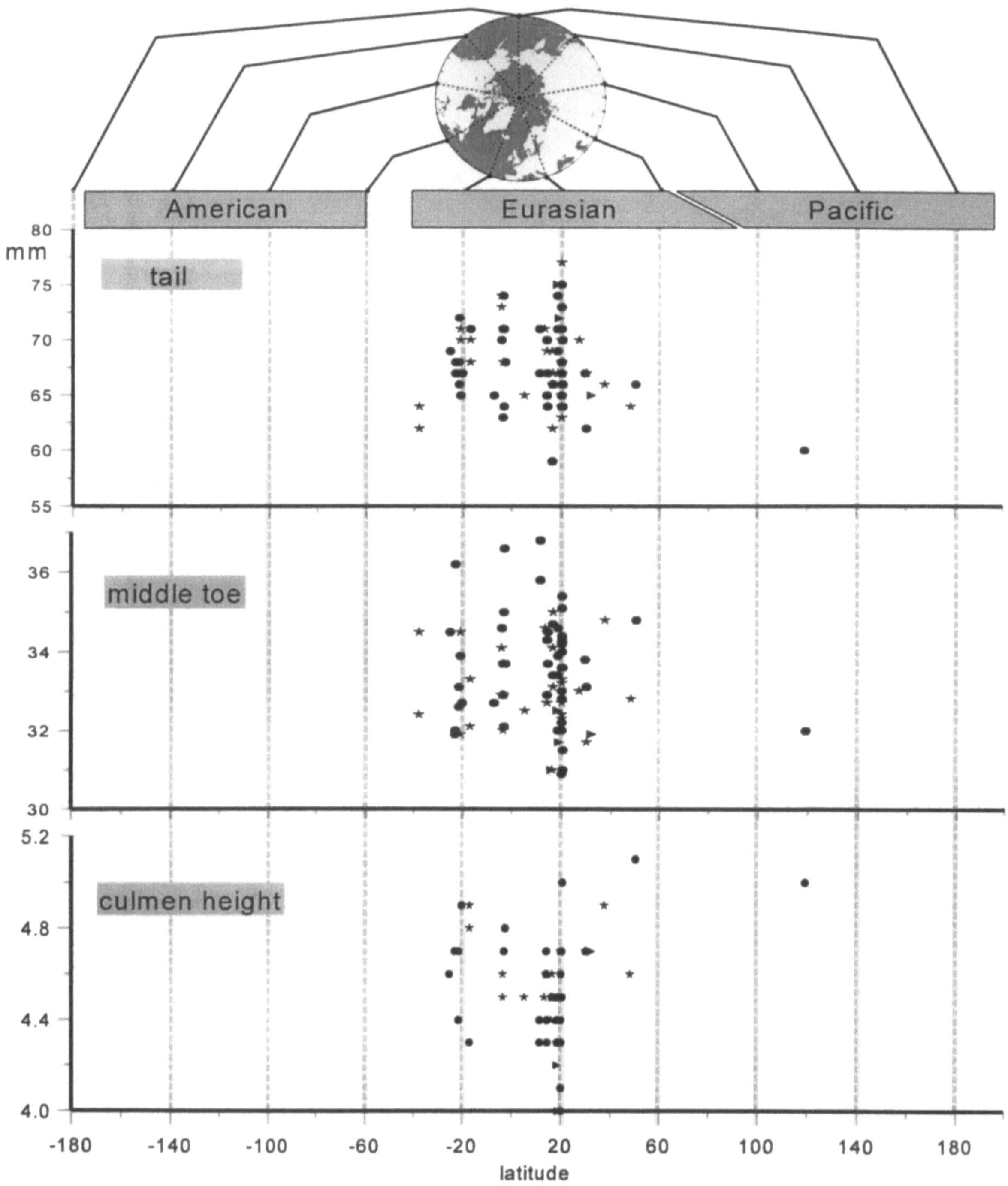

Fig. 28. *Measurements of Golden Plovers from the breeding grounds. The following measurements are presented from top to bottom: tail length, middle toe length and culmen height (description in chapter 2). Explanation of the used symbols:* ● ♂♂, ★ ♀♀, ▸ *sex unknown.*

Table 16. *Sexual size dimorphism of the Golden Plover. The mean percentage difference between the measurements of both sexes was calculated for every single variable and breeding population, according to: $(x_{♀}-x_{♂})/x_{♂}*100$. A negative value thus implies smaller measurements for ♀♀. The calculations were based upon the mean measurements in Table 14. The obtained values were averaged per variable and summed per breeding population.*

Locality	Wing	Culmen	Tarsus	Secondary	Tail	Middle toe	Culmen height	Σ
Iceland	2.1	-0.5	-1.3	-0.1	-2.2	-0.7	4.3	1.6
S Scandinavia	1.4	-0.9	-3.8	2.7	1.1	-4.0	4.5	1.0
N Scandinavia	0.4	2.7	-1.2	1.4	-1.0	-0.9	0.0	1.4
British isles	3.7	-0.4	-1.0	2.2	3.5	-2.6	-2.1	3.3
x	2.4	0.2	-1.8	1.5	0.4	-2.0	1.7	
SD	1.2	1.7	1.3	1.2	2.5	1.6	3.3	

Table 17. *Pearson's product-moment correlations between measurements of Golden Plovers. Both sexes are separately analysed. Two-tailed probabilities: * 0.01 < P < 0.05; ** 0.001 < P < 0.01; *** P < 0.001.*

	Wing	Culmen	Tarsus	Secondary	Tail	Middle toe	Culmen height	
Wing		0.322	0.258	0.353	0.492*	0.020	0.336	
Culmen	0.178		0.311	0.212	0.065	-0.116	-0.249	
Tarsus	0.093	0.184		0.193	-0.018	0.345	0.501	
Secondary	0.468**	0.184	-0.019		0.615**	-0.018	-0.163	♀♀
Tail	0.241	-0.020	-0.128	0.279		-0.071	-0.160	
Middle toe	-0.034	0.219	0.327	0.377	0.160		-0.031	
Culmen height	-0.369	0.259	-0.125	-0.241	-0.096	-0.022		
				♂♂				

half of May (Fig. 26). Most birds (63.2 %) examined were ♂♂.

Initially 6 breeding locations were distinguished (Fig. 25 & Table 14). The measurements are presented in the figs. 27 and 28. Since too few birds from Greenland and the Netherlands were measured, both these populations were excluded from data-analysis.

The Golden Plover shows distinct sexual variation and little or no geographical variation (Table 15). Females had longer wings (Table 16). Since wing length was regularly correlated with secondary- and tail length (Table 17), sexes may differ in other wing- and tail dimensions as well. On the other hand, ♀♀ had tarsi and middle toes systematically shorter than ♂♂ (Table 16). Significant sexual differences *within* breeding populations were found in the S Scandinavian population (4 measurements) and the Icelandic and British population (2 measurements - Table 18 shaded areas). Four of the 7 variables contributed significantly to the distinction between samples (Table 15 - analysis I). Sexes differed in wing length and middle toe length. Tarsus length, culmen height and middle toe length differed significantly over the breeding areas. Measuring length of culmen, secondary, and tail of the Golden Plover did not seem to be necessary for these populations (Table 15 - analyses I and II).

When only wing- and culmen lengths were analysed, sexes appeared to differ but breeding populations not (Table 15 - analysis III). The differences between both sexes completely relied upon wing length, not upon culmen length. This outcome is in line with the results presented above.

Table 18. *Results of MANOVA-analysis concerning the morphometric differentiation in the Golden Plover. Pillai's trace statistics are presented (Norušis 1992c). In these series of pairwise comparisons 8 groups (both sexes, 4 populations) were compared pairwise on the basis of 4 measurements in the upper table (wing-, tarsus-, middle toe length and culmen height) and 2 measurements in the lower table (wing- and culmen length). There were 6 and 60 df in the '6 variable' situation and 2 and 85 df in the second group of analyses. Significance levels: *0.01 < P < 0.05; **0.001 < P < 0.01; ***P < 0.001.*

Locality	n	♀ Iceland	♀ S Scandinavia	♀ N Scandinavia	♀ British isles	♂ Iceland	♂ S Scandinavia	♂ N Scandinavia
Iceland ♀	2							
S Scandinavia ♀	6	0.105						
N Scandinavia ♀	7	0.133	0.058					
British isles ♀	2	0.058	0.032	0.039				
Iceland ♂	7	**0.079**	0.107	0.077	0.080			
S Scandinavia ♂	6	0.339***	**0.245***	0.287**	0.229*	0.366***		
N Scandinavia ♂	23	0.188	0.105	**0.030**	0.081	0.135	0.289**	
British isles ♂	2	0.124	0.179	0.197*	**0.179**	0.107	0.307**	0.229*

Locality	n	♀ Iceland	♀ S Scandinavia	♀ N Scandinavia	♀ British isles	♂ Iceland	♂ S Scandinavia	♂ N Scandinavia
Iceland ♀	10							
S Scandinavia ♀	11	0.070*						
N Scandinavia ♀	7	0.091*	0.006					
British isles ♀	5	0.008	0.027	0.042				
Iceland ♂	24	**0.142***	0.021	0.016	0.071*			
S Scandinavia ♂	8	0.150***	**0.029**	0.009	0.082*	0.021		
N Scandinavia ♂	24	0.106**	0.021	**0.028**	0.057	0.010	0.044	
British isles ♂	8	0.151***	0.038	0.021	**0.091***	0.013	0.007	0.031

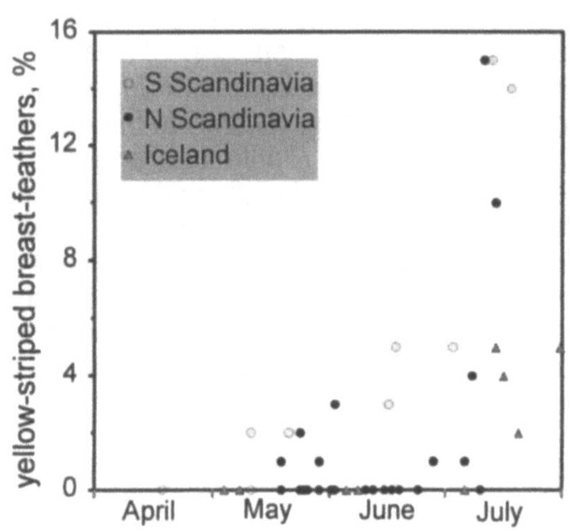

Fig. 29. *Seasonal occurrence of yellow-striped breast feathers in the Golden Plover.*

Yellow-striped breast-feathers occurred in low frequencies in all 3 breeding populations and in both sexes (Fig. 29). These feathers appeared in between the breeding plumage during the second part of the breeding season as was concluded earlier by Jukema & Piersma (1987). Yellow-striped feathers occurred in all samples, and can thus not be used to predict a bird's sex or breeding origin.

PREDICTION OF BREEDING ORIGIN

The results presented above do not permit a distinction between *apricaria* and *altifrons* on morphometrical grounds. To some extent, it is possible to predict a bird's sex with morphometric data. The degree of differentiation remains nevertheless small.

BODY MASS

Mass data of 17 ♂♂ and 5 ♀♀ from Iceland and Scandinavia were obtained from the skin's labels (Fig. 30): ♀♀ were 40 gm heavier than ♂♂ during May. Sexual differences explained up to 48 % of the variation (ANOVA). The breeding location did not influence mass differences.

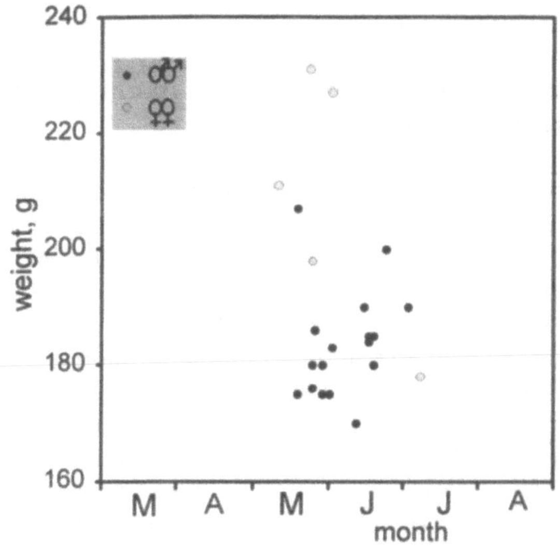

Fig. 30. *Body mass of Golden Plovers during the breeding season: ♂♂ - black circles; ♀♀ - open circles.*

PRIMARY MOULT

Eighteen out of 62 breeding birds were in active primary moult: 1 by late May, 3 in June, 11 in July and 3 in August (Fig. 31). More than half of the birds present during July and August were in primary moult. Primary moult started by shedding 3 to 4 primaries simultaneously. Moult was suspended when 6 to 7 primaries were renewed. The moult schedule apparently did not depend on the breeding location (Iceland, S or N Scandinavia) or a bird's sex.

DISCUSSION

The taxonomy of the Eurasian Golden Plover

This multivariate morphometric analysis did not result in a reliable separation between *apricaria* and *altifrons*. Perhaps both can be recognized when more details on plumage characters are collected.

The existence of a partial breast-feather moult during breeding

Jukema & Piersma (1987) were the first to describe an early start of post-breeding moult in the Golden Plover, in which yellow feathers with brown stripes start to grow in between the existing summer-plumage breast feathers. They noted that: (1) the later a breeding bird was collected the higher the probability that it had yellow-striped feathers on its breast; (2) this feather type was not observed in 533 non-breeding spring migrants, but was regularly present at arrival in the Netherlands after the breeding season (mainly August-September); and (3) 3 birds were trapped just before their departure to the breeding grounds without any yellow-striped feathers and retrapped at their return with several yellow-striped ones (Jukema & Piersma 1987). Our data of breeding birds are in line with these data: yellow-striped feathers gradually appear from May onwards in the Icelandic and Scandinavian samples (Fig. 29). As a consequence, the occurrence of these feathers cannot be used to separate populations or sexes. Jukema & Piersma (1987) argued, that this partial breeding moult might indicate a reduced need for a distinctive plumage in the course of the breeding season, when birds have to care for eggs and chicks. A distinctive plumage might be important early in the breeding season when couples are

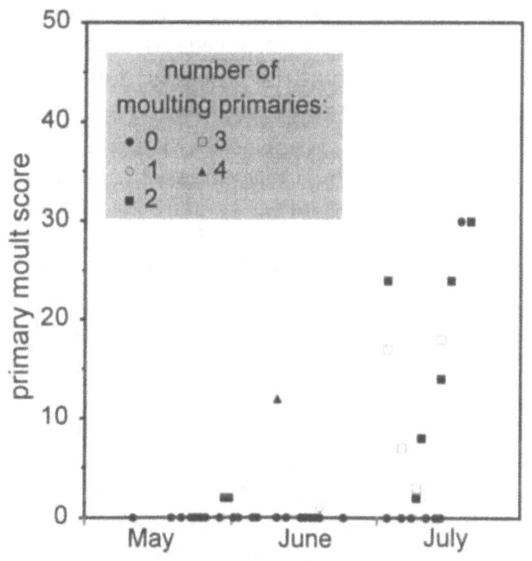

Fig. 31. *Primary moult of Golden Plovers on the breeding grounds. The different symbols represent the number of simultaneously missing or growing primaries.*

formed. However, from the timing of the moult (compare Fig. 29 and 31) it is clear that the moult to yellow-striped feathers is part of the early stages of post-breeding moult. As in the early stages of pre-breeding moult, when feathers intermediate in character between non-breeding and breeding plumage appear rather than full breeding-plumage feathers, the hormonal balance of the bird during the early stages of post-breeding leads to a feathering intermediate in character. Later on during post-breeding moult, these feathers are replaced again by normal non-breeding feathers. The appearance of scattered feathers intermediate between breeding and non-breeding in the early stages of a moult is a common feature in many waders, especially those with a protracted moult season and an opportunistic migration schedule, less so in long-distance migrants with very synchronized migration and moult periods. The small number of intermediate feathers grown by a bird and the occurrence of these feathers in a restricted number of birds in a sample does not favour the theory that a distinctive more cryptic plumage is needed during the egg- and chick-caring phase of the breeding cycle.

Morphometrical differentiation between both sexes

According to Jukema & Piersma (1992), both sexes of the Golden Plover are morphometrically inseparable. However, our data showed the opposite: ♀♀ were larger-sized than ♂♂. This different outcome is probably due to a different statistical treatment of the data. Jukema & Piersma (1992) performed several (33?) univariate *t*-tests with 2 independents (age and sex). The null hypothesis of no difference was rejected in none of these tests. By doing so, the overall power of the tests will be far too low and the outcome will be too prudent. Since 2 c.y. birds were not recognized amongst our breeding birds, it sufficed to analyse 7 variables simultaneously in a MANOVA with sex and type (*apricaria* and *altifrons*) being the independents: distinguishing sexes then appeared to be important for explaining the observed variation (Table 15). We obtained significant differences also when we contrasted (with MANOVA) both sexes within the S Scandinavian, Icelandic and British samples, and we feel confident that larger samples of other breeding populations would provide the same results. The mean morphometric values presented by Jukema & Piersma (1992) also show a slight but systematic size difference between ♂♂ and ♀♀, the latter again being somewhat larger. Multivariate analysis of their data might reveal a significant difference between the sexes as well. If not, absence of significant sexual difference may well be caused by mixing of different breeding populations on the wintering grounds.

The occurrence of primary moult on the breeding grounds

The primary moult could be completed in the second half of August, if the primary moult scores of breeding birds were used to estimate the timing of completion. However, Jukema (1982) already showed that completion occurs around 1 October with an estimated moult duration of 125 days. This is a very long interval, since primary moult in the related Grey Plover was estimated to take about 90-100 days (Boere 1976). This long duration can be explained by the occurrence of suspension of primary moult at the end of the breeding season when birds prepare for departure. This suspension is confirmed by Golden Plovers trapped in the Netherlands during August and September: several trapped adults had recently thrown out just one central primary (nr. 5, 6 or 7) with the inner primaries already being renewed (Jukema 1982). However, interruption does not seem to be the only explanation for a long moult duration. It also seems as if the number of simultaneously growing primaries is strongly reduced towards the outer primaries in order to maintain the flight capacity as much as possible.

SUMMARY

No morphometric differentiation between the northern (Iceland and N Scandinavia) and southern (S Scandinavia and British isles) samples could be found. On average, ♀♀ were larger and heavier than ♂♂. We agree with the findings of Jukema & Piersma (1987), that a partial breeding moult of breast feathers occurs.

Chapter 6

Grey Plover

-

Pluvialis squatarola

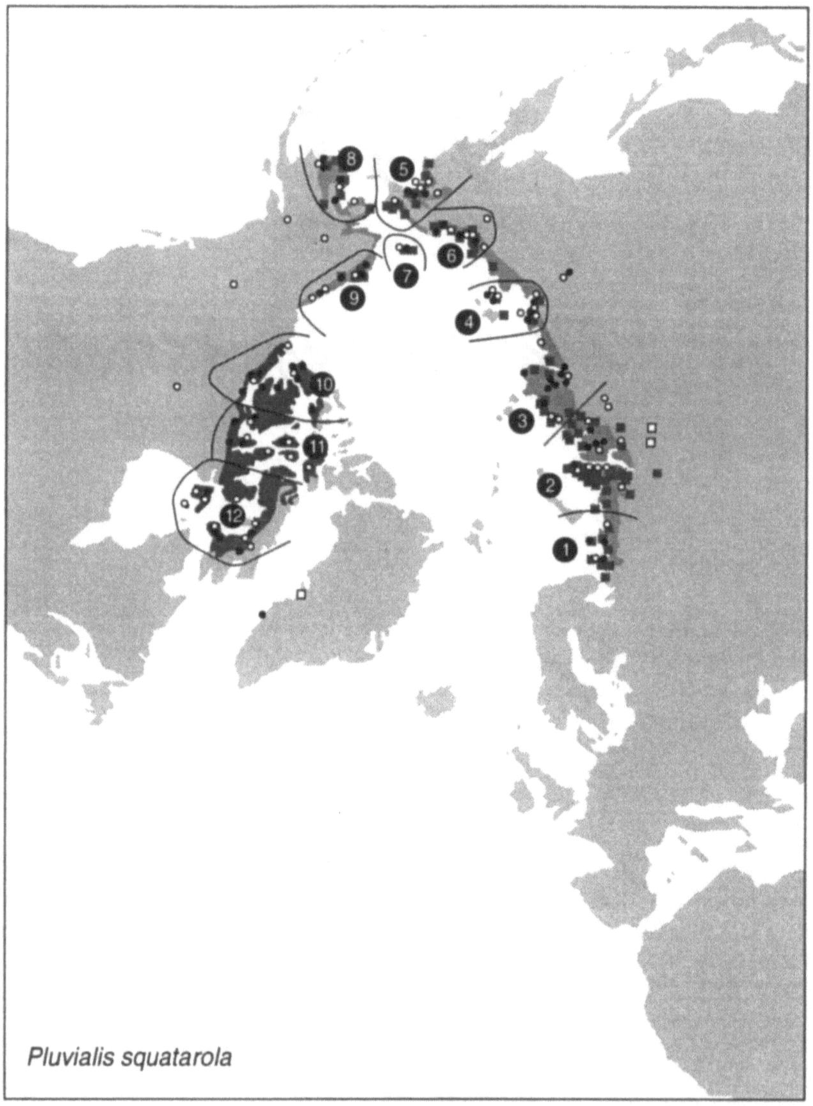

Fig. 32. *Breeding distribution of the Grey Plover. For explanation of the used symbols see Chapter 3: 'Introduction to the species' accounts'.*

INTRODUCTION

The breeding range of the Grey Plover extends from the Kanin Peninsula in N Europe throughout arctic Siberia and Alaska to Baffin Island in N Canada (Fig. 32). The species breeds on Wrangel Island, but not on most high arctic islands, nor on the northern New Siberian Islands and northern Melville Island. The breeding season starts in the second half of May or in the first half of June (Fig. 33).

Most authors considered the Grey Plover as a monotypic species with some clinal variation (Peters 1934, Low 1938, Dement'ev 1939, AOU 1957, Vaurie 1964, Glutz *et al.* 1975, Cramp & Simmons 1983, Hayman *et al.* 1986, Stepanyan 1990). Others recognized different subspecies. In all studies thus far 2 major problems were met: (1) no authors studied birds from the entire breeding range; and (2) clinal variations throughout the breeding range obscure discontinuities. Siberian breeding birds have been described as *hypomelaena* by Pallas in 1827, those of N Canada as *cynosurae* by Thayer & Bangs (1914). Hartert (1912-21) recognized *hypomelaena*, but had no material to check the validity of *cynosurae*. Peters (1934) agreed that some variation occurred, but considered the differences too small for recognition of different subspecies. According to Roselaar in Cramp & Simmons (1983) the variation in bill dimensions in Siberia might be discontinuous, Alaskan and

Table 19. *Mean measurements of Grey Plovers from the breeding grounds. The mean values with standard deviations and sample sizes are given. Numbers in the first column refer to the locations of Fig. 32.*

	Locality	Wing	Culmen	Tarsus	Secondary	Tail	Middle toe	Culmen height
				♂♂				
1	Archangelsk	201.2±1.9(5)	29.9±1.7(5)	47.2±1.8(5)	81.5±1.0(4)	71.7±1.0(4)	35.3±1.8(4)	7.1±0.3(5)
2	Yamal & Gydansk	204.7±3.7(10)	28.7±1.4(10)	46.9±2.1(10)	83.8±2.4(6)	73.2±1.7(6)	35.0±1.8(3)	6.9±0.5(10)
3	Taymyr	205.7±4.3(19)	29.0±1.4(19)	47.8±1.8(19)	83.4±3.3(14)	72.9±4.5(14)	34.6±1.6(12)	6.9±0.5(18)
4	Lena Delta & New Sib. I.	204.6±6.5(10)	30.5±1.6(10)	47.6±2.0(10)	85.2±2.3(6)	73.7±3.1(6)	35.7±0.8(5)	7.2±0.6(10)
5	Anadyr	204.3±6.1(13)	30.9±1.1(13)	49.8±1.4(12)	84.2±2.9(5)	74.2±5.1(5)	36.5±1.7(5)	7.7±0.3(12)
6	Indigirka & Kolyma	205.8±5.6(11)	31.1±1.3(11)	49.1±2.6(11)	83.8±1.1(5)	75.4±3.4(5)	36.3±1.4(5)	7.4±0.5(10)
7	Wrangel Island	208.6±4.2(20)	29.4±1.5(20)	46.4±1.4(20)	84.7±1.5(20)	72.7±3.0(20)	34.8±1.5(18)	7.0±0.4(20)
8	W Alaska	207.0±5.5(13)	31.2±1.5(13)	49.7±1.6(13)	85.0±2.3(9)	74.9±1.9(9)	34.7±1.1(8)	7.2±0.4(13)
9	N Alaska	202.6±5.1(16)	30.8±2.0(16)	48.4±1.9(16)	82.1±2.9(16)	72.9±2.9(16)	34.9±1.5(15)	7.4±0.5(11)
10	W Canada	197.1±3.0(15)	30.3±1.2(15)	45.7±1.6(15)	80.0±1.3(15)	68.6±3.2(15)	34.3±0.7(11)	7.1±0.3(15)
11	Central Canada	196.7±4.4(18)	29.5±1.7(18)	45.6±1.5(18)	80.2±2.3(15)	69.9±3.3(15)	33.7±1.0(15)	7.1±0.3(17)
12	E Canada	196.6±5.6(20)	30.6±1.4(20)	45.3±1.9(19)	81.2±2.3(9)	69.9±4.6(9)	35.0±1.2(8)	7.2±0.5(20)
				♀♀				
1	Archangelsk	200.7±2.9(3)	29.5±2.4(3)	47.7±2.9(3)	83.9±3.5(2)	67.5±4.9(2)	36.0±0.4(2)	6.6±0.4(3)
2	Yamal & Gydansk	204.9±3.7(9)	29.3±1.4(7)	47.2±1.9(9)	83.6±1.8(5)	70.2±1.9(5)	33.9±0.7(5)	6.9±0.5(7)
3	Taymyr	205.2±4.5(9)	28.7±0.8(9)	47.2±1.5(9)	83.2±2.6(6)	70.3±1.6(6)	35.3±0.8(6)	6.7±0.4(8)
4	Lena Delta & New Sib. I.	207.7±3.8(3)	29.5±1.8(3)	46.2±1.9(3)	83.5±0.7(2)	73.5±4.9(2)	33.7±0.9(2)	6.9±0.7(3)
5	Anadyr	201.2±1.4(2)	31.8±2.3(2)	46.8±4.9(2)	82.0 - (1)	73.0±1.4(2)	34.3±0.8(2)	-
6	Indigirka & Kolyma	208.8±5.4(5)	31.1±1.3(5)	48.5±0.9(5)	88.0 - (1)	74.0 - (1)	36.2 - (1)	7.7±0.4(5)
7	Wrangel Island	213.7±5.8(9)	28.9±2.2(9)	47.1±1.7(9)	85.0±2.2(7)	74.7±2.0(7)	34.5±1.6(7)	7.0±0.2(9)
8	W Alaska	206.5±5.6(7)	29.7±0.6(7)	48.9±2.9(7)	84.7±4.0(6)	72.2±3.1(6)	36.1±1.5(5)	7.1±0.5(6)
9	N Alaska	202.8±4.5(15)	30.0±1.2(15)	48.0±1.6(15)	83.3±2.1(15)	71.2±2.6(15)	34.4±1.3(10)	7.3±0.3(12)
10	W Canada	197.0±2.0(7)	30.1±0.8(7)	45.1±1.6(7)	79.3±1.5(6)	68.3±1.6(6)	34.0±1.1(5)	7.1±0.5(7)
11	Central Canada	198.1±5.8(7)	29.1±1.8(7)	45.2±2.1(7)	79.9±1.3(7)	67.7±4.5(7)	33.3±1.8(6)	7.0±0.4(7)
12	E Canada	195.7±3.6(14)	30.4±1.5(15)	45.0±1.6(15)	77.8±1.5(5)	65.8±4.6(5)	33.4±0.7(4)	7.1±0.4(15)

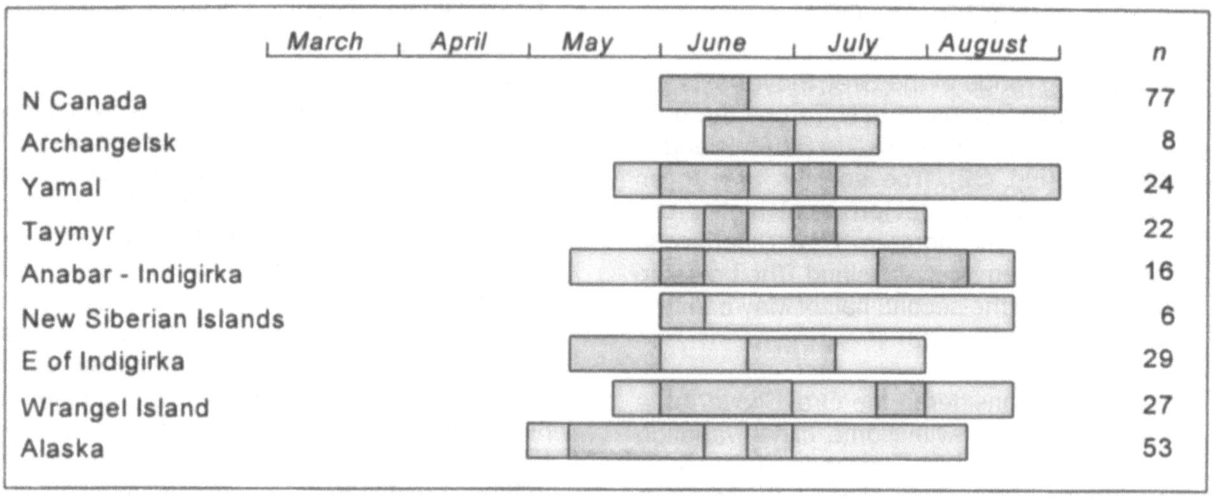

Fig. 33. *Collection dates of Grey Plovers. Most birds had been collected during the periods marked by the darkly shaded bars. The sample sizes are presented on the right side.*

Table 20. *Overall results of MANOVA for all distinguished breeding populations in the Grey Plover in combination with univariate ANOVA's in which the different measurements were analysed separately.*

Variables	Analysis I wing, culmen, tarsus, secondary, tail, middle toe, culmen height			Analysis II wing, culmen, tarsus			Analysis III wing, culmen		
n	150			256			258		
Multivariate analysis of variance									
Effects	Interaction	Area	Sex	Interaction	Area	Sex	Interaction	Area	Sex
Pillai's trace	0.468	1.495	0.076	0.120	0.949	0.026	0.079	0.712	0.014
F-value	0.910	3.135***	1.412	0.877	9.759***	2.056	0.876	11.771***	1.704
Hypoth. & error *df*	70/ 889	77/ 889	7/ 121	33/ 696	33/ 696	3/ 230	22/ 468	22/ 468	2/ 233
Univariate F tests									
df	10/127	11/127	1/127	11/232	11/232	1/232	11/234	11/234	1/234
wing length	0.817	12.157***	0.465	0.906	21.037***	0.605	0.902	21.279***	0.656
culmen length	0.513	3.645***	0.044	0.667	5.148***	1.970	0.671	5.150***	1.946
tarsus length	0.778	7.637***	0.689	0.730	11.281***	3.821			
secondary length	1.257	10.738***	0.231						
tail length	1.073	5.460***	4.267*						
middle toe length	1.183	3.070**	0.316						
culmen height	0.895	3.732***	0.753						

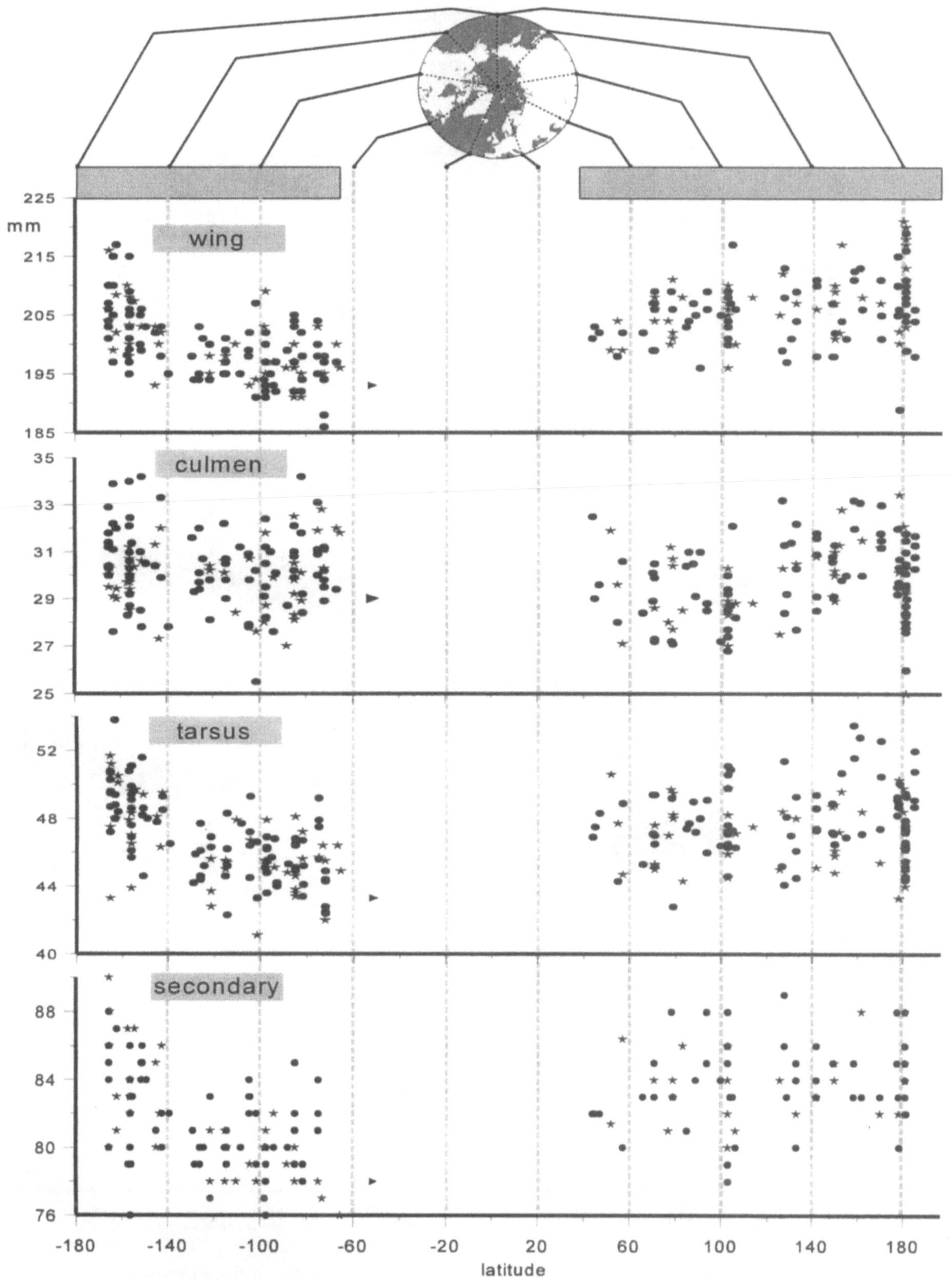

Fig. 34. *Measurements of Grey Plovers from the breeding grounds. The following measurements are presented from top to bottom: wing length, culmen length, tarsus length and secondary length (description in chapter 2). Explanation of the used symbols: ● ♂♂, ★ ♀♀, ▸ sex unknown.*

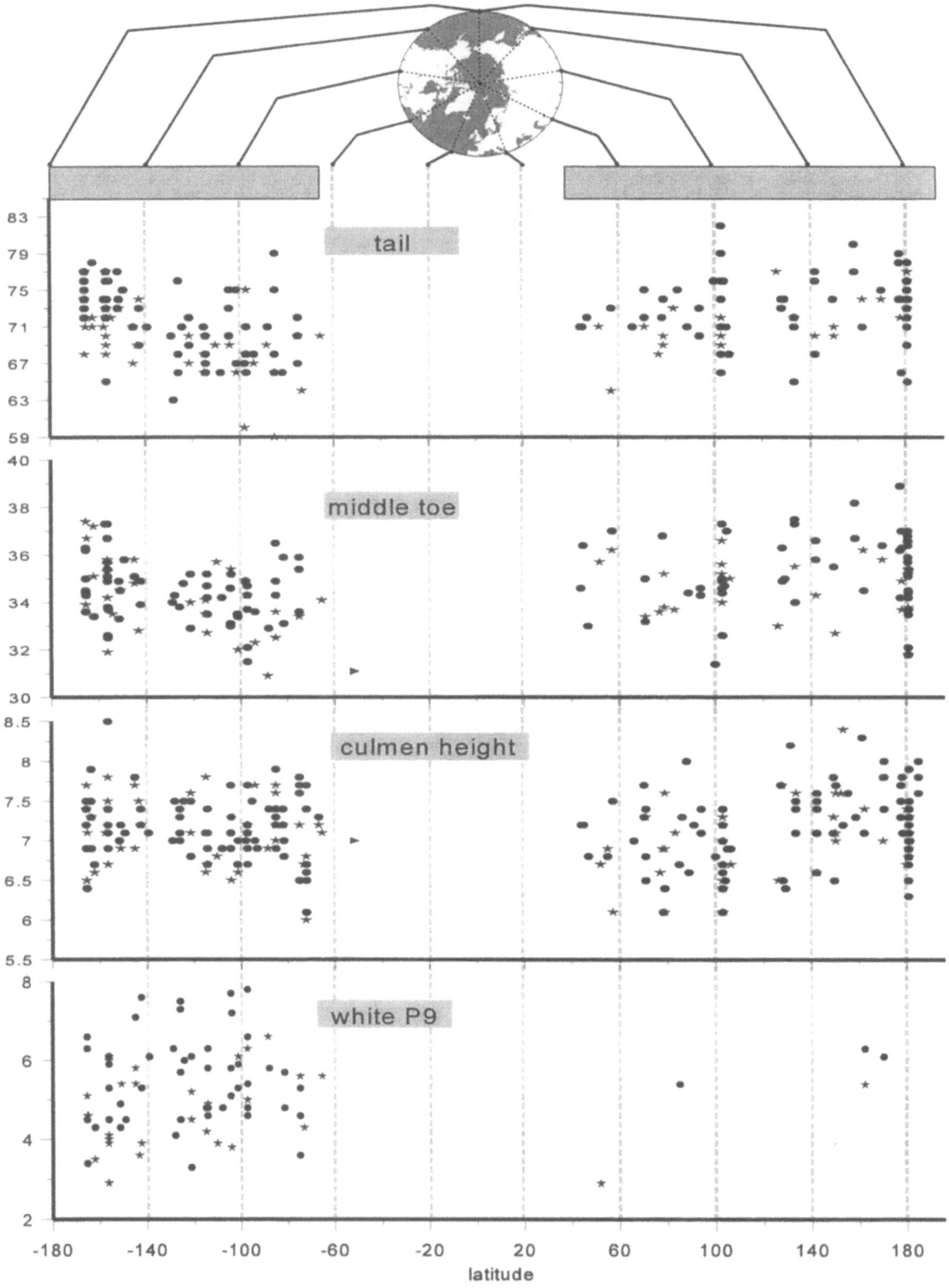

Fig. 35. *Measurements of Grey Plovers from the breeding grounds. The following measurements are presented from top to bottom: tail length, middle toe length, culmen height and the length of the white streak on the 9th primary (description in chapter 2). Explanation of the used symbols: ● ♂♂, ★ ♀♀, ▸ sex unknown.*

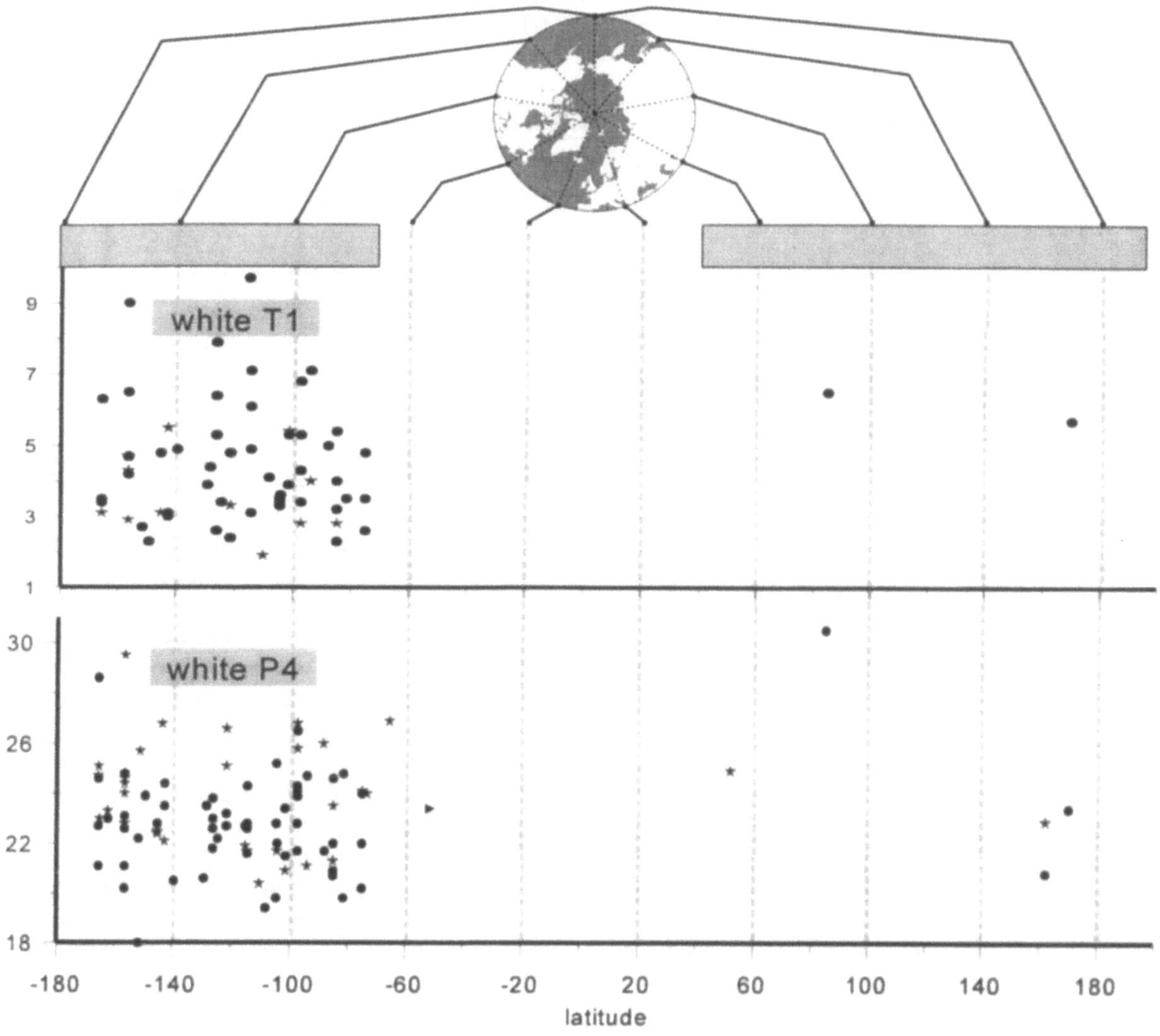

Fig. 36. *Measurements of Grey Plovers from the breeding grounds. Both following measurements are presented: (1) the width of the first white band on the innermost tail feather, and (2) the distance from the feather tip to the distal end of the white patch on the 4th primary (description in chapter 2). Explanation of the used symbols: ● ♂♂, ★ ♀♀, ▸ sex unknown.*

E Siberian samples were morphometrically similar and N Canadian breeders had distinctly smaller measurements. Cramp & Simmons (1983) followed Peters (1934) in not recognizing any subspecies in the Grey Plover, due to the overlap in size and to the lack of accurate data from the breeding grounds.

MORPHOMETRICAL DIFFERENTIATION

In total, 246 Grey Plovers were examined from the entire breeding range. Most birds had been collected soon after arrival on the breeding grounds or during breeding, but some as late as August (Fig. 33). Most of these (65.8 %) were ♂♂.

Initially, 12 breeding locations were distinguished (Fig. 32). Males from all samples were well represented, but fewer ♀♀ have been examined: none from Anadyr and few from Archangelsk, the Lena Delta and the New Siberian Islands (tables 19 - 22 and figures 34 - 37).

The Grey Plover showed distinct geographical variation (Table 20). In contrast, the samples of both sexes within a population did not differ in characters. The data of ♂♂ and ♀♀ were therefore combined.

Table 21. *Pearson's product-moment correlations between measurements in Grey Plovers. ♂and ♀ data were separately analysed. Two-tailed probabilities: *0.01 < P < 0.05; **0.001 < P < 0.01.*

	Wing	Culmen	Tarsus	Secondary	Tail	Middle toe	Bill height	
Wing		-0.029	0.415**	0.652**	0.681**	0.243	0.173	
Culmen	0.135		0.300*	-0.234	-0.091	0.254	0.373**	
Tarsus	0.408**	0.354**		0.331*	0.129	0.686**	0.131	
Secondary	0.687**	-0.006	0.249*		0.442**	0.396*	0.081	♀♀
Tail	0.580**	0.004	0.309**	0.525**		0.076	0.105	
Middle toe	0.312**	0.272*	0.502**	0.187	0.150		0.043	
Culmen height	0.108	0.449**	0.234*	-0.097	-0.086	0.127		
				♂♂				

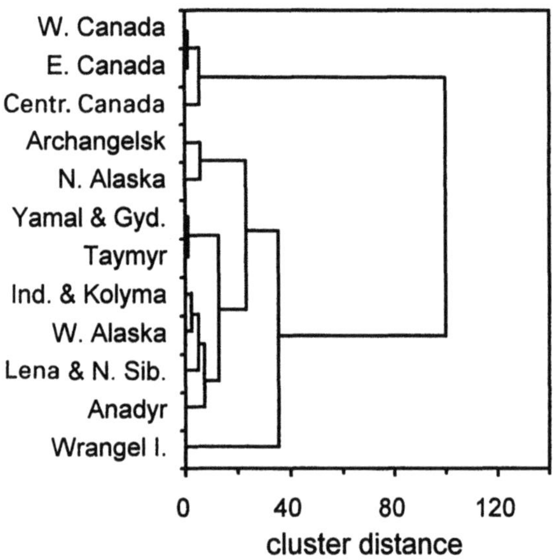

Fig. 37. *Dendrogram representing the results of cluster analysis in the Grey Plover and based upon the mean measurements per breeding population. Squared Euclidean distances were calculated while using average linkage between groups.*

Cluster analysis (Fig. 37) clearly showed the separation of the N Canadian sample from all other ones. These latter fall apart in 3 groups: (1) Wrangel Island, (2) Archangelsk and N Alaska, and (3) the remainder. Samples from Yamal & Gydansk and Taymyr differed from the E Palearctic ones (including W Alaska). The results of the cluster analysis were partly supported by the pairwise analyses in MANOVA (Table 22 - upper block):

- N Canadian samples clearly distinct from other samples, but not different from each other.
- Taymyr sample different from all other samples except neighbouring ones.
- E Siberian and W Alaskan samples could not be separated from each other.

There were some differences between the MANOVA- and cluster results:

- Wrangel Island birds were different from most of the other samples, but not from those of Yamal & Gydansk and of the Lena Delta & New Siberian Islands.
- The Archangelsk sample was significantly different only from the sample of Wrangel Island.
- The position of the Lena & New Siberian Islands sample is not clear, since it is like the E Siberian & W Alaskan samples according to cluster analysis, but it is also indistinct from the Yamal & Taymyr samples according to the MANOVA results.

From these data, we conclude that 4 'core' areas exist: (1) Canada, (2) Wrangel Island, (3) Yamal & Taymyr (incl. Gydansk), (4) E Siberia & W Alaska. A 5th core area might possibly be Archangelsk, but data were as yet too scarce. The morphometrics of birds from the Lena & New Siberian Islands and from N Alaska were studied in more detail in the section on contact zones.

All variables contributed significantly to the distinction between samples: wing-, secondary-,

Table 22. *Results of MANOVA analysis concerning the morphometric differentiation in the Grey Plover. Pillai's trace statistics are presented (Norušis 1992c). In these series of pairwise comparisons 12 groups (7 and 132 df) were compared on the basis of 7 measurements in the upper part and of 3 measurements in the lower part (3 and 242 df). Sexes were not separated. Significance levels: *0.01 < P < 0.05; **0.001 < P < 0.01; ***P < 0.001.*

Locality	n	Archangelsk	Yamal & Gydansk	Taymyr	Lena & New Sib. I.	Anadyr	Indigirka & Kolyma	Wrangel Island	W Alaska	N Alaska	W Canada	Central Canada
Wing-, culmen-, tarsus-, secondary-, tail-, middle toe length, culmen height												
Archangelsk	6											
Yamal & Gydansk	8	0.074										
Taymyr	17	0.109*	0.035									
Lena & New Sib. I.	7	0.081	0.050	0.085								
Anadyr	4	0.058	0.126*	0.133**	0.111*							
Indigirka & Kolyma	5	0.050	0.117*	0.131**	0.070	0.019						
Wrangel Island	25	0.154**	0.086	0.145**	0.088	0.145**	0.122*					
W Alaska	12	0.112*	0.145**	0.166***	0.143**	0.079	0.072	0.261***				
N Alaska	17	0.050	0.176***	0.246***	0.188***	0.050	0.059	0.295***	0.130**			
W Canada	16	0.090	0.228***	0.364***	0.253***	0.178***	0.182***	0.385***	0.360***	0.182***		
Central Canada	21	0.107*	0.211***	0.359***	0.267***	0.188***	0.207***	0.410***	0.365***	0.196***	0.034	
E Canada	12	0.098	0.249***	0.383***	0.255***	0.178***	0.178***	0.383***	0.358***	0.191***	0.020	0.080
wing-, culmen- and tarsus length												
Archangelsk	8											
Yamal & Gydansk	18	0.025										
Taymyr	28	0.041*	0.005									
Lena & New Sib. I.	13	0.019	0.028	0.046**								
Anadyr	14	0.030	0.079***	0.093***	0.044*							
Indigirka & Kolyma	16	0.041*	0.071***	0.084***	0.024	0.015						
Wrangel Island	29	0.109***	0.071***	0.088***	0.072***	0.186***	0.133***					
W Alaska	20	0.049**	0.072***	0.077***	0.044*	0.015	0.009	0.155***				
N Alaska	31	0.009	0.062***	0.097***	0.028	0.017	0.033*	0.223***	0.046**			
W Canada	22	0.049**	0.171***	0.256***	0.110***	0.170***	0.194***	0.332***	0.251***	0.153***		
Central Canada	25	0.035*	0.131***	0.213***	0.104***	0.169***	0.197***	0.312***	0.244***	0.144***	0.018	
E Canada	33	0.086***	0.252***	0.363***	0.165***	0.233***	0.262***	0.430***	0.336***	0.244***	0.011	0.058**

tarsus-, and tail length, culmen height, and culmen- and middle toe length (in decreasing order of importance, see Table 20). Sexes differed only in tail lengths.

Analysis with wing- and culmen length showed important geographical and no sexual variation (Table 20). Thus the data of both sexes were again combined per breeding location. MANOVA-analysis with wing-, culmen- and tarsus length produced similar results as the analysis with all measurements included (Table 22). The distinction was sometimes even clearer, e.g., the distinction of the Wrangel Island sample from all others was more evident; also the Archangelsk and the Lena & New Siberian Islands clearly differed from the Taymyr one. This better distinction was due to (1) larger sample sizes, and (2) the importance of wing lengths.

In general, the same 4 core areas were distinguished as with an analysis with all measurements: (1) N Canada, (2) Wrangel Island, (3) Yamal & Taymyr, and (4) E Siberia & W Alaska.

Three additional plumage characters were scored during the last stages of data collection in a search for better possibilities for distinguishing the sexes: (1) WHP9 - length of the white streak on the 9th primary; (2) WHP4 - distance from the feather tip to the distal end of the white on primary 4; (3) WTAIL - width of the first white band (starting from the top) of the innermost tail feather (see Introduction under Selected Variables). Since ♀♀ have a darker black-brown appearance than ♂♂ the size of the white areas on these feathers is likely to be affected also and may offer a possibility to separate the sexes. It appeared that ♂♂ had more white on P9 and ♀♀ more white on P4 (Table 23).

Table 23. *Results of T-tests on the differences of plumage characters between the sexes of the Alaskan and N Canadian breeding populations.*

Character	♂♂	♀♀	*T*-value	*df*	*P*
white P9	5.5±1.1(50)	4.9±1.2(29)	2.208	77	0.030
white P4	22.7±1.9(55)	24.1±2.1(32)	-3.058	85	0.003
white tail	4.6±1.7(48)	3.7±1.1(12)	1.728	58	0.089

The breeding plumage of N Canadian ♂♂ showed more contrast between black and white in the feather patterns than the plumage of Palearctic ♂♂. Several areas showed much more white, especially forehead, crown, nape, mantle, coverts, and scapulars. In Palearctic birds, the dark parts of the feathers in mantle, back and wings were brown-black while these were almost black in the N Canadian samples. The white streaks on the tail of N Canadian ♂♂ contained less of greyish and brownish wash. The ♀♀ in both samples were more or less similarly coloured with relatively more brownish feathers.

PREDICTION OF BREEDING ORIGIN

The MANOVA- and cluster analyses justified the distinction of 4 POSCON groups: (1) Yamal & Taymyr; (2) E Siberia & W Alaska; (3) Wrangel Island; and (4) N Canada. A 5th POSCON group was also included, *viz.* the Archangelsk breeding sample with its small wing- and tail dimensions. No distinction was made between the sexes. Basic data are summarized in Table 24. Two different sets of POSCON groups were constructed. In the first set, 5 variables were included in an analysis with 4 POSCON groups. These variables were the lengths of wing, culmen, tarsus, secondary and tail. In the second set, only wing- and culmen lengths were used in 5 POSCON groups. The Archangelsk sample was only included in this last set, since too few data were available to include it in the 5-variable set. The statistical differences between the groups were analysed with MANOVA (tables 25 and 26).

In a next step, all available measurements of any bird were used to calculate the typicality probability, that a bird resembled one of the constructed groups. In doing so, both the accuracy and the exclusivity of prediction is estimated (see Introduction under 'Statistical analysis of the data').

Table 24. *Mean measurements of the 5 POSCON groups of Grey Plovers. Presented are the means ± 1 SD (n).*

Locality	Wing	Culmen	Tarsus	Secondary	Tail	Middle toe	Bill height
Archangelsk	201.2±2.2(8)	29.8±1.8(8)	47.4±2.1(8)	82.3±2.2(6)	70.3±3.2(6)	35.5±1.5(6)	6.9±0.4(8)
Yamal & Taymyr	205.5±4.0(47)	28.9±1.3(45)	47.4±1.8(47)	83.5±2.7(31)	72.0±3.5(31)	34.7±1.3(26)	6.9±0.5(43)
E Siberia & Alaska	206.1±5.6(48)	30.9±1.3(48)	49.1±2.1(47)	84.8±2.7(25)	73.9±3.4(25)	35.6±1.5(21)	7.4±0.4(48)
Wrangel Island	210.4±5.0(29)	29.3±1.7(29)	46.6±1.5(29)	84.7±1.7(27)	73.2±2.9(27)	34.7±1.5(25)	7.0±0.4(29)
N Canada	196.9±4.4(79)	30.2±1.5(80)	45.4±1.7(79)	79.9±2.0(55)	69.3±4.2(55)	34.0±1.2(47)	7.1±0.4(79)

Table 25. *Differentiation between 4 POSCON samples of Grey Plovers (n = 120) when 5 measurements were included, or when analysis was restricted to wing-, culmen- and tarsus lengths. [1]The degrees of freedom in the multivariate tests of significance were 7 and 110. The degrees of freedom in the univariate tests were 1 and 116. [2]Including Yamal; [3]including Alaska.*

Contrast	F[1]	Wing	Culmen	Tarsus	Secondary	Tail	Middle toe	Culmen height
N Canada - Wrangel	20.875***	108.529***	4.725*	4.128*	70.882***	19.198***	4.344*	0.740
N Canada - Taymyr[2]	18.868***	45.962***	14.880***	20.875***	50.377***	8.281**	5.778*	19.447***
N Canada - E Siberia[3]	17.241***	49.211***	5.072*	72.872***	61.651***	23.698***	19.275***	0.457
Wrangel - Taymyr[2]	3.858**	11.511**	2.009	4.679*	1.778	1.972	0.054	9.421**
Wrangel - E Siberia[3]	8.868***	6.208*	14.553***	34.462***	0.006	0.426	4.623*	1.740
Taymyr[2] - E Siberia[3]	6.827***	0.492	27.386***	15.279***	1.422	4.013*	3.864	18.233***

Table 26. *Differentiation between the 5 POSCON groups of Grey Plovers (n = 211) when only wing- and culmen length were included. [1]The degrees of freedom in the multivariate tests of significance were 2 and 203. The degrees of freedom in the univariate tests were 1 and 204. [2]Including Yamal; [3]including Alaska.*

Contrast	F[1]	Wing	Culmen
Archangelsk - Taymyr[2]	5.180**	5.574*	2.420
Archangelsk - E Siberia[3]	4.443*	7.396**	3.504
Archangelsk - Wrangel	14.506***	24.093***	0.937
Archangelsk - N Canada	3.933*	6.244*	0.399
Taymyr[2] - E Siberia[3]	20.323***	0.422	39.961***
Taymyr[2] - Wrangel	9.818***	19.684***	0.780
Taymyr[2] - N Canada	72.832***	96.098***	19.812***
E Siberia[3] - Wrangel	24.560***	15.357***	21.932***
E Siberia[3] - N Canada	57.401***	115.367***	6.891**
Wrangel - N Canada	107.990***	176.845***	8.182**

Accuracy

Birds belonging to the N Canada population were predicted very well (Table 27). The accuracy of predicting N Canadian breeding birds was highest when wing-, culmen- and tarsus length were all included and strongly declined when only wing- and culmen length had been measured (Table 26). This was due to the additional inclusion of the Archangelsk POSCON group with its small wing dimensions, and thus resulting in much overlap with the N Canadian group (Table 24).

Grey Plovers from Yamal & Taymyr were best predicted when 5 measurements were used (79 % correctly predicted). However, they were poorly predicted when exclusively wing- and culmen lengths were used because of the large degree of similarity between the sample group of Yamal & Taymyr and several other POSCON groups with respect to the wing- and culmen lengths (Table 27). The Yamal & Taymyr birds were then often mistaken for birds from Wrangel Island (33.3%) or Archangelsk (15.6%).

The prediction of origin for the E Siberia & W Alaska group was best with the inclusion of 3 measurements (Table 27): > 80 % being predicted correctly. These results became worse when only wing- and culmen lengths were analysed, due to a close similarity to the Archangelsk sample. Prediction of the Wrangel Island birds was best when only wing- and culmen lengths had been used and became worse with more measurements. Most of the overlap occurred with the POSCON group of Yamal & Taymyr. The Archangelsk POSCON group was best predicted with use of wing- and culmen lengths.

Exclusivity

High exclusivity rates were reached in the N Canada group with wing-, culmen- and tarsus lengths: almost 40 % of the birds were predicted to resemble only the own POSCON group. In all other situations, exclusivity rates were lower. These rates varied between 0 (Yamal & Taymyr) and 15 % (N Canada) in the analyses with wing- and culmen lengths. In the analyses with 5 measurements, exclusivity rates varied between 12 (Yamal & Taymyr and E Siberia & Alaska) and 22 % (Wrangel Island). Exclusivity in both POSCON groups from Yamal & Taymyr and

Table 27. *Accuracy of prediction in Grey Plover. These predictions were based upon morphometric analysis and included (1) wing-, culmen-, tarsus-, secondary-, and tail length; (2) wing-, culmen- and tarsus length; and (3) wing- and culmen length. The columns show the predicted proportions in relation to the total number per breeding area (last column).*

	Predicted					*n*
Locality	Yamal & Taymyr	E Siberia & Alaska	Wrangel Island	N Canada	Archangelsk	
	Wing, culmen, tarsus, secondary and tail length					
Yamal & Taymyr	78.8	9.1	12.1	-	too few data	33
E Siberia & Alaska	20.0	76.0	4.0	-		25
Wrangel Island	25.9	11.1	55.6	7.4		27
N Canada	9.1	9.1	1.8	80.0		55
	Wing-, culmen- and tarsus length					
Yamal & Taymyr	55.3	23.4	14.9	6.4	too few data	47
E Siberia & Alaska	4.3	80.9	2.1	12.8		47
Wrangel Island	13.8	17.2	62.1	6.9		29
N Canada	3.8	7.7	2.6	85.9		78
	Wing- and culmen length					
Yamal & Taymyr	35.6	11.1	33.3	4.4	15.6	45
E Siberia & Alaska	10.4	54.2	6.3	8.3	20.8	48
Wrangel Island	6.9	6.9	75.9	3.4	6.9	29
N Canada	2.5	6.3	-	68.4	22.8	25
Archangelsk	25.0	-	-	12.5	62.5	8

Wrangel Island was highest with 5 measurements. For the E Siberia & Alaska- and the N Canada sample the highest exclusivity rates were found in the 3-variable situation.

CONTACT ZONES

The location of two possible contact zones between populations was studied with POSCON- analysis: (1) Central and E Siberia (Fig. 38) and (2) N Alaska (Fig. 39).

Central and E Siberia. Grey Plovers, resembling the Yamal & Taymyr-type, were predicted in a restricted part of the breeding range: northwest coast of Taymyr and around Lake Taymyr (Fig. 38). The Wrangel Island type was almost never predicted from the E Siberian mainland, whilst nearly 19 % of the Wrangel Island sample had the highest resemblance with the E Siberia & Alaska type. The E Siberia & Alaska type was most frequently predicted to occur even in the W Palearctic region (Fig. 38). Some migrants from the upper reaches of the Kolyma River closely resembled the Wrangel Island sample.

N Alaska. The N Alaskan sample appeared to be a mixture of different populations (Fig. 39). Breeding birds from this area resembled all 3 neighbouring POSCON groups: E Siberia & W Alaska (50 %; n = 24), N Canada (37.5 %) and Wrangel Island (12.5 %). Amongst these N Canadian predicted birds there were several (7 out of 12) not resembling both other POSCON groups.

The Mackenzie River Delta can be considered as the border separating the E Siberia & W Alaska population from the N Canada one (Fig 39). E Siberia & W Alaska-type birds were only predicted to occur in Canada along the south edge of the breeding range.

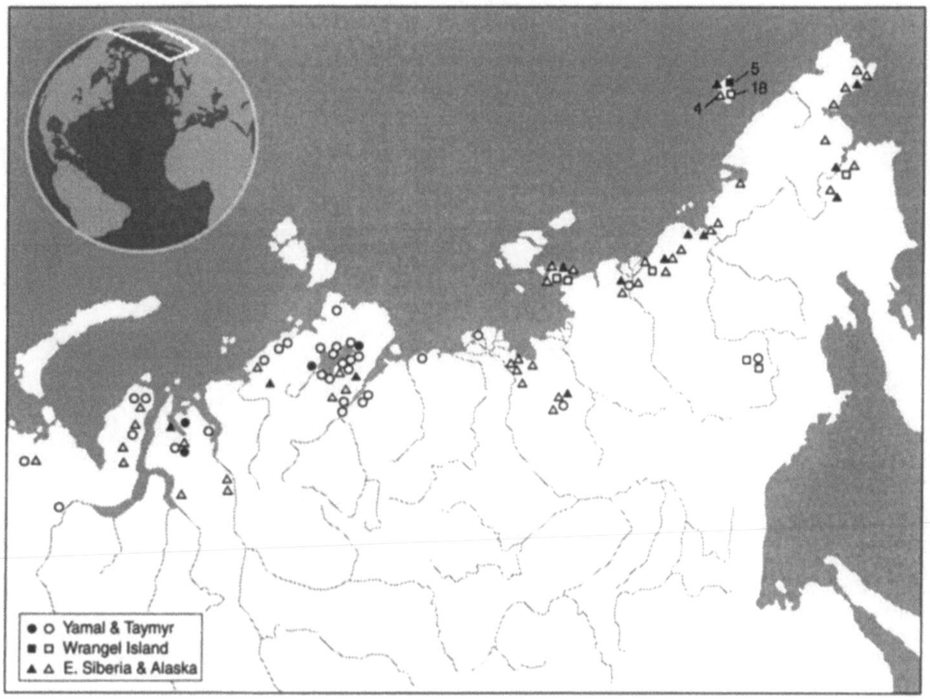

Fig. 38. *Distribution of Grey Plovers throughout the Palearctic. The Grey Plovers examined were predicted to resemble the following populations: (1) Yamal & Taymyr (circles), (2) E Siberia & Alaska (triangles) or (3) Wrangel Island (squares). Closed circles indicate that the typicality probability for E Siberia & Alaska was < 0.05, this probability was > 0.05 with an open circle. Closed triangles, west of the New Siberian Islands, indicate that the typicality probability for Yamal & Taymyr was < 0.05 and this probability was > 0.05 with an open triangle. From the New Siberian Islands eastwards, it implies a typicality probability for Wrangel Island < 0.05 and this probability was > 0.05 with an open triangle. Closed squares indicate that the typicality probability for E Siberia & Alaska was < 0.05 and this probability was > 0.05 with an open square.*

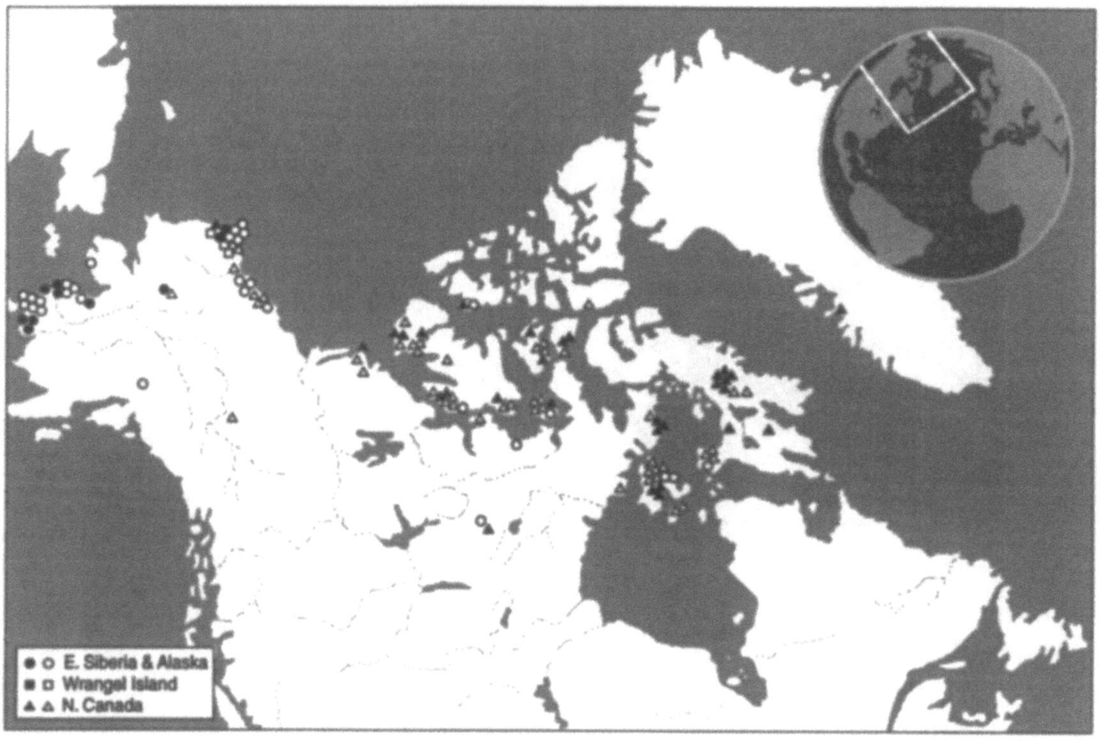

Fig. 39. *Distribution of Grey Plovers throughout the Nearctic. The Grey Plovers examined were predicted to resemble the following populations: (1) East Siberia & Alaska (circles), (2) North Canada (triangles) or (3) Wrangel Island (squares). Closed circles indicate that the typicality probability for North Canada or Wrangel Island was < 0.05 and this probability was > 0.05 with an open circle. Closed triangles indicate that the typicality probability for East Siberia & Alaska or Wrangel Island was < 0.05 and this probability was > 0.05 with an open triangle. Closed squares indicate that the typicality probability for East Siberia & Alaska or North Canada was < 0.05 and this probability was > 0.05 with an open square.*

BODY MASS

Mass data of 24 Grey Plovers were obtained from the skin's labels (Fig. 40). These varied significantly between regions but not between sexes (Table 28). The mass of birds breeding in N Canada varied between 180 and 220 gm (x = 199 gm), but the Palearctic and Alaskan breeding birds were heavier. No heavy individuals occurred among the N Canadian Grey Plovers, which might be caused by the occurrence of primary moult (see next chapter). Palearctic and/or Alaskan birds weigh about 225 gm at arrival (n = 5), which might be lower by mid-July (Fig. 40). W Palearctic birds increased their body mass between half July and early August, indicating a daily mass gain of 2-3 gm (1.0- 1.7 % per day), which is an intermediate estimate in comparison to other studies (Zwarts *et al.* 1990).

Table 28. *Results of analysis of variance (ANOVA) with respect to differences in body mass and primary moult in breeding populations of Grey Plover. The breeding location and sex of a bird were considered as independent factors. The breeding areas were divided over 3 groups: (1) Alaska, E Siberia and Wrangel Island; (2) N Canada; and (3) Palearctic E to the New Siberian islands. The E Siberian and Wrangel Island birds were not included in the first group but in the last one when analysing the moult-data. [1]The day number was used as a covariate in 2 analyses. [2]Analysis was restricted to those Grey Plovers collected on the Alaskan or N Canadian breeding grounds after May.*

	Covariate	Main effects		interaction	% explained	n
Variable	Date	Breeding location	Sex			
df		2	1	2		
Body mass		4.427*	0.169	1.059	35.0	24
Body mass[1]	0.120	4.177*	0.187	0.972	35.2	24
Moult		11.142***	0.000	1.122	11.1	203
Moult[1]	18.566***	13.085***	0.011	1.539	20.4***	195
Moult[2]	31.885***	0.024	0.002	1.47	36.1***	64

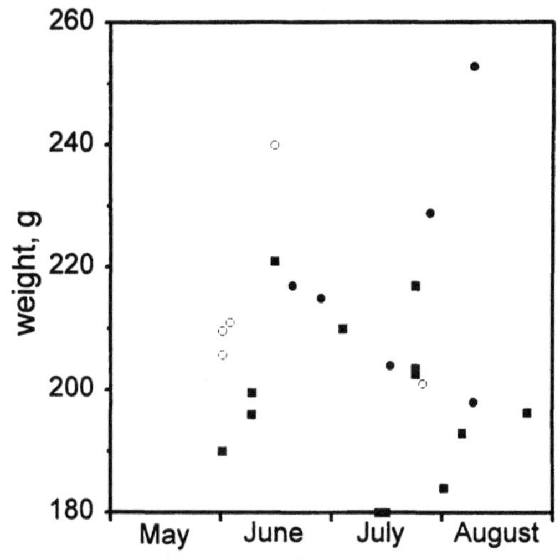

Fig. 40. *Body mass of Grey Plovers during breeding. Three breeding ranges were distinguished: (1) Palearctic E to the New Siberian Islands - ●; (2) E Siberia & Alaska - ○; Canada & Greenland - ■.*

PRIMARY MOULT

There were clear geographical differences in the occurrence and timing of primary moult on the breeding grounds (Table 28), but sexual differences were not significant. Grey Plovers breeding in the Nearctic started post-nuptial primary moult on the breeding grounds from 8 June onwards and nearly all were moulting by late June (Fig. 41). Our data did not show the occurrence of primary moult in Palearctic breeding populations, though it is reported to occur on the Taymyr breeding grounds from mid-July onwards (Hötker pers. comm. and Tomkovich pers. comm.). In the Palearctic samples primary moult must be a much rarer phenomenon than in the Nearctic, since none of the 130 Palearctic birds examined showed a sign of primary moult even though 38 of these 130 birds had been collected after mid-July. Thus these populations mostly defer their post-nuptial moult until after the breeding season.

Primary moult in Alaskan Grey Plovers was unexpected, since their measurements strongly resembled the ones of the E Siberian population, which were not moulting on the breeding grounds. Four of the 11 moulters in Alaska had

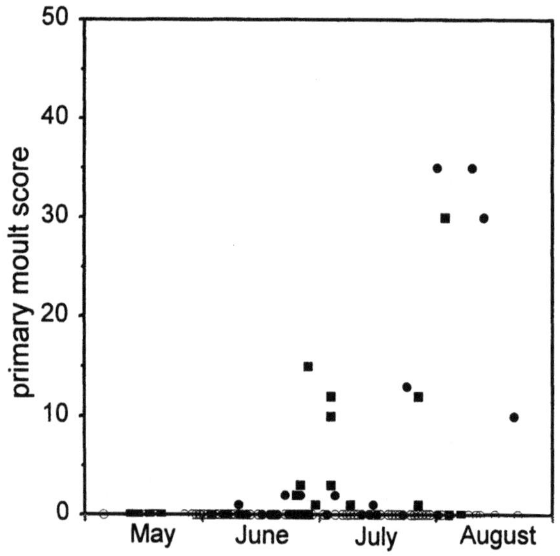

Fig. 41. *The occurrence of Grey Plovers on the breeding grounds with moulting primaries. Three areas were distinguished: ● Canada; ○ N Europe & Siberia; and ■ Alaska.*

been collected along the west coast, the others in N Alaska. The moulting schedules of the N Alaskan and N Canadian Grey Plovers were not different (ANOVA with date as covariate: F = 0.001, $P > 0.05$, $n = 66$).

Six out of 20 moulting Palearctic Grey Plovers suspended their primary moult with scores ranging between 10 and 35 at the end of the breeding season. Thus no Nearctic Grey Plovers finished primary moult on the breeding grounds.

One out of 20 breeding birds in primary moult was also moulting its secondaries.

DISCUSSION

The taxonomy of the Grey Plover

The information presented above indicated the existence of 4 distinct breeding populations. Most evident were the morphometric differences between the samples of N Canada and the ones of the Palearctic and Alaska. The Wrangel Island sample could also be well characterized. The Yamal & Taymyr sample differed from the E Siberia & W Alaska sample, but these differences were smaller and mainly concerned the birds breeding on N Taymyr. We therefore propose the distinction of 3 subspecies:

- *Pluvialis squatarola squatarola* (Linnaeus 1758). This subspecies breeds in the Palearctic (excluding Wrangel Island) and in Alaska. Nominate *squatarola* was originally described from Sweden. Birds from the type locality are Palearctic ones and thus the Palearctic and Alaskan breeding populations should be called by this name. This subspecies shows geographical variation with E Siberian birds being generally the largest-sized of all populations studied: culmen-, tarsus-, secondary-, middle toe- and tail length and culmen height. On average, nominate *squatarola* has an intermediate wing length, which ranges between the large-sized Wrangel Island population and the small-sized *cynosurae*.

- *Pluvialis squatarola cynosurae* (Thayer & Bangs 1914), based upon a type specimen from Baillie Island (70.35°N 128.12°W) in the NW Territories, Canada. The breeding birds of N Canada were clearly different from the Palearctic and Alaskan ones. They have characteristically short wings, tarsi, secondaries, tails and middle toes. They can easily be predicted to occur outside the breeding range. N Canadian ♂♂ also have a more contrastingly black-and-white breeding plumage and more extensive white parts of the body than the W Palearctic birds. They are not as brightly coloured as the E Palearctic ones. These colour differences were not evident in the ♀♀. This subspecies shows some geographical variation, Central Canadian populations having smaller culmen- and middle toe lengths.

- *Pluvialis squatarola tomkovichi* Engelmoer & Roselaar, nov. ssp. As holotypes are selected 2 individuals from the Zoological Museum of St. Petersburg nrs. 12213 (adult ♂) and 12230 (adult ♀). Both had been collected on Wrangel Island on 26 July and resp. 12 June 1939. Their wing lengths are 210 resp. 218 mm, culmen length 29.4 resp. 27.7, culmen height 6.9 resp. 6.9, first secondary length 84 resp. 88, tail length 72 resp. 74, and tarsus length 46.4 resp. 47.2. The breeding birds from Wrangel

Island clearly differ from the E Siberian & W Alaskan mainland populations. Most characteristic are the short lengths of culmen and tarsus, combined with a long wing. It differs from N Canadian *cynosurae* by its long wing- and tail dimensions. None of the subspecies' ames once in use belongs with certainty to the Wrangel Island population.

Small size at northern latitudes

From this study it appeared that the northern breeding populations had smaller measurements than the southern ones: it seems as if selectional forces caused size reduction in northern breeding populations. We present 3 examples:

(1) The Wrangel Island birds had short culmen- and tarsus lenghts and small culmen heights when compared with the E Siberia & Alaska samples (figs. 34 & 35).

(2) The breeding birds of N Taymyr were the most typical ones of the Yamal & Taymyr sample, since they had rather short culmen lengths and small culmen heights (figs. 34 & 35).

(3) Within the breeding population of N Canada, some birds had a closer resemblance to the large-sized E Siberia & Alaska population than with the own population. These birds mostly occurred south of or along the south edge of the breeding range in N Canada (Fig. 39).

Morphometrical variation in the Palearctic

As with other wader species (Knot, Dunlin, Bar-tailed Godwit, Curlew), Grey Plover measurements in the Palearctic range from small-sized populations in the west to large-sized ones in the E. It is difficult to understand the underlying mechanism resulting in a simultaneous size decrease of measurements in all these species when breeding further to the west. Tomkovich (pers. comm.) considers this variation -in very general terms- to be due to a process of gradual westward colonization since the last glaciation: the northern E Palearctic was ice-free, but the entire area from Scandinavia to central Taymyr was covered with ice. However, such processes of fragmentation, local extinction and recolonization might just as well have had different time scales (e.g., Dunlin - Wenink *et al.* 1993). We also argued in the introduction that processes of isolation might have been especially important during warm periods e.g., 8-5.000 B.P. because then there were no tundra areas west of the Taymyr Peninsula, whilst there were always possibilities to breed south of the ice sheets in the W Palearctic during the periods of glaciation.

Apart from the size differences, there are also differences in coloration between E and W Palearctic Grey Plovers with darker coloured W Palearctic birds and light coloured E Palearctic ones.

SUMMARY

The pattern of geographical variation of the Grey Plover was studied. We collected measurements of birds from the entire breeding range and these were analysed with multivariate statistics. The species is polytypic and 3 subspecies were recognized: (1) *P. s. squatarola*, breeding in the Palearctic and Alaska (but not on Wrangel Island), (2) *P. s. cynosurae*, breeding in N Canada, east of the Mackenzie River and (3) *P. s. tomkovichi* nov. ssp., breeding on Wrangel Island. The nominate subspecies was not quite uniform in size as N Taymyr birds had shorter bill and tarsus. *Cynosurae* was clearly recognizable because of its small size. It also had a lower body mass than nominate *squatarola*. *P. s. tomkovichi* had a long wing length, combined with short length of culmen and tarsus and a small culmen height.

W Palearctic birds increased their body mass prior to their departure from the breeding grounds, occurring between half July and early August. This pre-migratory mass increase was not observed in the N Canadian sample.

A few Palearctic birds started the primary moult on the breeding grounds, but this was regular in the Nearctic samples. Nearctic birds started primary moult from 8 June onwards, and suspended moult at the time of departure from the breeding grounds.

Chapter 7

Red Knot

-

Calidris canutus

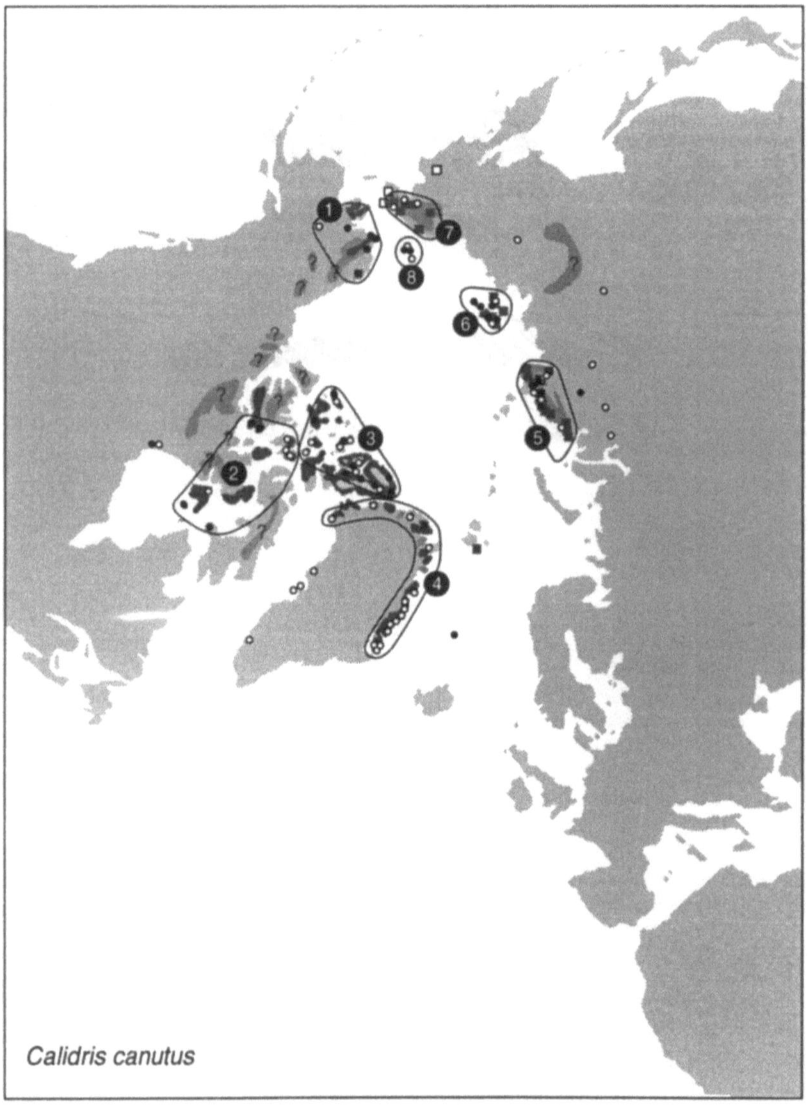

Fig. 42. *Breeding distribution of the Knot. For explanation of the used symbols see Chapter 3: 'Introduction to the species' accounts'.*

INTRODUCTION

The Red Knot is breeding scattered throughout the high arctic regions of the Nearctic and the Palearctic (Fig. 42). The species is mostly breeding on peninsulas and islands north of the continental mainland, near but not on the coast (Cramp & Simmons 1983). During the past decade much attention was focussed on the geographical variation in the Red Knot. Until the early 1980's much of the species' morphometrical variation was unknown or not well understood, since migrant birds from unknown breeding grounds were regularly used to describe a subspecies. Since then the discussion contained 2 elements: (1) the delineation of the regular breeding areas and (2) the degree of subspeciation. The debate is not yet finished and still includes most breeding populations.

Nowadays the following subspecies are generally distinguished:

- *C. c. rufa*, breeding in N Canada from Prince of Wales Island southwards.
- *C. c. islandica*, breeding on Greenland and in N Canada from Prince of Wales Island northwards.
- *C. c. canutus*, breeding in the north part of Taymyr and on the New Siberian Islands.
- *C. c. rogersi*, breeding on the Chukotsk Peninsula (Tomkovich 1990) and/or on the New Siberian Islands (Roselaar 1983).
- *C. c. roselaari*, breeding on Wrangel Island and possibly in Alaska.

Recent reviews concerning breeding distribution, taxonomy, morphometrics and migration system in this species were published by Barter *et al.* (1988), Harrington *et al.* (1988), Underhill *et al.* (1989), Tomkovich (1990, 1992a), Barter (1992), Davidson & Wilson (1992), Godfrey (1992), Morrison & Harrington (1992), Piersma *et al.* (1992), Baker *et al.* (1994), and Tomkovich & Soloviev (1996).

Barter *et al.* (1988) showed the occurrence of *rogersi* in Australia and New Zealand, although no corrections for museum shrinkage were undertaken and sex-ratios of the Australian and New Zealand birds were unknown. The subspecies name *rogersi* was used according to Roselaar (1983): a population breeding somewhere in Siberia, and recognizable by its short bill and rather a short wing.

Tomkovich (1990) showed that (1) the measurements of the populations from Greenland and Taymyr were identical, and (2) no *canutus*-like birds were recognized amongst the breeding birds of Taymyr. Roselaar (1983) and Tomkovich (1990, 1992a) suggested the presence of breeding grounds in the mountainous regions of Yakutia, where the long-billed '*canutus*' population should have its breeding grounds. Recently, Tomkovich & Soloviev (1996) showed that (1) the live-trapped Knots on Taymyr are larger-sized than the museum data suggest, and (2) Taymyr Knots occur in W Europe during autumn and spring.

The extent of the breeding range of *rufa* and the possible overlap with the range of *islandica* was analysed by Godfrey (1992).

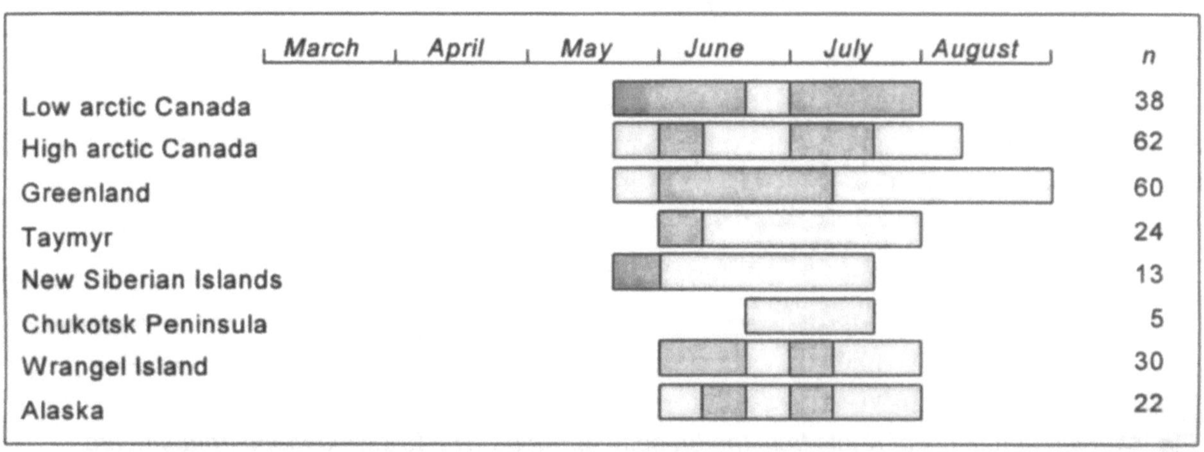

Fig. 43. *Collection dates of Red Knots. Most birds had been collected during the periods marked by the heavily shaded bars. The sample sizes are presented along the right side.*

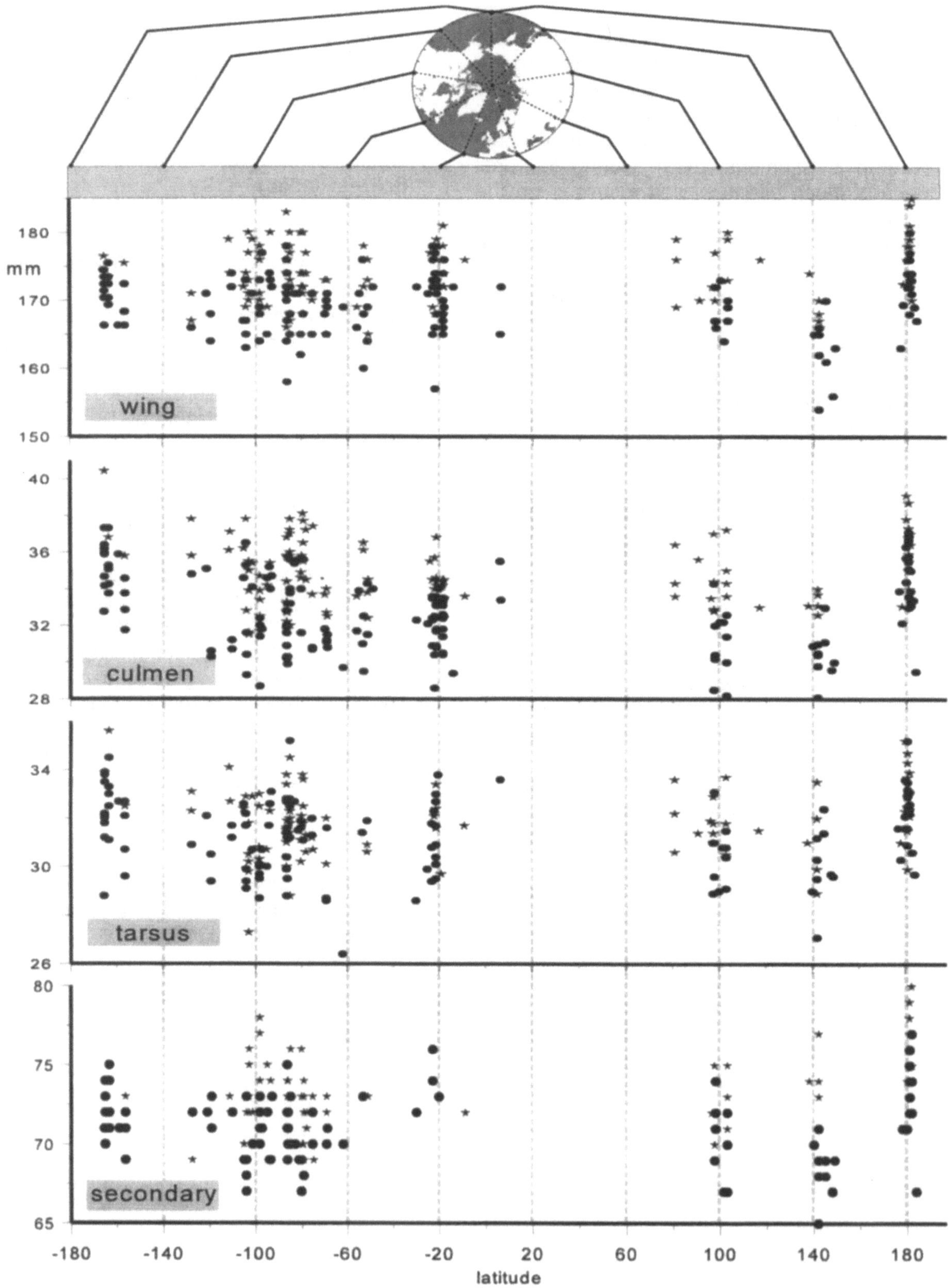

Fig. 44. *Measurements of Knots from the breeding grounds. The following measurements are presented from top to bottom: wing length, culmen length, tarsus length and secondary length (description in chapter 2). Explanation of the used symbols: ● ♂♂, ★ ♀♀.*

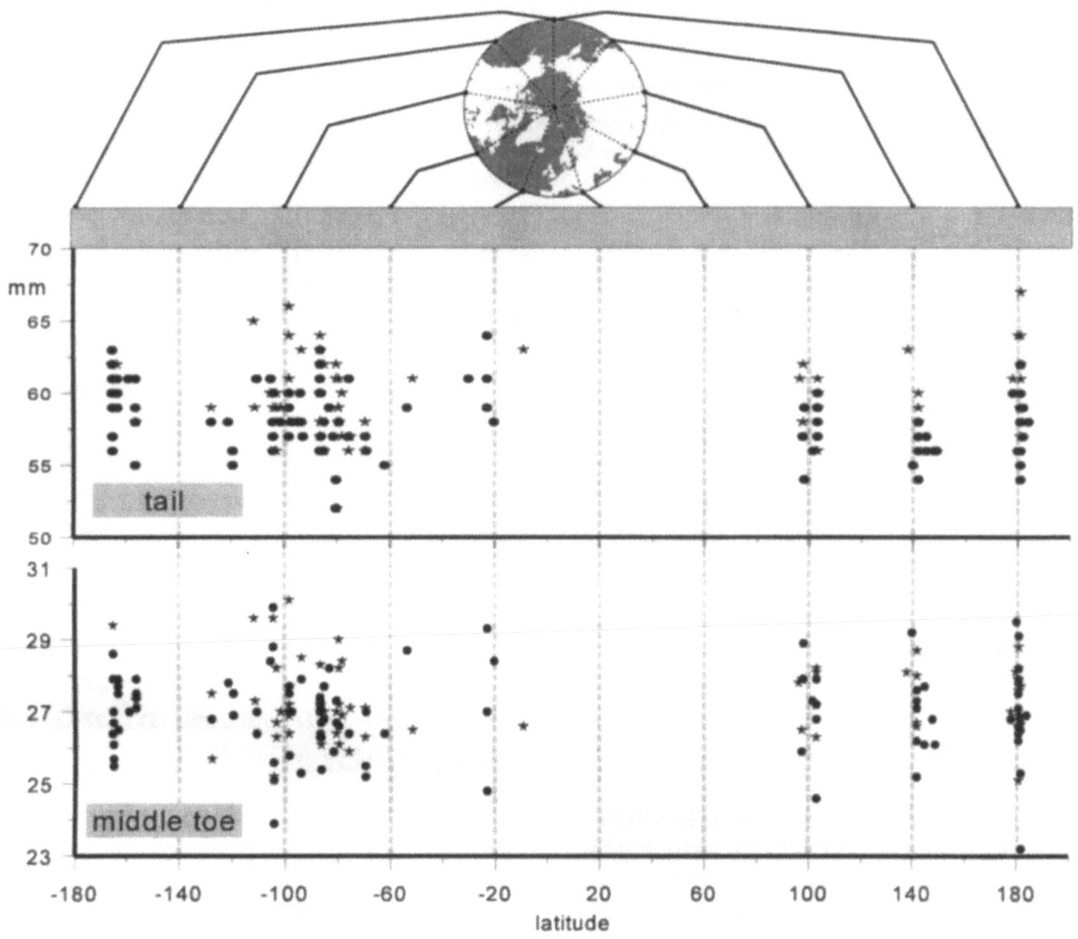

Fig. 45. *Measurements of Knots from the breeding grounds. Tail- and middle toe lengths are presented. Explanation of the used symbols:* ● ♂♂, ★ ♀♀.

Baker *et al.* (1994) found a low sequence divergence in the Red Knot when analysing the mtDNA-region sequences in this species. They concluded that the present-day populations have evolved from a numerically very small population about 10-20.000 years ago. This low variability did not allow intraspecific taxonomical conclusions. Research on nuclear DNA seemed to be better suited for taxonomical comparisons within this species.

MORPHOMETRICAL DIFFERENTIATION

In total, 240 Red Knots were sampled from the whole breeding range. Most breeding birds were present by late May - early June, although the low arctic Canada breeders arrived 1-2 weeks earlier (Fig. 43). Most birds had been collected during the first weeks after arrival or by the end of breeding. Most collected birds (58.7 %) were ♂♂.

Initially, 9 breeding locations were distinguished (Fig. 42, Table 29) with the inclusion of several small samples: ♂♂ from the New Siberian Islands and the Chukotsk Peninsula, and ♀♀ from the New Siberian Islands, the Chukotsk Peninsula and Alaska. In the Greenland sample nearly exclusively wing-, culmen- en tarsus lengths had been measured. The morphometrics of the Red Knot are presented and analysed in the tables 29-34 and the figures 44 - 46.

The species shows distinct sexual and geographical variation in measurements (Table 31, figs. 44 & 45).

Table 29. *Measurements of Red Knots, collected on the breeding grounds. The numbers in the first column refer to the numbers used in Fig. 42. Presented are the means ± 1 SD (n).*

Locality	Wing	Culmen	Tarsus	Secondary	Tail	Middle toe
♂♂						
2 Low arctic Canada	169.8±3.0(17)	34.8±0.8(17)	32.0±1.2(17)	70.5±1.8(15)	58.3±1.3(15)	27.5±1.3(12)
3 NE Canada	168.7±4.8(34)	31.1±1.1(34)	30.3±1.4(33)	70.7±1.7(31)	58.2±2.6(31)	26.5±0.9(29)
4 Greenland	170.4±4.3(38)	32.0±1.3(38)	31.0±1.2(17)	73.7±2.1(6)	60.0±2.7(6)	27.4±1.8(5)
5 Taymyr	168.5±2.9(11)	31.1±1.8(11)	30.5±1.3(11)	70.4±2.4(9)	57.7±1.9(9)	27.2±1.3(9)
6 New Siberian Islands	161.6±4.4(8)	30.0±0.9(8)	29.5±1.2(8)	68.5±1.9(8)	56.3±1.4(8)	26.9±1.2(8)
7 Chukotsk Peninsula	167.0±2.8(4)	32.2±2.0(4)	30.6±0.8(4)	69.0±1.0(2)	59.0±1.0(2)	26.8±0.1(2)
8 Wrangel Island	173.4±2.8(17)	35.1±1.4(17)	32.6±1.1(17)	74.0±1.6(15)	57.4±2.2(15)	27.0±1.5(15)
1 Alaska	171.1±2.9(19)	34.9±1.5(19)	32.2±1.4(19)	71.9±1.6(19)	59.4±2.1(19)	27.1±0.8(18)
♀♀						
2 Low arctic Canada	173.7±4.2(25)	36.7±1.3(25)	32.3±1.0(25)	71.9±2.1(19)	59.2±2.7(19)	27.6±1.1(16)
3 NE Canada	174.1±4.8(28)	33.6±1.4(28)	31.1±1.5(28)	73.7±2.0(26)	59.7±2.7(26)	27.0±1.1(18)
4 Greenland	173.5±3.8(23)	34.2±0.9(23)	31.5±1.1(8)	73.0 - (1)	61.0 - (1)	26.5± - (1)
5 Taymyr	173.9±4.3(13)	34.7±1.5(13)	31.8±1.1(13)	72.8±2.1(8)	59.4±2.0(8)	27.5±0.7(8)
6 New Sib. islands	169.0±3.2(5)	33.3±0.6(5)	31.1±1.8(5)	73.4±2.9(5)	59.6±2.1(5)	27.6±0.9(5)
7 Chukotsk Peninsula	172.0 - (1)	33.1 - (1)	31.0 - (1)	71.0 - (1)	61.0 - (1)	27.0 - (1)
8 Wrangel Island	178.5±4.2(13)	36.7±1.6(13)	32.8±1.5(13)	76.6±1.9(12)	61.2±2.9(12)	27.3±1.0(12)
1 Alaska	175.2±1.6(3)	37.7±2.4(3)	34.0±1.6(3)	73.7±0.6(3)	60.7±2.3(3)	28.2±1.1(3)

Sexual differences were larger than geographical differences. Sexual dimorphism was smallest in the Greenland samples and largest in the ones of the New Siberian Islands (Table 32). Wing length was correlated with tail length and culmen length with tarsus length (Table 33). Middle-toe lengths were least related to other measurements.

Cluster analysis clearly showed the distinctness of the samples from Wrangel Island and the New Siberian Islands, both in the ♂♂ and ♀♀ (Fig. 46). The measurements of the Alaskan population were intermediate between the ones of Wrangel and low arctic Canada. The samples of NE Canada and Taymyr were closely similar. The measurements of both these samples differed slightly from the ones of Greenland, especially in the ♂♂. The position of the Chukotsk sample remained unclear from this cluster analysis.

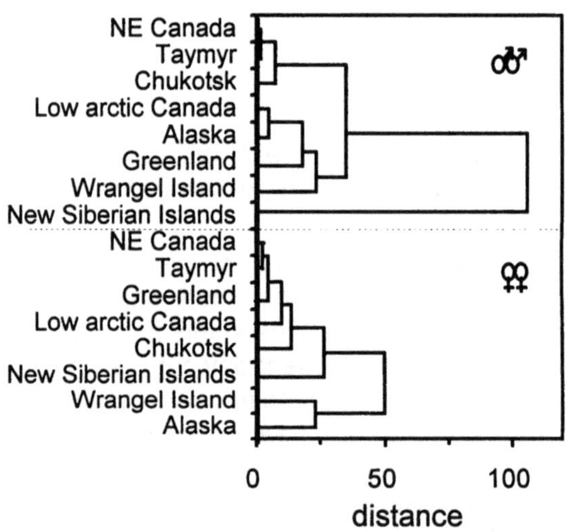

Fig. 46. *Dendrogram representing the results of cluster analysis. This analysis was based upon the mean measurements per breeding population. Squared Euclidean distances were calculated while using average linkage between groups.*

Table 30. *Overall results of MANOVA for all distinguished breeding populations in the Knot in combination with univariate ANOVA's in which the different measurements were analysed separately. Significance levels: *0.01 < P < 0.05; **0.001 < P < 0.01; ***P < 0.001.*

	Analysis I			Analysis II			Analysis III		
Variables	wing, culmen, tarsus, secondary, tail, middle toe length			wing, culmen, tarsus			wing, culmen		
n	161			222			267		
Multivariate analysis of variance									
Effects	Interaction	Area	Sex	Interaction	Area	Sex	Interaction	Area	Sex
Pillai's trace	0.237	1.246	0.370	0.879	0.810	0.348	0.061	0.704	0.310
F-value	0.852	5.428***	13.703***	0.888	10.888***	36.223	1.118	19.477***	56.231***
Hypoth. & error *df*	42/870	42/870	6/140	21/618	21/618	3/204	14/502	14/502	2/250
Univariate F tests									
df	7/145	7/145	1/145	7/206	7/206	1/206	7/251	7/251	1/251
wing length	0.297	9.616***	24.415***	0.658	10.402***	37.744***	0.629	9.177***	38.496***
culmen length	1.669	36.582***	51.049***	1.654	44.841***	72.509***	1.658	47.122***	78.936***
tarsus length	0.789	10.321***	8.315**	1.022	12.479***	10.921***			
secondary length	1.765	12.870***	21.893***						
tail length	0.910	1.090	12.524***						
middle toe length	0.408	1.753	1.126						

Table 31. *Sexual size dimorphism of the Knot. The mean percentage difference between the measurements of both sexes was calculated for every single variable and breeding population, according to: $(x_{♀}-x_{♂})/x_{♂} \ast 100$. A negative value thus implies smaller measurements for ♀♀. The calculations were based upon the means presented in Table 29. The obtained values were averaged per variable and summed per breeding population.*

Locality	Wing	Culmen	Tarsus	Secondary	Tail	Middle toe	Σ
Low arctic Canada	2.3	5.5	0.9	2.0	1.5	0.4	12.6
NE Canada	3.2	8.0	2.6	4.2	2.6	1.9	22.5
Greenland	1.8	6.9	1.6	(-0.9)	(-1.7)	(-3.3)	(4.4)
Taymyr	3.2	11.6	4.3	3.4	2.9	1.1	26.5
New Siberian Islands	4.6	11.0	5.4	7.2	5.9	2.6	36.7
Chukotsk Peninsula	3.0	2.8	1.3	2.9	3.4	0.7	14.1
Wrangel Island	2.9	4.6	0.6	3.5	6.6	1.1	19.3
Alaska	2.4	8.0	5.6	2.5	2.2	4.1	24.8
x	2.9	7.3	2.8	3.1	3.3	1.1	
SD	0.8	3.0	2.0	2.3	1.9	2.1	
n	8	8	8	8	8	8	

Table 32. *Pearson's product-moment correlations between measurements of the Knot. Both sexes are separately analysed. Two-tailed probabilities: * 0.01 < P < 0.05; ** P < 0.01.*

	Wing	Culmen	Tarsus	Secondary	Tail	Middle toe	
Wing		0.187	0.235	0.603**	0.499**	0.104	
Culmen	0.299**		0.636**	0.081	0.095	0.191	
Tarsus	0.319**	0.603**		0.094	0.299*	0.385*	♀♀
Secondary	0.566**	0.319*	0.365**		0.429**	0.080	
Tail	0.503**	0.047	0.143	0.289*		0.333*	
Middle toe	0.154	0.233	0.334*	0.120	0.064		
			♂♂				

The results of MANOVA-analysis, in which samples were pairwise contrasted, supported the findings of cluster analysis (Table 33):

(1) ♂♂ were nearly always distinguishable from ♀♀ of the same breeding population and nearly always from ♀♀ of other breeding populations. When differentiation was not evident, it was mostly caused by small sample sizes.

(2) Largely 2 groups of Knots can be distinguished amongst our samples. The first group includes the Knots from Wrangel Island, Alaska and low arctic Canada. And the other group includes the populations breeding between NE Canada and the Chukotsk Peninsula.

(3) The Wrangel Island population clearly differs from the low arctic Canada population, both in size and colouration. The Alaskan population cannot be distinguished from both others.

(4) Cluster analysis showed the distinct position of the breeding birds from the New Siberian Islands. This is also evident with MANOVA analysis even though the sample size is rather small, especially in the ♀♀.

(5) MANOVA also showed the lack of differentiation between the Taymyr- and high arctic Canadian sample.

(6) The Greenland ♀♀ could not be distinguished from the NE Canada or Taymyr ♀♀, but the Greenland ♂♂ differed nearly always from every other ♂ sample, due to the rather long wings and intermediately-sized bills.

(7) The small sample from Chukotsk is curious. The ♂♂ of this population nearly always differed from the neighbouring populations and show most resemblance with the populations of NE Canada, Greenland and Taymyr.

We concluded from these results that at least 4 distinct breeding populations can be characterized: (1) NE Canada, Greenland and Taymyr, (2) New Siberian Islands, (3) Wrangel Island, and (4) low arctic Canada. The position of the Alaskan population must then be considered as intermediate between groups 3 and 4. The sample size of the Chukotsk population is as yet to small to allow any conclusion on the distictiveness of the population from all others.

Nearly all variables contributed significantly to the distinction between samples, except for middle toe length (Table 30). Sexes differed in the following variables (in decreasing order of importance): culmen-, wing-, secondary-, tail-, and tarsus length. Culmen-, secondary-, tarsus-, and wing length (in decreasing order of importance) were important to distinguish between breeding locations (Table 30). When the middle toe lengths were ignored, the same results were obtained.

Both wing- and culmen lengths were important for the analysis of sexual and geographical variation (Table 30). The Greenland sample was also large enough to be included in the MANOVA analyses. The result corresponded with the all-measurements analyses.

Table 33. *Results of multivariate analysis of variance (MANOVA) concerning the morphometric differentiation in the Knot. The values presented are Pillai's trace values, which are presented with the level of significance concerning the difference between 2 groups (Norušis 1992c). Sixteen groups (2 sexes, 8 geographical groups) were compared on the basis of resp. 6 measurements (wing-, culmen-, tarsus-, secondary-, tail- and middle toe length), 3 measurements (wing-, culmen- and tarsus length), and 2 measurements (wing- and culmen length). Significance levels: *0.01 < P < 0.05; **0.001 < P < 0.01; ***P < 0.001.*

Locality	n	Low arctic Canada ♀	NE Canada ♀	Greenland ♀	Taymyr Peninsula ♀	New Siberian Islands ♀	Chukotsk Peninsula ♀	Wrangel Island ♀	Alaska ♀	Low arctic Canada ♂	NE Canada ♂	Greenland ♂	Taymyr Peninsula ♂	New Siberian Islands ♂	Chukotsk Peninsula ♂	Wrangel Island ♂
Wing, culmen, tarsus, secondary, tail & middle toe																
Low arctic Canada ♀	16															
NE Canada ♀	18	0.254***														
Greenland ♀	1	0.071	0.022													
Taymyr Peninsula ♀	8	0.123**	0.04	0.02												
New Siberian Islands ♀	5	0.227***	0.073	0.034	0.072											
Chukotsk Peninsula ♀	1	0.057	0.032	0.008	0.021	0.027										
Wrangel Island ♀	12	0.241***	0.237***	0.087*	0.217***	0.272***	0.108*									
Alaska ♀	3	0.042	0.138**	0.085*	0.104*	0.169***	0.08	0.064								
Low arctic Canada ♂	12	0.134**	0.572***	0.309***	0.496***	0.575***	0.229***	0.247***	0.142***							
NE Canada ♂	28	0.211***	0.398***	0.104*	0.346***	0.479***	0.160***	0.150***	0.116**	0.326***						
Greenland ♂	5	0.049	0.031	0.016	0.049	0.107*	0.037	0.057	0.035	0.205***	0.098*					
Taymyr Peninsula ♂	9	0.073	0.242***	0.093*	0.239***	0.374***	0.121**	0.105*	0.039	0.293***	0.044	0.092*				
New Siberian Islands ♂	8	0.127**	0.138**	0.054	0.139**	0.223***	0.089*	0.181***	0.100*	0.386***	0.181***	0.228***	0.089*			
Chukotsk Peninsula ♂	2	0.024	0.024	0.03	0.04	0.079	0.021	0.074	0.024	0.117**	0.021	0.075	0.022	0.041		
Wrangel Island ♂	15	0.381***	0.628***	0.312***	0.560***	0.641***	0.319***	0.198***	0.310***	0.250***	0.501***	0.202***	0.422***	0.539***	0.228***	
Alaska ♂	18	0.127**	0.348***	0.223***	0.346***	0.428***	0.231***	0.102*	0.101*	0.058	0.390***	0.175***	0.341***	0.456***	0.142**	0.180***
Wing, culmen & tarsus																
Low arctic Canada ♀	25															
NE Canada ♀	28	0.254***														
Greenland ♀	8	0.086***	0.009													
Taymyr Peninsula ♀	13	0.088***	0.029	0.002												
New Siberian Islands ♀	5	0.136***	0.036	0.036	0.044*											
Chukotsk Peninsula ♀	1	0.033	0.002	0.005	0.007	0.003										
Wrangel Island ♀	13	0.062**	0.228***	0.096***	0.106***	0.179***	0.043*									
Alaska ♀	3	0.024	0.122***	0.072**	0.065**	0.112***	0.046*	0.022								
Low arctic Canada ♂	17	0.135***	0.558***	0.363***	0.417***	0.500***	0.188***	0.088***	0.114***							
NE Canada ♂	33	0.109***	0.269***	0.074***	0.180***	0.342***	0.068**	0.089***	0.097***	0.290***						
Greenland ♂	17	0.044*	0.186***	0.072**	0.154***	0.287***	0.070**	0.022	0.031	0.152***	0.063**					
Taymyr Peninsula ♂	11	0.043*	0.273***	0.116***	0.207***	0.344***	0.086***	0.015	0.026	0.204***	0.002	0.047*				
New Siberian Islands ♂	8	0.026	0.051*	0.026	0.045*	0.124***	0.009	0.059**	0.035	0.312***	0.120***	0.208***	0.080**			
Chukotsk Peninsula ♂	4	0.011	0.013	0.002	0.014	0.052*	0.009	0.013	0.01	0.063**	0.015	0.027	0.014	0.057**		
Wrangel Island ♂	17	0.202***	0.510***	0.337***	0.412***	0.511***	0.227***	0.109***	0.172***	0.038*	0.353***	0.171***	0.256***	0.385***	0.105***	
Alaska ♂	19	0.077***	0.261***	0.180***	0.238***	0.324***	0.151***	0.047*	0.065**	0.005	0.324***	0.159***	0.227***	0.345***	0.077**	0.017

Continuation of Table 33: results of MANOVA-analysis based on the inclusion of wing- and culmen lengths.

Locality	n	Low arctic Canada ♀	NE Canada ♀	Greenland ♀	Taymyr Peninsula ♀	New Siberian Islands ♀	Chukotsk Peninsula ♀	Wrangel Island ♀	Alaska ♀	Low arctic Canada ♂	NE Canada ♂	Greenland ♂	Taymyr Peninsula ♂	New Siberian Islands ♂	Chukotsk Peninsula ♂	Wrangel Island ♂
Wing & culmen																
Low arctic Canada ♀	25															
NE Canada ♀	28	0.221***														
Greenland ♀	23	0.132***	0.02													
Taymyr Peninsula ♀	13	0.072***	0.023	0.002												
New Siberian Islands ♀	5	0.111***	0.026*	0.028*	0.034*											
Chukotsk Peninsula ♀	1	0.028*	0.001	0.004	0.006	0.002										
Wrangel Island ♀	13	0.045**	0.189***	0.138***	0.085***	0.143***	0.034*									
Alaska ♀	3	0.007	0.093***	0.064***	0.048**	0.087***	0.036*	0.011								
Low arctic Canada ♂	17	0.099***	0.513***	0.443***	0.364***	0.440***	0.154***	0.052**	0.081***							
NE Canada ♂	34	0.080***	0.232***	0.131***	0.140***	0.282***	0.071**	0.055**	0.069***	0.258***						
Greenland ♂	38	0.036**	0.289***	0.187***	0.182***	0.306***	0.061***	0.013	0.022	0.176***	0.040**					
Taymyr Peninsula ♂	11	0.031*	0.239***	0.155***	0.171***	0.287***	0.067***	0.004	0.017	0.175***	0	0.022				
New Siberian Islands ♂	8	0.02	0.044**	0.017	0.037**	0.099***	0.007	0.043**	0.027*	0.263***	0.085***	0.145***	0.059***			
Chukotsk Peninsula ♂	4	0.009	0.011	0.003	0.011	0.040**	0.006	0.009	0.008	0.051**	0.012	0.008	0.01	0.044**		
Wrangel Island ♂	17	0.158***	0.460***	0.391***	0.355***	0.447***	0.184***	0.079***	0.132***	0.027*	0.311***	0.221***	0.216***	0.324***	0.081***	
Alaska ♂	19	0.059***	0.223***	0.176***	0.200***	0.270***	0.121***	0.037**	0.050**	0.004	0.289***	0.200***	0.194***	0.291***	0.061***	0.012

Table 34. *The occurrence of brown feathers among the red breast feathers of Knot in summer plumage as an additional feature to distinguish ♂♂ from ♀♀. The distinction between 'some', 'frequent' or 'many' is qualitative: this scoring was merely aimed as a first step to unravel, as to whether the presence of brown feathers might be an indication for a ♀ and their absence for a ♂.*

		Frequency (in %)				
Locality	Sex	None	Few	Frequent	Many	*n*
Low arctic Canada	♂	53.8	38.5	7.7	-	13
	♀	25.0	25.0	50.0	-	16
NE Canada	♂	61.3	35.5	3.2	-	31
	♀	19.2	80.8	-	-	26
Greenland	♂	66.7	33.3	-	-	6
	♀	50.0	50.0	-	-	2
Taymyr	♂	75.0	12.5	12.5	-	7
	♀	-	57.1	14.3	28.6	7
New Siberian Islands	♂	62.5	37.5	-	-	8
	♀	20.0	-	60.0	20.0	5
Wrangel Island	♂	40.0	40.0	6.7	13.3	15
	♀	9.1	9.1	45.5	36.4	11

Females had more often brown breast feathers mixed between the red ones of the breeding plumage than ♂♂ (Table 34). Moreover, ♀♀ from Taymyr, the New Siberian Islands, and Wrangel Island had proportionally more brown feathers than ♀♀ from Canada.

The low arctic Canada population was also characterized by having palish-red breast feathers.

PREDICTION OF BREEDING ORIGIN

With the cluster- and MANOVA analyses 4 different POSCON groups could be identified: (1) low arctic Canada; (2) NE Canada, Greenland and Taymyr; (3) New Siberian Islands and (4) Wrangel Island. Since the breeding birds from Alaska were intermediate between the populations from Wrangel Island and low arctic Canada, they are left out of consideration. And since there was also a considerable degree of sexual dimorphism, 8 POSCON groups were recognized. Two different sets of groups were constructed, a set with 5 measurements (wing-, culmen-, tarsus-, secondary- and tail length) and a set with 2 measurements (wing- and culmen lengths).

In a next step, all available measurements of breeding birds were used to calculate the degree of resemblance (typicality probabilities) with the 8 constructed POSCON groups. By doing so, both the accuracy and exclusivity of prediction were estimated (see introduction under 'Statistical analysis of the data').

Accuracy

High accuracy rates of prediction could be reached when a bird's sex was known and up to 5 measurements were taken (Table 35). This was especially valid for the samples of low arctic Canada and Wrangel Island. Since the samples of NE Canada, Greenland and Taymyr had intermediate measurements, they also had a somewhat higher error rate. The estimated accuracy rates for the sample of the New Siberian Islands (5 measurements) did not include any predictions for the ♀♀, since too few ♀♀ could be sampled for the construction of a POSCON group.

When sexes were known and only wing- and culmen lengths were used, the error rate for the low arctic Canadian sample drastically increased because of a rather close resemblance with the Wrangel Island sample. The accuracy rate also decreased in the other samples, except for the New Siberian Islands one.

When sexes were *unknown* the situation became more complicated, since not 4 but 8 POSCON groups ought to be predicted and thus the accuracy rates correspondingly decreased. This occurred especially with the smallest-sized ♀♀ and the largest-sized ♂♂ (Table 36) Consequently, the best prediction results were obtained with large-sized ♀♀ (low arctic Canada, Wrangel Island) and small-sized ♂♂ (New Siberian Islands, NE Canada, Greenland, Taymyr).

Table 35. *Accuracy of prediction in Knot (sex* ***known*** *a priori). These predictions were based upon morphometric analysis and included resp. 5 measurements (wing-, culmen-, tarsus-, secondary- and tail length), and 2 measurements (wing- and culmen length). The columns show the predicted proportions in relation to the total number per breeding area (last column). [1]The sample group based on the ♀♀ from the New Siberian Islands, included too few individuals to allow predictions.*

	Predicted				
Locality	Low arctic Canada	NE Canada, Greenland, Taymyr	New Siberian Islands	Wrangel Island	*n*
	Wing-, culmen-, tarsus-, secondary- and tail length				
Low arctic Canada	80.0	6.7	-	13.3	30
NE Canada, Greenland, Taymyr	3.8	71.3	13.8	11.3	80
New Siberian Islands[1]	-	25.0	75.0	-	8
Wrangel Island	-	3.7	-	96.3	27
	Wing- and culmen length				
Low arctic Canada	47.4	2.6	2.6	47.4	38
NE Canada, Greenland, Taymyr	7.2	60.5	19.1	13.2	152
New Siberian Islands	-	7.7	92.3	-	13
Wrangel Island	6.7	3.3	-	90.0	30

Table 36. *Accuracy of prediction in Knot (sex* ***unknown*** *a priori). These predictions were based upon morphometric analysis and included resp. 5 measurements (wing-, culmen-, tarsus-, secondary- and tail length) and 2 measurements (wing- and culmen length). The columns show the predicted proportions in relation to the total number per breeding area (last column).*

	Predicted								
Locality	Low arctic Canada ♀	NE Canada, Greenland, Taymyr ♀	New Siberian Islands ♀	Wrangel Island ♀	Low arctic Canada ♂	NE Canada, Greenland, Taymyr ♂	New Siberian Islands ♂	Wrangel Island ♂	*n*
	Wing-, culmen-, tarsus-, secondary- and tail length								
Low arctic Canada ♀	70.6	11.8	?	11.8	-	-	-	2.2	17
NE Canada/ Greenland/ Taymyr ♀	5.7	51.4	?	20.0	5.7	5.7	2.9	8.6	35
New Siberian Islands ♀	-	60.0	?	20.0	20.0	-	-	-	5
Wrangel Island ♀	-	8.3	?	83.3	-	-	-	8.3	12
Low arctic Canada ♂	30.8	7.7	?	7.7	46.2	-	-	7.7	13
NE Canada/ Greenland/ Taymyr ♂	-	6.7	?	2.2	2.2	68.9	20.0	-	45
New Siberian Islands ♂	-	-	?	-	-	25.0	75.0	-	8
Wrangel Island ♂	6.7	6.7	?	13.3	-	-	-	73.3	15
	Wing- and culmen length								
Low arctic Canada ♀	52.2	-	-	39.1	4.3	-	-	4.3	23
NE Canada/ Greenland/ Taymyr ♀	6.3	40.6	14.1	6.3	9.4	4.7	-	10.9	64
New Siberian Islands ♀	-	20.0	60.0	-	20.0	-	-	-	5
Wrangel Island ♀	7.7	7.7	-	76.9	-	-	-	7.7	13
Low arctic Canada ♂	6.7	13.3	6.7	-	33.3	-	-	40.0	15
NE Canada/ Greenland/ Taymyr ♂	-	-	12.5	20.5	1.1	46.6	15.9	3.4	88
New Siberian Islands ♂	-	-	-	-	-	-	100.0	-	8
Wrangel Island ♂	35.3	29.4	-	11.8	-	-	-	23.5	17

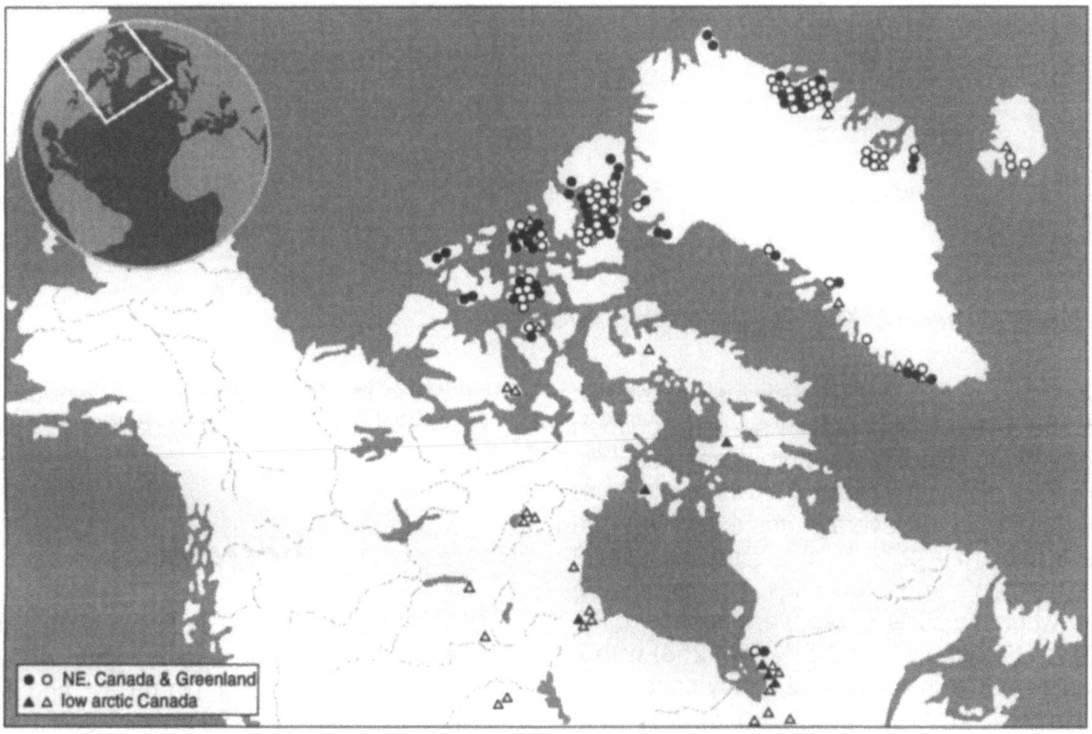

Fig. 47. *The predicted spatial distribution of both populations of Red Knot in arctic Canada as based on their measurements. These Knots were predicted to resemble one of both populations : (1) low arctic Canada, or (2) NE Canada & Greenland. Closed circles indicate that the typicality probability for NE Canada & Greenland was > 0.05, and that the typicality probability for the low arctic Canada population was < 0.05. An open circle indicates that both typicality probabilities were > 0.05, but that the NE Canada & Greenland populations was still more likely. An open square shows the opposite: the low arctic Canada population was most likely. With a closed square, the typicality probability for the NE Canada & Greenland sample was < 0.05 and for the low arctic Canada sample > 0.05.*

New Siberian Island birds were best predicted with only wing- and culmen length available, the other ones with more measurements available.

Exclusivity

On average, 6.3 % of the birds were predicted to resemble only 1 POSCON group when the birds' sexes were known and 5 measurements were used in the analyses. This degree of exclusivity was low in the POSCON groups of low arctic Canada (3.3 %, n = 30; 5 measurements and sex known) and NE Canada and Greenland and Taymyr (5.0 %, n = 80; 5 measurements and sex known). Higher exclusivity rates were estimated in the POSCON groups of the New Siberian Islands (15.4 %, n = 13) and Wrangel

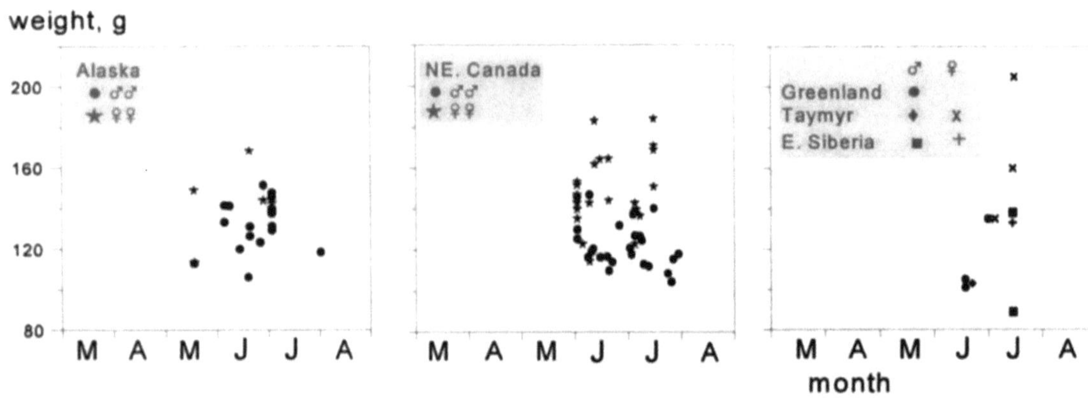

Fig. 48. *Body mass of Knots on the breeding grounds.*

Island (11.1 %; *n* = 27). The exclusivity rates in both increased when only wing- and culmen lengths were used, since both measurements were very distinctive (resp. 23.8 %, *n* = 13 and 20.0 %, *n* = 30). On average, the exclusivity rate was 5.9 % (*n* = 253) in this situation with 2 measurements and known sex. The rates were lower when the birds' sexes were unknown: 2.5 % (*n* = 160) with 5 and 2.7 % (*n* = 253) with 2 measurements included and it only concerned the small-sized ♂♂ or large-sized ♀♀.

CONTACT ZONES

The boundaries between the breeding areas of both populations in N Canada was investigated with help of our morphometric data and the constructed POSCON groups. Godfrey (1992) also considered the breeding ranges of both populations. His conclusion was that both occur on Prince of Wales Island. Our data and those of Godfrey mostly came from the same skins, but our analysis differed in approach, as we used the POSCON analyses to calculate typicality probabilities. Our findings support Godfrey's conclusions. The breeding range of the population breeding in NE Canada extends to Prince Patrick Island in the west and overlaps with the one of low arctic Canada on Somerset Island and Prince of Wales Island (Fig. 47).

Additionally, nearly all W Greenland migrants had the highest resemblance with the NE Canada & Greenland sample, whilst nearly all migrants in the NW Territories of Canada mostly resembled the low arctic Canadian sample (Fig. 47).

BODY MASS

Mass data of 21 Alaskan Knots were gathered (Fig. 48). Their average weights were resp. 132.4 (*SD* = 12.1, *n* = 18) for ♂♂ and 143.5 g (*SD* = 19.7, *n* = 3). Alaskan ♂♂ decreased their weights during the first 3 weeks of June. Their weight levels recovered during the second half of June and the first half of July.

On the breeding grounds in NE Canada mass data of 50 Knots were collected. Their average weights were resp. 122.5 (*SD* = 10.6, *n* = 26) for ♂♂ and 149.1 g (*SD* = 17.6, *n* = 24). These data showed that ♀ Knots increased mass with about 20 gm after arrival on the breeding grounds (Fig. 48), which is about the weight of a Knot's egg (on average 19 gm Cramp & Simmons 1983). Egg formation may therefore account for this observed mass increase. The ♀♀ approached the weight level of the ♂♂ by early July.

NE Canadian ♀♀, breeding west of Ellesmere Island, increased their body mass prior to departure during July, reaching to at least 180 gm. This information was lacking for the ♀♀ from Ellesmere Island.

The ♂♂ had a different pattern: their body mass steadily decreased throughout the breeding season. No weight increase prior to departure was observed.

Green (1978), De Korte *et al.* (1981), and Witts (1982) reported on the body mass of live-trapped adults. The body mass of these adults did not exceed 125 gm, which is lower than the ones from museum collections. De Korte *et al.* (1981) reported weight levels of Greenland Knots as low as 93 gm during starvation, caused by cold weather. The departure masses from the Canadian breeding grounds were more or less similar to those during spring migration (Prater 1974, Davidson & Wilson 1992).

PRIMARY MOULT

None of the birds collected showed primary moult on the breeding grounds.

DISCUSSION

The taxonomy of the Knot

The above presented morphometric data indicated the existence of 4 distinct breeding populations. The differences between the populations of low arctic Canada, Wrangel Island and the New Siberian Islands were most pronounced. The samples of NE Canada, Greenland and Taymyr were rather indistinguishable from each other, but differed from the first 3 mentioned.

These results mostly support the findings of Roselaar (1983), Tomkovich (1990, 1992a), Godfrey (1992) and Tomkovich & Soloviev (1996). Some additional remarks have to be made:

(1) *The lack of differentiation between the populations of Taymyr, NE Canada and Greenland.* Tomkovich (1992a) did not find morphometric differences between both breeding populations. We also failed to find such differences. Both were separated by Roselaar (1983). His analysis was primarily based upon 2 samples: (1) Greenland birds and (2) May-migrants passing through W Europe and coming from the W African wintering grounds. The latter sample was supposed to be representative for the Taymyr breeding population. Our new analysis was solely based upon breeding birds and revealed that (1) the measurements of Greenland ♂♂ differed somewhat from the ones of Taymyr and NE Canada, (2) the measurements of NE Canadian ♂♂ did not differ from the ones of Taymyr ♂♂, and (3) there were no differences at all between the ♀ samples of high arctic Canada, Greenland and Taymyr. Especially the high arctic Canadian breeding birds were morphometrically similar to the Taymyr Knots. These Taymyr Knots thus ought to be considered as *islandica* and not as *canutus*, since they did not resemble the May-migrants in W Europe (Table 37). However, recent field studies by Prokosch and co-workers (Prokosch & Hötker 1995) and Tomkovich & Soloviev (1996) clearly showed the N Taymyr birds to be larger-sized as indicated by our and Tomkovich's museum studies. The obtained field data of these Taymyr birds had close correspondence with the measurements of live-trapped birds staging in Guinea-Bissau, Mauritania, Germany (May) and with the older (!) museum data from W Europe during May, and are typical for nominate *canutus*. The discrepancy between the (older) museum- and (recent) field values on Taymyr is intriguing. It is probably not caused by wrong correction factors, since such would be an ad-hoc argument and not in line with other comparisons. It is also unlikely that spatial differences in sampling on Taymyr are responsible for this discrepancy, since both the museum- and field data were gathered on the breeding areas of N Taymyr. Temporal changes over the past 50 - 100 years might play a role with large-sized individuals having a selective advantage over small-sized ones, since nearly all museum skins had been collected in the first half of the 20th century. Such a shift in size was reported earlier from the Ringed Plover in Scandinavia (see chapter on Ringed Plover).

(2) *The subspecific status of the breeding population on the New Siberian Islands.* Tomkovich (1990) considered the breeding birds from the New Siberian Islands to be *canutus*, mainly because of a breeding plumage

Table 37. *Wing- and culmen lengths of Knots collected throughout the E Atlantic flyway. Information from both museum- and field studies are shown. The museum data represent uncorrected values in order to increase the comparability with other museum studies and to avoid potential errors due to wrong corrections. The wing lengths measured by Tomkovich (1992a) are shorter than later measurements because of methodological differences (Tomkovich 1992a).*

	Wing length		Culmen length		
Locality	♂♂	♀♀	♂♂	♀♀	Source
			Museum studies		
NE Canada	165.6±4.6(34)	170.6±4.4(28)	30.7±1.1(34)	33.2±1.4(28)	this study
E Greenland	167.2±4.4(32)	170.3±3.7(18)	31.6±1.3(32)	34.1±0.8(18)	this study
NE Canada & E Greenland	163.0±3.1(4)	165.8±6.2(4)	32.8±1.7(4)	33.9±0.1(3)	Tomkovich (1992a)
Taymyr	162.3±2.2(18)	168.6±4.2(6)	32.5±1.8(15)	34.9±1.8(7)	Tomkovich (1992a)
Taymyr	166.0±2.6(11)	171.0±3.9(12)	30.6±1.9(11)	34.1±1.6(12)	this study
W Europe - May	167.3±2.9(59)	170.5±2.7(53)	34.6±1.3(60)	36.5±1.3(53)	Roselaar (1983)
			Field studies		
NE Canada	-	-	31.8 - (22)	33.7 - (13)	Dick *et al.* (1976)
NE Canada	169.0 - (1)	173.5±0.5(2)	31.1 - (1)	35.0±0.0(2)	Witts (1982)
E Greenland	174.0±4.6(3)	-	-	-	De Korte *et al.* (1981)
Taymyr	169.0±4.2(38)	173.2±3.8(15)	33.7±1.6(38)	35.8±1.6(15)	Tomkovich & Soloviev (1996)
Taymyr	168.0±1.3(6)	169.9±5.4(9)	34.1±2.1(7)	34.8±1.5(9)	Prokosch & Hötker (1995)
Germany - March/April	172.4±3.6(111)	174.8±3.7(69)	32.4±4.6(109)	33.5±1.6(68)	Prokosch (1988)
Germany - May	170.4±3.7(208)	172.6±4.7(204)	34.7±1.6(209)	35.8±2.0(204)	Prokosch (1988)
Mauritania	168.1±3.2(11)	170.3±3.0(7)	35.0±1.3(11)	35.1±1.4(7)	Wymenga *et al.* (1990)
Guinea-Bissau	168.0±3.3(17)	173.2±2.8(5)	33.9±1.3(29)	36.1±2.0(11)	Wymenga *et al.* (1990)

similar to Taymyr birds. This plumage was rather different from the plumage of breeding birds on Wrangel Island or the Chukotsk Peninsula. Both latter also differed mutually in underparts, with the Wrangel Island birds being somewhat lighter (Paulson 1993) and having more often a completely rufous belly. Tomkovich considered the Chukotsk population as *rogersi* and separated the one of Wrangel Island as *roselaari* because of their different morphometrics and plumage characters. Tomkovich's morphometric information showed that the mean measurements of the breeding populations of the New Siberian Islands and Chukotsk were more similar to each other than to the measurements of the Taymyr breeding birds. Our few data of the Chukotsk population revealed more resemblance with the population of Taymyr than with the one of the New Siberian Islands. The small measurements of the birds from the New Siberian Islands are too different from the ones of Taymyr to consider them as one and the same subspecies. We thus prefer to consider the Knots from the New Siberian Islands as *rogersi*. Since the plumage characters of the birds from the New Siberian Islands really are too different, they should be considered as a separate subspecies and not as *canutus*.

(3) *The systematic position of the breeding population in Alaska.* The sampled Alaskan breeders could nearly never be differentiated from both the Wrangel Island and the low arctic Canada population, while they clearly differed from the populations breeding on the Chukotsk Peninsula and the New Siberian Islands (Table 33). This lack of differentiation with the neighbouring populations was mainly due to the occurrence of intermediate measurements. The morphometrics thus indicate that the Alaskan popu lation is influenced by both the Wrangel Island population and the population from low arctic Canada. When their measurements were analysed, they had mostly more resemblance with the low arctic Canada population than with the one from Wrangel

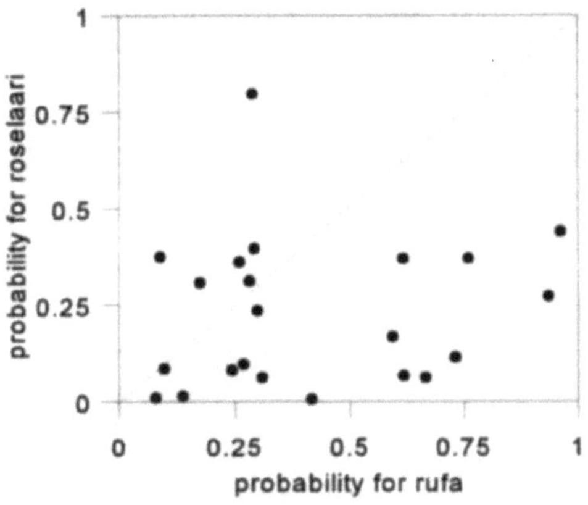

Fig. 49. *Estimating the degree of resemblance of 21 Alaskan breeding birds with (1) the breeding population of low arctic Canada - C. c. rufa (X-axis) and (2) the breeding population of Wrangel Island - C. c. roselaari (Y-axis). Typicality probabilities for these 2 breeding populations were calculated. These calculations were based upon 5 measurements per bird (wing-, culmen-, tarsus-, secondary- and tail length).*

Island (Fig. 49). However, the colouration of their summer plumage did not resemble the one of low arctic Canada but instead resembled more the one of Wrangel Island. We conclude that the Alaska population cannot be considered the same as the Wrangel Island- or the low arctic Canada population. On the other hand, the Alaska population can neither be distinguished as a separate subspecies, since it does not have clearly distinctional characters. It might be a population with mixed characters. It might also be that the populations of (1) low arctic Canada and Alaska, or (2) Wrangel Island and Alaska have common ancestors, whilst the present-day populations have little or no contact anymore which is resulting in an increasing differentiation.

The above presented data result in the recognition of 5 subspecies (Tables 37 & 38):

- *Calidris canutus canutus* (Linnaeus 1758) is breeding on N Taymyr. Nominate *canutus* was originally described from Sweden. Birds present on this type locality are most likely to be the long-billed breeding birds from Taymyr, which are also present in W Europe during autumn and spring. It is a dark-rufous coloured subspecies with intermediate measurements. It is in colouration and measurements most similar to the following subspecies, but it has significantly smaller wing- and longer culmen lengths.

- *Calidris canutus islandica* (Linnaeus 1767) is a breeding bird of NE Canada, NE Greenland, and Svalbard. It cannot be excluded to occur on Severnaya Zemlya or Taymyr. The breeding range of *islandica* in Arctic Canada extends as far west as Prince Patrick Island and is in contact with the low arctic *rufa* at Prince of Wales Island. The breeding plumage was described by Roselaar (1983) and Roselaar in Cramp & Simmons (1983). It has intermediate measurements. The subspecies shows some minor geographical variation.

- *Calidris canutus rogersi* (Mathews 1913) is breeding on the New Siberian Islands. The first description of *rogersi* was based upon a specimen collected in Japan during migration. The subspecies is characterized by its short measurements. Detailed breeding plumage descriptions were published by Roselaar (1983) and Tomkovich (1990 & 1992a).

- *Calidris canutus roselaari* (Tomkovich 1990) is breeding on Wrangel Island and possibly in Alaska. This subspecies is the largest one measured. The distinct breeding plumage is described by Tomkovich (1990 and 1992a). It is darker than *rufa*, and lighter than nominate *canutus* or *rogersi*. The subspecies does not show geographical variation.

- *Calidris canutus rufa* (Wilson 1813) is breeding in low arctic Canada. This subspecies has a pale breeding plumage and has long culmen- and tarsus lengths. However, the wing- and secondary lengths are intermediate and comparable with the lengths of *canutus*. It contacts *canutus* at Prince of Wales Island, but further breeds distinctly to the S. The subspecies does not show geographical variation.

The close similarity in the mitochondrial DNA of the different breeding populations, as studied by Baker *et al.* (1994), did not allow intraspecific taxonomical conclusions. However, the good

Table 38. *Measurements of 4 distinguished subspecies. No measurements of nominate canutus are given, since actual information from museum-data is lacking. The wing- and culmen lengths of this population, as derived from field studies, are presented in Table 37.*

Subspecies	Wing	Culmen	Tarsus	Secondary	Tail	Middle toe
			♂♂			
rufa	170.3±3.0(15)	34.4±0.9(15)	32.1±1.2(15)	70.2±1.9(13)	58.3±1.4(13)	27.6±1.4(10)
islandica	169.6±4.4(87)	31.2±1.4(87)	30.6±1.3(60)	71.1±2.1(45)	58.4±2.5(45)	26.7±1.1(42)
rogersi	163.3±4.4(13)	30.5±1.7(13)	30.1±1.4(13)	68.4±1.6(11)	56.5±1.3(11)	26.9±1.1(11)
roselaari	173.6±2.5(17)	34.7±1.4(17)	32.6±1.1(17)	74.0±1.6(15)	57.4±2.2(15)	27.0±1.5(15)
			♀♀			
rufa	174.4±3.6(22)	36.4±1.3(22)	32.3±1.0(22)	71.9±2.0(16)	59.3±2.9(16)	27.8±1.1(14)
islandica	173.9±4.0(64)	33.7±1.3(64)	31.3±1.3(46)	73.5±2.0(35)	59.7±2.5(35)	27.1±1.0(27)
rogersi	169.4±3.2(5)	32.9±0.6(5)	31.1±1.8(5)	73.4±2.9(5)	59.6±2.1(5)	27.6±0.9(5)
roselaari	178.2±4.0(13)	36.3±1.6(13)	32.8±1.5(13)	76.6±1.9(12)	61.3±2.9(12)	27.3±1.0(12)

possibilities to predict a Knot's breeding origin with morphometrics, allows us to recognize subspecies in the Red Knot. This is warranted if one compares the magnitude of the Knot's morphometric variation with the variation in other species. Starting from Baker *et al.*'s (1994) findings on the bottlenecking of the Red Knot 10-20.000 years ago, these morphometric and plumage differences seem to point to rapid evolution of morphometric and plumage characters when breeding areas are not in contact.

Is the Knot breeding in Yakutia?

It has been suggested that Knots were breeding in mountainous regions of Central- and E Siberia (Tomkovich 1992a). The proposal for the existence of such a breeding population was needed in order to find the breeding grounds of the long-billed Knots wintering in W Africa and migrating through W Europe. Nevertheless, no breeding grounds in Yakutia are identified up to now. We neither found possible breeding birds from Yakutia in the museum-collections. Our data are confined to 3 skins all from the Moscow museum, but neither of these were certain breeding birds:

- A ♀ was collected along the upper reaches of the Vilyuy River (62°N 117°E) on 7 June 1965. This bird had rather similar measurements in comparison with the Taymyr ♀♀ (wing 176, culmen 33.0, tarsus 31.5 mm).
- Two ♂♂ were collected along the mainland coast S of the New Siberian Islands (68.30°N 145°E). Collection dates are unknown. Both had small measurements like the birds from the New Siberian Islands. The first bird measured wing 161, culmen 33.0, tarsus 32.4, and secondary 69 mm and the second one had wing 170, culmen 31.1, tarsus 31.4, and secondary 68 mm. Both were probably migrants from the New Siberian Islands.

Thus these 3 birds do not indicate the presence of the expected long-billed breeding population in Yakutia. Moreover, the recent field information from Taymyr might well indicate a predominantly Taymyr breeding origin of the W African wintering population. However, there remain important discrepancies in the estimated population sizes as based upon the breeding esitimates from Taymyr in comparison with the estimates on the W African wintering grounds.

Different degrees of geographical variation in the ♂♂ and ♀♀

We agree with Tomkovich (1992a) that ♀♀ show less geographical variation than ♂♂ (Fig. 46 and Table 34). This cannot be due to the smaller sample size of ♀♀, since the cluster distances between the ♀ samples were also smaller than between the ♂ samples. These cluster distances are solely based upon the sample means.

SUMMARY

Measurements and plumage characters of Red Knots from all known breeding areas were examined and analysed with multivariate statistics. The species shows sexual and geographical variation in morphometrics and plumage characters. Sexual differences were larger than geographical differences, and ♂♂ were smaller than ♀♀. Sexual dimorphism was smallest in the breeding population of low arctic Canada (or possibly Greenland) and largest in the one of the New Siberian Islands. The species is polytypic and 5 subspecies are recognized: (1) nominate *canutus* breeding on the Taymyr Peninsula; (2) *islandica*, breeding in NE Canada, NE Greenland, Svalbard, (and possibly Severnaya Zemlya or parts of Taymyr); (3) *rogersi*, breeding on the New Siberian Islands,; (4) *roselaari*, breeding on Wrangel Island; and (5) *rufa*, breeding in low arctic Canada. The measurements of the Alaskan breeding population are intermediate between the ones of the Wrangel Island population and the ones of low Arctic Canadian, but have a generally higher resemblance with the latter than with the first. However, the darkish red bellies of Alaskan breeding birds are most similar to the ones of Wrangel Island. The population from Chukotsk has most in common with the ones on Taymyr, Greenland and NE Canada and differs from the populations on the New Siberian Islands, Wrangel Island, Alaska and low arctic Canada. Both first subspecies show some geographical variation; both last ones do not. *Rogersi* had the smallest measurements and *roselaari* the largest. Generally, culmen-, secondary-, wing-, tarsus-, and tail lengths were useful to be measured. It was possible to reach high accuracy rates when predicting a bird's breeding origin (at least as long as the intermediate populations of Chukotsk and Alaska were not considered). The exclusivity rates were low.

Chapter 8

Sanderling

-

Calidris alba

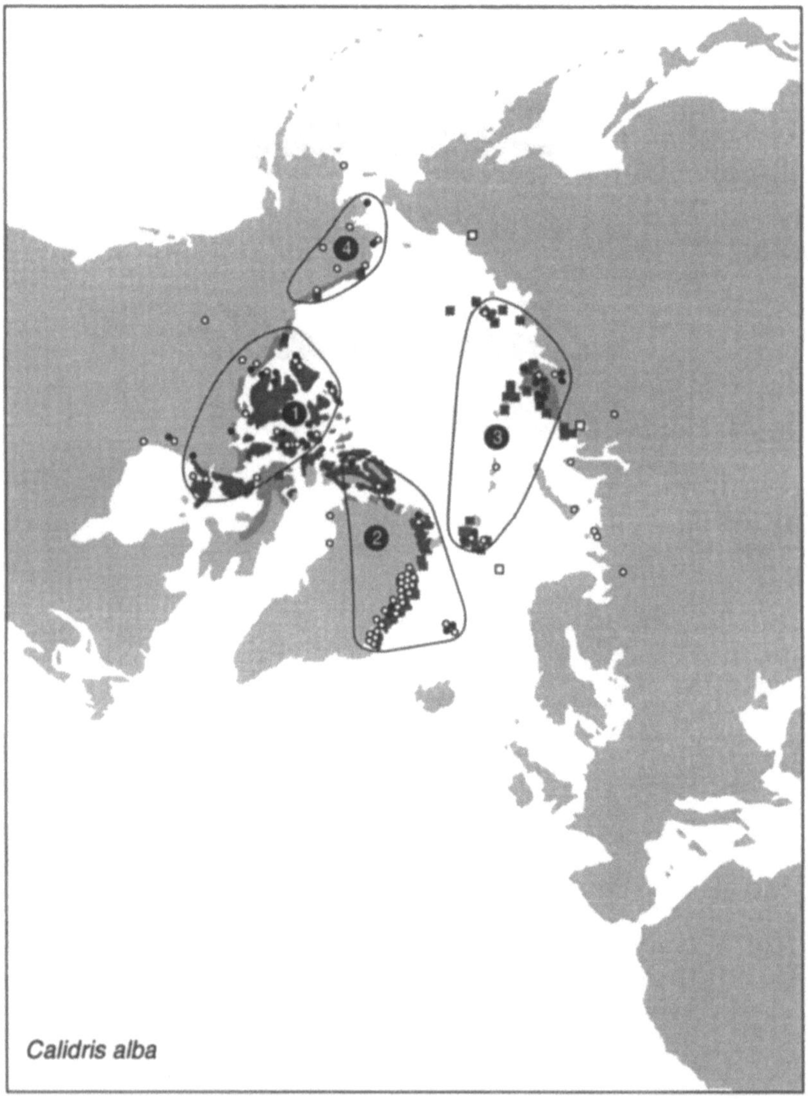

Fig. 50. *Breeding distribution of the Sanderling. For explanation of the used symbols see Chapter 3: 'Introduction to the species' accounts'.*

INTRODUCTION

Sanderlings are breeding in high arctic regions of both the Nearctic and the Palearctic (Fig. 50). The breeding distribution is scattered, with large areas in Siberia and Alaska being unoccupied. The major breeding concentrations are found in the Canadian Arctic, on N and E Greenland, on Svalbard and in Central and W Siberia: Taymyr, Severnaya Zemlya, and the New Siberian Islands.

Glutz *et al.* (1975) did not recognize any geographical variation. Cramp & Simmons (1983) showed the existence of a slight degree of geographical variation, which probably involved the colour of the adult breeding plumage. No different subspecies were recognized, since the morphometric similarity between populations and sexes was too large.

MORPHOMETRICAL DIFFERENTIATION

In total, 236 specimens were examined from nearly the whole breeding range (Fig. 50). Few birds from Svalbard, Franz Josephland and the New Siberian Islands were examined. Breeding

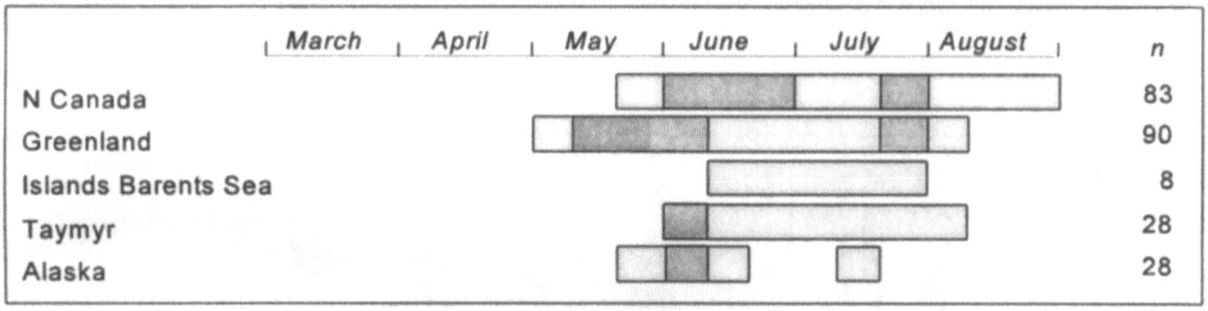

Fig. 51. *Collection dates of Sanderlings. Most birds had been collected during the periods marked by the heavily shaded bars. The sample sizes are presented on the right side.*

Table 39. *Measurements of Sanderlings, collected on the breeding grounds.* [1] *The measurements of the breeding birds from Ellesmere Island were combined with the ones of the Greenland population.* [2]*The measurements of birds collected just south of the regular breeding range on Taymyr (Fig. 50) are presented under S Taymyr. The numbers in the first column refer to the numbers used in Fig. 50. Presented are the means ± 1 SD (n).*

	Locality	Wing	Culmen	Tarsus	Secondary	Tail	Middle toe
				♂♂			
1	N Canada[1]	125.6±2.7(45)	24.8±0.9(45)	24.8±1.0(42)	56.4±1.7(36)	48.7±2.8(36)	20.1±0.8(32)
2	Greenland[1]	124.8±2.6(50)	23.9±1.1(49)	24.2±0.9(18)	54.6±1.3(8)	46.8±1.5(11)	19.4±0.9(8)
3	Svalbard	124.3±5.6(3)	24.0±0.6(3)	24.0±0.6(3)	52.5±0.7(2)	47.0 - (1)	19.7 - (1)
3	Franz Joseph Land	127.0 - (1)	22.7 - (1)	24.3 - (1)	-	-	-
3	Novaya Zemlya	125.3±2.6(3)	24.2±0.6(3)	24.5±1.6(3)	-	-	-
3	N Taymyr[2]	124.0±2.5(12)	23.5±1.4(12)	24.3±1.3(12)	54.4±1.7(12)	49.2±2.6(11)	20.1±0.9(11)
3	S Taymyr[2]	124.4±2.1(5)	22.7±0.8(5)	24.0±0.3(5)	55.0±1.2(5)	47.6±2.4(5)	19.6±0.3(5)
	New Siberian Isl.	126.0 - (1)	23.9 - (1)	24.5 - (1)	-	-	-
4	Alaska	125.3±2.9(17)	24.6±0.8(17)	24.5±0.8(17)	55.5±1.2(17)	48.5±1.8(17)	20.0±0.6(15)
				♀♀			
1	N Canada[1]	129.1±2.9(39)	26.6±1.1(39)	25.5±0.9(38)	57.2±1.7(37)	49.6±2.3(37)	20.0±0.9(34)
2	Greenland[1]	127.7±2.5(37)	25.7±1.2(36)	24.9±0.7(13)	56.2±0.5(4)	48.2±2.2(8)	19.7±0.9(5)
3	Svalbard	127.0 - (1)	25.5 - (1)	24.8 - (1)	-	-	-
3	N Taymyr[2]	128.3±3.6(7)	25.0±0.5(7)	25.0±0.8(7)	56.0±2.0(7)	50.3±2.4(7)	20.1±0.9(5)
3	S Taymyr[2]	130.3±3.3(4)	26.2±0.8(4)	25.4±0.3(4)	58.3±3.1(3)	48.0±1.7(3)	20.3±0.3(3)
4	Alaska	127.5±2.6(11)	27.1±1.0(11)	25.6±1.1(11)	56.7±1.0(11)	49.2±1.5(11)	20.2±0.9(11)

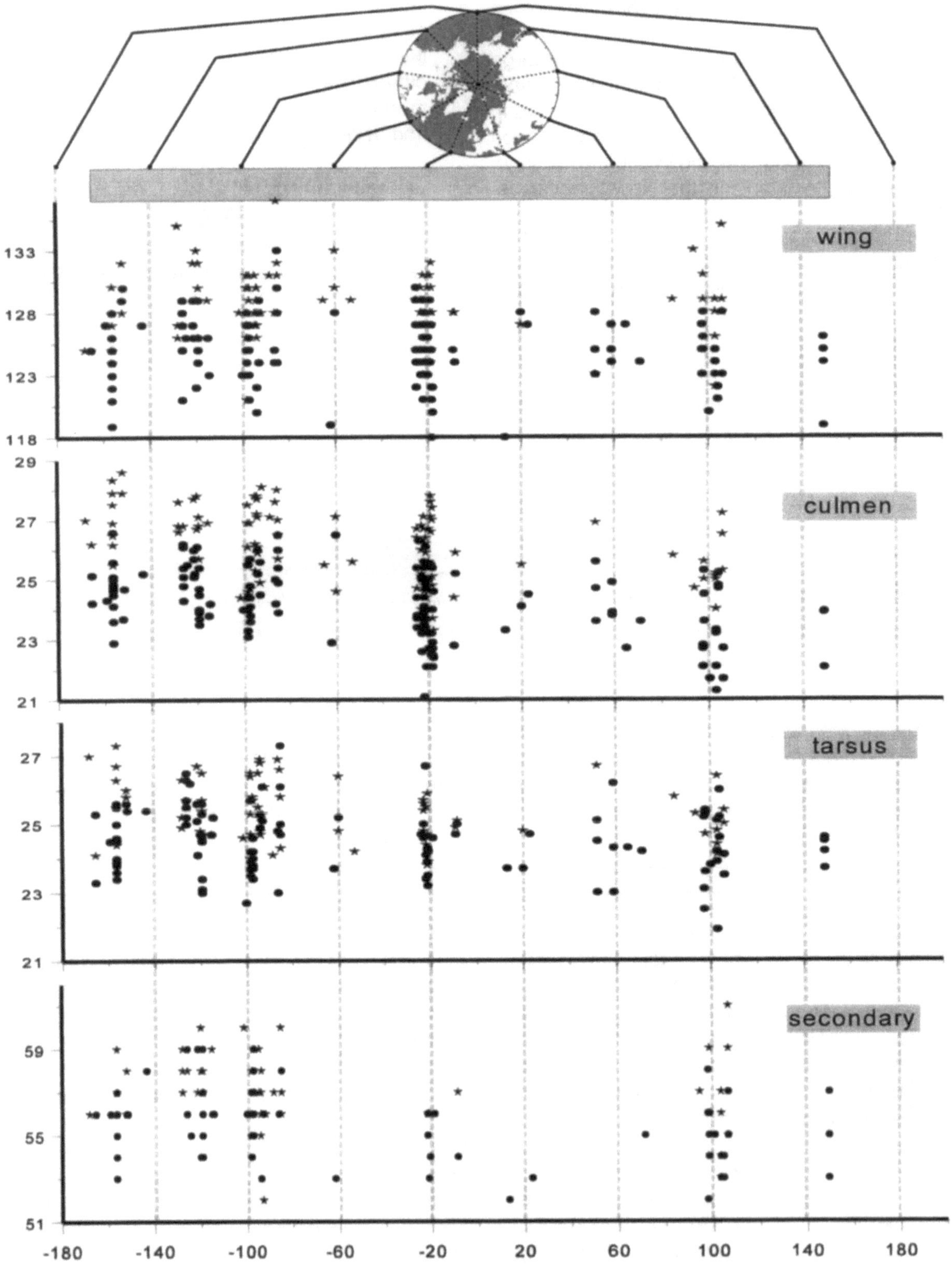

Fig. 52. *Measurements of Sanderlings from the breeding grounds. The following measurements are presented from top to bottom: wing length, culmen length, tarsus length and secondary length (description in chapter 2). Explanation of the used symbols: ● ♂♂ and ★ ♀♀.*

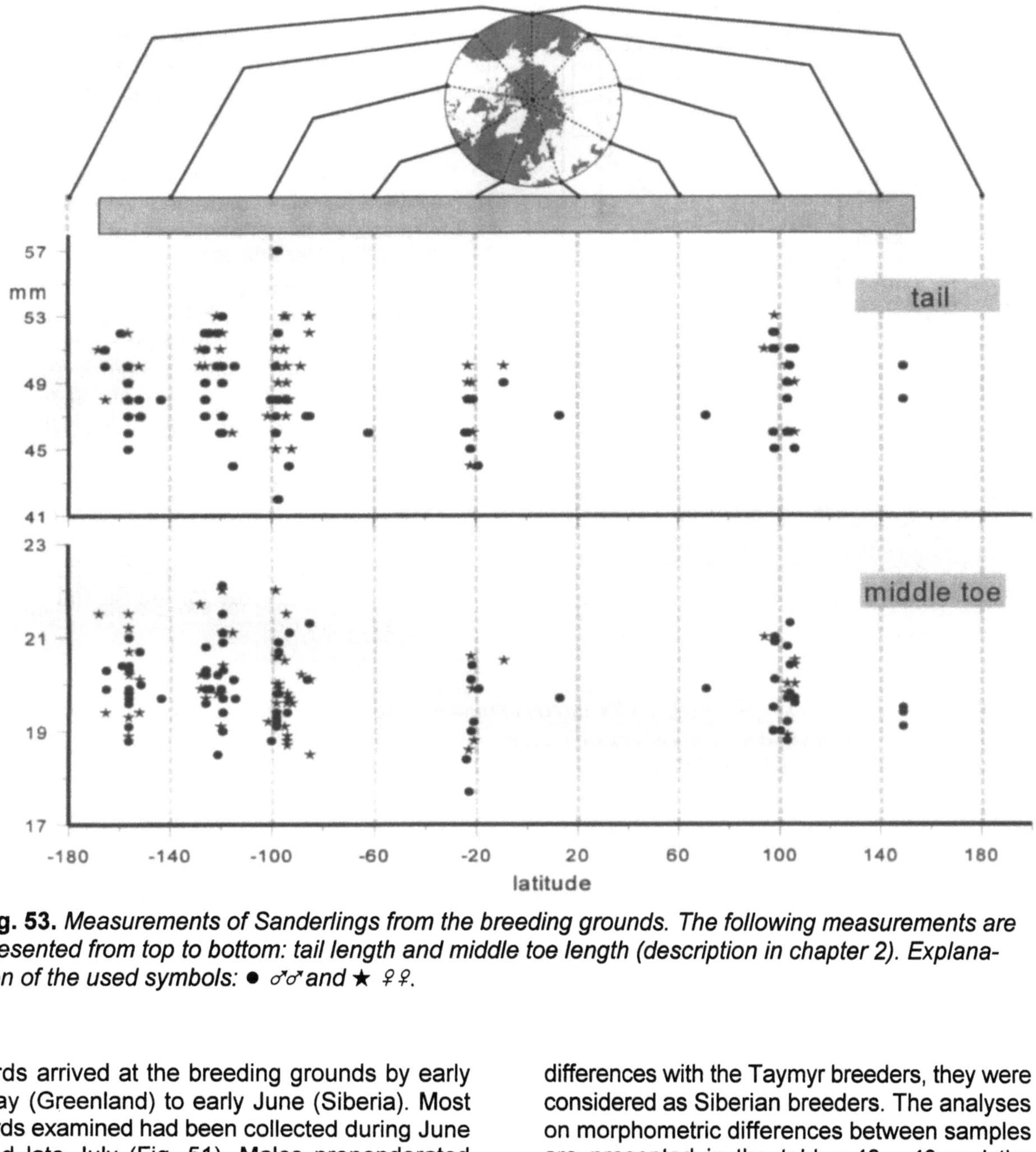

Fig. 53. *Measurements of Sanderlings from the breeding grounds. The following measurements are presented from top to bottom: tail length and middle toe length (description in chapter 2). Explanation of the used symbols: ● ♂♂ and ★ ♀♀.*

birds arrived at the breeding grounds by early May (Greenland) to early June (Siberia). Most birds examined had been collected during June and late July (Fig. 51). Males preponderated slightly (57.5 %) in the samples.

Initially, 4 breeding locations were recognized (Fig. 50): (1) N Canada (east of the Mackenzie River, excluding Ellesmere Island), (2) Greenland (Greenland, Ellesmere and Jan Mayen), (3) Siberia (Svalbard, Franz Josephland, Taymyr) and (4) Alaska. It resulted in the construction of 8 samples: 4 breeding locations of each sex. Several birds had been collected just south of the regular breeding range on Taymyr. Since they did not show significant morphometric differences with the Taymyr breeders, they were considered as Siberian breeders. The analyses on morphometric differences between samples are presented in the tables 40 - 43 and the figures 52 - 54.

The species showed sexual and geographical variation (Table 40). Sexual differences were larger than geographical ones. Sexual dimorphism was smallest in N Canada and largest in Siberia (Table 41). The wing- and tail variables were not always correlated with each other as in other species. Culmen- and tarsus lengths were correlated in both sexes (Table 42). Middle toe- and tail lengths correlated poorly with other measurements.

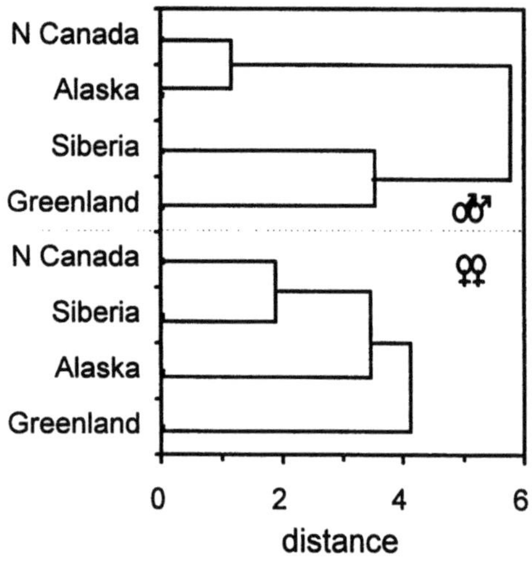

Fig. 54. *Dendrogram representing the results of cluster analysis in the Sanderling. The analysis was based upon the mean measurements per breeding population. Squared Euclidean cluster distances were calculated, with use of average linkage between groups.*

Cluster analysis in the ♂♂ showed similarities in measurements between the samples of Siberia and Greenland on the one side and N Canada and Alaska on the other side (Fig. 54). This was not so with the samples of the ♀♀.

The results of MANOVA-analysis, in which samples were pairwise contrasted, were in line with the results of cluster analysis (Table 43):

- Within a region, ♂♂ were nearly always distinguishable from ♀♀, but such was not always so with ♀♀ of other regions: when the differences between both sexes were not significant, small sample sizes were involved as is illustrated with the sample of Greenland ♀♀.
- When all 6 measurements were involved nearly no differences were evident between the 4 ♀ samples. The ♂♂ from N Alaska clearly differed from the ♂♂ of Siberia and the same was true for the Siberia ♂♂ and the Alaska ♂♂.

Table 40. *Overall results of MANOVA for the distinguished breeding populations in the Sanderling in combination with univariate ANOVA's in which the different measurements were analysed separately. Significance levels: *0.01 < P < 0.05; **0.001 < P < 0.01; ***P < 0.001.*

	Analysis I			Analysis II			Analysis III		
Variables	wing, culmen, tarsus, secondary, tail, middle toe length			wing, culmen, tarsus			wing, culmen		
n	129			183			241		
Multivariate analysis of variance									
Effects	Interaction	Area	Sex	Interaction	Area	Sex	Interaction	Area	Sex
Pillai's trace	0.110	0.360	0.484	0.058	0.234	0.436	0.025	0.211	0.437
F-value	0.751	2.686***	18.104***	1.150	4.935***	44.575***	0.989	9.155***	90.170***
Hypoth. & error *df*	18/354	18/354	6/116	35803	35803	3/173	6/466	6/466	2/232
Univariate F tests									
df	3/121	3/121	1/121	3/175	3/175	1/175	3/233	3/233	1/233
wing length	1.504	1.363	30.070***	0.913	1.813	42.843***	0.711	2.221	62.962***
culmen length	1.070	10.664***	93.893***	2.327	14.436***	117.923***	0.908	18.876***	159.701***
tarsus length	0.521	2.840*	10.315**	0.368	3.845*	33.892***			
secondary length	1.172	4.800**	13.888***						
tail length	0.065	2.966*	3.226						
middle toe length	0.334	0.165	0.857						

Table 41. *Sexual size dimorphism in the Sanderling. The mean percentage difference between the measurements of both sexes was calculated for every single variable and breeding population, according to: $(x_{♀}-x_{♂})/x_{♂}*100$. A negative value thus implies smaller measurements for ♀♀. The calculations were based upon the data presented in Table 39. The obtained values were averaged per variable and summed per breeding population.*

Locality	Wing	Culmen	Tarsus	Secondary	Tail	Middle toe	Σ
N Canada	2.8	7.3	3.2	1.4	1.9	-0.1	16.5
Greenland	2.3	7.3	2.9	3.0	3.1	1.7	20.3
Siberia	3.1	8.2	4.1	4.3	2.1	1.4	23.2
Alaska	1.8	10.2	4.5	2.2	1.4	1.0	21.1
x	2.5	8.2	3.7	2.7	2.1	1.0	
SD	0.6	1.4	0.8	1.2	0.7	0.8	
n	4	4	4	4	4	4	

- When 2 or 3 variables were used in the MANOVA-analyses there were regularly no significant differences between (1) the N Canada and Alaska populations, and (2) the Greenland and Siberia populations. And oppositely there were regularly differences between the N Canada or Alaska population on the one side and the Greenland or Siberia population on the other side.

We conclude from these results that 2 morphometrically different populations can be recognized: (1) N Canada and Alaska, and (2) Greenland and Siberia.

Nearly all variables contributed significantly to the distinction between samples except for middle toe length (Table 40). Sexes differed in culmen, wing, secondary, and tarsus (in decreasing order of importance). Culmen-, secondary-, tail-, and tarsus length (also in decreasing order of importance) were important to distinguish between breeding locations. These results did not change importantly when the middle toe lengths were left out of consideration.

Both wing- and culmen lengths were important for the analysis of sexual and geographical variation with the sexual variation being much larger than the geographical variation (Table 40). The results of MANOVA analysis with wing- and culmen length strengthened the conclusions of the all-measurements analyses, probably due to the larger sample sizes (of especially the Greenland population): the samples from N Canada and Alaska differed from the ones of Greenland and Siberia. This was also evident from the pairwise comparisons in the ♀♀, unlike the all-measurements analyses.

The breeding birds of N Canada and Alaska had a rather greyish-white appearance of the upperparts of the breeding plumage, while the breeding birds of Greenland and Taymyr had a more rufous colour. This difference was caused by more extensive greyish-white fringing along the mantle feathers of N Canadian and Alaskan birds. These 2 different colour types correspond with the differences found in measurements.

Table 42. *Pearson's product-moment correlations between measurements of the Sanderling. Both sexes are separately analysed. Two-tailed probabilities: * $0.01 < P < 0.05$; ** $0.001 < P < 0.01$; *** $P < 0.001$.*

	Wing	Culmen	Tarsus	Secondary	Tail	Middle toe	
Wing		0.307**	0.314**	0.465***	0.401***	0.295*	
Culmen	0.297***		0.485***	0.262*	0.137	0.249	
Tarsus	0.209*	0.478***		0.065	0.181	0.402**	
Secondary	0.529***	0.284**	0.192		0.275*	0.069	♀♀
Tail	0.099	0.045	-0.030	0.129		0.079	
Middle toe	0.112	0.191	0.330**	0.209	-0.041		
			♂♂				

Table 43. *Results of MANOVA analysis concerning the morphometric differentiation in the Sanderling. Pillai's trace statistics are presented (Norušis 1992c). In these series of pairwise comparisons 8 groups (2 sexes, 4 geographical groups) were compared on the basis of resp. 6, 3 and 2 measurements. Significance levels: *0.01 < P < 0.05; **0.001 < P < 0.01; ***P < 0.001.*

Locality	n	N Canada ♀	Greenland ♀	Siberia ♀	Alaska ♀	N Canada ♂	Greenland ♂	Siberia ♂
Wing, culmen, tarsus, secondary, tail & middle toe								
N Canada ♀	34							
Greenland ♀	4	0.075						
Siberia ♀	8	0.072	0.033					
Alaska ♀	11	0.066	0.092	0.124*				
N Canada ♂	32	0.414***	0.072	0.128*	0.325***			
Greenland ♂	6	0.312***	0.079	0.150**	0.310***	0.093		
Siberia ♂	19	0.572***	0.174***	0.268***	0.480***	0.219***	0.057	
Alaska ♂	15	0.333**	0.064	0.110*	0.290***	0.022	0.053	0.106*
Wing, culmen & tarsus								
N Canada ♀	38							
Greenland ♀	13	0.072**						
Siberia ♀	13	0.056*	0.007					
Alaska ♀	11	0.040	0.094***	0.094***				
N Canada ♂	42	0.314***	0.044*	0.070**	0.218***			
Greenland ♂	18	0.287***	0.062*	0.095***	0.230***	0.024		
Siberia ♂	31	0.488***	0.168***	0.200***	0.363***	0.119***	0.046*	
Alaska ♂	17	0.233***	0.041	0.068**	0.186***	0.004	0.008	0.067**
Wing & culmen								
N Canada ♀	39							
Greenland ♀	36	0.064***						
Siberia ♀	13	0.040**	0.003					
Alaska ♀	11	0.030*	0.077***	0.070***				
N Canada ♂	45	0.246***	0.084***	0.053**	0.158***			
Greenland ♂	49	0.403***	0.220***	0.125***	0.265***	0.065***		
Siberia ♂	31	0.403***	0.240***	0.149***	0.284***	0.095***	0.009	
Alaska ♂	17	0.176***	0.064***	0.049**	0.137***	0.001	0.027*	0.047**

PREDICTION OF BREEDING ORIGIN

The cluster- and MANOVA analyses result in the distinction of two different groups of breeding populations: (1) N Canada and Alaska; (2) Greenland and Siberia. Due to the sexual dimorphism, 4 POSCON groups were recognized. Two different sets of POSCON groups were constructed, a set with 4 measurements (wing-, culmen-, tarsus-, and secondary length) and a set with 2 measurements (wing- and culmen lengths). Tail- and middle toe lengths not used because of their insignificancy.

In a next step, all available measurements of breeding birds were used to calculate the degree of resemblance (typicality probabilities) with the POSCON groups.

Accuracy

Rather high accuracy rates could be reached when a bird's sex was known and 4 measurements were analysed: 75.6 % correct (n = 86) for N Canada / Alaska and 87.8 % (n = 41) for Greenland / Siberia. When only wing- and culmen lengths were used, these percentages were resp. 67.0 (n = 97) and 65.1 (n = 126).

Table 44. *Accuracy of prediction in the Sanderling with a bird's sex being* ***unknown****. These predictions were based upon morphometric analysis and included (A) wing-, culmen-, tarsus-, and secondary length; (B) wing- and culmen length. The columns show the predicted proportions in relation to the total number per breeding area (last column).*

Locality	Predicted				*n*
	N Canada & Alaska ♀	N Canada & Alaska ♂	Greenland & Siberia ♀	Greenland & Siberia ♂	
(A) Wing-, culmen-, tarsus- and secondary length					
N Canada & Alaska ♀	69.8	11.6	18.6	-	43
N Canada & Alaska ♂	4.7	55.8	18.6	20.9	43
Greenland & Siberia ♀	14.3	14.3	64.3	7.1	14
Greenland & Siberia ♂	-	11.1	7.4	81.5	27
(B) Wing- and culmen length					
N Canada & Alaska ♀	66.7	8.9	24.4	-	45
N Canada & Alaska ♂	1.9	32.7	32.7	32.7	52
Greenland & Siberia ♀	34.7	18.4	34.7	12.2	49
Greenland & Siberia ♂	-	26.0	9.1	64.9	77

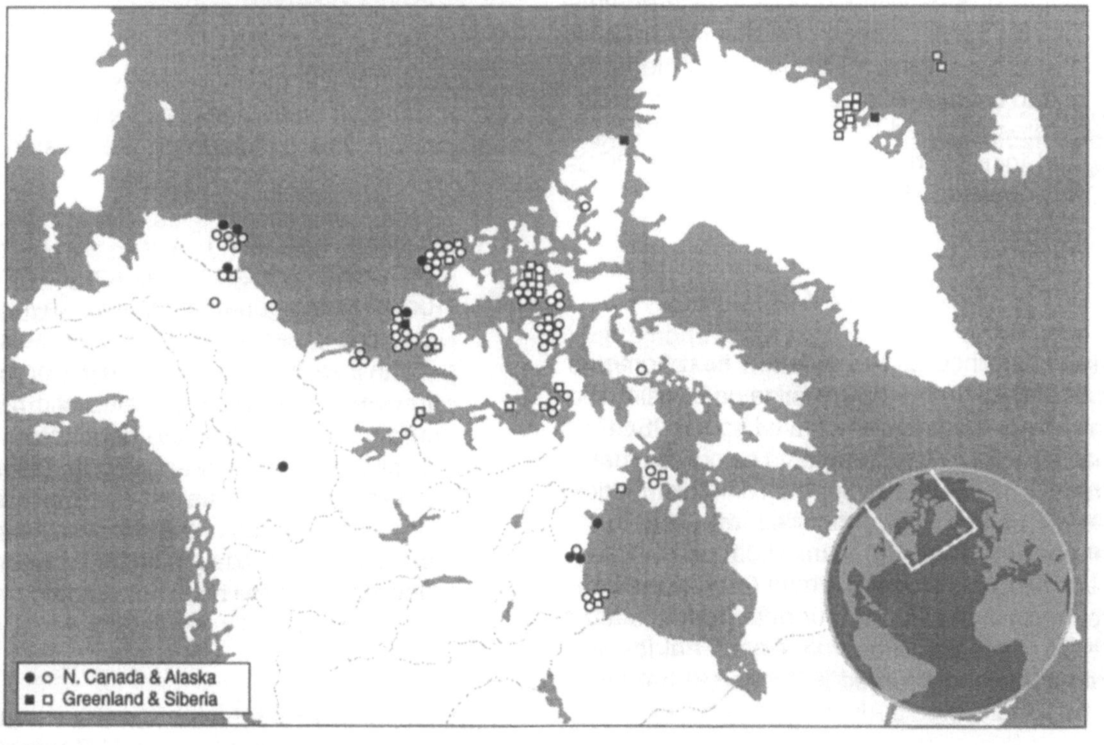

Fig. 55. *The predicted spatial distribution of both populations of Sanderling in Canada as based on their measurements. These Sanderling resemble 2 populations: (1) North Canada or Alaska (circles); and (2) Greenland or Siberia (squares). Closed symbols indicate that the typicality probability for the most resembling group was > 0.05 and that the typicality probability for the other was < 0.05. Open symbols indicate that the typicality probability for the most resembling group was again > 0.05 but that the typicality probability for the other was also > 0.05.*

When sexes were *unknown* not 2 but 4 typicality probabilities had to be analysed, resulting in lower accuracy rates. They were lowest for the N Canada / Alaska ♂♂ (55.8 %) and the Greenland / Siberia ♀♀ (64.3 %) when 4 measurements were analysed (Table 44). When sexes were unknown and only wing- and culmen lengths were used, the accuracy of prediction was reduced even more (Table 44). This makes it problematical to sex Sanderling when (1) only wing- and culmen lengths are measured and (2) different breeding populations can be involved.

Exclusivity

On average, 10.2 % of the birds were predicted to resemble only 1 POSCON group when the birds' sexes were known and 4 measurements were used in the analyses. This degree of exclusivity was smallest in both POSCON groups with intermediate measurements (2.3 % for the N Canada / Alaska ♂♂ and 0.0 % for the Greenland / Siberia ♀♀) and largest in both POSCON groups with extreme measurements (14.0 % for the N Canada / Alaska ♀♀ and 22.2 % for the Greenland / Siberia ♂♂). When only wing- and culmen lengths were used, the exclusivity rates dropped for both extreme (4.4-15.6 %) and increased for both intermediate POSCON groups (1.9-6.1 %).

CONTACT ZONES

The occurrence of two different morphometric types in N Canada was investigated by using the typicality probabilities as derived from POSCON-analysis with 4 measurements (wing-, culmen-, tarsus- and secondary length). The distribution pattern of both corresponded with the initial delineations (Fig. 55). The situation on Ellesmere and Axel Heiberg Island (both types contact here) is unclear as our data from this area were too scarce. Birds from Prince Patrick and Bathurst Island seemed to be somewhat more 'Greenlandic' than birds further south and west.

BODY MASS

Mass-data of 77 Sanderlings were taken from the specimen labels. Most of these originated from Canada (*n* = 32), Alaska (*n* = 21) and E Greenland (*n* = 12). The body masses ranged between 45 and 85 g and ♀♀ were mostly heavier than ♂♂ (Fig. 56). Unlike the situation in N Canada and Alaska, Greenland ♀♀ did not seem to be heavier than ♂♂. In N Canada both sexes seemed to decrease their mass from mid-July onwards.

PRIMARY MOULT

Only 1 Sanderling out of 236 birds examined was in primary moult: a ♀ breeding on NW Baffin Island (Canada) with score 4 on 22 July (the 4 innermost primaries recently shed). The occurrence of primary moult on the breeding grounds is thus a rare event.

DISCUSSION

The taxonomy of the Sanderling

The data presented above show the existence of 2 distinct breeding populations, one in the Nearctic and the other in the Palearctic and on Greenland. The measurements of both populations differed enough to reach adequate predictions. We thus conclude on the existence of 2 subspecies:

(1) *Calidris alba alba* (Pallas, 1764) breeds in NE and E Greenland, on Jan Mayen, Svalbard, Franz Josephland and Taymyr. The species is described from the coast of the North Sea. Most wintering birds or migrants from this area have a Palearctic breeding origin and therefore the nominate name is applied to the Palearctic population. The subspecies is characterized by its short bill- and tarsus length. The ♂♂ also have short wing- and secondary lengths. Moreover, the upperparts of the birds in breeding plumage are predominantly reddish-brown or brownish.

(2) *Calidris alba rubidus* (Gmelin, 1789) was described from the Hudson Strait in Canada. It breeds in Alaska and N Canada and possibly also in the Delta of the Lena River and on the New Siberian Islands. It has long culmen- and tarsus lengths and the ♂♂ also have long wing- and secondary lengths. The

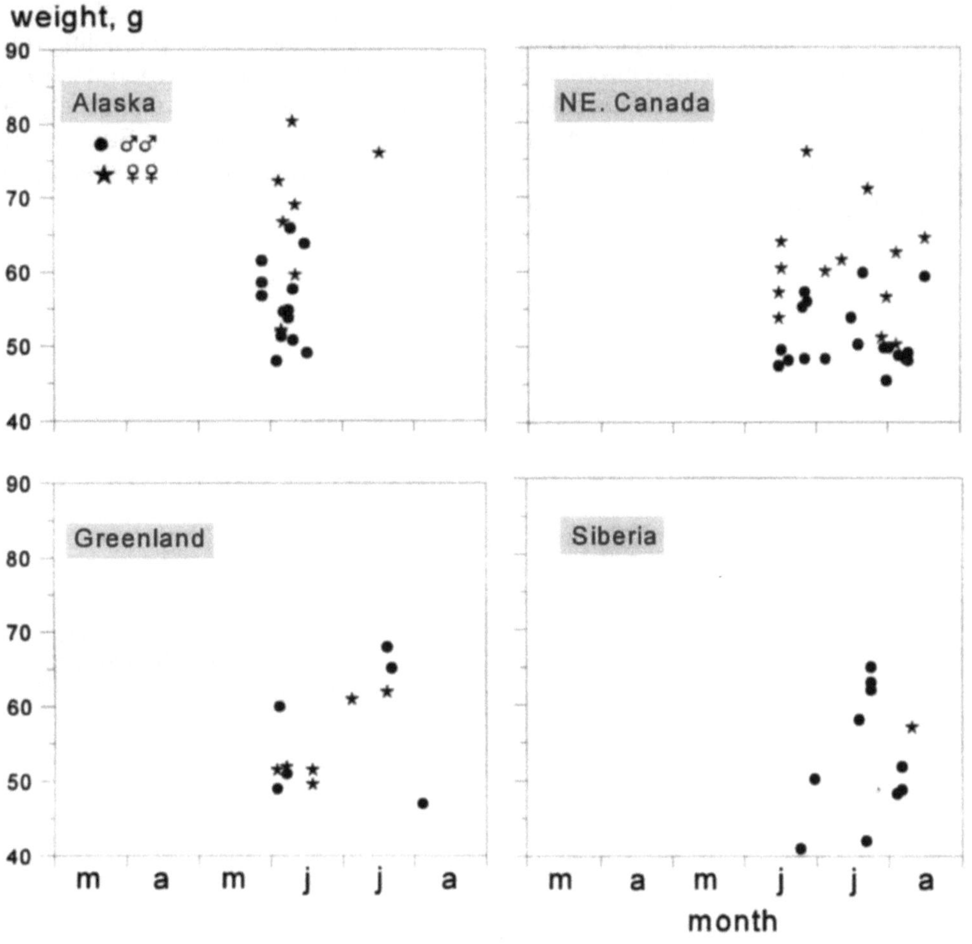

Fig. 56. *Body mass of Sanderlings on the breeding grounds in Alaska, N Canada, Greenland and Siberia.*

birds from this subspecies have more greyish-white upperparts than the Palearctic ones. Within this subspecies, birds from Alaska have slightly different measurements from those of N Canada (figs. 52 & 53).

Sexing Sanderlings

Two other studies have been dealing with sex differentiation in the Sanderling. One concerned Nearctic birds (Bodega Bay, California; Maron & Myers 1984), the other Palearctic ones (Teesmouth, NE England; Wood 1987). The data from England can be compared with the ones presented here, but such is not the case with the data from California as different measurements were taken. The English winter data strongly correspond with our data for the Palearctic breeding population: mean wing length of ♂♂ in England 124.7 mm (SD 4.5, $n = 22$), of ♀♀ 130.6 mm (SD 2.6, $n = 20$); mean culmen length of ♂♂ in England 23.3 mm (SD 1.2, $n = 22$), and of ♀♀ 25.5 mm (SD 1.4, $n = 20$) (Wood 1987).

SUMMARY

Measurements and plumage characters of Sanderling from nearly all breeding areas had been analysed with multivariate statistics. The species shows both geographical and sexual variation but the variation between sexes is larger than between populations. Sexual dimorphism was smallest for the Alaska population and largest in Siberia. The species is polytypic and 2 subspecies are recognized: *C. a. alba*, breeding in Siberia and on Greenland; and (2) *C. a. rubidus* (Gmelin, 1789), breeding in Alaska and in N Canada. Both differ in measurements and in plumage characters: nearctic breeders are larger-sized and have generally a more greyish breeding plumage. *Rubidus* shows some minor geographical variation in contrast to nominate *alba*. Generally, culmen-, wing-, secondary-, and tarsus lengths were useful to be measured. It was possible to reach good accurracy rates when predicting a bird's breeding origin, especially when its sex was known. The mean estimated exclusivity rates did not exceed 11%. Almost no Sanderlings started to moult primaries on the breeding grounds.

Chapter 9

Curlew Sandpiper
-
Calidris ferruginea

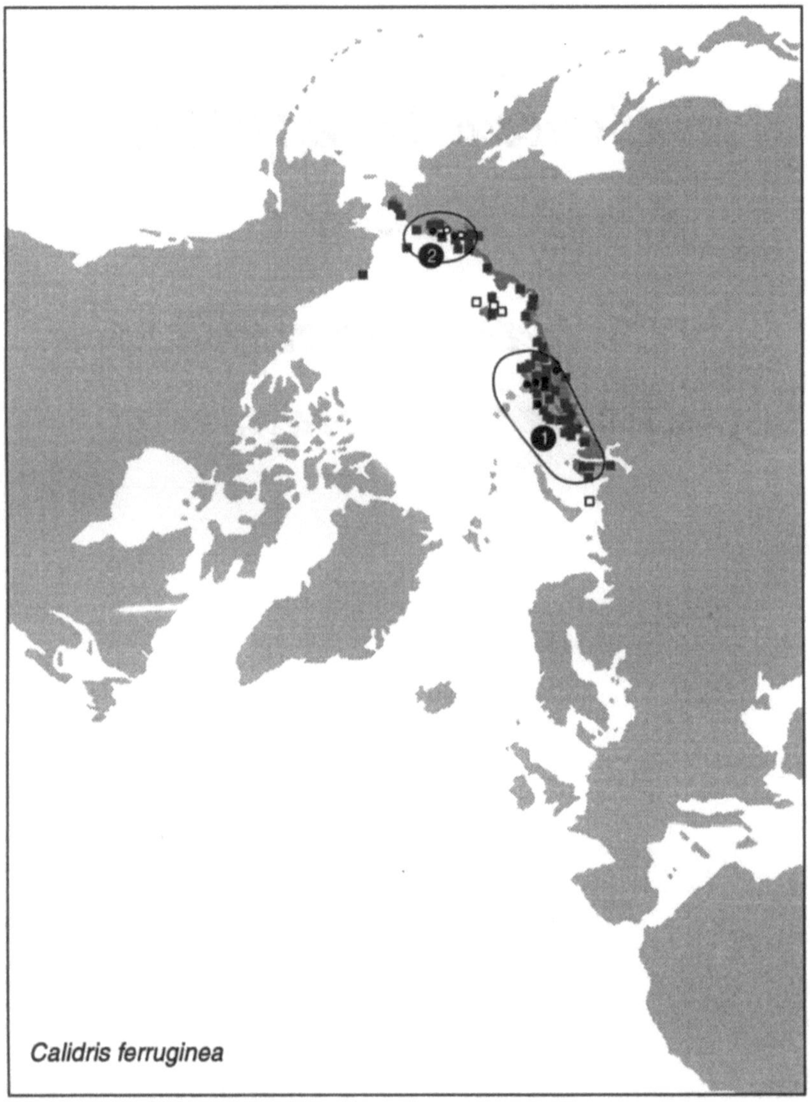

Fig. 57. *Breeding distribution of the Curlew Sandpiper. For explanation of the used symbols see Chapter 3: 'Introduction to the species' accounts'.*

INTRODUCTION

The Curlew Sandpiper is breeding in high arctic Siberia from the Gydansk Tundra in the west eastwards to the Chukotsk Peninsula. Curlew Sandpipers breed irregularly in NW Alaska (Kessel & Gibson 1978).

No geographical variation in measurements is known to occur in this species (Glutz *et al.* 1975, Prater *et al.* 1977, Cramp & Simmons 1983). However, there is some variation in breeding plumage of ♀♀ (Roselaar in Cramp & Simmons 1983): eastern ♀♀ show less dark barring on the underparts than western ♀♀.

MORPHOMETRICAL DIFFERENTIATION

In total, 68 specimen from two breeding locations were examined: Taymyr and the region near the Indigirka and Kolyma (Fig. 57). Breeding birds on Taymyr were present by early June and most had been collected between mid-June and late July (Fig. 58). The breeding birds from the Indigirka and Kolyma region were mostly collected between late May and early June. Males predominated amongst the birds examined (61.8 %).

Initially, 4 samples (both sexes and two breeding regions) were distinguished. These regions were (1) Taymyr and (2) Indigirka and Kolyma (Fig. 57). The mean measurements per sample are presented in Table 45. The morphometric differences between samples were analysed and presented in the figures 59 & 60 and the tables 46 - 48.

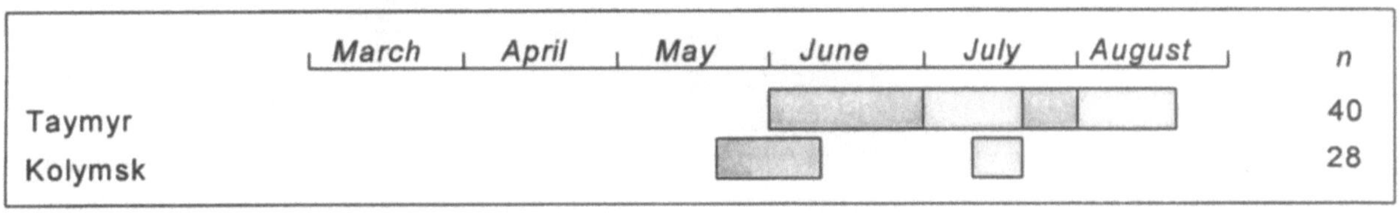

Fig. 58. *Collection dates of Curlew Sandpipers. Most birds had been collected during the periods marked by the heavily shaded bars. The sample sizes are presented on the right side.*

Table 45. *Measurements of Curlew Sandpipers, collected on the breeding grounds. Presented are the means ± 1 SD (n).*

Locality	Wing	Culmen	Tarsus	Secondary	Tail	Middle toe
			♂♂			
Taymyr	132.9±2.6(23)	36.3±0.9(23)	29.2±1.5(23)	56.6±1.5(23)	46.1±1.8(23)	23.9±0.8(21)
Kolymsk	132.1±3.3(19)	36.5±1.7(19)	29.3±1.2(19)	57.5±1.7(19)	47.3±4.6(19)	24.4±0.8(17)
all ♂♂	132.5±2.9(42)	36.4±1.8(42)	29.3±1.3(42)	57.0±1.6(42)	46.7±3.4(42)	24.2±0.8(38)
			♀♀			
Taymyr	135.6±3.1(17)	40.7±1.1(17)	31.1±1.2(17)	57.8±1.3(17)	45.9±2.5(17)	25.3±1.1(17)
Kolymsk	133.6±4.5(9)	39.9±2.1(9)	30.6±1.6(9)	57.9±1.6(9)	46.0±4.4(9)	24.9±1.0(8)
all ♀♀	134.9±3.7(26)	40.4±1.5(26)	30.9±1.3(26)	57.8±1.4(26)	46.0±3.2(26)	25.2±1.1(25)

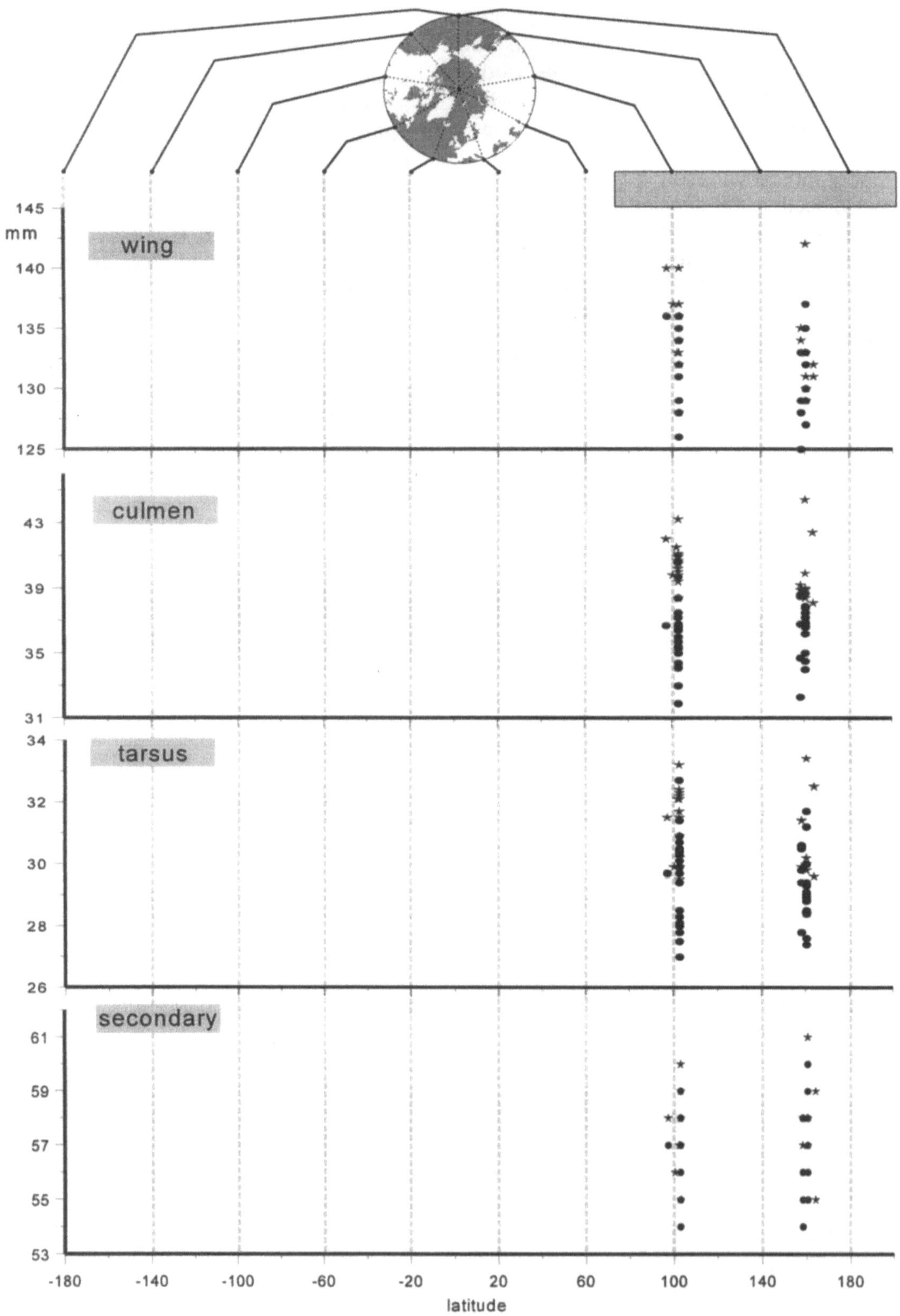

Fig. 59. *Measurements of Curlew Sandpipers from the breeding grounds. The following measurements are presented from top to bottom: wing length, culmen length, tarsus length and secondary length (description in chapter 2). Explanation of the used symbols: ● ♂♂ and ★ ♀♀.*

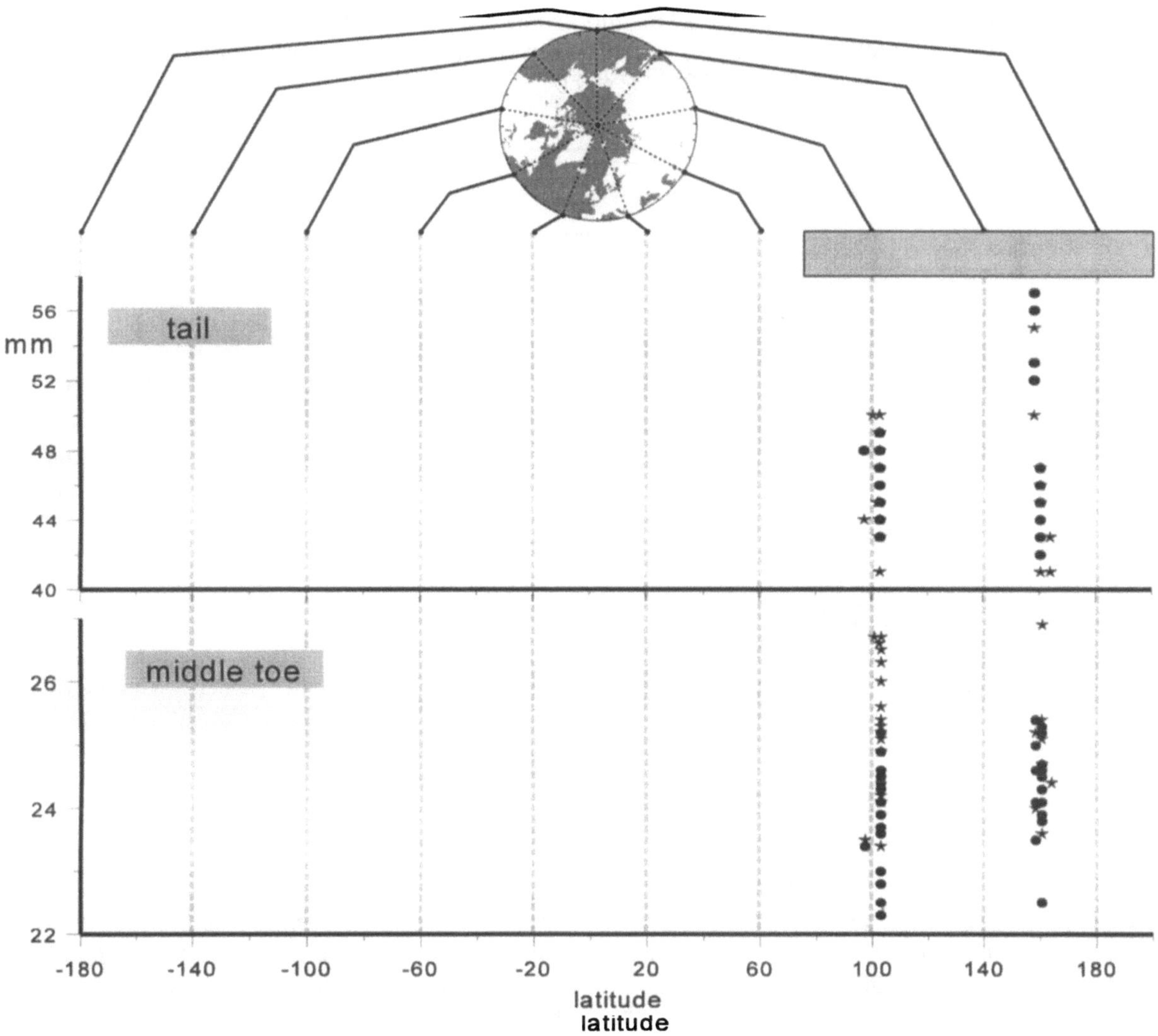

Fig. 60. *Measurements of Curlew Sandpipers from the breeding grounds. The following measurements are presented from top to bottom: tail length and middle toe length (description in chapter 2). Explanation of the used symbols:* ● ♂♂ *and* ★ ♀♀.

The Curlew Sandpiper shows a strong sexual variation and hardly any geographical variation. Size difference between the sexes was smaller in the E Siberian sample than in the Taymyr one (Table 47). Most of the variables were not correlated, perhaps due to small sample sizes (Table 48). Sexes are well-separated: ♂♂ are smaller than ♀♀. This was recognized already by e.g., Thomas & Dartnall 1970, Elliott *et al.* 1976 and Prater *et al.* 1977. Since (1) geographical variation in this species was negligible, and (2) the sexual differentiation was marked, we did not perform a cluster analysis or a MANOVA, in which samples were pairwise contrasted.

Four variables contributed significantly to separation of the sexes (Table 46). Generally, the culmen lengths were most distinctive, followed by the tarsus-, middle toe- and wing lengths (in decreasing order of importance).

When only wing- and culmen lengths are available, the sexes could still be separated well. Especially culmen length was important (Table 46).

Table 46. *Overall results of MANOVA for all distinguished breeding populations in the Knot in combination with univariate ANOVA's in which the different measurements were analysed separately. Significance levels: *0.01 < P < 0.05; **0.001 < P < 0.01; ***P < 0.001.*

	Analysis I			Analysis II			Analysis III		
Variables	**wing, culmen, tarsus, secondary, tail, middle toe length**			**wing, culmen, tarsus**			**wing, culmen**		
n	63			68			68		
	Multivariate analysis of variance								
Effects	Interaction	Area	Sex	Interaction	Area	Sex	Interaction	Area	Sex
Pillai's trace	0.097	0.121	0.566	0.022	0.042	0.555	0.020	0.042	0.551
F-value	0.969	1.241	11.736***	0.455	0.906	25.816***	0.649	1.380	38.639***
Hypoth. & error *df*	6/54	6/54	6/54	3/62	3/62	3/62	2/63	2/63	2/63
	Univariate F tests								
df	1/59	1/59	1/59	1/64	1/64	1/64	1/64	1/64	1/64
wing length	0.298	2.116	7.887**	0.444	2.757	6.882*	0.444	2.757	6.882*
culmen length	3.392	0.189	67.396***	1.106	0.305	77.793***	1.106	0.305	77.793***
tarsus length	2.159	0.510	15.764***	0.916	0.423	20.567***			
secondary length	0.937	0.869	3.737						
tail length	0.372	1.183	0.528						
middle toe length	3.467	0.018	14.929***						

Table 47. *Sexual size dimorphism in the Curlew Sandpiper. The mean percentage difference between the measurements of both sexes was calculated for every single variable and breeding population, according to: $(x_♀-x_♂)/x_♂*100$. A negative value thus implies smaller measurements for ♀♀. The calculations were based upon the means presented in Table 45. The obtained values were averaged per variable and summed per breeding population.*

Locality	Wing	Culmen	Tarsus	Secondary	Tail	Middle toe	Σ
Taymyr	2.0	12.1	6.5	2.1	-0.4	5.8	28.1
Kolymsk	1.1	9.3	4.4	0.7	-2.7	2.0	14.8
x	1.5	10.7	5.5	1.4	-1.5	3.9	

Table 48. *Pearson's product-moment correlations between measurements of Curlew Sandpipers. Both sexes are separately analysed. Two-tailed probabilities: * 0.01 < P < 0.05; ** P < 0.01.*

	Wing	Culmen	Tarsus	Secondary	Tail	Middle toe	
Wing		0.431	0.613**	0.442	0.244	0.376	
Culmen	0.074		0.666**	0.583**	0.011	0.260	
Tarsus	0.106	0.580		0.669**	0.086	0.610**	♀♀
Secondary	0.439*	0.082	0.056		0.219	0.317	
Tail	-0.109	-0.067	0.199	-0.031		0.140	
Middle toe	0.094	0.274	0.375	0.119	0.055		
				♂♂			

Prediction of a bird's sex

Two different sets of POSCON groups were constructed, a set with 4 measurements (wing-, culmen-, tarsus- and middle toe length) and one with 2 measurements (wing- and culmen length). In both sets, sexes were separated. Secondary- and tail lengths could be excluded from the analyses because of their insignificant contribution to the distinction between groups.

In a next step, all available measurements of breeding birds were used to calculate the degree of resemblance (typicality probabilities) with the constructed POSCON groups. By doing so, both the accuracy and exclusivity of prediction were estimated (see introduction under 'Statistical analysis of the data').

Accuracy

A high accuracy rate of 92 % correct was reached in predicting a bird's sex when 4 measurements were analysed (n = 93). The accuracy rates were not different for both sexes. When 2 measurements were used, the accuracy rates were resp. 90.5 % (n = 42) for ♂♂ and 88.5 % (n = 26) for ♀♀.

Exclusivity

The estimated exclusivity rates for sexing were high, since only 2 POSCON groups were involved: 41.3 % when 4 measurements were used and 44.1 % when 2 measurements were used.

BODY MASS

Mass-data of 13 breeding Curlew Sandpipers were taken from specimen labels. Two exceptionally heavy birds had been on Taymyr around mid-July (Fig. 61). These may have been birds near to departure from the breeding grounds, since the first Curlew Sandpipers are generally known to arrive at the moulting grounds from this time onwards. ♀♀ were heavier than ♂♂ throughout June and July (Fig. 61).

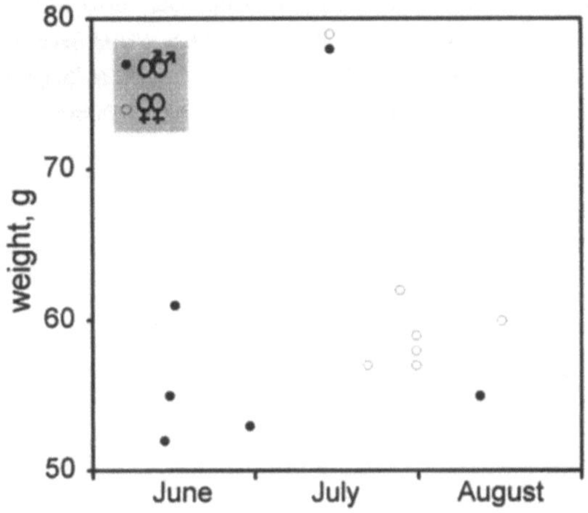

Fig. 61. *Body mass of Curlew Sandpipers on the breeding grounds.*

PRIMARY MOULT

No birds with primary moult were found on the breeding grounds.

DISCUSSION

The taxonomy of the Curlew Sandpiper

The results presented above showed that no geographical variation was detectable in this species. It was also not expected since (1) the species has a limited breeding range, and (2) it shows little breeding site tenacity (Underhill *et al.* 1993). The species is thus considered as monotypic.

SUMMARY

Measurements of Curlew Sandpiper from 2 breeding localities in W and E Siberia were analysed with help of multivariate statistics. The species shows much sexual variation, but almost no geographical variation. Females were larger than ♂♂. The degree of sexual dimorphism differed between both sites, with larger differences on Taymyr and smaller ones in E Siberia. Most measurements and several primary lengths were not correlated. Four measurements were especially important for the distinguishing the sexes: culmen-, tarsus-, middle toe- and wing. High accuracy and exclusivity rates were reached when predicting a bird's sex. Curlew Sandpipers did not moult primaries on the breeding grounds. It was concluded that the Curlew Sandpiper is a monotypic species with a strong sexual dimorphism.

Chapter 10

Purple Sandpiper

-

Calidris maritima

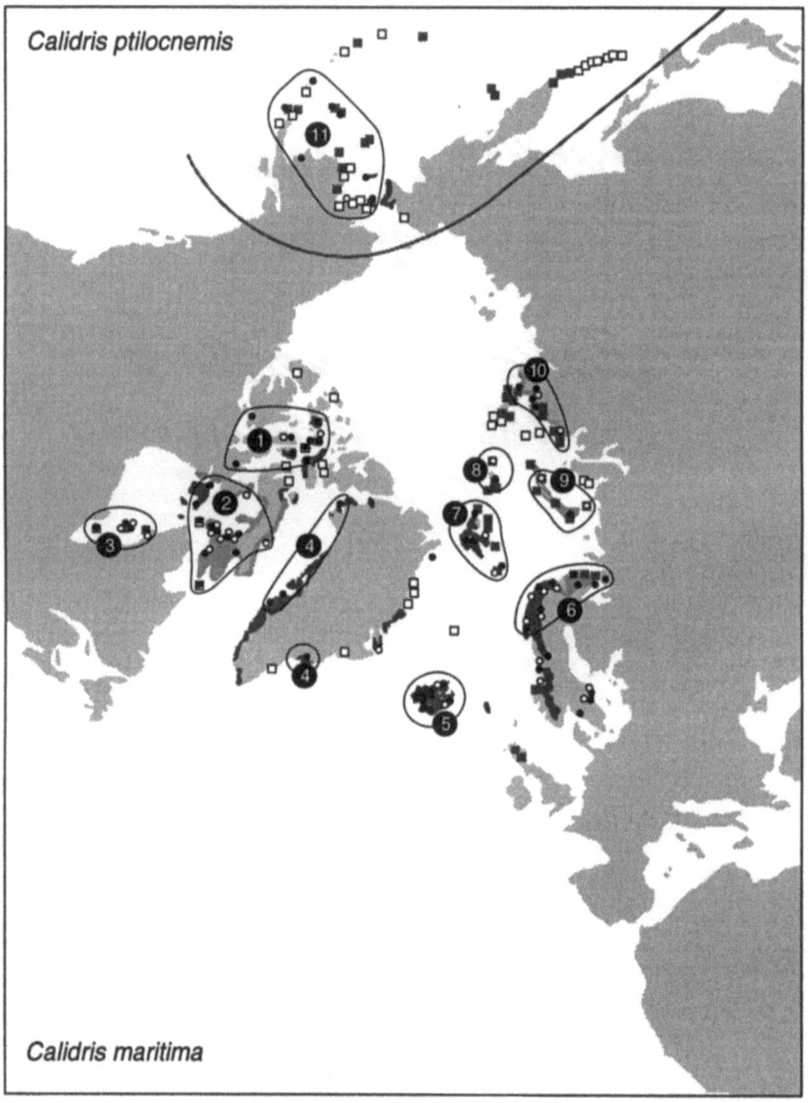

Fig. 62. *Breeding distribution of the Purple- and the Rock Sandpiper. For explanation of the used symbols see Chapter 3: 'Introduction to the species' accounts'.*

INTRODUCTION

The Purple Sandpiper is breeding along arctic coasts and in alpine areas at the border of permanent snow in N Canada, Greenland, Iceland, Jan Mayen, the Færoes, Scotland, Scandinavia, Svalbard, Bear Island, Novaya Zemlya, Franz Jozephland, Taymyr and Severnaya Zemlya (Fig. 62). It is replaced by the Rock Sandpiper *Calidris ptilocnemis* in the Bering Sea region.

Several authors recognized geographical variation of measurements in this species (Løvenskiold 1950, Salomonsen 1950, Glutz *et al.* 1975, Prater *et al.* 1977, Innes 1980, Atkinson *et al.* 1981, Cramp & Simmons 1983, Boere *et al.* 1984, Tomkovich 1985, Rae *et al.* 1986, Summers *et al.* 1987, Summers *et al.* 1988a, Nicoll *et al.* 1988, Stepanyan 1990, Nicoll *et al.* 1991). In general, two to four morphometrically distinct populations were recognized by these authors:

- very small-sized birds, breeding in S Norway and on Franz Josef Land;
- small-sized birds breeding in N Scandinavia, Bear Island, Svalbard and Greenland;
- intermediately-sized birds breeding on the Kola Peninsula, Novaya Zemlya and Vaygach;
- large-sized birds, breeding in N Canada, Iceland and Taymyr.

Of these latter, Icelandic birds were clearly the largest of all. Glutz *et al.* (1975) distinguished 2 different colour forms with (1) palish birds breeding in Scandinavia, Svalbard and N Russia, and (2) birds with a brighter breeding plumage on Greenland and Iceland. Brehm (1831) distin-

Table 49. *Measurements of Purple Sandpipers, collected on the breeding grounds. [1]The breeding birds of Melville Island, Somerset Island, Prince of Wales Island, Frederick Island and Southampton Island. (east of 80°W). [2]The breeding birds of the Belcher islands in the east part of the Hudson Bay. [3]Bear Island included. Presented are the means ± 1 SD (n).*

	Locality	Wing	Culmen	Tarsus	Secondary	Tail	Middle toe
				♂♂			
1	N Canada[1]	131.3±2.3(16)	29.6±1.3(16)	22.9±0.8(16)	64.3±1.1(12)	54.3±3.3(12)	27.5±0.9(11)
2	Baffin Island	132.2±2.3(26)	31.3±1.8(26)	22.8±0.8(26)	63.9±1.4(25)	55.8±1.9(25)	27.5±0.9(25)
3	Hudson Bay[2]	125.7±3.6(12)	29.2±1.3(12)	21.4±0.6(11)	60.8±1.6(12)	53.6±2.0(12)	26.5±0.9(9)
4	Greenland	127.4±2.9(11)	27.8±1.6(11)	22.4±0.7(11)	62.4±1.1(5)	53.5±2.1(6)	27.2±0.6(6)
5	Iceland	135.9±2.6(27)	30.2±1.1(27)	23.1±0.8(26)	65.0±1.7(21)	59.0±2.0(25)	27.4±0.9(21)
6	N Fennoscandia	127.7±2.5(20)	27.6±1.8(20)	22.1±1.1(19)	60.5±1.4(11)	57.0±5.7(9)	26.7±1.3(11)
7	Svalbard[3]	128.5±2.3(55)	27.9±1.3(54)	22.4±0.8(46)	61.4±1.3(16)	56.5±3.4(27)	25.9±1.1(31)
8	Franz Josef Land	128.4±1.5(5)	28.7±1.0(5)	22.3±0.2(5)	-	-	-
9	Novaya Zemlya	130.0±2.4(12)	29.5±1.6(12)	22.3±0.5(12)	60.5±0.7(2)	54.0±4.2 (2)	27.2 - (1)
10	Taymyr	127.3±2.9(4)	28.1±1.1(4)	22.0±0.6(4)	61.5±1.3(4)	-	27.0±1.0(4)
11	Rock Sandpiper	126.0±3.4(16)	28.4±1.6(16)	23.1±0.7(16)	61.1±2.9(16)	52.0±2.3(15)	27.7±1.0(13)
				♀♀			
1	N Canada[1]	136.0±3.8(9)	34.9±1.6(9)	23.6±0.6(9)	66.1±1.4(8)	55.6±4.6(8)	27.5±0.6(7)
2	Baffin Island	135.5±3.0(15)	35.3±1.3(15)	23.4±0.7(15)	66.0±1.1(15)	56.9±2.5(15)	28.4±0.9(13)
3	Hudson Bay[2]	129.9±3.0(11)	33.5±1.3(11)	22.8±0.7(11)	63.7±2.0(11)	54.5±2.2(11)	27.4±0.9(11)
4	Greenland	132.4±3.0(5)	31.7±1.7(4)	23.6±0.9(5)	63.7±2.3(3)	58.5±2.6(4)	27.3±1.2(4)
5	Iceland	141.2±4.2(20)	34.2±2.0(19)	24.1±1.0(20)	67.9±2.3(15)	60.1±3.0(18)	28.6±1.3(17)
6	N Fennoscandia	132.8±2.2(13)	32.5±1.7(11)	22.8±0.5(12)	64.5±1.0 (4)	53.4±4.2(5)	27.2±1.3(5)
7	Svalbard[3]	133.9±2.2(42)	32.5±1.5(42)	23.0±0.6(36)	64.2±1.6(10)	59.2±3.6(15)	26.9±1.2(23)
8	Franz Josef Land	131.3±1.5(3)	32.9±0.7(3)	23.0±0.3(3)	66.0 - (1)	49.0 - (1)	27.5 - (1)
9	Novaya Zemlya	136.0±2.5(9)	33.4±1.6(8)	23.5±0.7(9)	-	-	-
10	Taymyr	135.2±3.7(5)	31.5±1.5(5)	22.4±0.3(5)	66.0±1.2(5)	-	27.4±0.5(5)
11	Rock Sandpiper	132.7±3.5(12)	32.7±2.2(13)	23.8±1.1(13)	64.9±2.4(12)	54.2±4.8(13)	28.7±1.2(12)

Table 50. *Overall results of MANOVA for the distinguished breeding populations of the Purple Sandpiper and the Alaskan population of the Rock Sandpiper (Table 49) in combination with univariate ANOVA's in which the different measurements were analysed separately. Significance levels: *0.01 < P < 0.05; **0.001 < P < 0.01; ***P < 0.001. In the first analysis the populations from Franz Josef Land and Novaya Zemlya were not included.*

	Analysis I			Analysis II			Analysis III		
Variables	wing, culmen, tarsus, secondary, middle toe length			wing, culmen, tarsus length			wing, culmen length		
n	186			323			342		
Multivariate analysis of variance									
Effects	Interaction	Area	Sex	Interaction	Area	Sex	Interaction	Area	Sex
Pillai's trace	0.227	1.263	0.631	0.106	1.001	0.579	0.064	0.830	0.575
F-value	0.998	7.096***	56.174***	1.099	15.085***	137.226***	1.063	22.721***	216.071***
Hypoth. & error *df*	40/840	40/840	5/164	30/903	30/903	3/299	20/640	22/640	2/319
Univariate F tests									
df	8/168	8/168	1/168	10/301	10/301	1/301	10/320	10/320	1/320
wing length	1.236	26.937***	108.016***	1.158	36.868***	161.048***	1.215	38.539***	170.460***
culmen length	0.744	14.336***	246.019***	0.617	16.849***	377.308***	0.623	18.231***	391.534***
tarsus length	0.743	6.514***	40.397***	0.824	10.614***	59.737***			
secondary length	1.703	19.093***	117.086***						
middle toe length	0.594	5.007***	16.201***						

guished the Icelandic population as *C. a. littoralis* and Løvenskiold (1950) recognized a separate Greenlandic subspecies *C. m. groenlandica*. Both claims were generally not accepted, since the amount of overlap in measurements and in colour pattern between the different populations was considered too large (Salomonsen 1950, Glutz *et al.* 1975). Cramp & Simmons (1983) thought it possible to recognize at least 2 races. They suggested to re-introduce the Icelandic *C. m. littoralis* (Brehm 1831) to be separated from the nominate form. However, such was only possible when the Icelandic population was clearly recognizable from every

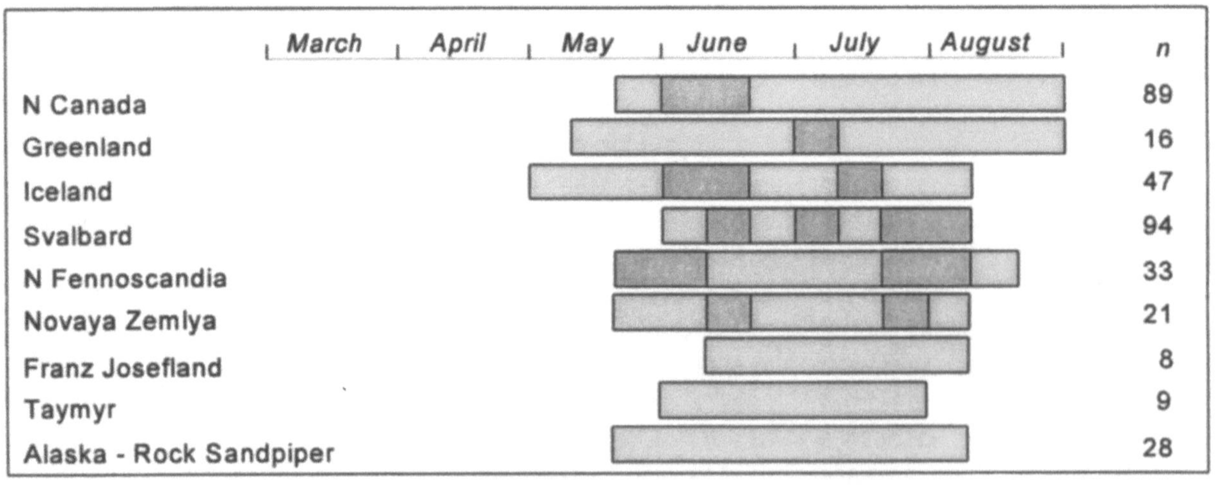

Fig. 63. *Collection dates of Purple- and Rock Sandpipers. Most birds had been collected during the periods marked by the heavily shaded bars. The sample sizes are presented on the right side.*

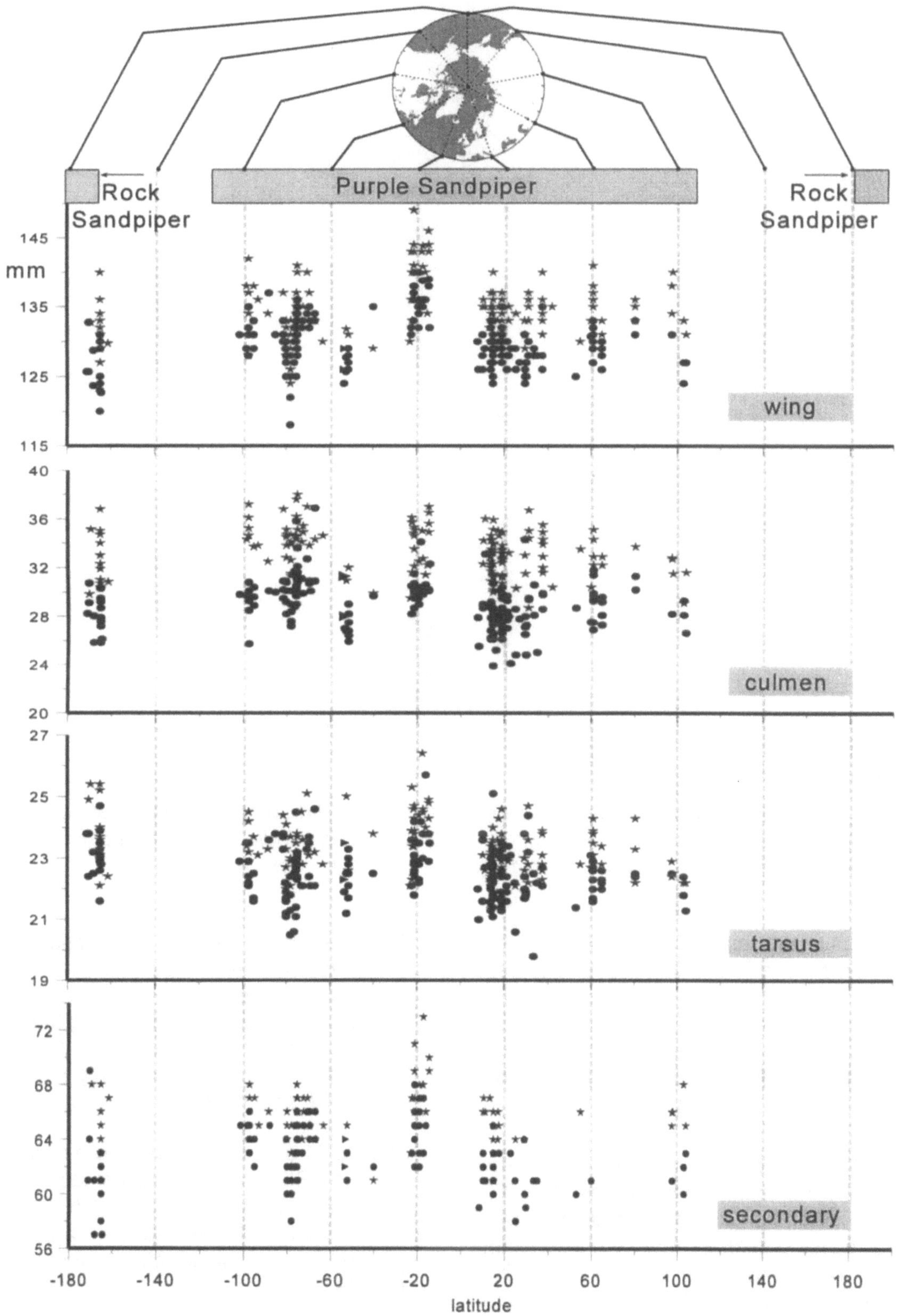

Fig. 64. *Measurements of Purple- and Rock Sandpipers from the breeding grounds. The following measurements are presented from top to bottom: wing length, culmen length, tarsus length and secondary length (description in chapter 2). Explanation of the used symbols:* ● ♂♂, ★ ♀♀, ▸ *sexe unknown.*

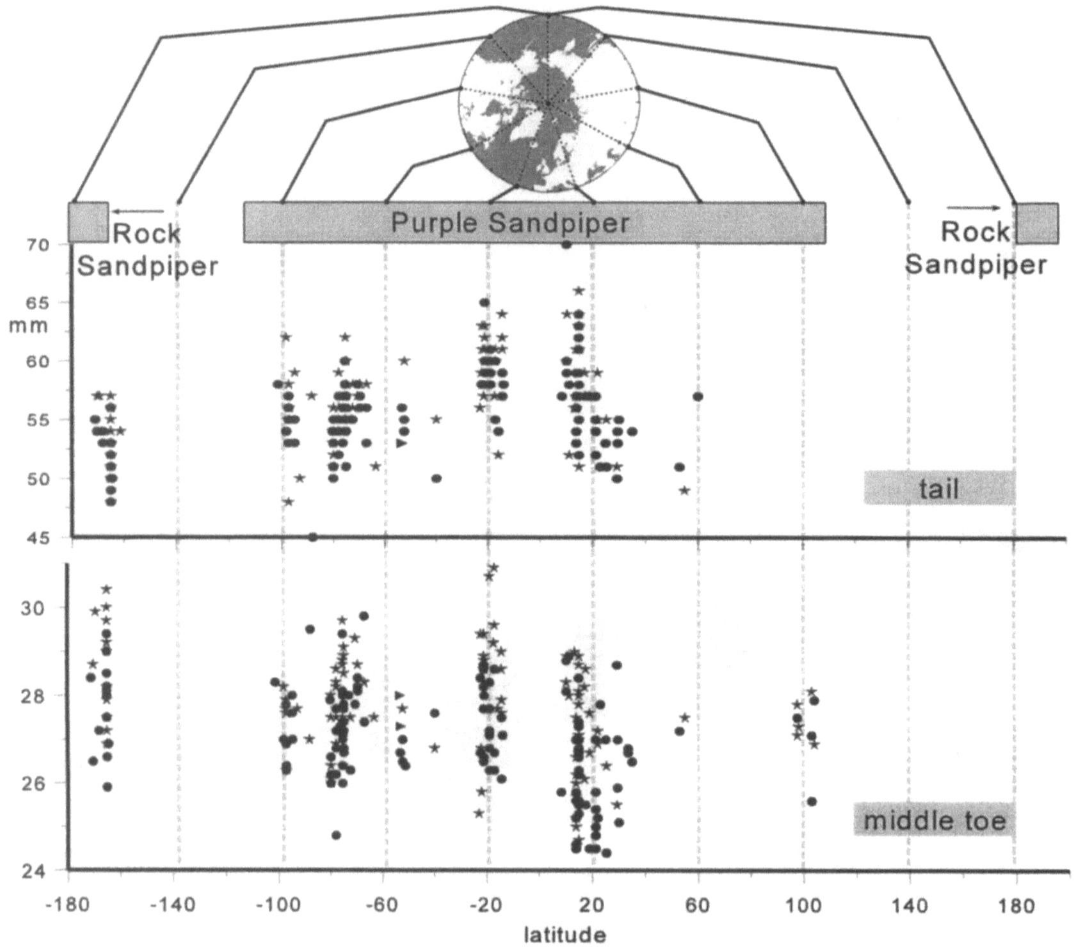

Fig. 65. *Measurements of Purple Sandpipers from the breeding grounds. The following measurements are presented from top to bottom: tail- and middle toe length (description in chapter 2). Explanation of the used symbols: ● ♂♂, ★ ♀♀, ▸ sexe unknown.*

other population, and accurate data of the intermediately-sized Canadian populations were lacking.

MORPHOMETRICAL DIFFERENTIATION

In total, 320 Purple Sandpipers were examined from nearly the entire breeding range (Fig. 62, Table 49). Another 29 Rock Sandpipers from Alaska subsp. *tschuktschorum* were also examined for comparative purposes. Since Nicoll *et al.* (1991) warned against the use of measurement data of Purple Sandpipers, present in N Fennoscandia during May, only those individuals with brood patches in May, June birds and birds collected from inland breeding sites were used for N Fennoscandia. No Purple Sandpipers from S Norway were examined. The available samples of birds breeding on Franz Josef Land and Taymyr were small. Most breeding birds arrived by late May or early June at the breeding grounds (Fig. 63). Most examined Purple Sandpipers (57.3 %) were ♂♂.

Initially, 11 breeding regions were distinguished (Fig. 64 & 65). Three different breeding locations in Canada were distinguished (Fig. 62). The morphometric differences are analysed and presented in the tables 49 - 53 and the figures 64 - 66.

Rock Sandpipers had somewhat shorter wing- and tail lengths and a bit longer tarsus- and

Table 51. *Differences between the sexes of the various populations of Purple- and Rock Sandpipers. The mean percentage difference between the measurements of both sexes was calculated for every variable and breeding population, according to: $(x_♀-x_♂)/x_♂*100$. A negative value thus implies smaller measurements for ♀♀. The calculations were based upon the means of Table 49. The obtained values were averaged per variable and summed per breeding population. [1]Summation concerned the first 3 variables. . [2]Summation concerned all 6 variables.*

Locality	Wing	Culmen	Tarsus	Secondary	Tail	Middle toe	Σ_3 [1]	Σ_6 [2]
N Canada	3.7	17.9	3.1	2.8	2.4	0.0	24.7	29.9
Baffin Island	2.5	12.8	2.6	3.3	2.0	3.3	17.9	26.5
Hudson Bay	3.3	14.7	6.5	4.8	1.7	3.4	24.5	34.4
Greenland	3.9	14.0	5.4	2.1	9.3	0.4	23.3	35.1
Iceland	3.9	13.2	4.3	4.5	1.9	3.3	21.4	31.1
N Fennoscandia	4.0	17.8	3.2	6.6	-6.3	1.9	25.0	27.2
Svalbard	4.2	16.5	2.7	4.6	4.8	3.9	23.4	36.7
Franz Josef Land	2.1	14.6	3.1	-	-	-	19.8	-
Novaya Zemlya	4.6	13.2	5.4	-	-	-	23.4	-
Taymyr	6.2	12.1	1.8	7.3	-	1.5	20.1	-
Rock Sandpiper	5.3	15.1	3.0	6.2	4.2	3.6	23.4	37.4
x	4.0	14.7	3.7	4.7	2.5	2.4		
SD	1.2	2.0	1.5	1.8	4.4	1.5		
n	11	11	11	9	8	9		

middle toe lengths than Purple Sandpipers, but the similarities between both were more remarkable than the differences (figs. 64 and 65).

In this species-complex there was more sexual than geographical variation (Table 50). Sexual dimorphism was smallest in the samples from Baffin Island, Franz Josef Land, and Taymyr and largest in the samples from N Fennoscandia, N Canada and the Hudson Bay (Table 51).

The squared euclidean distances as calculated with cluster analysis were relatively small (Fig. 66) and comparable with those of Knot and Turnstone. The distinctness of the Icelandic sample was striking. The sample of the Hudson Bay also had a somewhat distinct position, although less evident as the Iceland sample. The other samples were quite similar.

Rock Sandpiper ♂♂ showed a high morphometric resemblance with Purple Sandpiper ♂♂ from the Hudson Bay, but ♀♀ closely resembled the samples from the small-sized N Atlantic populations of the Purple Sandpiper.

Table 52. *Pearson's product-moment correlations between measurements of the Purple Sandpiper. Both sexes are separately analysed. Two-tailed probabilities: * $0.01 < P < 0.05$; ** $0.001 < P < 0.01$; *** $P < 0.001$.*

	Wing	Culmen	Tarsus	Secondary	Tail	Middle toe	
Wing		0.410***	0.458***	0.731***	0.531***	0.386***	
Culmen	0.506***		0.353***	0.414***	0.056	0.257**	
Tarsus	0.368***	0.400***		0.328**	0.176	0.507***	
Secondary	0.781***	0.516***	0.351***		0.360**	0.435***	♀♀
Tail	0.476***	0.078	0.048	0.333***		0.220*	
Middle toe	0.243**	0.480***	0.466***	0.330***	-0.036		
			♂♂				

Table 53 - *continuation*

Locality	n	N Canada ♀	Baffin Island ♀	Hudson Bay ♀	Greenland ♀	Iceland ♀	N Fennoscandia ♀	Svalbard ♀	Fr. Joseph Land ♀	Nov. Zemlya ♀	Taymyr ♀	Rock Sandpiper ♀	N Canada ♂	Baffin Island ♂	Hudson Bay ♂	Greenland ♂	Iceland ♂	N Fennoscandia ♂	Svalbard ♂	Fr. Joseph Land ♂	Nov. Zemlya ♂	Taymyr ♂
N Canada ♀	9																					
Baffin Island ♀	15	0.01																				
Hudson Bay ♀	11	0.076***	0.080***																			
Greenland ♀	4	0.048**	0.071***	0.049**																		
Iceland ♀	19	0.094***	0.172***	0.299***	0.104***																	
Fennoscandia ♀	10	0.047**	0.062***	0.024	0.021	0.198***																
Svalbard ♀	36	0.060***	0.110***	0.091***	0.02	0.240***	0.013															
Fr. Joseph Land ♀	3	0.025	0.028*	0.01	0.012	0.100***	0	0.01														
Nov. Zemlya ♀	8	0.016	0.040**	0.092***	0.028*	0.055***	0.036*	0.023	0.024													
Taymyr ♀	5	0.059***	0.081***	0.084***	0.033*	0.091***	0.032*	0.02	0.027*	0.025												
Rock Sandp. ♀	12	0.044**	0.078***	0.060***	0.01	0.184***	0.027*	0.042**	0.01	0.034*	0.060***											
N Canada ♂	16	**0.183*** **	0.262***	0.166***	0.022	0.308***	0.084***	0.114***	0.041**	0.109***	0.047**	0.101***										
Baffin Island ♂	26	0.114***	**0.178*** **	0.097***	0.015	0.293***	0.023	0.031*	0.013	0.060***	0.023	0.058***	0.043**									
Hudson Bay ♂	11	0.263***	0.311***	**0.137*** **	0.097***	0.451***	0.128***	0.230***	0.062***	0.222***	0.118***	0.205***	0.129***	0.150***								
Greenland ♂	11	0.275***	0.347***	0.208***	**0.066*** **	0.411***	0.155***	0.229***	0.078***	0.208***	0.113***	0.185***	0.050***	0.135***	0.072***							
Iceland ♂	26	0.186***	0.293***	0.274***	0.047**	**0.214*** **	0.146***	0.171***	0.072***	0.085***	0.033*	0.159***	0.082***	0.133***	0.295***	0.194***						
Fennoscandia ♂	19	0.324***	0.416***	0.257***	0.084***	0.489***	**0.190*** **	0.306***	0.089***	0.244***	0.120***	0.242***	0.070***	0.183***	0.080***	0.01	0.248***					
Svalbard ♂	45	0.347***	0.468***	0.296***	0.074***	0.543***	0.209***	**0.380*** **	0.087***	0.253***	0.119***	0.259***	0.060***	0.218***	0.125***	0.01	0.281***	0.01				
Fr. Joseph Land ♂	5	0.160***	0.195***	0.105***	0.038**	0.249***	0.071***	0.093***	**0.043** **	0.116***	0.058***	0.094***	0.017	0.046**	0.032*	0.01	0.092***	0.01	0.01			
Nov. Zemlya ♂	12	0.185***	0.245***	0.132***	0.036*	0.323***	0.072***	0.110***	0.038**	**0.123*** **	0.047**	0.115***	0.015	0.041**	0.061***	0.034*	0.119***	0.038**	0.039**	0.01		
Taymyr ♂	4	0.170***	0.199***	0.110***	0.050**	0.251***	0.082***	0.105***	0.053***	0.129***	**0.069*** **	0.108***	0.029*	0.059***	0.022	0	0.103***	0	0.01	0	0.013	
Rock Sandp. ♂	16	0.310***	0.398***	0.231***	0.068***	0.488***	0.193***	0.315***	0.084***	0.251***	0.168***	**0.197*** **	0.107***	0.211***	0.129***	0.034*	0.305***	0.078***	0.092***	0.033*	0.096***	0.030*

Table 53. *Results of MANOVA-analysis concerning the morphometric differentiation in Purple- and Rock Sandpiper. Pillai's trace values are presented together with the degree of significance as based upon F-values (Norušis 1992c). Two tables are presented. In the first 5 measurements and 9 populations were included (5 and 194 df). In the second one 3 measurements and 11 populations were included (3 and 299 df). Significance levels: *0.01 < P < 0.05; **0.001 < P < 0.01; ***P < 0.001.*

Locality	n	N Canada ♀	Baffin Island ♀	Hudson Bay ♀	Greenland ♀	Iceland ♀	N Fennoscandia ♀	Svalbard ♀	Fr. Josef Land ♀	Nov. Zemlya ♀	Taymyr ♀	Rock Sandpiper ♀	N Canada ♂	Baffin Island ♂	Hudson Bay ♂	Greenland ♂	Iceland ♂	N Fennoscandia ♂
N Canada ♀	7																	
Baffin Island ♀	13	0.034																
Hudson Bay ♀	11	0.091**	0.127***															
Greenland ♀	3	0.083*	0.127***	0.090**														
Iceland ♀	14	0.159***	0.232***	0.374***	0.127***													
N Fennoscandia ♀	4	0.035	0.072*	0.045	0.028	0.124***												
Svalbard ♀	10	0.087**	0.138***	0.096**	0.034	0.199***	0.008											
Taymyr ♀	5	0.092**	0.134***	0.141***	0.076*	0.111**	0.032	0.046										
Rock Sandpiper ♀	11	0.099**	0.127***	0.114***	0.033	0.232***	0.045	0.062	0.120***									
N Canada ♂	11	0.211***	0.306***	0.206***	0.055	0.302***	0.056	0.076*	0.055	0.157***								
Baffin Island ♂	25	0.156***	0.263***	0.148***	0.037	0.331***	0.017	0.014	0.059	0.112**	0.054							
Hudson Bay ♂	8	0.325***	0.407***	0.210***	0.139***	0.502***	0.148***	0.210***	0.219***	0.292***	0.179***	0.216***						
Greenland ♂	5	0.232***	0.298***	0.185***	0.06	0.304***	0.086*	0.101**	0.107**	0.160***	0.031	0.078*	0.072*					
Iceland ♂	16	0.255***	0.382***	0.354***	0.072*	0.227***	0.090**	0.129***	0.062	0.243***	0.087*	0.168***	0.334***	0.105**				
N Fennoscandia ♂	11	0.404***	0.513***	0.363***	0.132***	0.523***	0.187***	0.259***	0.235***	0.358***	0.171***	0.274***	0.089**	0.035	0.278***			
Svalbard ♂	15	0.409***	0.543***	0.399***	0.106**	0.530***	0.180***	0.269***	0.234***	0.365***	0.162***	0.294***	0.156***	0.032	0.259***	0.029		
Taymyr ♂	4	0.243***	0.299***	0.180***	0.080*	0.319***	0.102**	0.118***	0.125***	0.178***	0.055	0.097**	0.039	0.006	0.132***	0.013	0.026	
Rock Sandpiper ♂	13	0.417***	0.524***	0.355***	0.111**	0.566***	0.208***	0.296***	0.299***	0.314***	0.225***	0.328***	0.139***	0.069*	0.386***	0.119***	0.131***	0.055

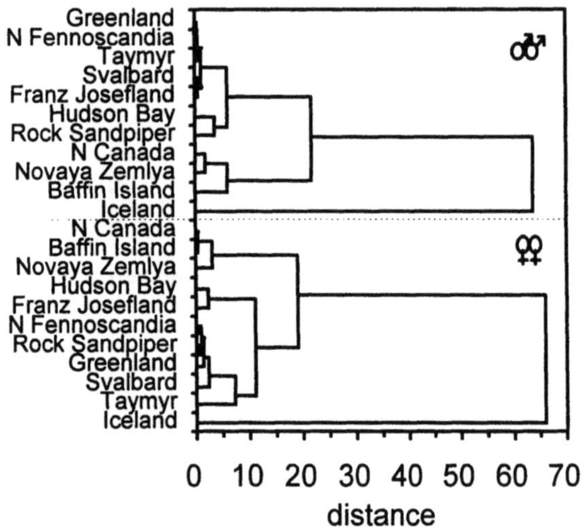

Fig. 66. *Results of a cluster analysis in the Purple- and Rock Sandpiper. The analysis was based upon the mean wing-, culmen- and tarsus lengths of the populations presented in Table 49.*

Two MANOVA-analyses were performed. The first included 5 measurements and thus 2 populations (Franz Joseph Land, Novaya Zemlya) had to be excluded from analysis. In the second analysis both these populations were included, but then only 3 measurements could be used.

The following conclusions could be drawn from the obtained results (Table 53):

(1) Nearly all ♂ samples differed statistically significant from the ♀ samples in measurements. If this was not the case, it was nearly always due to small sample sizes (e.g., ♀♀ Greenland, ♀♀ N Fennoscandia).

(2) Rock Sandpiper ♀♀ were indistinguishable from Purple Sandpiper ♀♀ on Franz Joseph Land and Greenland. Rock Sandpiper ♂♂ were distinguishable from every other ♂ group.

(3) Iceland birds were the largest measured, and thus Iceland samples were distinct from any other sample.

(4) The Hudson Bay population was distinct from nearly every other population. They were characterized by their small measurements.

(5) The birds from N Canada and Baffin Island were also separable from every other population except the Novaya Zemlya population.

(6) There were only minor morphometric differences between the populations of Greenland, N Fennoscandia, Svalbard, Franz Joseph Land and Taymyr, if any at all. From here on, these samples will be grouped and referred to as the 'N Atlantic' breeding population. The status of the Novaya Zemlya breeding birds with respect to this N Atlantic population is unclear.

Culmen-, wing-, and tarsus lengths were important for the distinction between samples (Table 50). Wing lengths were important to separate breeding regions and culmen lengths to identify sexes. Tarsus lengths were less important than wing- and culmen lengths.

MANOVA analysis with exclusively wing- and culmen lengths produced practically identical results as one with 3 measurements. Therefore, the results from pairwise contrasting with wing- and culmen lengths were not separately presented.

PREDICTION OF BREEDING ORIGIN

Due to much sexual dimorphism, both sexes were treated separately. Moreover, the cluster- and MANOVA analyses resulted in the identification of 4 distinct breeding populations in the Purple Sandpiper: (1) Hudson Bay, (2) Baffin Island and N Canada, (3) Iceland, and (4) N Atlantic (Greenland, N Fennoscandia, Svalbard, Franz Joseph Land, Novaya Zemlya and Taymyr). This resulted in the construction of 8 POSCON groups. When 3 measurements are used, at least 11 birds per POSCON group are needed. This need was fulfilled in all samples. However, when 6 measurements would be used, 14 birds per POSCON group were minimally required. This sample size could not be reached in 3 of the 8 POSCON groups. It was thus decided to use 3 measurements when predicting a bird's breeding origin or sex: wing-, culmen- and tarsus length.

In a next step, the available measurements of breeding birds were used to calculate the degree of resemblance (typicality probabilities) with these 8 constructed POSCON groups. By doing so, both the accuracy and exclusivity of prediction were estimated (see introduction under 'Statistical analysis of the data').

Accuracy

The estimated accuracy rates clearly showed the distinctness of the samples from the Hudson Bay and Iceland. The estimated accuracy rates for the Hudson Bay sample were nearly 82 % and they ranged for the Iceland sample between 76 and 88 % (Table 54). Differentiation between the intermediately-sized samples from N Canada and the N Atlantic was somewhat more problematical and the accuracy rates were accordingly lower. The quality of the predictions only slightly improved with measuring tarsus length (tables 54 and 55).

The mean % of 'correctly' predicted birds improved with the additional inclusion of tarsus lengths from 68.2 to 71.8 %, when the birds' sexes were known, and from 62.3 to 66.1 % when they were unknown. When the sexe was unknown, this only had serious consequences for predictions in which the N Atlantic samples were involved. The N Atlantic ♀♀ were often predicted as ♂♂ from other populations.

Exclusivity

When the birds' sexes were known, the estimated exclusivity rates were about 18 % with 2 measurements being used (Table 56). This percentage reduced somewhat when more measurements were used: the predictions become more accurate but less certain. When a bird's sex was unknown, the estimated exclusivity rates were about 8 % lower in comparison with the situation that sexes were. The inclusion of more measurements resulted in lower exclusivity rates but higher accuracy rates.

Table 54. *Accuracy of prediction in Purple Sandpipers (sex* **known** *a priori). These predictions have been based upon morphometric analysis and included (A) wing-, culmen- and tarsus length and (B) wing- and culmen length. The columns show the predicted proportions in relation to the total number per breeding area (last column).*

	Predicted								
Locality	Hudson Bay ♂	N Canada ♂	Iceland ♂	N Atlantic ♂	Hudson Bay ♀	N Canada ♀	Iceland ♀	N Atlantic ♀	*n*
	(A) Wing-, culmen- and tarsus length								
Hudson Bay ♂	81.8	-	-	18.2	-	-	-	-	11
N Canada ♂	7.5	57.5	15.0	20.0	-	-	-	-	40
Iceland ♂	-	20.0	76.0	4.0	-	-	-	-	25
N Atlantic ♂	11.2	10.1	3.4	75.3	-	-	-	-	89
Hudson Bay ♀	-	-	-	-	81.8	9.1	-	9.1	11
N Canada ♀	-	-	-	-	8.7	60.9	17.4	13.0	23
Iceland ♀	-	-	-	-	-	5.9	88.2	5.9	17
N Atlantic ♀	-	-	-	-	13.9	12.5	20.8	52.8	72
	(B) Wing- and culmen length								
Hudson Bay ♂	75.0	8.3	-	16.7	-	-	-	-	12
N Canada ♂	17.5	65.0	12.5	5.0	-	-	-	-	40
Iceland ♂	-	26.9	73.1	-	-	-	-	-	26
N Atlantic ♂	24.2	13.1	3.0	59.6	-	-	-	-	99
Hudson Bay ♀	-	-	-	-	72.7	18.2	-	9.1	11
N Canada ♀	-	-	-	-	8.7	60.9	17.4	13.0	23
Iceland ♀	-	-	-	-	-	11.8	76.5	11.8	17
N Atlantic ♀	-	-	-	-	7.6	15.2	13.9	63.3	79

Table 55. *Accuracy of prediction in Purple Sandpipers (sex* **unknown** *a priori). These predictions have been based upon morphometric analysis and included (A) wing-, culmen- and tarsus length and (B) wing- and culmen length. The columns show the predicted proportions in relation to the total number per breeding area (last column).*

	Predicted								
Locality	Hudson Bay ♂	N Canada ♂	Iceland ♂	N Atlantic ♂	Hudson Bay ♀	N Canada ♀	Iceland ♀	N Atlantic ♀	*n*
	(A) Wing-, culmen- and tarsus length								
Hudson Bay ♂	81.8	-	-	18.2	-	-	-	-	11
N Canada ♂	7.5	45.0	10.0	20.0	-	5.0	-	12.5	40
Iceland ♂	-	16.0	68.0	4.0	-	-	12.0	-	25
N Atlantic ♂	11.2	6.7	3.4	75.3	1.1	-	-	2.2	89
Hudson Bay ♀	-	9.1	-	-	81.8	9.1	-	-	11
N Canada ♀	-	-	-	-	8.7	60.9	17.4	13.0	23
Iceland ♀	-	-	-	5.9	-	5.9	88.2	-	17
N Atlantic ♀	2.8	18.1	8.3	4.2	9.7	12.5	16.7	27.8	72
	(B) Wing- and culmen length								
Hudson Bay ♂	75.0	8.3	-	16.7	-	-	-	-	12
N Canada ♂	17.5	52.5	12.5	5.0	-	5.0	-	7.5	40
Iceland ♂	-	19.2	69.2	-	-	-	3.8	7.7	26
N Atlantic ♂	24.2	11.1	3.0	59.6	1.0	-	-	1.0	99
Hudson Bay ♀	-	9.1	-	-	72.7	18.2	-	-	11
N Canada ♀	-	-	-	-	8.7	60.9	17.4	13.0	23
Iceland ♀	-	-	-	5.9	-	11.8	76.5	5.9	17
N Atlantic ♀	1.3	21.5	8.9	-	7.6	15.2	13.9	31.6	79

Table 56. *Estimated exclusivity rates of predictions, derived from morphometric data, of the breeding origin and sex of Purple Sandpipers. [1]Wing- and culmen lengths; [2]Wing-, culmen- and tarsus lengths.*

Locality and sex	Sex known		Sex unknown	
n of variables	2[1]	3[2]	2[1]	3[2]
n	340	321	340	321
Hudson Bay ♂	4.6	2.3	3.1	1.1
N Canada ♂	0.3	0.3	-	-
Iceland ♂	5.2	4	0.5	-
N Atlantic ♂	0.9	1	0.9	1
Hudson Bay ♀	2.3	2.3	1.1	1.1
N Canada ♀	-	-	-	-
Iceland ♀	4.4	5.9	3.7	5.2
N Atlantic ♀	0.2	-	-	-
Total %	17.9	15.8	9.3	8.4

BODY MASS

Mass-data of 61 Purple Sandpipers were taken from specimen labels. On average, ♀♀ were heavier than ♂♂: 78.9 gm ± 9.8 SD (n = 27), and 67.1 gm ± 5.6 SD (n = 43), respectively. Most of the mass data came from the N Canada and N Atlantic samples. No mass data were obtained from Iceland breeders. In the N Atlantic samples, masses of both sexes stayed rather constant throughout the breeding season (Fig. 67), without an indication of pre-migratory fattening at the end of the breeding season.

In the N Canada samples most mass data had been collected during the first half of June, and by that time these birds were generally heavier than the birds from the N Atlantic.

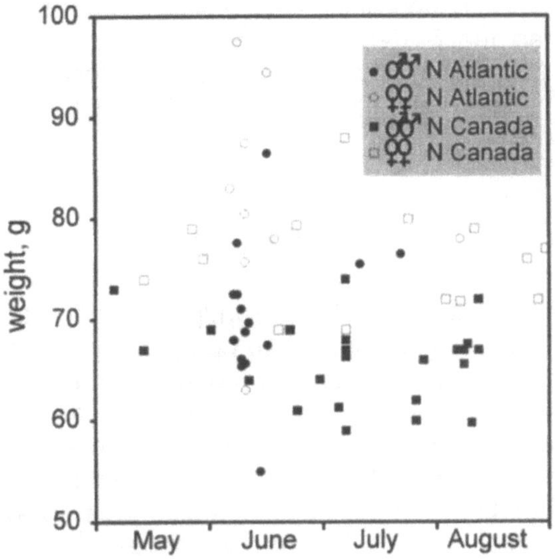

Fig. 67. *Body masses of Purple Sandpipers from the breeding grounds in the N Atlantic region and in N Canada.*

PRIMARY MOULT

Among 312 birds examined, 9 showed primary moult. These latter were all collected after 20 July. However, some must have started primary moult much earlier, since a bird with score 22 (3 fully grown new primaries) was collected on 22 July and another one with score 42 (7 fully grown new primaries) on 2 August. However, another bird had a moult score of 2 only on 27 August. Eight of the 9 moulting birds were from the N Atlantic sample.

DISCUSSION.

The taxonomy of the Purple Sandpiper

Our data largely supported the earlier findings of Atkinson *et al.* (1981), Boere *et al.* (1984), Nicoll *et al.* (1988), Summers *et al.* (1988a) and Nicoll *et al.* (1991). No birds from the Hardangervidda population in S Norway have been examined. The measurements of the N Fennoscandian population averaged somewhat smaller than published earlier (Boere *et al.* 1984), but not as small as measured by Nicoll *et al.* (1991). Iceland birds were largest (as was known) and clearly differed from other N Atlantic populations. We did not recognize other morphometrically distinct populations in the N Atlantic region. None of the Franz Joseph Land birds examined was as small as published by Tomkovich (1985). According to our data, this population was identical to other N Atlantic populations.

N Canadian breeders (north of the Hudson Bay) were intermediate in size between the Icelandic and N Atlantic birds, but had more resemblance with the N Atlantic samples than with the Icelandic ones. Icelandic birds were distinct from the N Canada and N Atlantic ones. The largest surprise concerned the samples from the east part of the Hudson Bay. Here birds with very small measurements were found, which hitherto were not considered separable. This population has much smaller measurements than the other N Canada populations, and it is also spatially separated from them. Accordingly, we recognize 3 subspecies in the Purple Sandpiper:

- *Calidris maritima maritima* (Brünnich, 1764) breeds in N Canada (incl. Baffin Island), Greenland, Jan Mayen, Bear Island, Svalbard, N Fennoscandia, Franz Joseph Land, Novaya Zemlya, and W Siberia. It was originally described from Christiansö (off Bornholm, Denmark) and Norway. Birds belonging to this subspecies have smaller measurements than those breeding on Iceland and larger than the birds of the Hudson Bay population. Within this subspecies, N Canadian birds have on average larger measurements than those of Greenland and the N Atlantic.

- *Calidris maritima littoralis* (Brehm, 1831) is breeding on Iceland. It was originally described from "Island und andere nördliche länder'. Here it is restricted to the breeding population of Iceland. It has the largest measurements (especially wing-, tarsus-, and secondary lengths) of all the analysed breeding populations of Purple Sandpiper.

- *Calidris maritima belcheri* (Engelmoer & Roselaar, nov. ssp.) is breeding in the east part of the Hudson Bay and in the James Bay: the Belcher Islands, the Nastapooka Islands, on North Twin Island and possibly also along the east coast of the Hudson Bay Type specimen is at the Carnegie Museum of Natural History (Pittsburgh), coll. nr. 122.627. This adult ♂ had brood patches and was collected at Les Iles Belcher (56.30°N 78.00°W, E Hudson Bay, Canada) on 28 June 1938, and had wing 122, culmen 29.0, secondary 58, tail 54, tarsus 21.3 and middle toe 27.2 mm. They have characteristically short wings and small tarsi (Table 49). The subspecies was also collected on the Nastapooka Islands in the Hudson Bay and on North Twin Island in the James Bay. No specimen were available from the breeding site NW of Inoucdjouac along the east coast of the Hudson Bay, but 2 specimen collected on Southampton Island had most resemblance with the breeding birds of Baffin Island. According to Manning (1981) only 100 Purple Sandpipers breed on North Twin Island.

Finally, there is published information on the measurements of the S Norwegian population, which is also characterized by small and short measurements (Innes 1980, Nicoll *et al.* 1991) Wing lengths of this S Norwegian breeding population were comparable with *C. m. belcheri*, but culmen lengths were shorter: wing of ♂♂ 125.6 mm ± 2.5 SD (n = 12), of ♀♀ 130.3 ± 4.4 (4); culmen of ♂♂ 25.0 ± 1.1 (12), of ♀♀ 29.8 ± 1.3 (4) (Innes 1980). Nicoll *et al.* (1991) obtained larger samples of the S Norwegian population and their data are similar with the ones of Innes (1980): wing of ♂♂ 125.8 mm ± 2.5 SD (n = 49), of ♀♀ 129.7 ± 3.3 (18); culmen of ♂♂ 24.7 ± 1.0 (49), of ♀♀ 29.2 ± 1.0 (18). Since our data set did not contain a complete data set of measurements of the S Norwegian population, it was not considered furthermore.

The N Fennoscandian population is larger-sized than the S Fennoscandian one. It might be influenced by both the small-sized S Norwegian population and the larger-sized Atlantic population.

SUMMARY

Measurements of Purple Sandpipers from nearly all breeding locations had been analysed with multivariate statistics. Several morphometrically different breeding populations could be allocated. However, the differences in measurements between both sexes were larger than the ones between populations. Males were always smaller than ♀♀. The amount of sexual dimorphism was not the same over the whole breeding range. The species is polytypic and 3 different subspecies were recognized: (1) *C. m. maritima*, breeding in the N Atlantic (except Iceland), on Greenland and in N Canada (except coasts of Hudson Bay); (2) *C. m. littoralis*, breeding on Iceland; and (3) *C. m. belcheri* nov. ssp., breeding along the east coast of the Hudson Bay. The first subspecies shows some geographical variation, the N Canadian population having larger measurements than the Greenland and N Atlantic one. *Littoralis* is largest, *belcheri* smallest. To predict a bird's breeding origin, it is useful to measure wing-, culmen-, secondary-, tarsus- and middle toe lengths. We had too limited data to assess the usefulness of secondary-, middle toe- and tail lengths.

Chapter 11

Dunlin

-

Calidris alpina

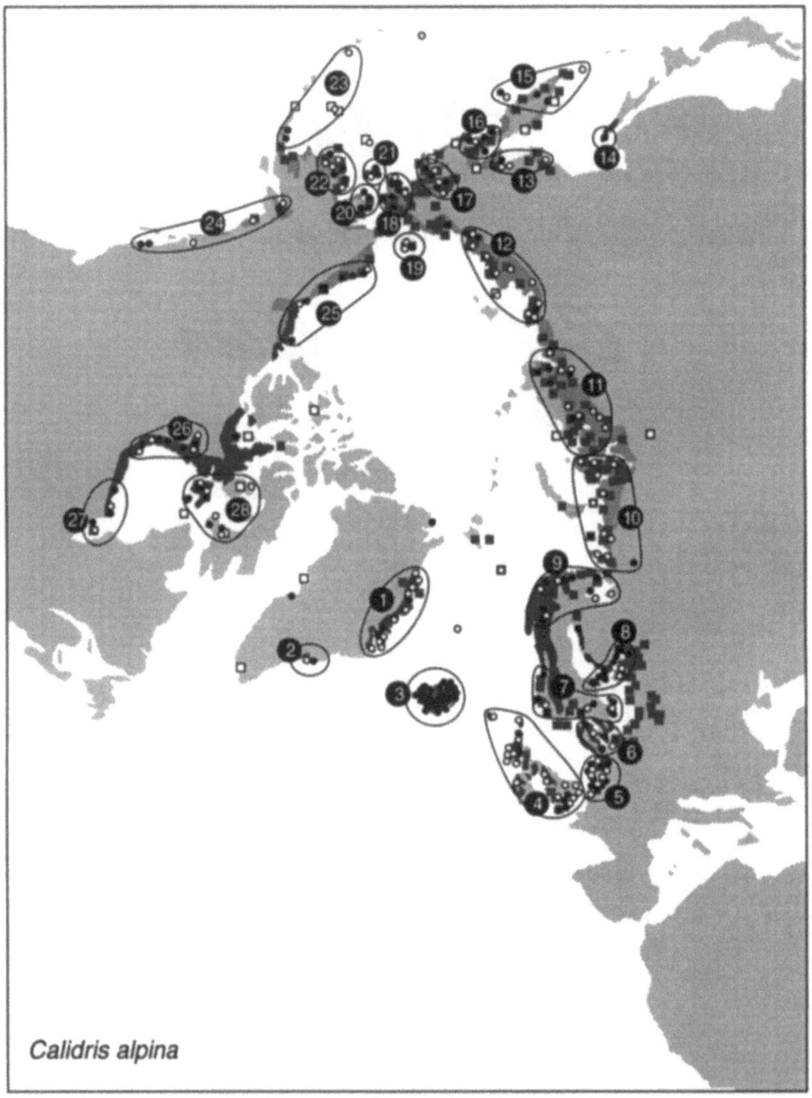

Fig. 68. *Breeding distribution of the Dunlin. For explanation of the used symbols see Chapter 3: 'Introduction to the species' accounts'.*

INTRODUCTION

The Dunlin is the most widespread of all species analysed. The breeding areas cover the coastal grasslands, saltmarshes, and upland moors of the temperate zone, mires in forests of the boreal zone and tundras of the subarctic and arctic (Fig. 68).

The geographical variation in the Dunlin has received much attention. Glutz *et al.* (1975) summarized the diverging views of the geographical variation of the Dunlin. Since then, Hale (1980), Cramp & Simmons (1983), Greenwood (1986), Tomkovich (1986), Nechaev & Tomkovich (1987 and 1988), Browning (1991) and Wenink *et al.* (1993) undertook further studies of the complex patterns of variation. Cramp & Simmons, Browning, Nechaev and Tomkovich studied both plumage characters and morphometrics, Greenwood predominantly morphometrics, and Wenink *et al.* differences in mtDNA. Their efforts resulted in the separation of 10 subspecies, which however are not all commonly recognized by all authors. The described subspecies are:

- *C. a. arctica*, breeding in NE Greenland;
- *C. a. schinzii*, breeding on Iceland, the Faroes, the British isles, W Europe, the Baltic and S Scandinavia, and occasionally S and SE Greenland;
- *C. a. alpina*, breeding in N Fennoscandia and NW Russia, east to the Lena River valley (127°E);
- *C. a. centralis* breeding in E Siberia in between the Lena River and the Chaunsk Bay (170°E);
- *C. a. sakhalina*, breeding on the Chukotsk Peninsula (E Siberia), north of 62°N and west to the Chaunsk Bay, and on Wrangel Island;
- *C. a. kistchinskii*, breeding in the SW Koryak highlands, the northeast coast of the Sea of Ochotsk, Kamchatka, and the northern Kurilen, all in E Siberia south of 62°N. This subspecies has recently been separated from *sakhalina* (Tomkovich 1986);
- *C. a. actites*, breeding on N Sakhalin Island. The subspecies was recently described by Nechaev & Tomkovich (1987 and 1988);
- *C. a. arcticola*, breeding in Alaska north of 65°N, east to the Mackenzie river delta in NW Canada;
- *C. a. pacifica*, breeding in Alaska south of 65°N;
- *C. a. hudsonia*, breeding in N Canada.

The discussion on the geographical variation in this species concentrated on:

(1) The systematic position of *centralis*. Hale (1980) considered *centralis* as a hybrid between *alpina* and *sakhalina*. Cramp & Simmons (1983) rejected the recognition of this subspecies because of the lack of evidence for a zone of secondary contact in Siberia between *sakhalina* and *alpina*. Greenwood (1986) considered the Central Siberian populations as *alpina*. Tomkovich (1986) considered *centralis* as a separate subspecies, which intergraded into *sakhalina* around the Chaunsk Bay.

(2) Subspeciation in the N Pacific. Some years ago, Tomkovich (1986) described the existence of 2 subspecies in E Siberia: *sakhalina* breeding on Wrangel Island, the Chukotsk Peninsula and Anadyr, and *kistchinski* mainly breeding around the Ochotsk Sea. This recognition was based on measurements and plumage characters. Browning (1991) analysed several other specimen including N Alaskan ones, and arrived at the same conclusions. Also, the existence of the N Alaskan subspecies *arcticola* was established. Browning's arguments were primarily based upon plumage characters but also upon measurements. Greenwood (1986) recognized this subspecies mainly because of its spatially segregated breeding range. Tomkovich (1986) noticed that the measurements of *arcticola* were largely similar to those of *sakhalina*. Moreover, the schedules of primary moult in the N Pacific populations are closely similar (Holmes 1966, Holmes 1971, Tomkovich 1986, Browning 1991, this study). Recently, Wenink *et al.* (1996) showed that 4 birds from

Anadyr had clearly different mtDNA patterns from those of W and N Alaska (*n* resp. 7 and 15), whereas the Alaskan populations appeared to have more or less the same patterns. They concluded that Alaska and Siberia were populated by different subspecies. This conclusion posed a new problem, since the W Alaskan breeding population was already considered to form a separate subspecies for long, named *pacifica*, and separable from the N Alaska population by longer average bill- and tarsus length (Browning 1977).

The separation of the 2 new subspecies from the E Siberian breeding grounds has stimulated new research. The first studies, involving mtDNA suggested important differences in genetic structure between breeding populations in combination with a large degree of genetical homogeneity within breeding areas (Wenink 1994, Wenink *et al.* 1993, 1996). Their results indicate a maximal time of splitting between supposed *hudsonia* and all other breeding populations of about 220.000 years ago. They distinguished 5 so-called haplotypes: (1) the '*alpina*' type, including *arctica*, *schinzii*, and *alpina* (east to Yamal); (2) the '*centralis*' type, including the breeding populations between Yamal and the Lena River Delta; (3) the '*sakhalina*' type, including breeding birds from Anadyr ; (4) the '*pacifica*' type, including breeding birds from N and W Alaska and (5) the '*hudsonia*' type, including *hudsonia*. No mtDNA from the recently detected breeding population on Sakhalin Island was available.

MORPHOMETRICAL DIFFERENTIATION

In total, 925 specimens were examined from all over the breeding range (Fig. 68). Some birds with brood patches occurred in regions unknown as regular breeding sites: (1) W Greenland near the Disko Bugt, (2) NE Greenland in the Independence Fjord, and (3) Archangelsk near

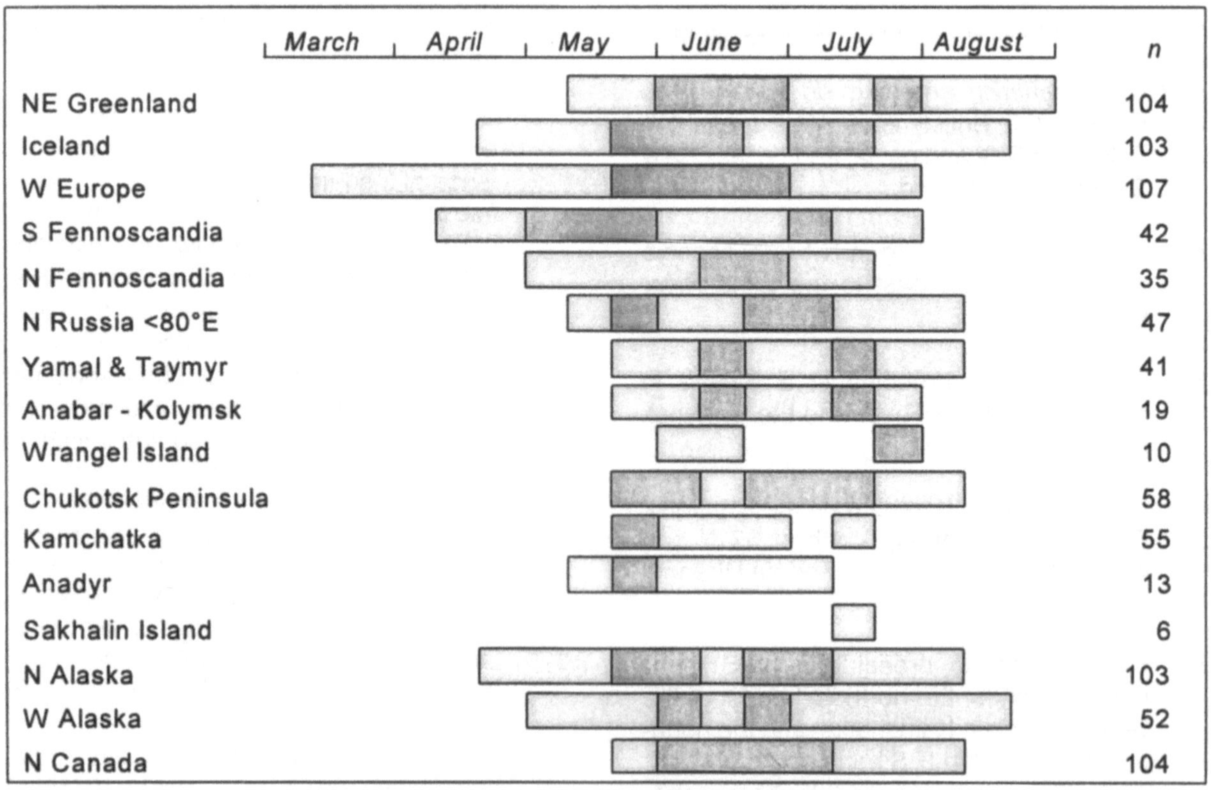

Fig. 69. *Collection dates of Dunlins. Most birds had been collected during the periods marked by the heavily shaded bars. The sample sizes are presented on the right side.*

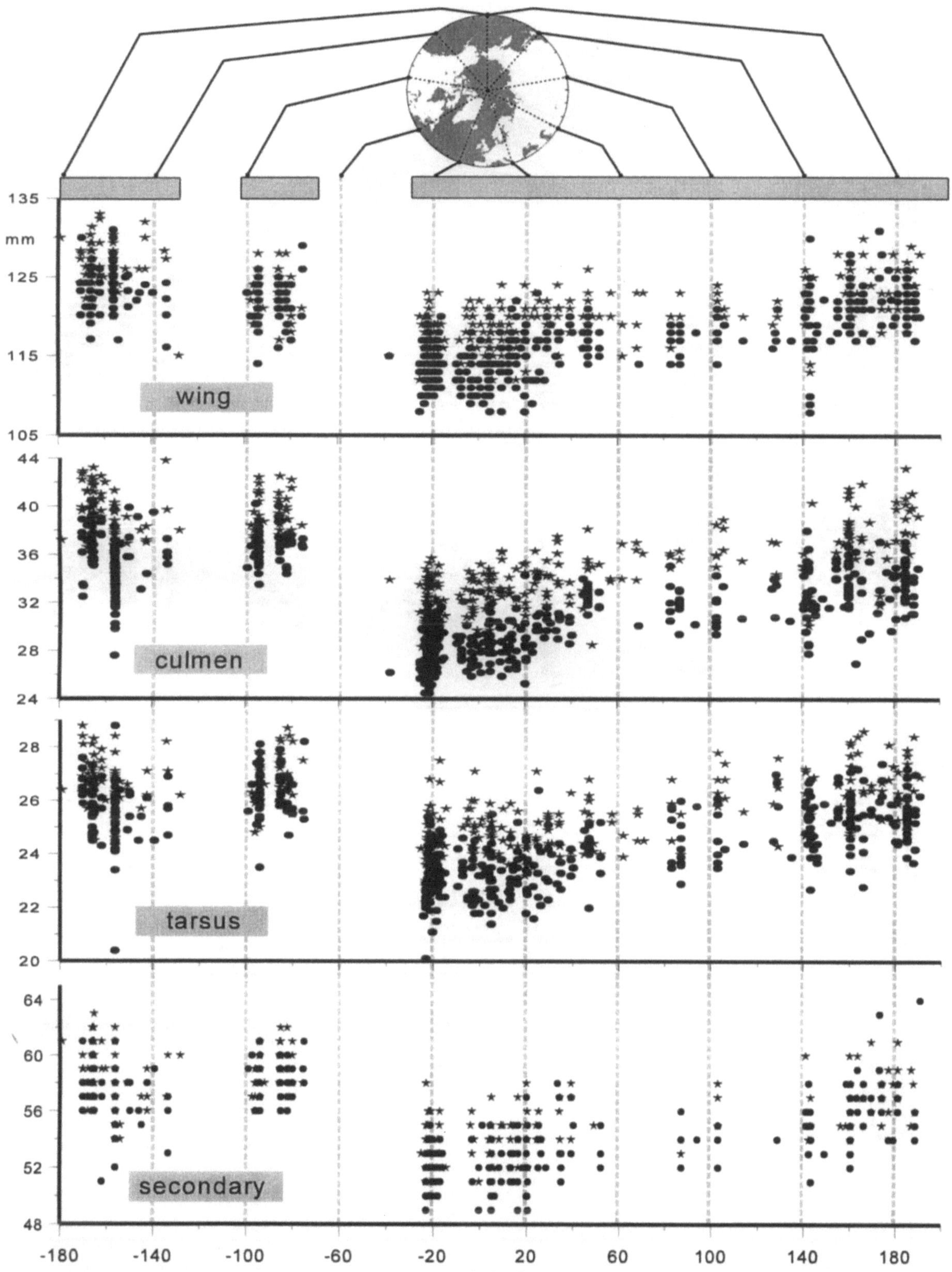

Fig. 70. *Measurements of Dunlins from the breeding grounds. The following measurements are presented from top to bottom: wing length, culmen length, tarsus length and secondary length (description in chapter 2). Explanation of the used symbols: ● ♂♂, ★ ♀♀, ▸ sexe unknown.*

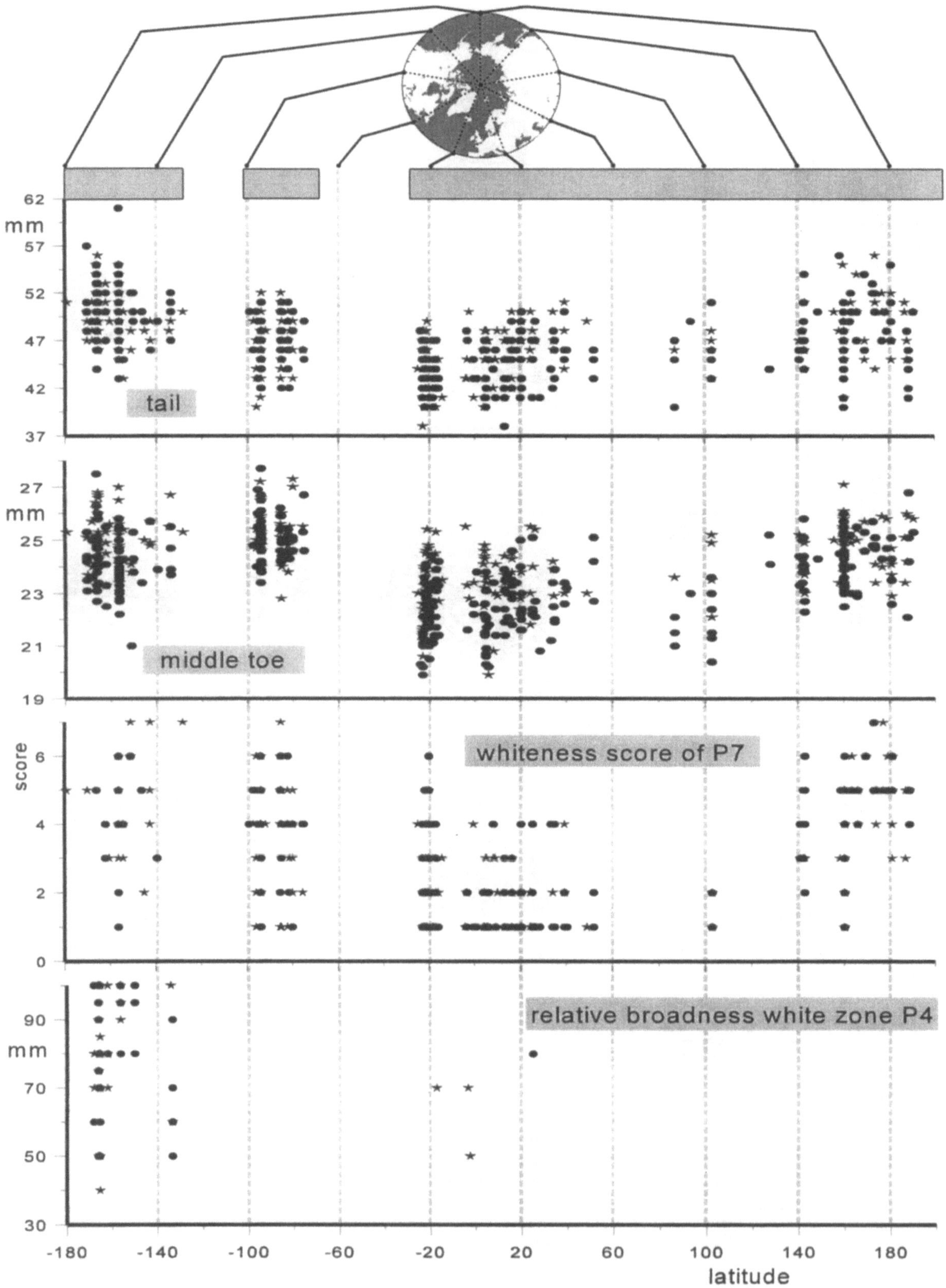

Fig. 71. *Measurements of Dunlins from the breeding grounds. The following measurements are presented from top to bottom: tail length, middle toe length, Nieboer's score on P4 and Greenwood's score on P7 (description in chapter 2). Explanation of the used symbols: ● ♂♂, ★ ♀♀, ▸ sexe unknown.*

Table 57. *Measurements of breeding populations of Dunlins. The delineation of the various populations is presented in Fig. 68. Presented are the means ± 1 SD (n).*

Locality	Wing	Culmen	Tarsus	Secondary	Tail	Middle toe
			♂♂			
1 E & NE Greenland	113.2±2.2(63)	26.9±1.0(61)	22.5±0.6(26)	51.0±1.5(12)	44.4±2.2(15)	21.1±0.7(12)
2 SE Greenland	115.0 - (1)	26.2 - (1)	-	-	-	-
3 Iceland	114.0±2.4(58)	29.1±1.5(58)	23.5±1.0(54)	52.2±1.3(48)	43.3±1.9(46)	22.5±1.0(42)
4 British Isles	112.4±2.4(34)	28.1±1.5(34)	23.0±0.8(33)	52.2±2.4(5)	45.2±2.3(5)	22.6±0.9(4)
5 Cont. W Europe	113.6±1.3(19)	28.9±1.2(19)	23.0±0.8(19)	51.6±1.3(16)	44.1±2.5(15)	21.8±1.0(15)
6 S Baltic	112.7±2.9(7)	28.7±1.5(7)	22.0 - (1)	-	-	-
7 S Fennoscandia	115.2±3.7(22)	28.8±1.9(22)	23.6±0.9(22)	52.1±1.8(22)	44.4±2.9(22)	22.8±1.0(22)
8 Gulf of Finland	113.6±3.2(14)	28.8±1.6(13)	23.2±0.9(14)	52.4±1.9(11)	44.9±3.4(11)	22.4±1.0(11)
9 N Fennoscandia	117.3±2.6(22)	30.6±1.7(22)	23.6±1.0(22)	54.1±2.4(17)	46.1±2.1(17)	22.7±1.0(17)
10 Kanin Peninsula-Yamal	117.1±2.1(18)	32.4±0.9(18)	24.3±0.8(17)	53.3±1.5(3)	44.7±1.5(3)	24.0±1.2(3)
11 Taymyr	117.6±1.9(24)	31.8±1.5(24)	24.6±1.0(23)	53.9±1.1(11)	45.4±3.0(11)	22.1±1.0(11)
12 Anabar - Kolymsk	117.6±1.9(14)	31.6±1.7(15)	25.0±1.0(15)	55.0±1.4(2)	46.0±3.5(3)	24.1±1.1(3)
13 Gizhiga	121.9±2.6(26)	33.8±1.7(26)	25.5±0.8(26)	54.5±1.3(26)	46.2±2.9(26)	24.3±0.9(26)
14 Sakhalin Island	109.0±1.0(3)	28.3±0.4(3)	23.7±1.0(3)	53.3±2.1(3)	48.3±5.1(3)	22.9±0.8(3)
15 S Kamchatka	120.2±2.0(10)	33.6±2.1(10)	25.4±1.3(10)	57.5±1.7(4)	51.2±3.8(4)	23.6±0.8(4)
16 N Kamchatka	119.5±3.0(4)	33.4±2.7(4)	25.8±0.8(4)	56.7±0.5(4)	49.0±3.6(4)	25.2±0.4(4)
17 Anadyr	124.0±4.3(8)	32.7±2.3(8)	26.0±0.8(8)	57.3±3.1(7)	50.3±2.1(8)	24.9±0.4(7)
18 Chukotsk Peninsula	121.6±2.0(31)	33.7±1.6(32)	25.5±0.9(31)	56.8±4.1(5)	44.6±3.3(8)	24.7±1.7(5)
19 Wrangel Island	122.5±1.6(6)	33.6±1.6(6)	25.1±0.8(6)	56.8±1.0(4)	50.5±3.4(4)	23.9±1.1(4)
20 Seward Peninsula	123.8±2.4(3)	36.9±1.6(3)	25.8±1.2(3)	58.3±1.5(3)	51.0±3.5(3)	24.3±0.7(3)
21 St. Lawrence Island	122.7±2.0(6)	36.1±2.5(6)	26.8±0.5(6)	57.7±1.9(6)	50.8±3.3(6)	24.1±0.8(6)
22 Yukon Delta	123.2±2.4(29)	37.4±1.3(29)	25.9±0.8(29)	57.7±1.8(29)	49.7±2.6(29)	24.5±1.1(27)
23 SW Alaska	130.0 - (1)	33.4 - (1)	25.9 - (1)	-	-	
24 S & SE Alaska	121.9±2.7(9)	37.4±1.8(9)	25.6±0.9(9)	56.6±1.7(9)	50.0±1.6(9)	24.3±0.7(9)
25 N Alaska	124.2±2.7(46)	34.1±2.2(47)	25.5±1.2(47)	56.6±1.7(43)	49.6±3.4(46)	23.9±1.0(36)
26 W coast Hudson Bay	121.4±2.5(33)	36.6±1.4(33)	26.2±0.9(33)	58.2±1.5(32)	47.2±2.3(33)	25.1±1.0(28)
27 James Bay	121.6±2.6(10)	36.6±1.4(10)	26.1±0.7(10)	58.0±1.2(10)	47.6±2.1(10)	25.0±0.4(9)
28 Foxe Basin	121.6±2.9(18)	37.3±1.1(18)	26.5±0.9(18)	58.5±1.2(17)	46.9±2.5(18)	25.1±0.8(15)
			♀♀			
1 E & NE Greenland	117.7±2.5(44)	30.1±1.3(43)	23.5±0.9(24)	53.9±2.7(7)	45.1±3.2(10)	21.7±0.6(10)
2 SE Greenland	115.0 - (1)	33.9 - (1)	-	-	-	-
3 Iceland	117.3±2.3(41)	33.0±1.5(41)	24.7±1.0(39)	53.6±1.6(36)	43.9±2.4(34)	23.6±0.9(29)
4 British isles	117.9±2.5(23)	32.5±1.5(23)	24.7±1.1(23)	53.8±1.9(5)	44.4±3.4(5)	23.8±1.2(4)
5 Cont. W Europe	117.2±2.4(23)	33.2±1.0(22)	24.5±1.0(23)	53.6±1.3(18)	45.2±2.7(18)	23.3±1.4(16)
6 S Baltic	118.4±3.5(7)	34.0±1.8(6)	24.2±0.5(2)	55.0 - (1)	45.0 - (1)	23.0 - (1)
7 S Fennoscandia	118.3±3.5(7)	31.9±1.8(7)	24.6±0.9(7)	54.1±1.8(7)	45.1±2.5(7)	23.7±0.9(6)
8 Gulf of Finland	119.6±3.1(10)	33.6±1.5(10)	24.0±0.6(10)	55.4±1.9(9)	47.4±1.6(8)	22.8±1.1(9)
9 N Fennoscandia	120.5±2.1(20)	34.3±1.7(20)	25.0±0.8(20)	55.5±1.5(10)	47.9±2.1(10)	23.7±0.8(10)
10 Kanin Penin. - Yamal	120.2±2.8(20)	34.6±2.2(19)	25.1±0.8(20)	55.0 - (1)	49.0 - (1)	23.0 - (1)
11 Taymyr	120.8±2.0(13)	36.4±1.3(13)	26.1±1.1(13)	55.6±1.9(5)	46.0±1.9(5)	23.9±1.2(5)
12 Anabar - Kolymsk	121.3±2.0(8)	36.4±2.1(8)	26.4±1.0(8)	55.0 - (1)	48.0 - (1)	23.4 - (1)
13 Gizhiga	123.6±2.4(13)	38.2±1.9(13)	26.7±1.1(13)	56.2±1.5(13)	46.6±2.2(13)	25.1±1.0(13)
14 Sakhalin Island	113.4±0.6(3)	31.0±1.3(3)	24.5±0.2(3)	55.7±1.2(3)	47.0±3.6(3)	23.9±1.0(3)
15 S Kamchatka	125.9±3.0(6)	38.8±2.4(6)	27.0±1.6(6)	57.7±2.0(6)	51.8±2.1(6)	25.2±0.2(6)
16 N Kamchatka	125.0±1.4(2)	37.5±1.8(2)	26.5±0.1(2)	58.5±3.5(2)	46.0±1.4(2)	25.2±0.8(2)
17 Anadyr	122.3±3.3(6)	34.7±2.2(6)	26.8±0.9(6)	57.2±1.5(6)	50.5±3.9(6)	24.4±1.1(6)
18 Chukotsk Peninsula	124.6±2.5(19)	38.2±2.4(19)	26.6±0.9(18)	57.0±1.8(4)	48.6±2.5(5)	25.1±1.1(6)
19 Wrangel Island	124.0±2.4(4)	38.0±2.7(4)	26.1±0.8(4)	58.8±1.7(4)	48.0±3.6(3)	23.8±0.8(4)
20 Seward Peninsula	125.3±1.8(5)	39.6±1.9(5)	26.6±0.7(5)	57.5±1.3(4)	48.8±1.5(5)	24.9±0.9(3)

Locality	Wing	Culmen	Tarsus	Secondary	Tail	Middle toe
21 St. Lawrence Island	127.0±1.6(5)	41.5±1.6(5)	27.7±1.1(5)	59.2±0.4(5)	48.8±1.3(5)	24.4±1.0(2)
22 Yukon Delta	127.2±2.8(29)	40.5±1.3(29)	26.8±0.8(29)	59.6±1.7(27)	50.7±2.7(27)	25.1±1.0(27)
23 SW Alaska	127.5±3.5(2)	39.6±3.4(2)	26.2±0.3(2)	60.0±1.4(2)	50.0±1.4(2)	25.5±0.2(2)
24 S & SE Alaska	127.8±0.7(2)	41.8±2.9(2)	27.6±0.8(2)	58.5±2.1(2)	48.5±0.7(2)	26.1±0.8(2)
25 N Alaska	125.3±3.6(27)	37.4±1.9(29)	26.3±0.9(29)	58.0±2.2(27)	49.1±2.9(27)	25.0±0.8(24)
26 W coast Hudson Bay	122.9±2.5(20)	39.3±1.9(20)	26.6±0.9(20)	58.5±1.4(20)	46.5±3.1(20)	25.3±0.9(19)
27 James Bay	122.5±3.1(11)	39.4±1.7(11)	27.0±1.1(10)	59.5±1.2(11)	47.2±2.4(11)	25.0±1.6(7)
28 Foxe Basin	123.2±2.9(12)	39.5±1.3(12)	27.2±0.7(12)	59.9±1.3(11)	47.2±2.7(12)	25.1±0.6(11)

Kotlas. Small samples were available from the region between Taymyr and the Chaunsk Bay, which is due to scarcity of the species in this region (Tomkovich pers. comm.).

Birds did not arrive at the northernmost breeding grounds until late May, whereas the southern breeding populations were present by March (Fig. 69). Most birds had been collected during June and the first half of July. Males were more common (58.6 %) than ♀♀ in the samples.

Initially, 28 breeding regions were distinguished (Fig. 68, Table 57). The morphometric differences between populations were analysed and are presented in the tables 57 - 64 and in the figures 70 - 73.

Reduction in the number of samples.

Some preliminary analyses were carried out to reduce the number of initially distinguished samples in 4 breeding regions: (1) Europe as far as *schinzii* is involved; (2) N Canada concerning *hudsonia*; (3) N Fennoscandia east to the Chaunsk Bay in E Siberia; and (4) E Siberia. No significant geographical variation could be shown to exist in the samples from the first 2 regions, and these samples were combined. There was geographical variation between the samples in both latter regions:

(1) *N Europe and Siberia*. The geographical variation within this region showed differences between 2 western and 2 eastern populations (Fig. 72). The Taymyr ♂♂ were distinguishable from both their western and eastern neighbours, but the Taymyr ♀♀ were only distinct from their western neighbours.
The samples of N Fennoscandia and Kanin - Yamal were indistinguishable and could be combined in further analyses.

(2) *E Siberia*. The geographical variation in this region is difficult to understand. Since southern populations start to breed earlier than the northern ones, the Dunlins examined from the shores of the Sea of Ochotsk might include some migrants. Several late May birds at Amur, Gizhiga, or Koryak might still be migrants, since most birds on Wrangel Island or Chukotsk did not arrive prior to early June. In order to reduce the probability for using the wrong birds (migrants) in the southern breeding areas, only those birds were selected, which (1) had been collected between early June and 10 July; or (2) had brood patches already in May. For Anadyr, Wrangel and Chukotsk also other July birds with brood patches were included.

Table 58. *Overall results of MANOVA analysis of the E Siberian breeding populations of the Dunlin. Three different analyses per sex were performed These analyses differed in relation to the number of variables included.*

Variables	♂♂				♀♀			
	n	*df*	*F*-value	*P*	*n*	*df*	*F*-value	*P*
Wing, culmen	86	10/160	1.735	0.08	30	10/88	1.565	0.13
Wing, culmen, tarsus	85	15/237	1.556	0.09	49	15/129	1.186	0.29
Wing, culmen, tarsus, secondary, tail, middle toe	50	30/215	1.962**	0	32	30/125	1.461	0.08

Table 59. *Results of pairwise MANOVA analysis in samples of E Siberian ♂♂. This analysis included 6 measurements (n = 50; 6 and 39 df): wing-, culmen-, tarsus-, secondary-, tail- and middle toe length. Pillai's trace values are presented together with the level of significance as based upon F-values (Norušis 1992c). Significance levels: *0.01 < P < 0.05; **0.001 < P < 0.01; ***P < 0.001.*

Locality	*n*	Gizhiga	S Kamchatka	N Kamchatka	Anadyr	Chukotsk
Gizhiga	25					
S Kamchatka	4	0.294*				
N Kamchatka	4	0.233	0.171			
Anadyr	7	0.333*	0.251	0.179		
Chukotsk	6	0.307*	0.288*	0.100	0.218	
Wrangel Island	4	0.183	0.043	0.142	0.161	0.229

In general, the MANOVA-analyses did not lead to the recognition of significant geographical variation in this region (Table 58), except for the ♂ samples when all 6 measurements were included. Pairwise comparison of the ♂ samples showed small but significant differences (Table 59), especially between the Gizhiga ♂♂ and the other ♂ samples. Thus the Gizhiga sample was separated from the other NE Siberian ones. These latter were combined.

Sexual dimorphism.

The Dunlin showed much sexual and geographical variation (Table 60). Sexual differences were larger than geographical differences, but the amount of geographical variation was the highest of all birds examined in this study. Sexual dimorphism was relatively small in the samples of N Canada and Alaska and large in the ones of Greenland, Europe and W Siberia (Table 61). This difference is caused by 2 factors:

Table 60. *Overall results of MANOVA for all distinguished breeding populations in the Dunlin in combination with univariate ANOVA's in which the different measurements were analysed separately. Significance levels: *0.01 < P < 0.05; **0.001 < P < 0.01; ***P < 0.001.*

	Analysis I			Analysis II			Analysis III		
Variables	**wing, culmen, tarsus, secondary, tail, middle toe length**			**wing, culmen, tarsus**			**wing, culmen**		
n	508			814			892		
Multivariate analysis of variance									
Effects	Interaction	Area	Sex	Interaction	Area	Sex	Interaction	Area	Sex
Pillai's trace	0.168	1.600	0.252	0.064	1.132	0.322	0.076	1.042	0.304
F-value	1.268	16.001***	26.900***	2.076***	43.492***	124.876***	3.123***	85.903***	189.518***
Hypoth. & error *df*	66/2904	66/2904	6/479	33/2370	33/2370	3/788	22/1736	22/1736	2/867
Univariate F tests									
df	11/484	11/484	1/484	11/790	11/790	1/790	11/868	11/868	1/868
wing length	1.877*	95.029***	38.124***	3.391***	151.155***	96.530***	3.713***	173.739***	99.851***
culmen length	1.677	162.553***	140.975***	2.461***	261.102***	321.322***	2.713**	319.285***	340.036***
tarsus length	1.137	64.241***	35.952***	1.547***	112.291***	94.291***			
secondary length	1.427	94.246***	20.659***						
tail length	0.738	34.550***	0.762						
middle toe length	1.938*	39.953***	12.804***						

Table 61. *Sexual size dimorphism in the Dunlin. The mean percentage difference between the measurements of both sexes was calculated for every single variable and breeding population, according to: $(x_♀-x_♂)/x_♂*100$. A negative value thus implies smaller measurements for ♀♀. The calculations were based upon the means of Table 57. The obtained values were averaged per variable and summed per breeding population.*

Locality	Wing	Culmen	Tarsus	Secondary	Tail	Middle toe	Σ 3 variables	Σ 6 variables
E & NE Greenland	3.9	11.9	4.4	5.7	1.6	2.8	20.2	30.3
Iceland	2.9	13.4	5.1	2.7	1.4	4.9	21.4	30.4
British Isles	4.9	15.7	7.4	3.1	-1.8	5.3	28.0	34.6
Cont. W Europe	3.2	14.9	6.5	3.9	2.5	6.9	24.6	37.9
S Baltic	5.1	18.5	-	-	-	-	-	-
S Fennoscandia	2.7	10.8	4.2	3.8	1.6	3.9	17.7	27.0
Gulf of Finland	5.3	16.7	3.4	5.7	5.6	1.8	25.4	38.5
N Fennoscandia	2.7	12.1	5.9	2.6	3.9	4.4	20.7	31.6
Kanin Penin.-Yamal	2.6	6.8	3.3	-	-	-	12.7	-
Taymyr	2.7	14.5	6.1	3.2	1.3	8.1	23.3	35.9
Anabar-Kolymsk	3.1	15.2	5.6	-	-	-	23.9	-
Gizhiga	1.4	13.0	4.7	3.1	0.9	3.3	19.1	26.4
Sakhalin Island	4.0	9.5	3.4	4.5	-2.7	4.4	16.9	23.1
S Kamchatka	4.7	15.5	6.3	0.3	1.2	6.8	26.5	34.8
N Kamchatka	4.6	12.3	2.7	3.2	-6.1	0.0	19.6	16.7
Anadyr	-1.4	6.1	3.1	-0.2	0.4	-2.0	7.8	6.0
Chukotsk Peninsula	2.5	13.3	4.3	0.4	9.0	1.6	20.1	31.1
Wrangel Island	1.2	13.1	4.0	4.0	-4.9	-0.4	18.3	17.0
N Alaska	0.9	9.7	3.1	2.5	-1.0	4.6	13.7	19.8
Seward Peninsula	1.2	7.3	3.1	-1.4	-4.3	2.5	11.6	8.4
St. Lawrence Island	3.5	15.0	3.4	2.6	-3.9	1.2	21.9	21.8
Yukon Delta	3.2	8.3	1.2	3.3	2.0	2.4	12.7	20.4
S & SE Alaska	4.8	11.7	7.8	2.5	-3.0	7.4	24.3	31.2
W coast Hudson Bay	1.2	7.4	1.5	0.5	-1.5	0.8	10.1	9.9
James Bay	0.7	7.7	3.4	2.6	-0.8	0.0	11.8	13.6
Foxe Basin	1.3	5.9	2.6	2.4	0.6	0.0	9.8	12.8
x	2.8	11.8	4.3	2.7	0.1	3.1		
SD	1.7	3.6	1.7	1.8	3.5	2.7		
n	26	26	25	23	23	23		

Table 62. *Pearson's product-moment correlations between measurements of the Dunlin. Data of ♂♂ and ♀♀ were analysed separately. Two-tailed probabilities: * 0.01 < P < 0.05; ** 0.001 < P < 0.01; *** P < 0.001.*

	Wing	Culmen	Tarsus	Secondary	Tail	Middle toe	
Wing		0.717***	0.638***	0.750***	0.635***	0.531***	
Culmen	0.761***		0.732***	0.731***	0.443***	0.620***	
Tarsus	0.714***	0.775***		0.622***	0.370***	0.664***	♀♀
Secondary	0.738***	0.766***	0.698***		0.592***	0.491***	
Tail	0.613***	0.485***	0.446***	0.571***		0.278***	
Middle toe	0.594***	0.667***	0.728***	0.619***	0.345***		
				♂♂			

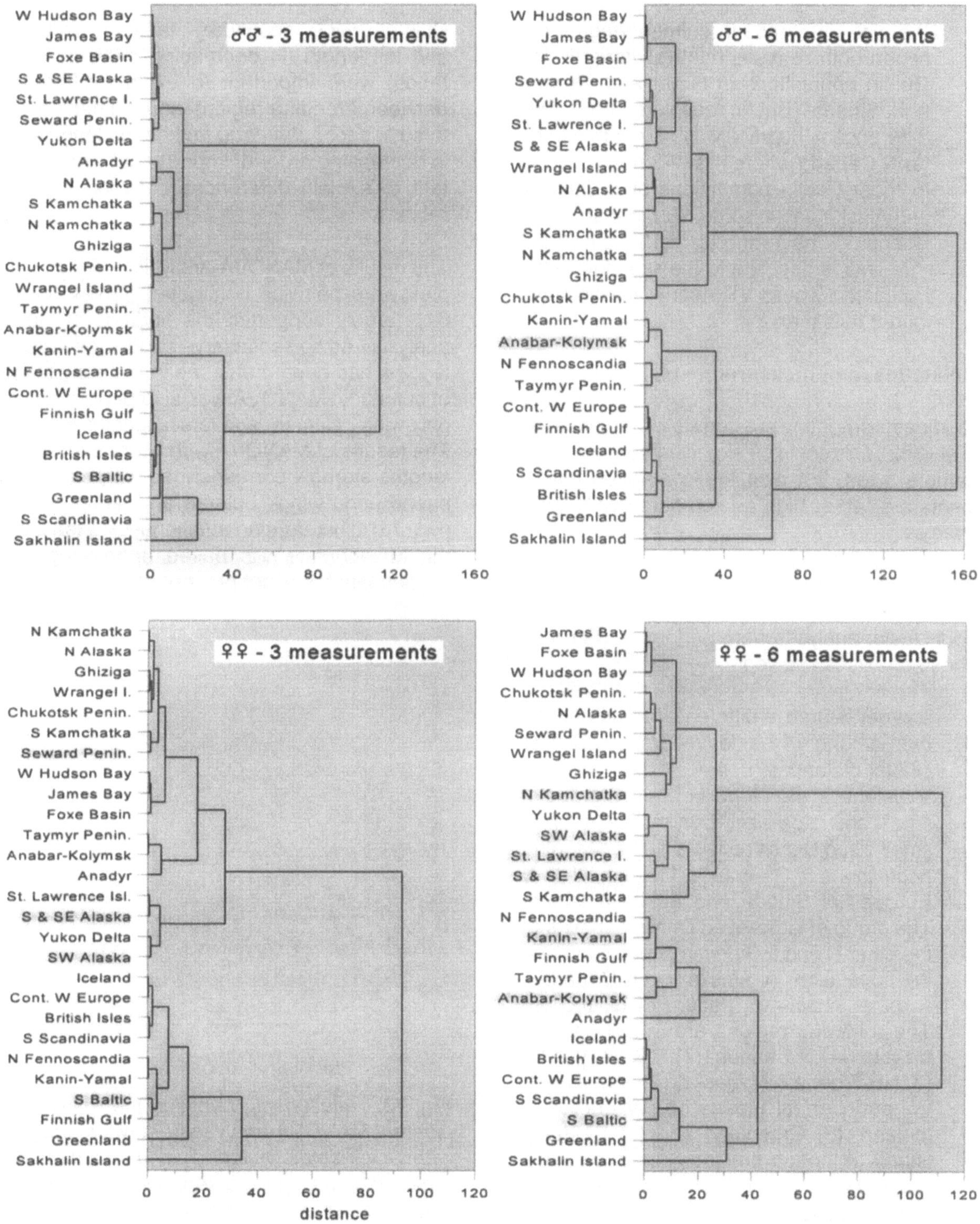

Fig. 72. *Cluster analysis in the Dunlin based upon the measurements presented in Table 57. The samples of both sexes were analysed separately. The results in both left figures were based on the wing-, culmen- and tarsus lengths and in both right figures also secondary-, middle toe and tail lengths were used in the analyses. Sometimes mean values in a sample were derived from only 1 or 2 birds. Such samples are recognizable by their shaded background.*

- the size difference in culmen lengths between both sexes is relatively large in the Dunlin populations of Greenland, Europe and Siberia. But in absolute terms this difference is nearly always 3-4 mm except for N Canada where it is 2-3 mm.
- In most Dunlin populations of N Canada and Alaska and in some NE Siberian populations ♀♀ have shorter tail lengths than ♂♂. This is opposite to the situation in the Palearctic where ♀♀ nearly always have longer tails than ♂♂.

Relatedness of measurements.

All measurements were correlated significantly (Table 62).This could be expected from the large sample sizes, but also the correlation coefficients are rather high as compared with other species.

Cluster analysis.

Four cluster analyses were performed (Fig. 72). These analyses showed:

- a larger degree of differentiation among ♂♂ than among ♀♀ as derived from the larger cluster distances;
- a systematic dichotomy between a group of small- and large-sized Dunlin. The group of small-sized Dunlin consists of the breeding populations of Greenland, Iceland, W- & N Europe, W Siberia and Sakhalin Island. The group of large-sized Dunlin consists of the other N and E Siberian populations and the populations of Alaska and N Canada.
- a rather consistent pattern of distinction. The following regions are usually recognizable: (1) N Canada, (2) S & W Alaska, (3) NE Siberia & N Alaska, (4) N Europe & W- and Central Siberia, (5) W Europe & Iceland, (6) Greenland and (7) Sakhalin Island. Although the recognition of these groups is rather evident, individual samples sometimes interchange between groups.

MANOVA-analysis.

All variables contributed significantly to the distinction between samples (Table 60). Culmen-, wing-, secondary-, tarsus-, middle toe- and tail length (in decreasing order of importance) were important to explain differences between breeding regions and culmen-, wing-, tarsus-, secondary- and middle toe length (also in decreasing order of importance) were important to explain differences between breeding regions.

The results of MANOVA-analysis, in which samples were pairwise contrasted (tables 63 and 64), largely supported the findings of cluster analysis . As no secondary-, middle toe- and tail lengths had been measured in the early stages of our study, MANOVA was also performed with only wing-, culmen- and tarsus lengths included. The results of MANOVA with wing- and culmen lengths strongly corresponded with the results based upon wing-, culmen and tarsus lengths (Fig. 73). Therefore an additional third table with the MANOVA-results based upon wing- and culmen lengths is not presented.

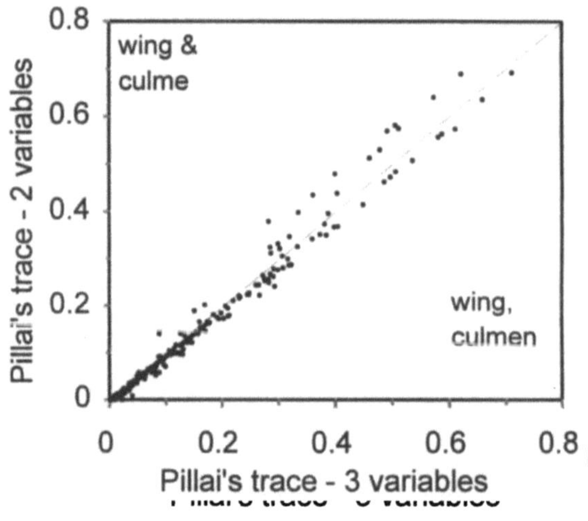

Fig. 73. *Relationship between the results of 2 different MANOVA-analyses. Pairwise analyses with 3 variables included wing-, culmen and tarsus lengths and are represented on the X-axis. Pairwise analyses based upon 2 variables (wing- and culmen length) were related to the '3 variable'-analyses .*

Table 63. *Results of MANOVA-analysis concerning the morphometric differentiation in the Dunlin. Pillai's trace values are presented together with the level of significance as based upon F-values (Norušis 1992c). In this series of analyses 6 measurements and 12 populations were included (6 and 479 df). Significance levels: *0.01 < P < 0.05; **0.001 < P < 0.01; ***P < 0.001.*

Locality	n	E & NE Greenland ♀	W Europe ♀	N.Fenn.-Yamal ♀	Taymyr ♀	Anabar-Kolymsk ♀	Gizhiga ♀	Sakhalin Island ♀	NE Siberia ♀	Seward Penin. ♀	W & S Alaska ♀	N Alaska ♀	N Canada ♀	E & NE Greenland ♂	W Europe ♂	N.Fenn.-Yamal ♂	Taymyr ♂	Anabar-Kolymsk ♂	Gizhiga ♂	Sakhalin Island ♂	NE Siberia ♂	Seward Penin. ♂	W & S Alaska ♂	N Alaska ♂
E & NE Greenl. ♀	6																							
W Europe ♀	56	0.068***																						
N.Fenn.-Yamal ♀	11	0.067***	0.048***																					
Taymyr ♀	5	0.105***	0.065***	0.038**																				
Anabar-Kol. ♀	1	0.041**	0.020	0.013	0.003																			
Gizhiga ♀	13	0.225***	0.233***	0.118***	0.021	0.008																		
Sakhalin I. ♀	3	0.050***	0.052***	0.044***	0.079***	0.042**	0.135***																	
NE Siberia ♀	19	0.224***	0.295***	0.097***	0.027*	0.008	0.041**	0.117***																
Seward Pen. ♀	2	0.130***	0.088***	0.057***	0.022	0.009	0.011	0.068***	0.012															
W & S Alaska ♀	33	0.367***	0.582***	0.278***	0.103***	0.024	0.105***	0.213***	0.123***	0.010														
N Alaska ♀	22	0.246***	0.360***	0.129***	0.039**	0.011	0.041**	0.143***	0.019	0.016	0.091***													
N Canada ♀	36	0.321***	0.512***	0.237***	0.071***	0.024	0.060***	0.160***	0.117***	0.017	0.127***	0.114***												
E & NE Greenl. ♂	9	**0.040****	0.209***	0.185***	0.217***	0.075***	0.393***	0.088***	0.409***	0.204***	0.562***	0.442***	0.529***											
W Europe ♂	78	0.027*	**0.308*****	0.188***	0.201***	0.057***	0.474***	0.058***	0.536***	0.179***	0.757***	0.595***	0.726***	0.044***										
N.Fenn.-Yamal ♂	20	0.019	0.081***	**0.055*****	0.116***	0.035**	0.294***	0.037**	0.307***	0.122***	0.531***	0.353***	0.481***	0.096***	0.077***									
Taymyr ♂	11	0.019	0.057***	0.048***	**0.072*****	0.024	0.219***	0.053***	0.219***	0.102***	0.411***	0.260***	0.359***	0.103***	0.092***	0.037**								
Anabar-Kol. ♂	2	0.032*	0.048***	0.038**	0.055***	**0.033***	0.104***	0.035**	0.085***	0.085***	0.172***	0.100***	0.144***	0.062***	0.046***	0.038**	0.031*							
Gizhiga ♂	26	0.110***	0.109***	0.035**	0.033*	0.011	**0.119*****	0.067***	0.143***	0.061***	0.395***	0.172***	0.356***	0.270***	0.378***	0.151***	0.104***	0.044**						
Sakhalin I. ♂	3	0.076***	0.115***	0.107***	0.149***	0.072***	0.231***	**0.020**	0.213***	0.145***	0.321***	0.247***	0.276***	0.073***	0.070***	0.076***	0.093***	0.055***	0.166***					
NE Siberia ♂	23	0.135***	0.212***	0.036**	0.065***	0.022	0.167***	0.055***	**0.114*****	0.063***	0.366***	0.154***	0.319***	0.304***	0.414***	0.168***	0.145***	0.038**	0.100***	0.136***				
Seward Pen. ♂	3	0.108***	0.081***	0.031*	0.024	0.010	0.028*	0.065***	0.008	**0.013**	0.034*	0.011	0.037**	0.197***	0.176***	0.097***	0.087***	0.064***	0.049***	0.127***	0.029*			
W & S Alaska ♂	42	0.236***	0.381***	0.096***	0.039**	0.009	0.068***	0.111***	0.026*	0.015	**0.187*****	0.056***	0.188***	0.435***	0.653***	0.346***	0.243***	0.099***	0.201***	0.208***	0.158***	0.003		
N Alaska ♂	34	0.147***	0.286***	0.046***	0.053***	0.012	0.149***	0.104***	0.109***	0.053***	0.353***	**0.122*****	0.369***	0.334***	0.524***	0.222***	0.156***	0.056***	0.063***	0.190***	0.074***	0.024	0.155***	
N Canada ♂	50	0.241***	0.413***	0.135***	0.049***	0.020	0.069***	0.097***	0.081***	0.029*	0.262***	0.102***	**0.104*****	0.459***	0.682***	0.375***	0.273***	0.095***	0.252***	0.209***	0.170***	0.021	0.131***	0.268***

Table 64. *Results of MANOVA-analysis concerning the morphometric differentiation in the Dunlin. Pillai's trace values are presented together with the level of significance as based upon F-values (Norušis 1992c). In this series of analyses 3 measurements and 12 populations were included (3 and 788 df): wing-, culmen- and tarsus lengths. Significance levels: *0.01 < P < 0.05; **0.001 < P < 0.01; ***P < 0.001.*

Locality	n	E & NE Greenland ♀	W Europe ♀	N.Fenn.-Yamal ♀	Taymyr ♀	Anabar-Kolymsk ♀	Gizhiga ♀	Sakhalin Island ♀	NE Siberia ♀	Seward Penin. ♀	W & S Alaska ♀	N Alaska ♀	N Canada ♀	E & NE Greenland ♂	W Europe ♂	N.Fenn.-Yamal ♂	Taymyr ♂	Anabar-Kolymsk ♂	Gizhiga ♂	Sakhalin Island ♂	NE Siberia ♂	Seward Penin. ♂	W & S Alaska ♂	N Alaska ♂
E & NE Greenl. ♀	24																							
W Europe ♀	93	0.090***																						
N.Fenn.-Yamal ♀	39	0.126***	0.060***																					
Taymyr ♀	13	0.157***	0.080***	0.026***																				
Anabar-Kol. ♀	8	0.126***	0.062***	0.023***	0.001																			
Gizhiga ♀	13	0.230***	0.170***	0.075***	0.016**	0.010*																		
Sakhalin I. ♀	3	0.023***	0.012*	0.034***	0.045***	0.044***	0.080***																	
NE Siberia ♀	36	0.319***	0.307***	0.121***	0.024***	0.013*	0.003	0.089***																
Seward Pen. ♀	5	0.160***	0.114***	0.059***	0.024***	0.020**	0.005	0.084***	0.008															
W & S Alaska ♀	38	0.461***	0.506***	0.298***	0.113***	0.076***	0.041***	0.151***	0.083***	0.004														
N Alaska ♀	27	0.286***	0.276***	0.109***	0.035***	0.023***	0.010*	0.093***	0.007	0.009	0.067***													
N Canada ♀	42	0.403***	0.398***	0.202***	0.040***	0.025***	0.008	0.102***	0.038***	0.006	0.074***	0.057***												
E & NE Greenl. ♂	26	0.088***	0.283***	0.335***	0.306***	0.248***	0.388***	0.034***	0.512***	0.272***	0.622***	0.479***	0.574***											
W Europe ♂	129	0.087***	0.333***	0.373***	0.283***	0.213***	0.384***	0.012*	0.588***	0.245***	0.711***	0.536***	0.659***	0.052***										
N.Fenn.-Yamal ♂	39	0.021**	0.036***	0.096***	0.124***	0.099***	0.210***	0.012*	0.322***	0.142***	0.486***	0.291***	0.405***	0.150***	0.125***									
Taymyr ♂	23	0.035***	0.008	0.049***	0.079***	0.065***	0.152***	0.010*	0.220***	0.115***	0.381***	0.206***	0.291***	0.159***	0.120***	0.010*								
Anabar-Kol. ♂	14	0.040***	0.016**	0.048***	0.069***	0.056***	0.132***	0.009	0.173***	0.110***	0.316***	0.170***	0.232***	0.131***	0.088***	0.022***	0.005							
Gizhiga ♂	26	0.106***	0.076***	0.015**	0.035***	0.026***	0.074***	0.039***	0.099***	0.061***	0.260***	0.087***	0.190***	0.299***	0.318***	0.096***	0.050***	0.042***						
Sakhalin I. ♂	3	0.049***	0.053***	0.088***	0.102***	0.096***	0.146***	0.009	0.161***	0.142***	0.231***	0.165***	0.177***	0.027***	0.016**	0.043***	0.044***	0.039***	0.093***					
NE Siberia ♂	58	0.133***	0.117***	0.027***	0.046***	0.032***	0.096***	0.039***	0.152***	0.072***	0.359***	0.130***	0.275***	0.361***	0.449***	0.130***	0.059***	0.043***	0.001	0.094***				
Seward Pen. ♂	3	0.061***	0.035***	0.011*	0.005	0.005	0.004	0.043***	0.003	0.006	0.020***	0.002	0.012*	0.136***	0.110***	0.050***	0.038***	0.040***	0.012*	0.085***	0.015**			
W & S Alaska ♂	45	0.285***	0.265***	0.077***	0.014*	0.010*	0.010*	0.074***	0.016**	0.014*	0.132***	0.018**	0.062***	0.492***	0.581***	0.282***	0.184***	0.149***	0.076***	0.144***	0.127***	0.001		
N Alaska ♂	46	0.169***	0.210***	0.061***	0.060***	0.041***	0.083***	0.062***	0.116***	0.058***	0.285***	0.082***	0.247***	0.400***	0.497***	0.194***	0.118***	0.094***	0.016**	0.124***	0.033***	0.011*	0.101***	
N Canada ♂	61	0.302***	0.266***	0.082***	0.002	0.001	0.015**	0.062***	0.034***	0.023***	0.198***	0.053***	0.068***	0.506***	0.611***	0.293***	0.176***	0.132***	0.083***	0.129***	0.138***	0.005	0.023***	0.146***

The following conclusions were derived from the MANOVA-results presented in Table 63:

- There was a large degree of differentiation: within each sex nearly all samples differed from each other. And if there was no significancy, it was often due to small sample size (♀♀ Anabar-Kolymsk, ♀♀ Seward Peninsula, ♂♂ Seward Peninsula). There are 2 exceptions on this rule: (1) Gizhiga ♀♀ did not differ from Taymyr ♀♀, and (2) N Alaska ♀♀ did not differ from NE Siberia ♀♀.
- Both samples from Gizhiga differed significantly from the ones of NE Siberia.
- The distances between the ♂-samples is nearly always larger -as calculated by Pillai's trace value- in comparison with the ones between the ♀-samples.

When less measurements were included (and thus more birds per sample could be used), we arrived largely at the same findings (Table 64). There are some supplementary conclusions:

- The ♀♀ from Gizhiga can still badly be differentiated. They were not different from the NE Siberia ♀♀, Seward Peninsula ♀♀ and N Canada ♀♀. Since the Gizhiga population is mainly characterized by its small secondaries, there will be a lack of recognition when such a character is lacking in the analyses.
- The same conclusion is valid for the Seward Peninsula ♀♀. These cannot be differentiated from the Gizhiga ♀♀, NE Siberia ♀♀, W & S Alaska ♀♀, N Alaska ♀♀ and N Canada ♀♀.
- The measurements of Taymyr birds (both ♀♀ and ♂♂) had more in common with the ones of Anabar-Kolymsk than with the ones of N Fennoscandia - Yamal.

Plumage characters

Two plumage characters were quantified in order to study their usefulness for the distinction between samples: (1) the occurrence of white fringed primaries (Horvath & Keve 1956, Greenwood 1986) and (2) the presence of buff-fringed inner wing coverts in adults (Clark 1984, Gromadzka & Przystupa 1984, Gromadzka 1989). For quantification of these data see chapter 3 - 'The selected variables'.

(1) *the occurrence of white fringed primaries.* Horvath & Keve (1956) already noticed that migrant Hungarian Dunlins of eastern breeding origin had more white on the leading edge of the 4th primary. Greenwood (1986) confirmed this, but he erroneously published it as having quantified the amount of white on P7 instead of P4. In order to obtain comparable results with Greenwood (1986), we estimated the amount of white on P7 and thus erroneously obtained data, which could *not* be compared with earlier studies. Moreover, the white edges on P7 are more vulnerable to wear than the white edges of P4 (Miltenburg 1990). Nevertheless, the Dunlins sampled west of Taymyr appeared to have little or no white on the outer vane of P7 (Fig. 71), while samples from E Siberia, Alaska and N Canada mostly had much more. The extremely abraded state of P7 of these latter samples caused a lot of variation and made the amount of white a useless character for prediction purposes.

In a later stage of the project some data on P4 were collected by estimating the relative broadness of the white zone on the outer vane of this primary. See also chapter 2 for a detailed description of this variable. This measure aims to introduce a normally distributed variable instead of the categorical one used above, since categorical data require too large amounts of data in predictional analyses.
There are as yet too few data to conclude on the existence of geographic variation in this character.

(2) *The occurrence of buff-fringed inner coverts in adults.* Dunlin from Central and E Siberia show buff-brown fringes on the fresh inner median wing coverts in summer plumage, in contrast to those of northern Europe (Gromadzka 1989). The occurrence of these so-called 'adult buff' feathers is important when the breeding origin of migrants has to be predicted.We found adult buff feathers to be present in the samples

Table 65. *The occurrence of buff-fringed inner coverts in different breeding populations of Dunlin.*

Locality	Buff-fringed coverts Absent	Buff-fringed coverts Present	*n*	% present	Locality	Buff-fringed coverts Absent	Buff-fringed coverts Present	*n*	% present
E & NE Greenland	19	-	19	0.0	S Kamchatka	6	-	6	0.0
Iceland	80	-	80	0.0	Anadyr	2	1	3	33.3
British Isles	10	-	10	0.0	Chukotsk Peninsula	4	10	14	71.4
Cont. W Europe	22	-	22	0.0	N Alaska	67	6	73	8.2
S Baltic	1	-	1	0.0	Seward Peninsula	6	1	7	14.3
S Fennoscandia	2	-	2	0.0	St. Lawrence Island	2	-	2	0.0
Gulf of Finland	6	-	6	0.0	Yukon Delta	56	-	56	0.0
N Fennoscandia	4	-	4	0.0	SW Alaska	2	-	2	0.0
Kanin Penin.-Yamal	2	1	3	33.3	S & SE Alaska	24	-	24	0.0
Anabar-Kolymsk	2	-	2	0.0	west coast Hudson Bay	50	3	53	5.7
Gizhiga	38	1	39	2.6	James Bay	20	-	20	0.0
Sakhalin Island	6	-	6	0.0	Foxe Basin	28	-	28	0.0
					total	*446*	*23*	*469*	*4.9*

of Kanin-Yamal, Gizhiga, Anadyr, Chukotsk, N Alaska, the Seward Peninsula and the west coast of the Hudson Bay (Table 65). They were absent in the samples from the other areas. Since these data are discrete, for predictive purposes large sample sizes are needed.

PREDICTION OF BREEDING ORIGIN

In line with the results presented above 10 samples per sex were constructed:

1. E & NE Greenland (locality 1, Table 57);
2. W Europe & Iceland (localities 3 to 7, Table 57);
3. N Fennoscandia - Yamal (localities 9 and 10, Table 57);
4. Taymyr - Kolymsk (localities 11 and 12, Table 57);
5. Gizhiga (locality 13, Table 57);
6. Sakhalin Island (locality 14, Table 57);
7. NE Siberia (localities 15 to 19, Table 57);
8. N Alaska (locality 25, Table 57);
9. W & S Alaska (localities 22 to 24, Table 57);
10. N Canada (localities 26 to 28, Table 57).

In a next step all available measurements of breeding birds were used to calculate the degree of resemblance (typicality probabilities) with the constructed POSCON sample groups. By doing so, both the accuracy and exclusivity of prediction were estimated (see introduction under 'Statistical analysis of the data'). Three different designs were used, which varied in respect of the number of variables included: (1) wing-, culmen-, tarsus-, secondary-, tail- and middle toe lengths; (2) wing-, culmen-, and tarsus lengths; (3) wing- and culmen lengths.

The analyses could only include the small samples of Sakhalin Island when equal covariance matrices were assumed. These predictions evidently showed the distinctive character of the Sakhalin Island birds: they were always predicted to resemble their own group. Some W and N European Dunlin were also predicted as Dunlin from Sakhalin Island: resp. 7.5 and 3.2 % of the samples.

Since the assumption of equal covariance matrices may not hold true, the Sakhalin Island sample was further left out of consideration. The prediction results based on the remaining 9 samples per sex are presented in the tables 66 and 67.

Accuracy

The sample groups of E & NE Greenland, Taymyr-Kolymsk and N Canada had accuracy rates higher than 70 % under the best possible conditions (sex known; all measurements included - Table 66). Two of these samples (E & NE Greenland and N Canada) also have rather extreme measurements. However, the same is

Table 66. *Accuracy of prediction in Dunlin (sex* ***known*** *a priori). These predictions have been based upon 3 different analyses, including resp. (1) all 6 measurements; (2) wing-, culmen- and tarsus length; and (3) wing- and culmen length. The columns show the predicted proportions in relation to the total number per breeding area (last column).*

	Predicted									
Locality	E & NE Greenland	W Europe & Iceland	N Fennoscandia - Yamal	Taymyr - Kolymsk	Gizhiga	NE Siberia	N Alaska	W & S Alaska	N Canada	*n*
			All measurements							
E & NE Greenland	93.3	-	6.7	-	-	-	-	-	-	15
W Europe & Iceland	19.4	59.7	15.7	4.5	0.7	-	-	-	-	134
N Fennoscandia - Yamal	9.7	16.1	54.8	12.9	6.5	-	-	-	-	31
Taymyr - Kolymsk	-	-	15.8	78.9	-	-	5.3	-	-	19
Gizhiga	-	-	5.1	7.7	66.7	5.1	12.8	-	2.6	39
NE Siberia	-	2.4	11.9	4.8	19.0	31.0	16.7	7.1	7.1	42
N Alaska	-	-	5.5	3.6	14.5	12.7	47.3	7.3	9.1	55
W & S Alaska	-	-	-	-	7.5	1.5	11.9	61.2	17.9	67
N Canada	-	-	-	-	5.8	4.7	-	16.3	73.3	86
			Wing-, culmen- and tarsus length							
E & NE Greenland	90.0	10.0	-	-	-	-	-	-	-	50
W Europe & Iceland	17.1	64.9	13.5	3.2	0.5	0.5	0.5	-	-	222
N Fennoscandia - Yamal	2.6	15.4	57.7	11.5	7.7	3.8	1.3	-	-	78
Taymyr - Kolymsk	-	3.4	34.5	37.9	6.9	10.3	-	1.7	5.2	58
Gizhiga	-	-	7.7	5.1	25.6	23.1	25.6	7.7	5.1	39
NE Siberia	-	1.1	6.4	7.4	23.4	22.3	27.7	6.4	5.3	94
N Alaska	-	-	4.1	1.4	9.6	15.1	60.3	5.5	4.1	73
W & S Alaska	-	-	1.4	-	4.2	4.2	15.3	55.6	19.4	72
N Canada	-	-	1.9	1.9	12.6	5.8	10.7	20.4	46.6	103
			Wing-, and culmen length							
E & NE Greenland	76.9	22.1	1.0	-	-	-	-	-	-	104
W Europe & Iceland	16.7	61.9	19.2	0.4	0.4	1.3	-	-	-	239
N Fennoscandia - Yamal	1.3	15.2	60.8	3.8	7.6	8.9	1.3	-	1.3	79
Taymyr - Kolymsk	-	6.8	44.1	18.6	11.9	10.2	-	1.7	6.8	59
Gizhiga	-	-	7.7	2.6	17.9	25.6	23.1	5.1	17.9	39
NE Siberia	-	1.0	12.5	3.1	12.5	28.1	21.9	9.4	11.5	96
N Alaska	-	-	2.7	1.4	5.5	26.0	52.1	5.5	6.8	73
W & S Alaska	-	-	-	-	-	5.6	13.9	58.3	22.2	72
N Canada	-	-	1.0	2.9	4.8	6.7	9.6	28.8	46.2	104

true for the W & S Alaska population and the sample of this population stuck at an accuracy rate of 61 %.
When less measurements were included the accuracy rates of the samples from Greenland, Taymyr-Kolymsk, Gizhiga and N Canada decreased much, whereas the rates of the other samples remained more or less the same.

These prediction results clearly reveal the degree of resemblance between samples. For instance, the breeding populations of Gizhiga, NE Siberia and N Alaska have slightly different measurements: the prediction results correspondingly show most confusion between these samples.

Comparison of Table 67 with Table 66 learns that the prediction results become generally worse when sexes are unknown. When only wing- and culmen lengths are available and sexes are unknown, some samples have an accuracy rate lower than 30 %: N Fennoscandia - Yamal, Taymyr - Kolymsk, Gizhiga and NE

Table 67. *Accuracy of prediction in Dunlin (sex* **unknown** *a priori). These predictions have been based upon 3 different analyses, including resp. (1) all 6 measurements; (2) wing-, culmen- and tarsus length; and (3) wing- and culmen length. The columns show the predicted proportions in relation to the total number per breeding area (last column).*

	Predicted									
Locality	E & NE Greenland	W Europe & Iceland	N Fennoscandia - Yamal	Taymyr - Kolymsk	Gizhiga	NE Siberia	N Alaska	W & S Alaska	N Canada	*n*
			All measurements							
E & NE Greenland	80.0	6.7	-	13.3	-	-	-	-	-	15
W Europe & Iceland	19.4	58.2	10.5	7.5	3.7	0.8	-	-	-	134
N Fennoscandia - Yamal	16.1	25.8	38.7	6.5	9.7	3.2	-	-	-	31
Taymyr - Kolymsk	10.5	15.8	-	68.4	-	5.3	-	-	-	19
Gizhiga	2.6	5.1	12.8	12.8	48.7	5.1	7.7	2.6	2.6	39
NE Siberia	-	4.8	9.5	14.3	11.9	31.0	11.9	9.5	7.1	42
N Alaska	1.8	-	1.8	1.8	14.6	18.2	45.5	7.3	9.1	55
W & S Alaska	-	1.5	-	4.5	6.0	6.0	11.9	58.2	11.9	67
N Canada	-	-	-	1.2	2.3	4.7	1.2	17.4	73.3	86
			Wing-, culmen- and tarsus length							
E & NE Greenland	82.0	6.0	6.0	-	4.0	2.0	-	-	-	50
W Europe & Iceland	22.5	52.7	12.2	5.0	3.6	2.7	1.4	-	-	222
N Fennoscandia - Yamal	16.7	25.6	24.4	7.7	9.0	6.4	3.9	-	-	78
Taymyr - Kolymsk	-	22.4	25.9	25.9	3.5	8.6	5.2	1.7	6.9	58
Gizhiga	2.6	5.1	7.7	10.3	20.5	18.0	25.6	7.7	2.6	39
NE Siberia	3.2	3.2	8.5	12.8	12.8	19.2	28.7	8.5	3.2	94
N Alaska	-	1.4	2.7	2.7	8.2	11.0	63.0	6.9	4.1	73
W & S Alaska	-	1.4	1.4	4.2	11.1	4.2	20.8	38.9	18.1	72
N Canada	-	1.0	2.9	14.6	16.5	11.7	7.8	15.5	30.1	103
			Wing-, and culmen length							
E & NE Greenland	66.4	21.1	6.7	-	1.9	3.8	-	-	-	104
W Europe & Iceland	25.1	52.3	12.6	2.5	2.9	4.6	-	-	-	239
N Fennoscandia - Yamal	15.2	26.6	27.9	2.5	8.9	13.9	2.5	-	2.5	79
Taymyr - Kolymsk	8.5	30.5	23.7	11.9	3.4	6.8	1.7	3.4	10.2	59
Gizhiga	-	7.7	10.3	5.1	12.8	25.6	23.1	5.1	10.3	39
NE Siberia	3.1	4.2	11.5	9.4	4.2	27.1	22.9	11.5	5.2	96
N Alaska	-	-	5.5	2.7	1.4	20.6	53.4	11.0	5.5	73
W & S Alaska	-	1.4	1.4	5.6	5.6	8.3	19.4	33.3	25.0	72
N Canada	-	1.0	3.9	8.6	1.0	11.5	9.6	18.3	46.2	104

Siberia. But comparison with the results in Table 66 demonstrated that an average improvement of 26 % can be reached with knowing a bird's sex and measuring 6 in stead of 2 morphometric characters.

Exclusivity

It was impossible to reach exclusive predictions when birds of unknown sex were analysed. With known sex only 2 % of the birds was predicted to resemble 1 POSCON group exclusively. This did not improve when more variables were included. A higher degree of exclusivity could only be reached by combining POSCON groups: (1) Greenland with W Europe; (2) N Fennoscandia-Yamal with Taymyr-Kolymsk; (3) NE Siberia and Gizhiga with N Alaska; (4) W & S Alaska; (5) N Canada. Then, the exclusivity increased to 6 % in the analyses with 5 measurements and to 20 % in the analyses with 2 measurements.

CONTACT ZONES

The contact zones between different populations could be studied by calculating typicality probabilities. We analysed 3 regions: (1) Scandinavia, (2) N Russia and W Siberia, and (3) E Siberia.

(1) *Scandinavia* (Fig. 74). The typicality probabilities were calculated with use of 5 measurements, but if these 5 were not all available, only wing-, and culmen lengths were used. The distribution of the W European type of Dunlin in Scandinavia corresponds with the present-day views, although the absence along the S Finnish coast was unexpected. Most breeders from the Baltic Islands resembled the W European type closely. The distribution of the N Fennoscandian - Yamal type reached to coastal W Norway, Finland, and Karelia. There were no data from the alpine regions in Scandinavia and from the coast of W Finland. Both types were predicted to occur on Åland.

(2) *N Russia and W Siberia* (Fig. 75). On the peninsulas of Yamal and Taymyr, 2 morphometrically different populations are in contact with each other: (1) N Fennoscandia-Yamal (here called 'N Europe' sample), and (2) Taymyr-Kolymsk (here called 'Central Siberian' sample). The character of the contact zone was studied with POSCON-analysis. The calculations were based upon the inclusion of 3 that

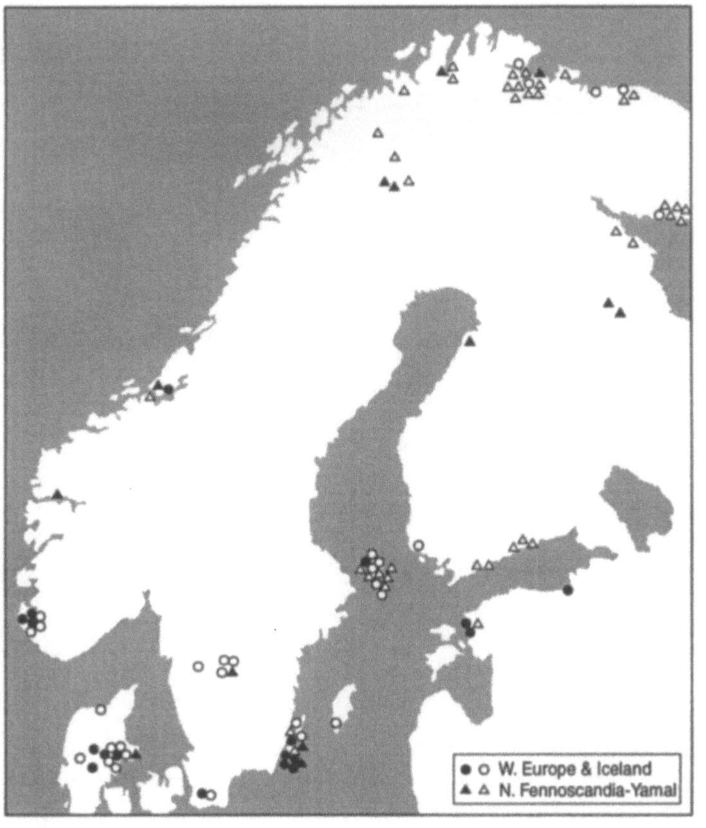

Fig. 74. *The predicted distribution of 2 groups of Dunlin in Fennoscandia: (1) W Europe & Iceland; (2) N Fennoscandia - Yamal. The predictions were based upon the calculated typicality probabilities for these 2 groups. Circles indicate the occurrence of W European Dunlin; a closed circle indicates that the typicality probability for N Fennoscandian Dunlin was < 0.05, and an open circle > 0.05. Squares indicate the occurrence of N Fennoscandian Dunlin; a closed square indicates that the typicality probability for W European Dunlin was < 0.05, and an open square > 0.05.*

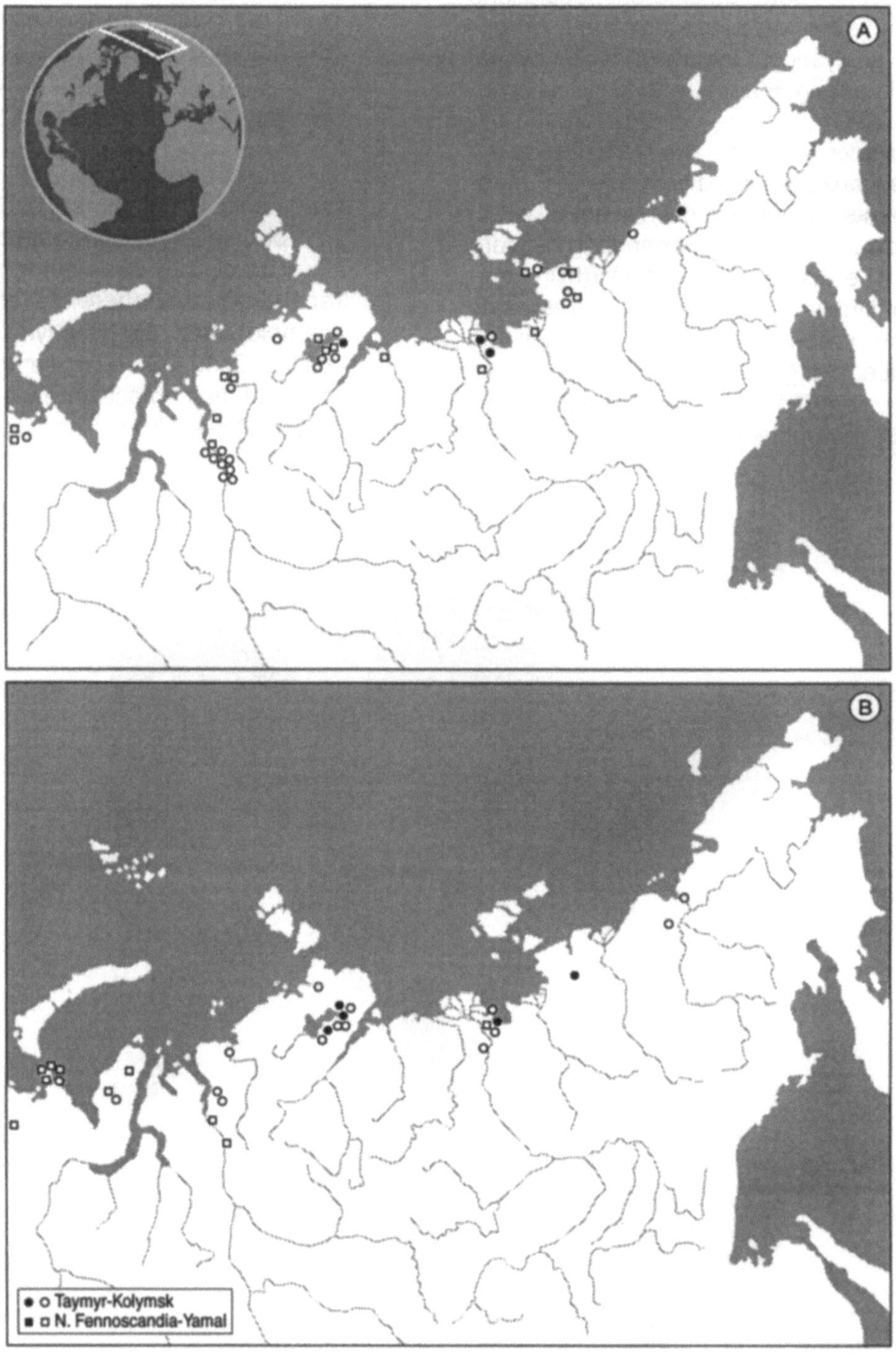

Fig. 75. *The predicted distribution of 2 groups of Dunlin in Siberia: (1) N Fennoscandia - Yamal (squares); and (2) Taymyr - Kolymsk (circles). 'A' refers to ♂♂ and 'B' to ♀♀. Predictions have been based upon the calculated typicality probabilities. A closed square indicates that the typicality probability for Taymyr - Kolymsk was < 0.05, an open square > 0.05. A closed circle indicates that the typicality probability for N Fennoscandia - Yamal was < 0.05, an open circle > 0.05.*

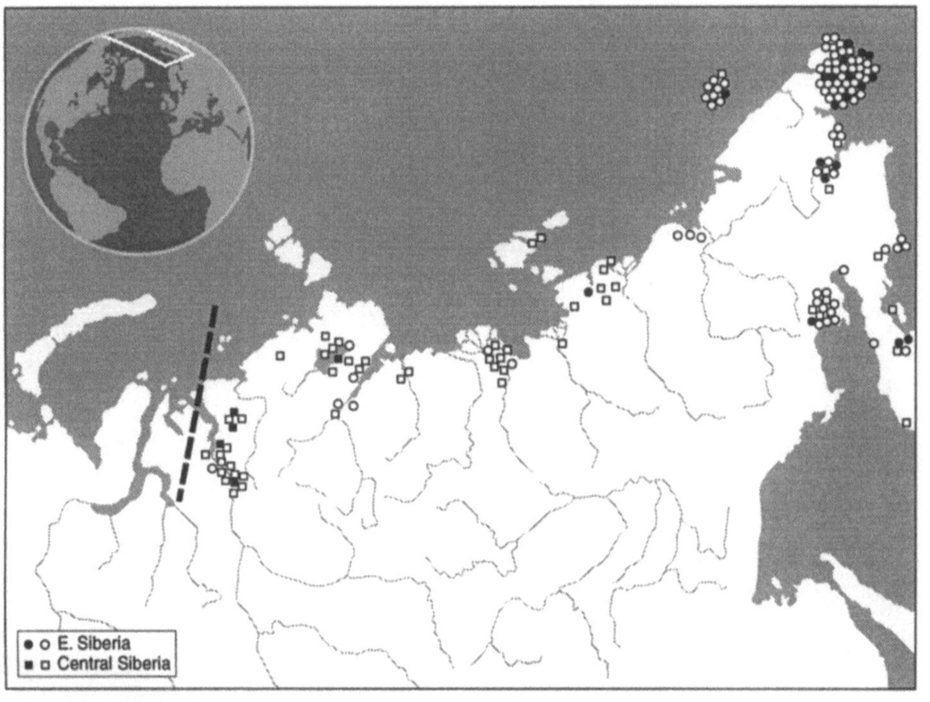

Fig. 76. *The predicted distribution of 2 groups of Dunlin in Central and E Siberia. Predictions have been based upon the calculated typicality probabilities of 3 sample groups (details in text). Circles indicate that a bird resembled mostly the Central Siberian population: a closed circle means that the typicality probability for E Siberian Dunlin was < 0.05, and an open circle > 0.05. Squares indicate that a bird resembled mostly the E Siberian population: a closed square means that the typicality probability for Central Siberian Dunlin was < 0.05, and with an open square > 0.05. These probabilities were based upon 5 or 2 measurements and thus sample sizes differ from an analysis with 3 measurements (previous figure).*

measurements since 5 measurements were taken from too few birds. There was a clear segregation between both ♀ samples. Central Siberian ♀♀ mostly occurred from Taymyr eastwards and N European ♀♀ from the Yenisei westwards. However, several Yenisei breeding ♂♂ had highest resemblance with the Central Siberian sample and several Taymyr breeding ♂♂ with the N European sample. This discrepancy between ♂♂ and ♀♀ does not seem to be caused by wrong predictions: Central Siberian breeding birds moult primaries in July and this also happened with the 5 predicted Central Siberian ones collected west of 85°. Three of the 4 Dunlins predicted to be N European, and collected east of 85°E during July were not moulting when all predicted Central Siberian ones were in moult. Therefore, some degree of intermixing between both populations is likely.

(3) *E Siberia*. Three geographical regions were included in the analyses: (1) Taymyr - Kolymsk (here called 'Central Siberia' population), (2) Gizhiga, and (3) NE Siberia. Since the morphometrical differences between the second and third sample were small, the highest typicality probability for any of both was taken as representative for the combined 'E Siberian population'. Typicality probabilities were calculated while using 5 measurements. If not all 5 were available, only wing-, and culmen lengths were used.

The Central Siberian population contacts the E Siberian one between the Indigirka and the Kolyma River (Fig. 76), as nearly all birds breeding from the Indigirka west-

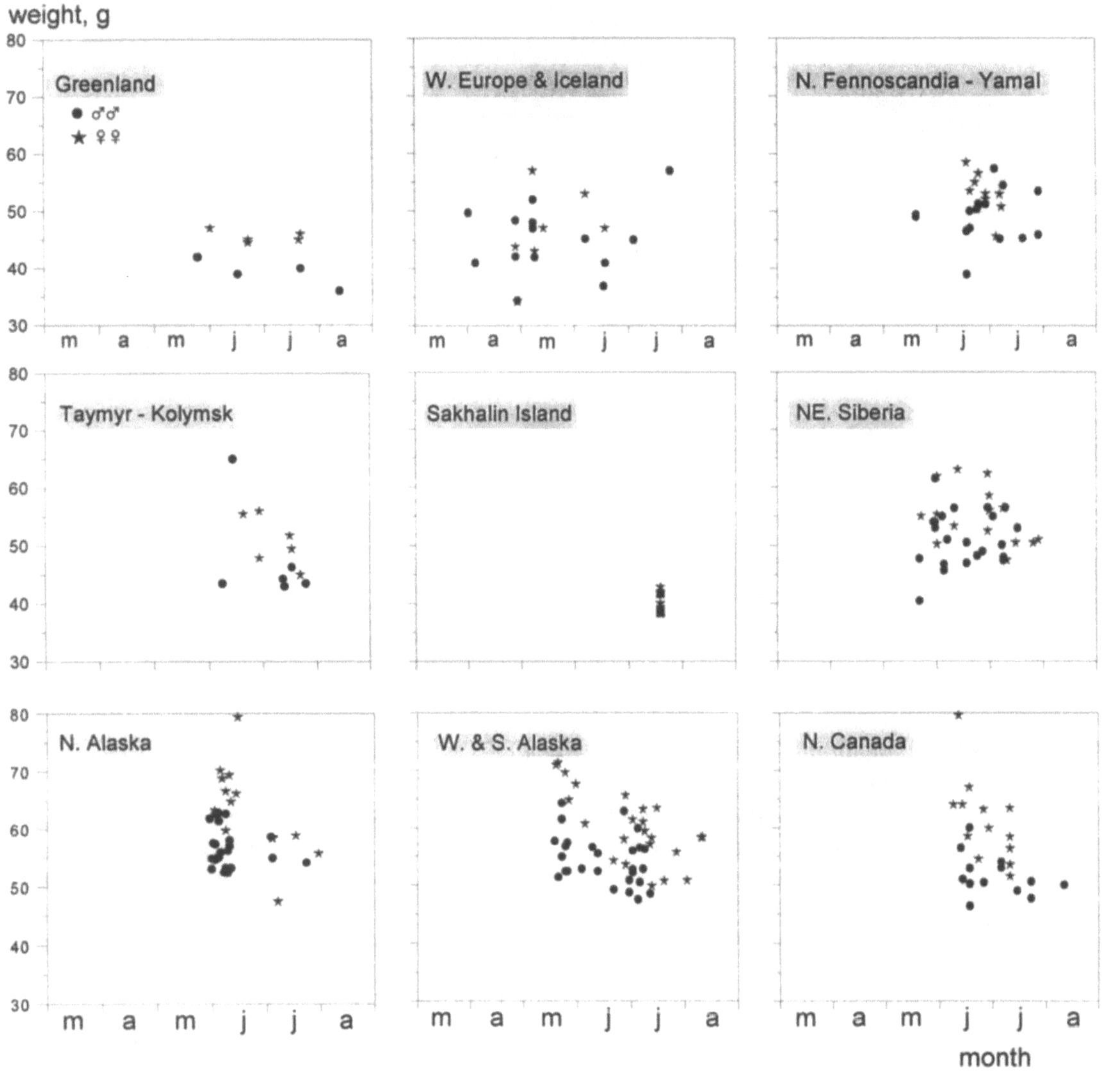

Fig. 77. *Body masses of Dunlin in 9 breeding populations. Circles represent ♂♂ and stars ♀♀.*

wards were predicted to resemble the Central Siberian population, and those from the Kolyma Delta eastwards were assigned to the E Siberian population.

BODY MASS

The average mass level increased from west to east with 0.3 gm per 10 degrees longitude (Table 68). Mass declined during June-July in N European ♂♂, Central Siberian ♀♀, and N Canadian ♂♂ and ♀♀ (Fig. 77). S Scandinavian ♂♂ showed a mass gain of 0.9 g/day prior to departure from the breeding grounds. This increase corresponded with data collected elsewhere (Zwarts *et al.* 1990). The mass reached by these ♂♂ in the second half of July allows a flight of about 2.300 - 2.700 km, according tothe formulae presented by Castro & Myers (1989), which is the distance between S Scandinavia and the Iberian Peninsula or S France. We had no other data on pre-migratory fattening on the breeding grounds.

Table 68. *Mean body mass of breeding Dunlin. Presented are the mean body masses ± 1 SD and the total number of birds involved.*

Locality	♂♂	♀♀
Greenland	39.3±2.5(4)	45.5±1.0(5)
W Europe	45.0±6.0(14)	46.4±7.3(7)
N Fennoscandia-Yamal	49.2±4.4(16)	52.3±4.3(10)
Taymyr-Kolymsk	47.6±8.6(6)	52.1±3.6(5)
Gizhiga	49.0 - (1)	-
Sakhalin Island	39.7±1.8(3)	41.6±1.4(3)
NE Siberia	51.2±4.7(22)	54.9±4.8(15)
N Alaska	56.8±3.4(23)	63.5±7.9(14)
W & S Alaska	54.5±4.4(27)	60.2±6.3(23)
N Canada	51.8±3.5(14)	61.1±7.3(13)

PRIMARY MOULT

The various breeding populations show clear differences in primary moult schedule. Analysis of variance showed -not unexpectedly- date and breeding region to be the key factors in explaining the variation in primary moult (F = 37.383; P = 0.000; df = 8). Most of the variance was explained by the breeding region (56.4 % of the explained variance with F = 47.433, P = 0.000, df = 8). The date, expressed as day number, explained 42 % of the variation (F = 282.307; P = 0.000; df = 1). A bird's sex did not play a role in explaining (part of) the variation in moult. The moulting schedules on the breeding grounds are described in more detail below.

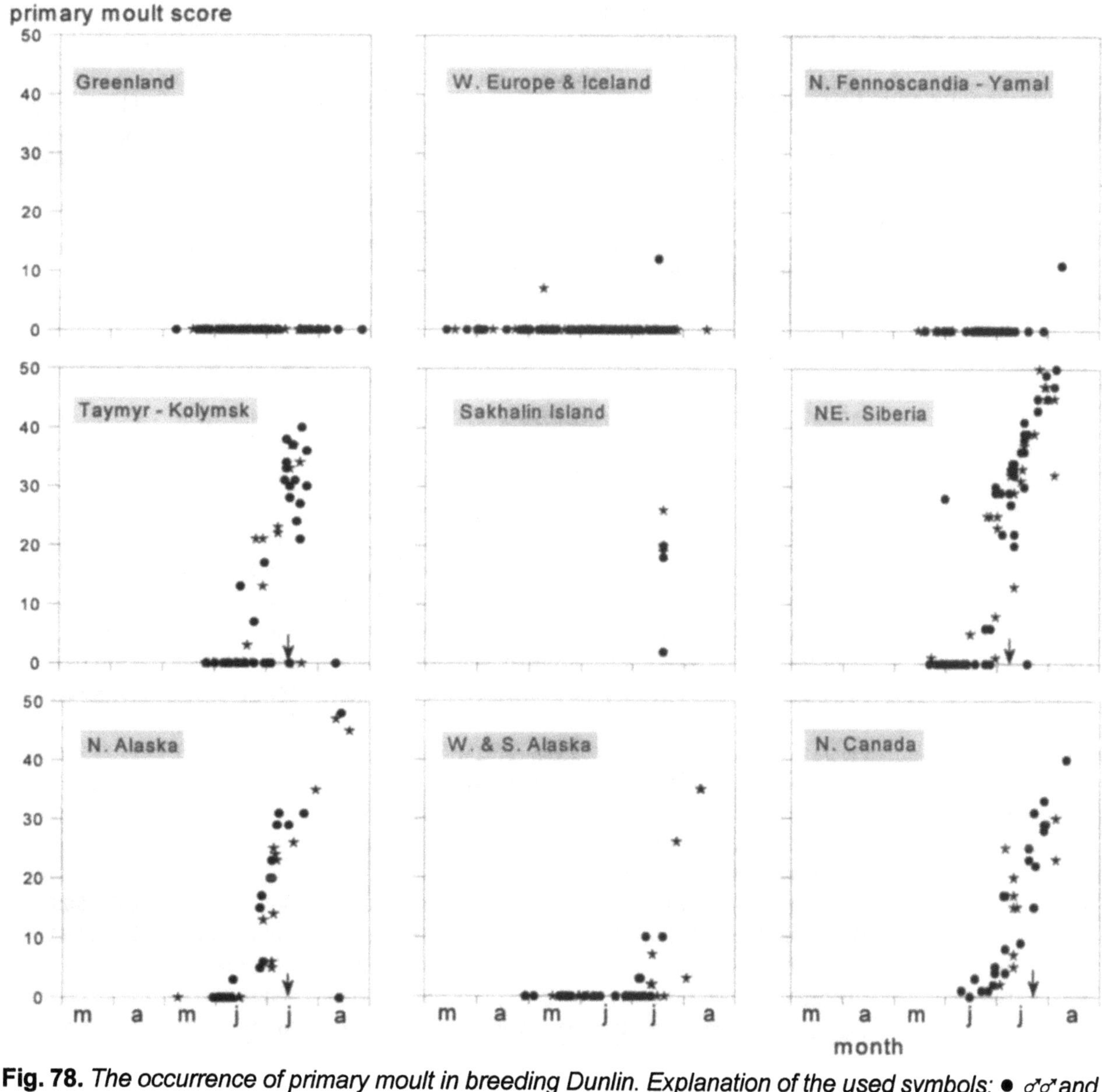

Fig. 78. *The occurrence of primary moult in breeding Dunlin. Explanation of the used symbols:* ● *♂♂ and* ★ *♀♀. The arrows represent the date at which an average moult score 25 was reached.*

The sampled breeding birds from Greenland, W Europe and N Europe did not show primary moult. The other Palearctic and Nearctic samples included birds in active primary moult during breeding (Fig. 78), although some Taymyr breeding birds (27.6 % of the ones studied; *n* = 29) were not yet in primary moult in July. For moult in Central Siberian birds, see the section on contact zones. Birds from populations, which moult primaries on the breeding grounds, had very worn primaries at arrival, and primary moult already started soon. Birds from N Canada and Central Siberia had a slower moulting rate than those of NE Siberia and N Alaska. This was expressed by calculating the average day number, at which primary moult score 25 was reached (arrows in Fig. 78). No birds on the breeding grounds of N Canada and Central Siberia had all primaries renewed: scores of over 40 were not recorded. According to Holmes (1966, 1971) not all breeding birds from N Alaska finish their primary moult on the breeding grounds.

DISCUSSION

The separation of N European and Central Siberian populations

The populations breeding between N Fennoscandia and the Kolyma Delta are evidently polymorphic. The N Fennoscandia - Yamal population differed from both the Taymyr- and the Anabar-Kolymsk population, whilst both latter were nearly always indistinguishable in the morphometric analyses (tables 63 & 64). We conclude on the recognition of 2 taxa in this region: the western '*alpina*' type is characterized by (1) smaller dimensions (Table 69), (2) lack of 'adult buff' wing coverts, (3) narrow white edges along the outer vanes of the primaries and (4) the absence of primary moult on the breeding grounds; the eastern '*centralis*' type is larger, has 'adult buff' coverts, more white on the vanes and starts primary moult during breeding. This difference is supported by mtDNA research (Wenink *et al.* 1996). These are good reasons to

Table 69. *Summary of the measurements of 10 taxa in the Dunlin. This summary is based upon the data presented in Table 57. Presented are the means ± 1 SD (n).*

Locality	Wing	Culmen	Tarsus	Secondary	Tail	Middle toe
			♂♂			
Greenland	113.2±2.2(63)	26.9±1.0(61)	22.5±0.6(26)	51.0±1.5(12)	44.4±2.2(15)	21.1±0.7(12)
W Europe	113.7±2.7(140)	28.8±1.6(140)	23.3±0.9(129)	52.1±1.5(91)	43.8±2.3(88)	22.5±1.0(83)
N Fennoscandia-Yamal	117.3±2.3(39)	31.4±1.7(39)	23.9±0.9(39)	54.0±2.3(20)	45.9±2.0(20)	22.9±1.1(20)
Taymyr-Kolymsk	117.5±1.9(39)	31.7±1.6(40)	24.8±1.0(38)	54.1±1.2(13)	45.5±3.0(14)	22.5±1.3(14)
Gizhiga	121.9±2.6(26)	33.8±1.7(26)	25.5±0.8(26)	54.5±1.3(26)	46.2±2.9(26)	24.3±0.9(26)
NE Siberia	121.6±2.7(59)	33.5±1.8(60)	25.5±1.0(59)	57.0±2.5(24)	48.6±3.9(28)	24.5±1.1(24)
N Alaska	124.2±2.7(46)	34.1±2.2(47)	25.5±1.2(47)	56.6±1.7(43)	49.6±3.4(46)	23.9±1.0(36)
Sakhalin Island	109.0±1.0(3)	28.3±0.4(3)	23.7±1.0(3)	53.3±2.1(3)	48.3±5.1(3)	22.9±0.8(3)
W & S Alaska	123.1±2.7(39)	37.3±1.5(39)	25.8±0.8(39)	57.4±1.9(38)	49.8±2.4(38)	24.4±1.0(36)
N Canada	121.5±2.6(61)	36.8±1.3(61)	26.3±0.9(61)	58.3±1.3(59)	47.1±2.3(61)	25.1±0.9(52)
			♀♀			
Greenland	117.7±2.5(44)	30.1±1.2(43)	23.5±0.9(24)	53.8±2.7(7)	45.1±3.2(10)	21.7±0.6(10)
W Europe	117.6±2.5(101)	32.9±1.5(99)	24.6±1.0(94)	53.7±1.5(67)	44.4±2.6(65)	23.5±1.1(56)
N Fennoscandia-Yamal	120.3±2.4(36)	34.3±1.9(35)	25.0±0.8(36)	55.5±1.4(11)	48.0±2.0(11)	23.6±0.8(11)
Taymyr-Kolymsk	120.8±2.2(25)	36.3±1.6(25)	26.0±1.1(25)	55.5±1.8(6)	46.3±1.9(6)	23.8±1.1(6)
Gizhiga	123.6±2.4(13)	38.2±1.9(13)	26.7±1.1(13)	56.2±1.5(13)	46.6±2.2(13)	25.1±1.0(13)
NE Siberia	124.4±2.8(37)	37.7±2.6(37)	26.6±1.0(36)	57.7±1.8(22)	49.7±3.3(22)	24.7±0.9(24)
N Alaska	125.3±3.6(27)	37.4±1.9(29)	26.3±0.9(29)	58.0±2.2(27)	49.1±2.9(27)	25.0±0.8(24)
Sakhalin Island	113.4±0.6(3)	31.0±1.3(3)	24.5±0.2(3)	55.7±1.2(3)	47.0±3.6(3)	23.9±1.0(3)
W & S Alaska	127.2±2.7(33)	40.5±1.5(33)	26.8±0.8(33)	59.5±1.7(39)	50.5±2.6(31)	25.2±1.0(31)
N Canada	122.9±2.7(43)	39.4±1.7(43)	26.9±0.9(42)	59.1±1.5(42)	46.8±2.7(43)	25.2±1.0(37)

distinguish both as separate subspecies. As shown earlier , some degree of intermixing may occur between both.

Subspeciation in the N Pacific

The breeding populations of Sakhalin Island and W & S Alaska clearly differed from the breeding populations of Gizhiga, NE Siberia and N Alaska. The Sakhalin Island birds had very small measurements for the E Palearctic as was shown earlier by Nechaev & Tomkovich (1987 & 1988). The breeding birds of W & S Alaska had characteristically long culmen- and tail lengths. It is more difficult to recognize the current subspecies *kistchinski*, *sakhalina* and *arcticola* from our morphometric material. Wenink (1994) showed that the mtDNA patterns of *sakhalina* (Anadyr) and *arcticola* (Barrow) were evidently different. Also Tomkovich (1986) claimed *kistchinski* to breed on the southwest part of the Koryak Highland, the east coast of the Okhotsk Sea, Kamchatka and the northern Kuril Islands. Our morphometric data did not reveal important morphometric differences between these populations (Gizhiga, N- and S Kamchatka) and the *sakhalina*-type populations from Anadyr, Chukotsk and Wrangel (Table 59, Fig. 72).There was only 1 exception on this rule: a MANOVA-analysis with all 6 variables including the ♂♂ of the different populations (Table 58). Other than morphometric arguments have thus to be used for the recognition of these subspecies. Tomkovich (1986) and Browning (1991) also described these subspecies mainly in terms of plumage characters (details in chapter on the taxonomy of the Dunlin). Here we adopt their findings with the addition that our morphometric data could not be a basis for the distinction between *sakhalina* and *kistchinski*.

In future analyses concerning the subspeciation in the N Pacific, it is necessary to obtain ordinal in stead of categorical data on plumage characters. The plumage characters used at present are too qualitative and generalized to allow comparison of birds in different collections. They also largely ignore the existence of variation in colour patterns within a population. An approach as was used in the Golden Plover (Jukema & Piersma 1987, Jukema 1989) might serve as an example for future work.

Wenink (1994) also showed the large degree of resemblance between the W- and N Alaskan population, which is different from all morphometric findings including ours. His W Alaskan sample originated from Chevak in the Yukon Delta: our Yukon sample (*pacifica*) evidently differed from the N Alaska sample (*arcticola*): Fig. 72, tables 63 & 64.

The taxonomy of the Dunlin

Ten subspecies are recognized. Their measurements are presented in Table 69. They are described below:

- *C. a. arctica* (Schiøler 1922) is breeding on E and NE Greenland. The type locality is the east coast of Greenland. *Arctica* is most similar to *schinzii* and *actites*. Differentiating characters are both measurements and plumage characters (Ferns & Green 1979, this study). It is the smallest subspecies studied, although it has on average longer wings than *actites* (Table 69). Their wing- and tail lengths are comparable with *schinzii*.

- *C. a. schinzii* (Brehm 1822) is breeding on Iceland, the Færoes, the British isles, W Europe, the Baltic, S Scandinavia and (occasionally) S and SE Greenland. The type locality is Rügen Island, Germany. *Schinzii* is most similar to *arctica* and *actites*. It has longer culmen lengths than *arctica* and longer wings and shorter tails than *actites*. It is smaller-sized than *alpina*. There are important differences in breeding plumage between *schinzii*, *arctica* and *alpina* (Ferns & Green 1979, Cramp & Simmons 1983).

- *C. a. alpina* (Linnaeus 1758) is breeding from N Fennoscandia east to the Yenisei Delta. The type locality is Lappland. *Alpina* is most similar to *centralis*. It intergrades with *centralis* E of the Yenisei. However, *alpina* and *centralis* are separated by 4 characters: (1) *alpina* is not moulting primaries on the breeding grounds as *centralis* does; (2) *alpina* does not have 'adult buff' upper wing coverts; (3) *alpina* has small white edges along the outer vanes of the primaries; and (4) *alpina* is on average smaller than *centralis*. *Alpina* has on average larger measurements than *arctica* and *schinzii* and differs from *actites* by its longer wing- and culmen lengths. *Arctica* and *schinzii* are neither moulting their primaries on the breeding grounds. There are also important differences in breeding plumage with *arctica* and *schinzii* (Ferns & Green 1979, Cramp & Simmons 1983).

- *C. a. centralis* (Buturlin 1932) is breeding in Central Siberia from the Taymyr Peninsula east to the Indigirka River. The type locality is Yakutsk along the Lena River. It is most resembling *alpina* and is intergrading with this subspecies in the west part of its range. The separating characters with *alpina* are given under the description of *alpina*. It has generally smaller measurements than *sakhalina* or *kistchinski* and longer wing-, culmen- and tarsus lengths than *actites* (Table 69). *Centralis*, *sakhalina* and *kistchinski* have in common that they start their primary moult on the breeding grounds.

- *C. a. actites* (Nechaev & Tomkovich 1988) is breeding on N Sakhalin Island. The type locality is the shore of the Chayvo Bay, Sakhalin Island. *Actites* has the shortest wing lengths of all the populations studied (Table 69). It has for the E Palearctic characteristically small measurements. However, these values are still larger than in *arctica* and (mostly) in *schinzii*. Its plumage characters are described by Nechaev & Tomkovich (1987 & 1988) and Browning (1991).

- *C. a. kistchinski* (Tomkovich 1986) is breeding (according to Tomkovich 1986) on the southwest part of the Koryak Highland, the northeast coast of the Ochotsk Sea, the Kamchatka Peninsula and the northern Kuril Islands. The type locality is the Moroshechnaya River on the west part of the Kamchatka Peninsula. *Kistchinski* is most similar to *sakhalina* and their measurements do nearly not differ. *Kistchinski* is (according to Browning 1991) less black above and larger than *actites* and is more black dorsally with the edges of the back feathers more yellow than in *arcticola* and *sakhalina*.

- *C. a. sakhalina* (Vieillot 1816) is breeding on Wrangel Island, the Chukotsk Peninsula and at Anadyr. The type locality of *sakhalina* is the headwaters of the Kanchalan River (75 km west of Egvekinot), Chukotsk Autonomous Region (Tomkovich 1986 after Browning 1991). *Sakhalina* is mostly resembling *arcticola* and *kistchinski*. It is intergrading with *arcticola* and *pacifica* in the extreme east part of the Chukotsk Peninsula (Tomkovich 1986). The subspecies distinction is primarily based on plumage characters. Browning (1991) produced the following description: *sakhalina* differs from *kistchinski* by 'its darker (less golden) rusty back and reduced amount of black on the crown and upper breast' and *sakhalina* differs from *arcticola* (among some other characters) by 'the slightly less reddish and less dark rust colour dorsally'.

- *C. a. arcticola* (Todd 1953) is breeding north of the Brooks Range in Alaska. Its type locality is Point Barrow, Alaska. It has much morphometric resemblance with both E Siberian subspecies *sakhalina* and *kistchinski*. It is the longest-winged subspecies of the three. It differs from *kistchinski* by its more reddish back and fewer black feathers on the back and crown and from *sakhalina* by its greater weight, greater amount of white on the leading edge of the eighth primary and slightly more reddish and darker rust colour dorsally (Browning 1991).

- *C. a. pacifica* (Coues 1861) is breeding in Alaska south of 65°N. The type locality is Simiahmoo, Washington State, which is outside the breeding range. It is the largest subspecies studied here and has most resemblance with *hudsonia* and *arcticola*. It has generally larger measurements than *arcticola*. It has longer wing-, culmen- and tail lengths and shorter tarsus lengths than *hudsonia* (Table 69). Plumage characters of this subspecies are described by Browning (1977).

- *C. a. hudsonia* (Todd 1953) is breeding in the NW Territories of Canada. The type locality is Southampton Island, Hudson Bay. *Hudsonia* has most resemblance with the large-sized subspecies of the N Pacific. It has shorter wing-, culmen- and tail lengths than *pacifica*, but somewhat longer tarsus lengths (Table 69). It has shorter wing- and tail lengths and longer culmen-, tarsus-, secondary-, and middle toe lengths than *arcticola*.

Comparison with mtDNA-patterns

Our results differed in 3 major aspects from the mtDNA studies described in Wenink *et al.* (1993) and Wenink (1994) (see Fig. 79):

(1) The haplotypes as found in the DNA studies did not reveal differentiation between *arctica*, *schinzii* and *alpina* in contrast to - morphometric and plumage studies (Ferns & Green 1979, Greenwood 1986, Meininger & Van Swelm 1989, this study). It is beyond doubt that both *arctica* and *schinzii* can be recognized as subspecies, since both are clearly different from other populations, and from each other. The separate position is clearly illustrated in Fig. 72. The small size of *arctica* and *schinzii* might be the result of a relatively recent process in contrast to the separation of the *alpina* haplotype. If so, this process of subspeciation might be explained as due to character displacement. An ecological process may select in favour of small individuals.

(2) The mtDNA studies showed *hudsonia* to be the oldest subspecies at present existing (Wenink *et al.* 1993). In morphometrics, the N Canadian population is well distinguishable from others, but not as deviating as can be derived from the mtDNA studies.

(3) The N Alaskan population *arcticola* was morphometrically rather similar with *sakhalina* and *kistchinski*, while mtDNA studies showed a clear pattern of segregation between the Dunlin populations on both sides of the Bering Strait.

(4) Wenink (1994) showed a large degree of similarity in observed haplotypes between the N- and W Alaskan populations. In contrast, our morphometric study supported earlier findings of Browning (1977), which showed clear differences between *arcticola* and *pacifica*.

SUMMARY

Measurements and plumage characters of all breeding populations of Dunlin were analysed with multivariate statistics in order to describe the geographical variation in this species. The species showed much geographical and sexual

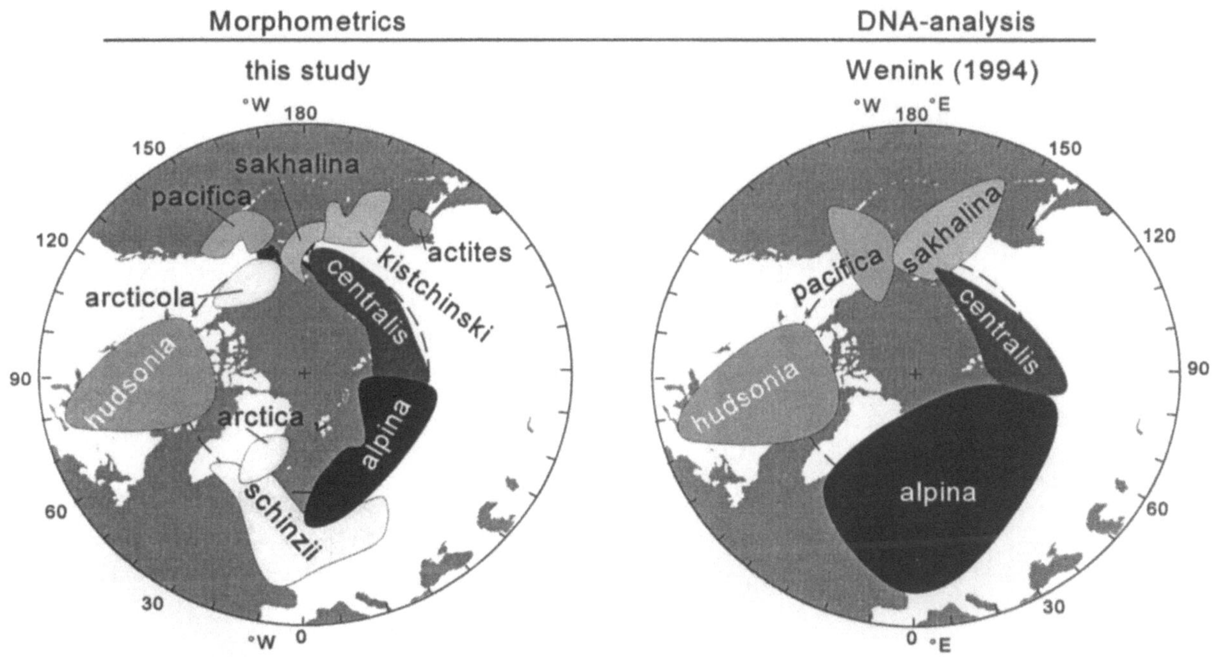

Fig. 79. *Comparison of the taxonomy of Dunlin resulting from morphometric research (this study) with the typology resulting from mtDNA sequence analysis (Wenink 1994).*

variation. The degree of subspeciation was high, and 10 subspecies were recognized: *arctica*, *schinzii*, *alpina*, *centralis*, *actites*, *kistchinski*, *sakhalina*, *arcticola*, *pacifica* and *hudsonia*. Most populations moulted primaries on the breeding grounds, though with differences in speed and timing. The atlantic populations of the Dunlin did not start primary moult on the breeding grounds. Birds from N Canada had broader wings than birds from other populations. The differences in taxonomical interpretation with earlier morphometric and mitochondrial DNA studies were discussed.

Chapter 12

Black-tailed Godwit

-

Limosa limosa

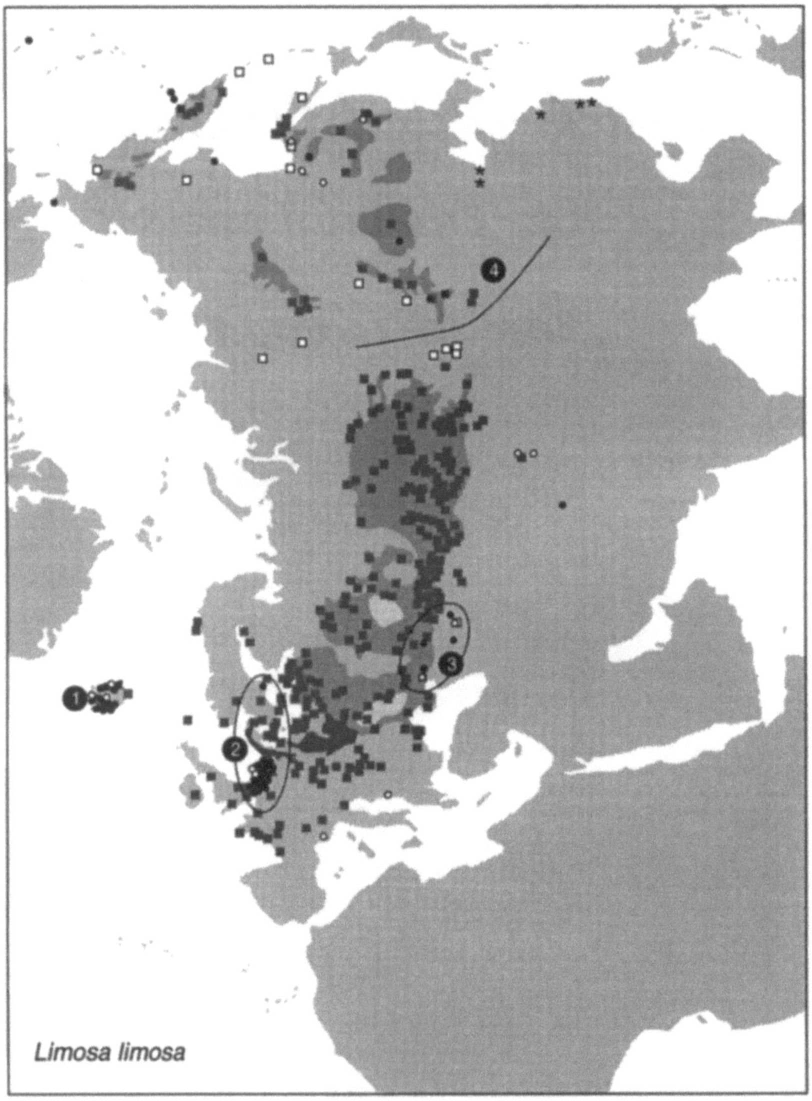

Fig. 80. *Breeding distribution of the Black-tailed Godwit. Explanation of the used symbols in chapter 3: 'Introduction to the species' accounts'. Some stars are used in the upper right part of this figure. These concern some non-breeding Godwits collected in coastal areas of E China during March, April and May.*

INTRODUCTION

The Black-tailed Godwit has a scattered breeding distribution in marshy and wet freshwater habitat in the temperate zone of the W Palearctic, from Iceland in the west to the Baraba and Kulunda steppes and the western limits of the Altai Mountains in the east (Cramp & Simmons 1983, Stepanyan 1990 - Fig. 80). It has a fragmented breeding distribution in the E Palearctic, where it is known to breed near the middle reaches of the Lena River, from Lake Baikal to Amur, Sakhalin Island, Kamchatka and Anadyr (Fig. 80).

Three distinct populations are recognized, based upon morphometrics and plumage characters (Vaurie 1965, Glutz *et al.* 1977, Hale 1980, Cramp & Simmons 1983, Stepanyan 1990):

- *L. l. limosa*, breeding from continental W Europe east to the Baraba and Kulunda steppes and the western Altai Mountains;
- *L. l. islandica*, breeding on Iceland, the Færoes and in NW Norway;
- *L. l. melanuroides*, breeding in the E Palearctic as far west as the upper Vilyuy River and Lake Baikal.

Prater *et al.* (1977), Hale (1980) and Cramp & Simmons (1983) noticed the existence of clinal variation in the nominate subspecies with increasing measurements from west to E. The species strongly resembles the Nearctic Hudsonian Godwit *Limosa haemastica*. Hale (1980) considered both to be separate species, since they had different wing patterns and a different habitat choice. Cramp & Simmons (1983) considered both as separate allospecies belonging to 1 superspecies, with the Hudsonian Godwit approaching the measurements of *L. l. islandica* but having different plumage characters.

MORPHOMETRICAL DIFFERENTIATION

In total, 176 Black-tailed Godwits from different breeding populations were examined (Table 70). About 80 % of these birds had been collected on Iceland and in the Netherlands. Males were more common (60.8 %) than ♀♀ in the samples. Most of the Iceland breeders had been collected during May and June, and the W European ones between half March and late August. The sample of E Palearctic birds is rather fragmented and therefore some non-breeders, collected in China and Alaska outside the regular breeding range were included in the statistical analyses.

Table 70. *Measurements of breeding populations of Black-tailed Godwits. The numbers in the first column refer to the numbers used in Fig. 70. Presented are the means ± 1 SD (n). [1]Including 7 non-breeding specimen from E China (stars in Fig. 70), 1 bird from Little Diomede Island (end of May) and 1 bird from Unalaska (Fox Islands, Aleutians - June).*

	Locality	Wing	Culmen	Tarsus	Secondary	Tail	Middle toe
				♂♂			
1	Iceland	210.3±4.9(22)	79.9±3.6(21)	65.9±3.4(22)	-	-	-
2	W Europe	213.6±5.5(63)	91.1±5.5(62)	72.6±5.2(61)	97.5±0.7(2)	74.5±3.2(8)	46.1±1.8(8)
3	E Europe & W Asia	219.6±6.1(7)	102.2±4.5(7)	78.2±1.6(7)	96.0 - (1)	74.6±3.2(7)	47.5±1.9(5)
4	E Palearctic[1]	199.4±7.3(15)	76.8±5.0(15)	62.0±3.6(15)	90.1±3.9(15)	66.1±5.1(15)	43.1±1.8(13)
				♀♀			
1	Iceland	223.6±5.1(9)	96.6±4.4(7)	74.0±3.3(9)	-	-	-
2	W Europe	225.1±5.3(42)	108.0±4.6(41)	80.3±4.4(39)	109.0 - (1)	79.0±5.0(3)	48.0±3.5(2)
3	E Europe & W Asia	229.9±5.4(10)	113.7±9.7(10)	83.0±6.7(10)	-	77.9±3.7(8)	48.9±2.9(7)
4	E Palearctic[1]	203.0±5.5(8)	92.7±9.6(8)	68.3±4.8(8)	94.0±3.3(8)	70.0±3.3(8)	44.2±2.2(8)

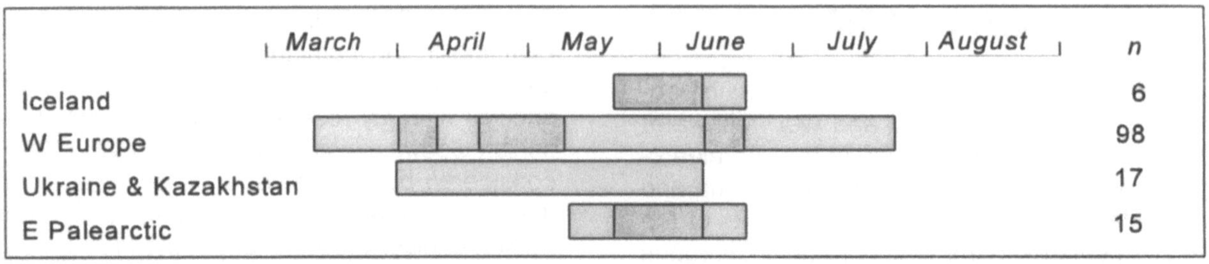

Fig. 81. *Collection dates of Black-tailed Godwits. Most birds had been trapped during the periods marked by the heavily shaded bars. The sample sizes are presented on the right side.*

Initially, 4 geographic regions were distinguished (Fig. 80, Table 70). The various locations of the E Palearctic *melanuroides* were combined per sex, due to scarcity of data. With these 4 regions and both sexes 8 samples were constructed. The analyses concerning the morphometric differences in this species were mostly based upon these samples. The measurements and results are presented in the tables 70-74 and in the figures 81 - 84.

There was a high degree of sexual and geographical variation in this species (Table 71). Sexual differences were larger than geographical differences. Sexual dimorphism was smallest in the sample from E Europe & SW Asia and largest in the Iceland sample (Table 72). Both sexes mostly differed in culmen lengths, and least in wing lengths.

When the sample sizes were reasonably large, the measurements were correlated to a large degree (Table 73). Lower significance of the correlations in ♀♀ was probably because of scarcity of data. Culmen lengths were most important to be measured when analysing sexual and geographical variation in this species, being followed by resp. wing- and tarsus length. Secondary-, tail- and middle toe lengths could not be included in the analyses due to scarcity of data.

Table 71. *Overall results of MANOVA for all distinguished breeding populations in the Black-tailed Godwit in combination with univariate ANOVA's in which the different measurements were analysed separately. Significance levels: *0.01 < P < 0.05; **0.001 < P < 0.01; ***P < 0.001.*

	Analysis I			Analysis II		
Variables	wing, culmen, tarsus			wing, culmen		
n	166			171		
Multivariate analysis of variance						
Effects	Interaction	Area	Sex	Interaction	Area	Sex
Pillai's trace	0.090	0.949	0.545	0.078	0.857	0.522
F-value	1.624	24.366***	62.332***	2.204*	40.760***	88.204***
Hypoth. & error *df*	9/ 474	9/ 474	3/ 156	6/ 326	6/ 326	2/ 162
Univariate F tests						
df	3/ 158	3/ 158	1/ 158	3/ 163	3/ 163	1/ 163
wing length	3.572*	72.545***	74.452***	3.474*	70.976***	73.246***
culmen length	1.356	77.605***	146.473***	1.510	79.850***	149.136***
tarsus length	0.470	51.390***	48.749***			

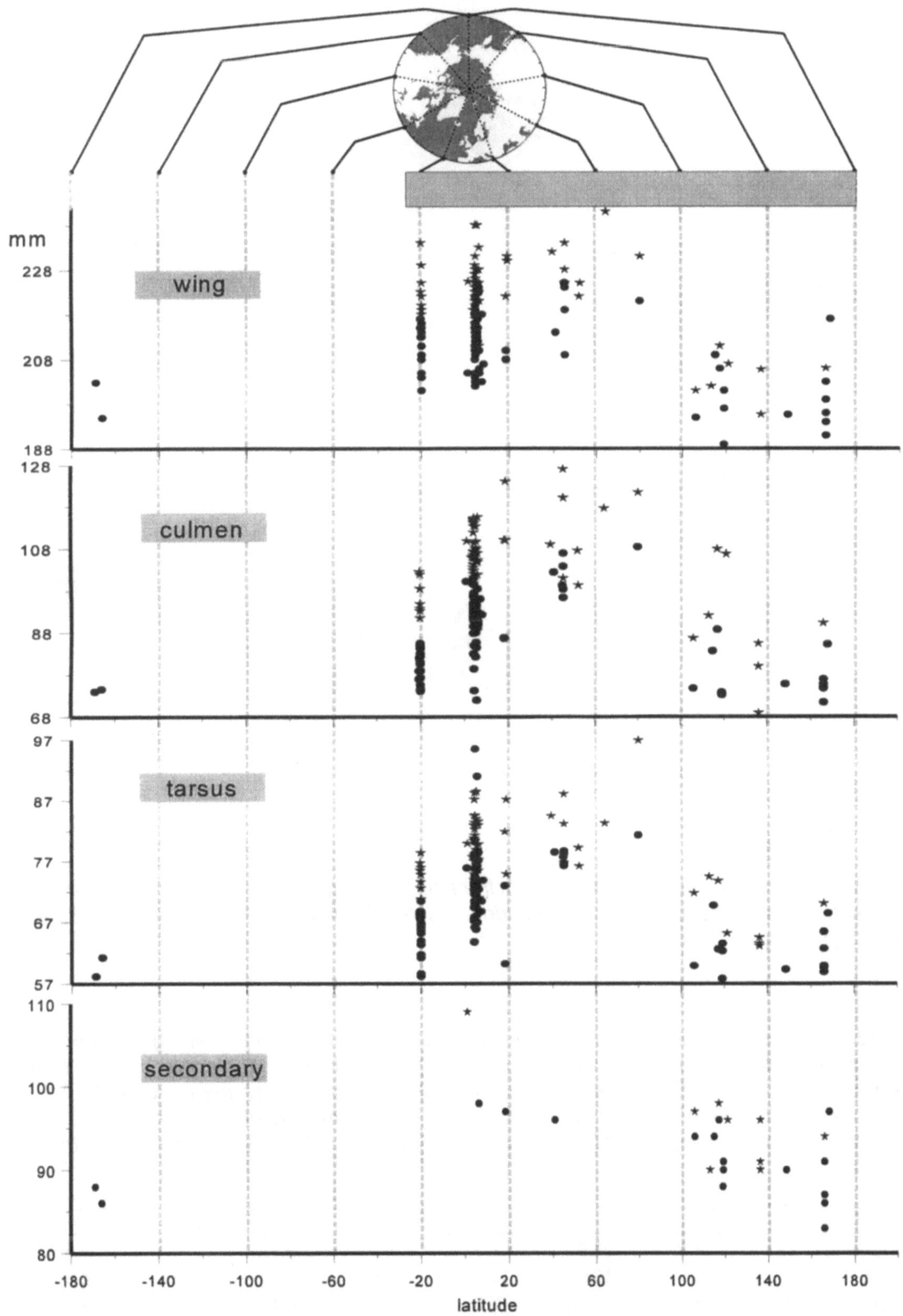

Fig. 82. *Measurements of Black-tailed Godwits from the breeding grounds. The following measurements are presented from top to bottom: wing length, culmen length, tarsus length and secondary length (description in chapter 2). Explanation of the used symbols: ● ♂♂, ★ ♀♀.*

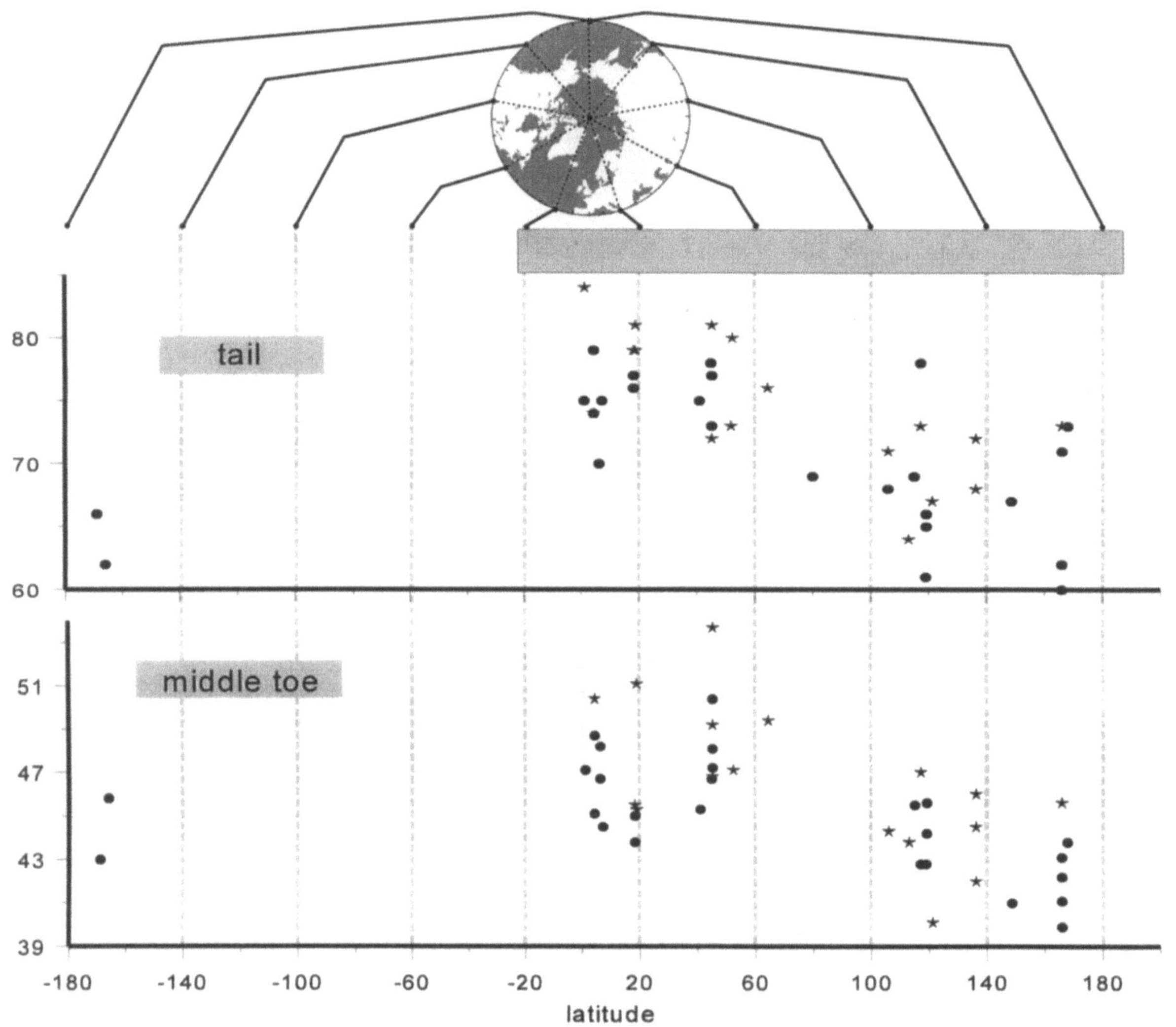

Fig. 83. *Measurements of Black-tailed Godwits from the breeding grounds. The following measurements are presented from top to bottom: tail- and middle toe length (description in chapter 2). Explanation of the used symbols: ● ♂♂, ★ ♀♀.*

Table 72. *Sexual size dimorphism of the Black-tailed Godwit. The mean percentage difference between the measurements of both sexes was calculated for every single variable and breeding population, according to: $(x_{♀}-x_{♂})/x_{♂}*100$. A negative value thus implies smaller measurements for ♀♀. The calculations were based upon the means of Table 84. The obtained values were averaged per variable and summed per breeding population ($\sum_3$ represents the sum of the first 3 variables, $\sum_6$ the sum of all 6 variables).*

Locality	Wing	Culmen	Tarsus	Secondary	Tail	Middle toe	$\sum_3$	$\sum_6$[1]
Iceland	6.3	20.9	12.3	-	-	-	39.5	-
W Europe	5.4	18.6	10.6	11.7	6.0	4.1	34.6	56.4
E Europe & SW Asia	4.7	11.3	6.1	-	4.4	2.9	22.1	-
E Palearctic	1.8	20.7	10.2	4.3	5.9	2.5	32.7	45.4
x	4.6	17.9	9.8	8.0	5.4	3.2		
SD	1.9	4.5	2.6	5.2	0.9	0.8		

Table 87. *Pearson's product-moment correlations between measurements of Black-tailed Godwits. Data of ♂♂ and ♀♀ were analysed separately. Two-tailed probabilities: * 0.01 < P < 0.05; ** 0.001 < P < 0.01; *** P < 0.001.*

	Wing	Culmen	Tarsus	Secondary	Tail	Middle toe	
Wing		0.700***	0.692***	0.928***	0.731***	0.698**	
Culmen	0.569***		0.611***	0.690***	0.485*	0.435	
Tarsus	0.644***	0.565***		0.635	0.699**	0.763***	♀♀
Secondary	0.765***	0.777***	0.738**		0.862**	-0.150	
Tail	0.736***	0.701***	0.632***	0.823***		0.600*	
Middle toe	0.628***	0.682***	0.743***	0.503*	0.587**		
				♂♂			

Table 88. *Results of MANOVA analysis concerning the morphometric differentiation of the Black-tailed Godwit. Pillai's trace statistics are presented and significant differences as derived from F-tests) are marked (Norušis 1992c). In these series of pairwise comparisons 8 groups (2 sexes, 4 geographical groups) were compared on the basis of 3 measurements in the upper part (wing-, culmen- and tarsus length). These 8 groups were compared in the lower part on the basis of wing- and culmen lengths. There were 3 and 156 df in the '3 variable' analyses and 2 and 159 df in the second group of analyses. Significance levels: *0.01 < P < 0.05; **0.001 < P < 0.01; ***P < 0.001.*

Locality	*n*	Iceland ♀	W Europe ♀	E Europe & SW Asia ♀	E Palearctic ♀	Iceland ♂	W Europe ♂	E Europe & SW Asia ♂
wing-, culmen- & tarsus length								
Iceland ♀	7							
W Europe ♀	38	0.178***						
E Europe & SW Asia ♀	10	0.237***	0.069*					
E Palearctic ♀	8	0.249***	0.467***	0.480***				
Iceland ♂	21	0.279***	0.716***	0.653***	0.220***			
W Europe ♂	60	0.130***	0.608***	0.519***	0.148***	0.349***		
E Europe & SW Asia ♂	7	0.086**	0.053*	0.129***	0.215***	0.399***	0.158***	
E Palearctic ♂	15	0.425***	0.747***	0.698***	0.175***	0.168***	0.469***	0.476***
wing- & culmen length								
Iceland ♀	7							
W Europe ♀	41	0.125***						
E Europe & SW Asia ♀	10	0.180***	0.061**					
E Palearctic ♀	8	0.239***	0.449***	0.457***				
Iceland ♂	21	0.265***	0.682***	0.610***	0.177***			
W Europe ♂	62	0.120***	0.604***	0.496***	0.137***	0.263***		
E Europe & SW Asia ♂	7	0.043*	0.053*	0.125***	0.194***	0.328***	0.133***	
E Palearctic ♂	15	0.413***	0.723***	0.667***	0.136***	0.162***	0.417***	0.423***

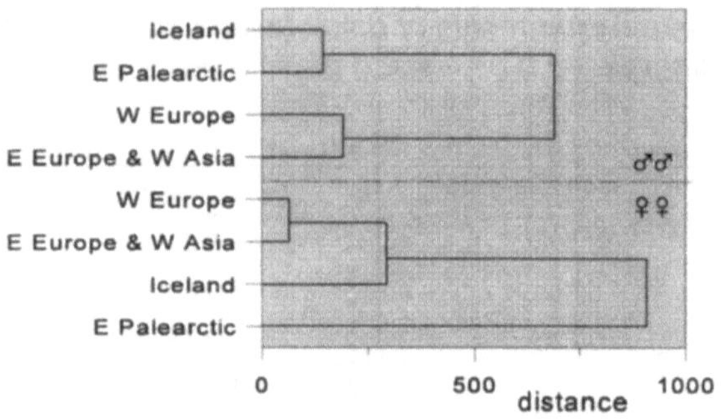

Fig. 84. *Cluster analysis in the Black-tailed Godwit based upon the wing-, culmen- and tarsus lengths presented in Table 70. The samples of both sexes were analysed separately. Warning: the values of the ♀-sample from the E Palearctic (Commander Islands) are based upon just 1 individual.*

luster analysis clearly showed the large degree of differentiation between the samples (Fig. 84). The ♂-samples were split in two: the Iceland ♂♂ were most similar to the E Palearctic ♂♂ and the W Europe ♂♂ to the ♂♂ from E Europe & SW Asia. The E Palearctic ♀♀ were clearly distinct from the other 3 samples. The W Europe ♀♀ were most similar to the ♀♀ from E Europe & SW Asia as was the case in the ♂♂.

The results from MANOVA analysis corresponded closely with the ones of cluster analysis (Table 74). All ♂ samples and all ♀ samples differed significantly from each other. The largest differences existed between the E Palearctic populations on one side and the samples from W Europe and E Europe & SW Asia on the other side. The measurements of the E Palearctic birds had most in common with the Icelandic ones.

When only wing- and culmen lengths were analysed with MANOVA (Table 74 - lower part), the results were nearly the same as compared with the 3-measurement analyses (Table 74 - upper part). The most obvious differences were the smaller estimated distances between the various samples in the 2-measurement analysis.

PREDICTION OF BREEDING ORIGIN

The above presented findings resulted in the construction of 8 POSCON groups: both sexes of the following 4 geographic regions: (1) Iceland, (2) W Europe, (3) E Europe & SW Asia, and (4) E Palearctic. A set of POSCON groups with 3 measurements (wing-, culmen- and tarsus length) was constructed. All breeding birds were used to calculate the degree of resemblance (typicality probabilities) with these POSCON groups. By doing so, both the accuracy and exclusivity of prediction were estimated (see introduction under 'Statistical analysis of the data'). Estimates of the accuracy rates could only be produced by calculating the typicality probabilities under the assumption that the covariance matrices of the 8 sample groups were equal.

When sexes are known, the predictionary power is rather high (Table 75 - upper part). In the ♂♂ there is some confusion between the Icelandic ♂♂ and the E Palearctic ♂♂. Some W European ♂♂ are classified as Icelandic ones, others as E European ones as would be expected in such an intermediately-sized population. There is no confusion between the Icelandic ♀♀ and the E Palearctic ♀♀ as there was between both corresponding ♂-populations. The only wrongly classified E Palearctic ♀ was one of the 2 ♀♀ with relatively long culmen lengths, presented in Fig. 82. Both were non-breeders collected along the E Chinese coast and both had a low resemblance with any other known population. They might (1) originate from an as yet unknown breeding population, or (2) be the result of a process of intermixing with the nominate subspecies.

Obviously, the prediction results became worse when sexes were unknown (Table 75 - lower part). This is especially true for the small-sized ♀♀ (e.g. Iceland ♀♀) and the large-sized ♂♂ (e.g. W- & E European ♂♂). And in the E Palearctic population some birds had a higher probability of belonging to the opposite sex than to the own sex.

Exclusion of tarsus lengths resulted in generally worse predictions since the accuracy rates were on average 8.9 % lower.

Table 75. *Accuracy of prediction in Black-tailed Godwits. These predictions have been based upon morphometric analysis and included wing-, culmen- and tarsus length. The columns show the predicted proportions in relation to the total numbers per breeding area (last column).*

	Predicted								
Locality	Iceland ♀	W Europe ♀	E Europe & SW Asia ♀	E Palearctic ♀	Iceland ♂	W Europe ♂	E Europe & SW Asia ♂	E Palearctic ♂	*n*
				Sex known					
Iceland ♀	100.0	-	-	-	-	-	-	-	7
W Europe ♀	10.5	63.2	26.3	-	-	-	-	-	38
E Europe & SW Asia ♀	10.0	30.0	60.0	-	-	-	-	-	10
E Palearctic ♀	-	12.5	-	87.5	-	-	-	-	8
Iceland ♂	-	-	-	-	71.4	-	-	28.6	21
W Europe ♂	-	-	-	-	10.3	77.6	10.3	1.7	58
E Europe & SW Asia ♂	-	-	-	-	-	-	100.0	-	7
E Palearctic ♂	-	-	-	-	20.0	-	-	80.0	15
				Sex unknown					
Iceland ♀	71.4	-	-	-	-	14.3	14.3	-	7
W Europe ♀	7.9	42.1	26.3	-	-	-	23.7	-	38
E Europe & SW Asia ♀	10.0	30.0	60.0	-	-	-	-	-	10
E Palearctic ♀	-	-	-	50.0	12.5	-	12.5	25.0	8
Iceland ♂	-	-	-	4.8	71.4	-	-	23.8	21
W Europe ♂	21.7	-	-	13.3	10.0	48.3	5.0	1.7	60
E Europe & SW Asia ♂	28.6	28.6	-	-	-	-	42.9	-	7
E Palearctic ♂	-	-	-	6.7	13.3	-	-	80.0	15

DISCUSSION

The taxonomy of the Black-tailed Godwit

There is a high degree of geographical variation in this species which we can only partly understand due to the incompleteness of the data-set. In general, we agree with the recognition of 3 separate groups: (1) Icelandic breeders - usually known as *L. l. islandica* , (2) breeding birds from mainland Europe & S Asia - usually known as *L. l. limosa*, and (3) E Palearctic breeding birds - usually known as *L. l. melanuroides*. We will give some additional remarks in order to facilitate a systematic revision for this species:

1. The differences between the W and E Palearctic populations are large and concern morphometric as well as plumage differences of both the summer- and winter plumage. The average differences in wing length vary between 11 and 26 mm, in culmen length between 3 and 25 mm and in tarsus length between 4 and 15 mm. The morphometric differences between the E Palearctic population and the neighbouring one of the nominate subspecies are the largest, whereas the differences between the E Palearctic and the Icelandic population are smallest but still evident. Moreover, there are important differences in plumage characters (described and/or shown by Prater *et al.* 1977, Cramp & Simmons 1983, Hayman *et al.*1986 - Plate 48, Paulson 1993 and Hoyo *et al.* 1996 - Plate 42). Because of these important differences, it

might be better to consider the W and E Palearctic Black-tailed Godwits as separate allospecies forming a superspecies with the Hudsonian Godwit *L. haemastica* as well. For such a revision it would be useful to include the morphometrics and plumage characters of the Hudsonian Godwit.

2. We agree with the findings of Prater *et al.* (1977), Hale (1980) and Cramp & Simmons (1983) that the easterly breeding birds of the nominate subspecies are larger-sized than the W European breeders. However, the size-differences between these 2 groups are of the same magnitude as between W Europe and Iceland. Such a size-difference might be an indication for the existence of clinal variation in this subspecies, but it might also indicate the presence of 2 separate subspecies. Subspeciation in E Europe and S Asia was also evident in the Whimbrel, Curlew and Redshank (see species' chapters). Our data-set of E European Black-tailed Godwits is too small to favour any of both opinions. However, we can conclude from this data-set that especially the breeding region between 10 and 40°E is interesting since this is the contact zone between the W andE European breeding populations. Moreover when the measurements would change clinally, birds of the nominate subspecies breeding E of 60°E ought to show a gradual increase in size.

3. The small dimensions of the Iceland population are remarkable. They are larger than the E Palearctic *melanuroides* and their coloration is totally different with Icelandic breeding birds having more reddish and more coarsely striped underparts. It is intriguing to speculate on the causation of the small dimensions of the Iceland population. There might be selection against large size in the Icelandic Godwits somewhere in the year's cycle. Or the population might have an evolutionary background making it more related to the E Palearctic *melanuroides* than to the W Palearctic *limosa*. It might also be that the population is influenced by the Hudsonian Godwit having its nearest breeding grounds south of the Hudson Bay.

4. Our data show some indications on the existence of geographical variation within the E Palearctic *melanuroides*. For instance, some migrants along the E Chinese coast were (with respect to this subspecies) very long-billed. There were no breeding specimen found with such long bills, but neither all breeding populations were sampled (e.g. the isolated population breeding at the upper reaches of the Vilyuy and Lena rivers). Moreover, Black-tailed Godwits from Amur seemed to be smaller-sized than the ones breeding at Kamchatka and on the Commander Islands. The scattered breeding distribution of this species in the E Palearctic with basically 5 isolated breeding areas might well result in some degree of morphometric variation.

SUMMARY

A marked degree of sexual and geographical variation exists in the Black-tailed Godwit. Sexual differences were larger than geographical differences: ♂♂ had smaller measurements than ♀♀. Sexual dimorphism was smallest in the samples from E Europe & SW Asia and largest in the Icelandic sample.
All ♂ samples and all ♀ samples differed significantly from each other. The W Palearctic samples (especially W Europe and E Europe & SW Asia) differed from the E Palearctic ones. The measurements of the E Palearctic birds had most in common with the Icelandic ones.
Four geographical regions were recognized: (1) Iceland, (2) W Europe, (3) E Europe & SW Asia and (4) E Palearctic. By separating both sexes 8 POSCON sample-groups were constructed. The accuracy of prediction increased with nearly 10 % when more measurements were included. The species is polytypic and 3 subspecies are recognized: *L. l. limosa*, *L. l. islandica*, and *L. l. melanuroides*. *L. l. limosa* is a polymorphic subspecies, but it might also be that the E European and S Asian populations ought to be recognized as a separate subspecies. There were too few data to analyse the possible occurrence of geographical variation in the E Palearctic region.

Chapter 13

Bar-tailed Godwit

-

Limosa lapponica

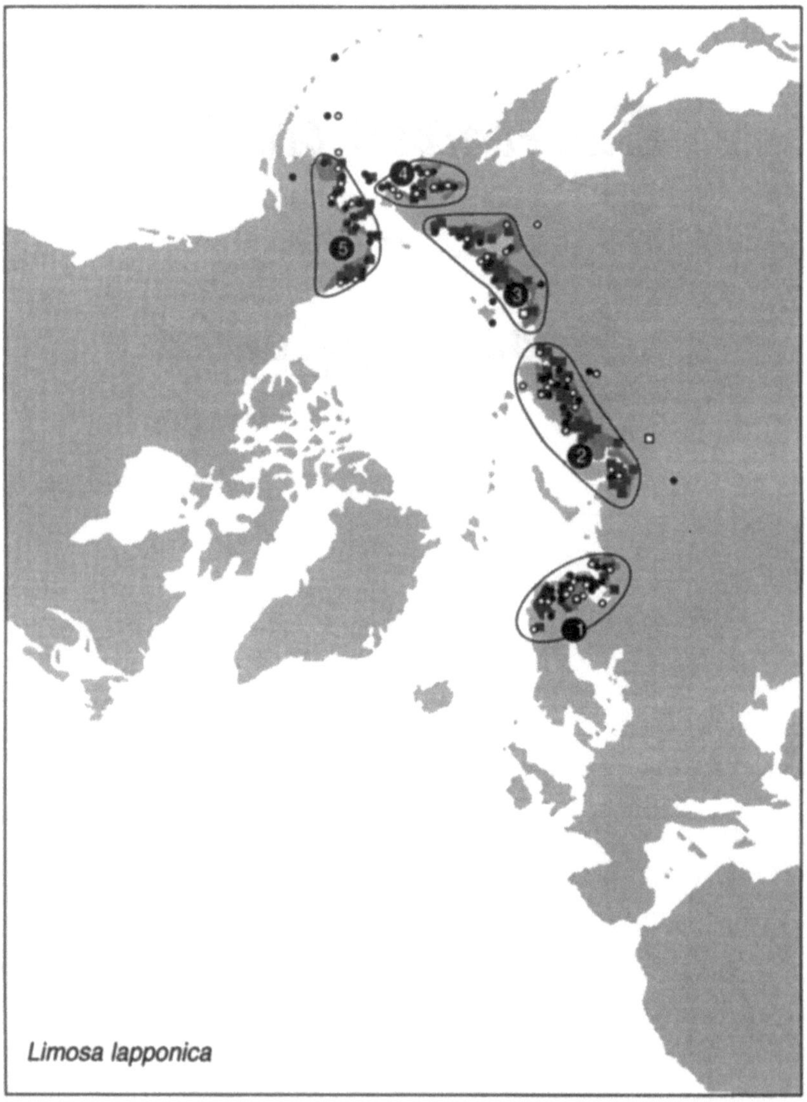

Fig. 85. *Breeding distribution of the Bar-tailed Godwit.*

INTRODUCTION

The Bar-tailed Godwit breeds in the arctic regions of the Palearctic and Alaska. Five disjunct breeding ranges are currently known (Fig. 85): (1) N Fennoscandia, the north part of the White Sea and the Kanin Peninsula; (2) from Yamal east to the delta of the Anabar River; (3) from the Lena Delta east to the Chaunsk Bay; (4) Anadyr and the south part of the Chukotsk Peninsula; and (5) N and W Alaska. The birds breed in lowland tundra and rolling uplands, both on coastal and inland sites (Hayman *et al.* 1986). Scattered trees are sometimes tolerated.

There is a fairly large amount of geographical variation in the Bar-tailed Godwit (Naumann 1836, Portenko 1936, Vaurie 1965, Prater *et al.* 1977, Glutz *et al.* 1977, Hale 1980, Cramp & Simmons 1983, Stepanyan 1990). This variation involves small-sized individuals on the western and large-sized ones on the eastern breeding grounds. It also involves plumage differences with easterly birds having a less contrasting flight pattern, a brown lower back, and heavily brown barred underwings, axillaries, tail and rump (Cramp & Simmons 1983, Paulson 1993). In the present-day literature 2 subspecies are recognized (Vaurie 1965, Glutz *et al.* 1977, Prater *et al.* 1977, Cramp & Simmons 1983, Paulson 1993):

- *L. l. lapponica*, breeding from N Europe east to the delta of the Khatanga river (110 °E), including the whole Taymyr Peninsula.
- *L. l. baueri*, breeding east of the Khatanga River in E Siberia and in N and W Alaska.

Portenko (1936) described a third subspecies *L. l. menzbieri*, occupying the range between the rivers Khatanga and Kolyma. This subspecies was not accepted by most later authors, because *lapponica* and *baueri* graded clinally into each other. *Baueri* was considered as a variable subspecies breeding as far west as the Khatanga River. Hale (1980) recognized *menzbieri* as a highly variable intermediate form between both other subspecies, rather than as a step on the cline, not warranting inclusion in either *lapponica* or *baueri*. True *lapponica* thus occur west of the Khatanga River and true *baueri* at Anadyr, Chukotsk and in N and W Alaska. Paulson (1993) noticed that *baueri* varies more in rump color than evident from literature with some easterly ones being quite *lapponica*-like.

Table 76. *Measurements of breeding populations of Bar-tailed Godwits. [1]The north part of the White Sea and the Kanin Peninsula are included. [2]This region is situated between the Lena River and the Chaunsk Bay. [3]This region is restricted to the breeding grounds at Anadyr and the south part of the Chukotsk Peninsula. Presented are the means ± 1 SD (n).*

Locality	Wing	Culmen	Tarsus	Secondary	Tail	Middle toe
			♂♂			
1 N Fennoscandia[1]	218.1±5.8(33)	81.4±4.2(33)	51.6±2.0(30)	90.8±2.6(17)	70.6±4.9(18)	36.2±1.8(17)
2 Yamal & Taymyr	214.9±4.5(41)	78.8±3.9(41)	50.3±2.0(42)	90.0±2.8(29)	67.2±2.8(29)	35.8±1.3(29)
3 Central Siberia[2]	219.8±4.2(13)	85.3±3.8(12)	52.4±1.5(12)	92.0±6.0(6)	69.8±2.0(6)	37.3±1.7(5)
4 E Siberia[3]	225.8±4.2(16)	84.7±4.9(16)	54.3±1.9(21)	92.7±3.1(13)	72.8±3.5(13)	36.7±0.9(13)
5 N & W Alaska	231.4±5.9(52)	85.2±5.5(52)	53.6±3.2(36)	91.9±2.3(32)	73.0±3.9(32)	37.6±1.5(31)
			♀♀			
1 N Fennoscandia[1]	230.8±6.1(23)	101.6±5.3(23)	55.9±2.8(20)	98.5±3.8(13)	71.6±3.8(13)	38.2±2.0(14)
2 Yamal & Taymyr	227.0±4.4(29)	96.0±4.4(29)	54.0±2.4(29)	95.6±2.4(25)	70.3±2.9(27)	37.8±1.6(26)
3 Central Siberia[2]	231.2±5.3(16)	108.6±4.2(16)	55.6±1.6(16)	97.0±0.7(5)	74.3±1.9(6)	38.6±1.4(6)
4 E Siberia[3]	236.3±7.7(7)	107.8±6.6(7)	58.3±2.5(7)	99.4±3.3(7)	74.9±3.4(7)	39.8±0.6(7)
5 N & W Alaska	246.4±6.4(44)	108.5±6.9(43)	59.2±2.2(44)	97.4±2.9(18)	76.0±3.2(21)	40.5±1.7(17)

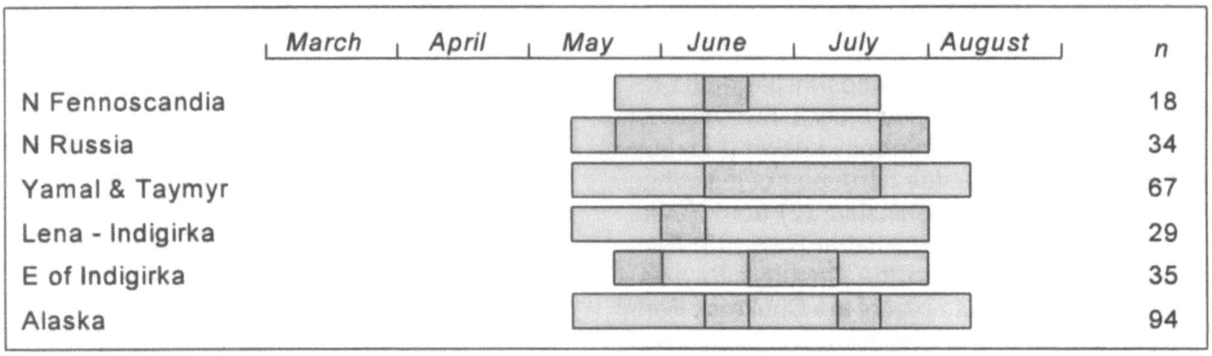

Fig. 86. *Collection dates of Bar-tailed Godwits. Most birds had been collected during the periods marked by the heavily shaded bars. The sample sizes are presented on the right side.*

MORPHOMETRICAL DIFFERENTIATION

In total, 277 Bar-tailed Godwits (58.3% ♂♂) were examined from all known breeding grounds except the Lena River Delta. However, the breeding of Bar-tailed Godwits in the delta of the Lena River is doubtful (Tomkovich, pers. comm.). The breeding grounds were reached from mid-May onwards (Fig. 86). Most skins had been collected during June. Birds collected after July were excluded from the analyses. Five distinct breeding locations were distinguished (Fig. 85). Males from all 5 locations and ♀♀ from 4 of these were well-represented, but few ♀♀ of the Anadyr and Chukotsk population had been measured. Therefore, 7 non-breeding ♀♀ collected near Koryakskaya, south of Anadyr, were included in this sample, as MANOVA tests showed a lack of morphometric differentiation between these breeding ♀♀ and non-breeding ♀♀. The morphometric and plumage differences between samples were analysed and presented in the tables 76 - 81 and figures 87 - 90.

The Bar-tailed Godwit showed much sexual and geographical variation (Table 77, figs. 87 & 88).

Table 77. *Overall results of MANOVA for all distinguished breeding populations in the Bar-tailed Godwit in combination with univariate ANOVA's in which the different measurements were analysed separately. Significance levels: *0.01 < P < 0.05; **0.001 < P < 0.01; ***P < 0.001.*

	Analysis I			Analysis II			Analysis III		
Variables	wing, culmen, tarsus, secondary, tail, middle toe length			wing, culmen, tarsus			wing, culmen		
n	183			277			284		
Multivariate analysis of variance									
Effects	Interaction	Area	Sex	Interaction	Area	Sex	Interaction	Area	Sex
Pillai's trace	0.163	0.910	0.252	0.088	0.821	0.785	0.065	0.790	0.785
F-value	1.211	8.395***	26.900***	2.028*	25.140***	322.525***	2.310*	44.754***	499.543***
Hypoth. & error *df*	24/684	24/684	1/173	12/801	12/801	3/735	8/548	8/548	2/273
Univariate F tests									
df	4/173	4/173	1/173	4/267	4/267	1/267	4/274	4/274	1/274
wing length	1.018	93.936***	179.542***	1.313	123.425***	283.504***	1.213	124.957***	289.533***
culmen length	1.134	21.823***	423.564***	3.508**	37.527***	884.845***	3.801**	37.795***	912.108***
tarsus length	1.582	24.330***	96.552***	2.961*	39.861***	187.232***			
secondary length	1.222	5.003**	119.688***						
tail length	0.514	23.233***	16.794***						
middle toe length	2.682*	17.140***	59.613***						

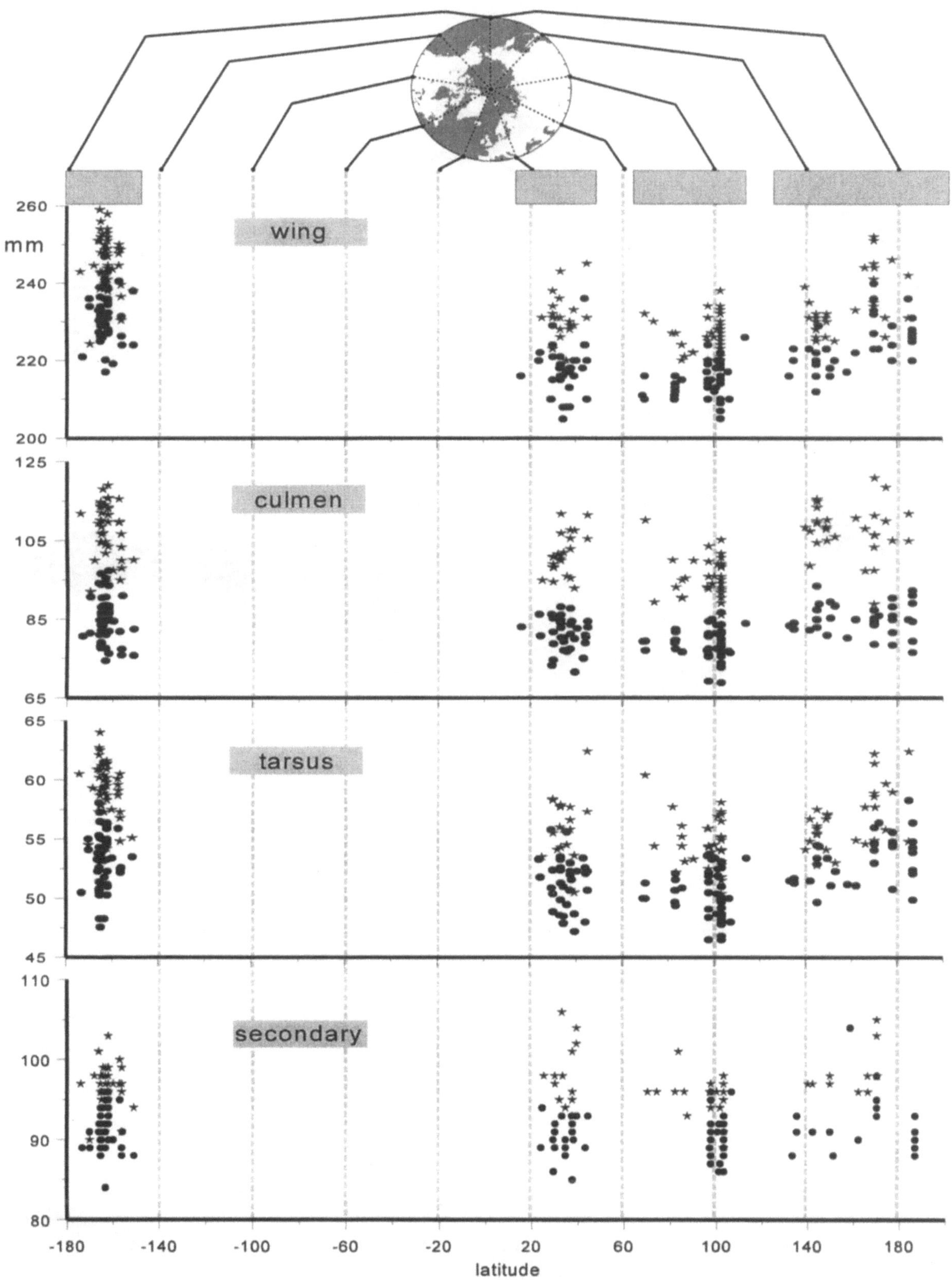

Fig. 87. *Measurements of Bar-tailed Godwits from the breeding grounds. The following measurements are presented from top to bottom: wing length, culmen length, tarsus length and secondary length (description in chapter 2). Explanation of the used symbols: ● ♂♂, ★ ♀♀.*

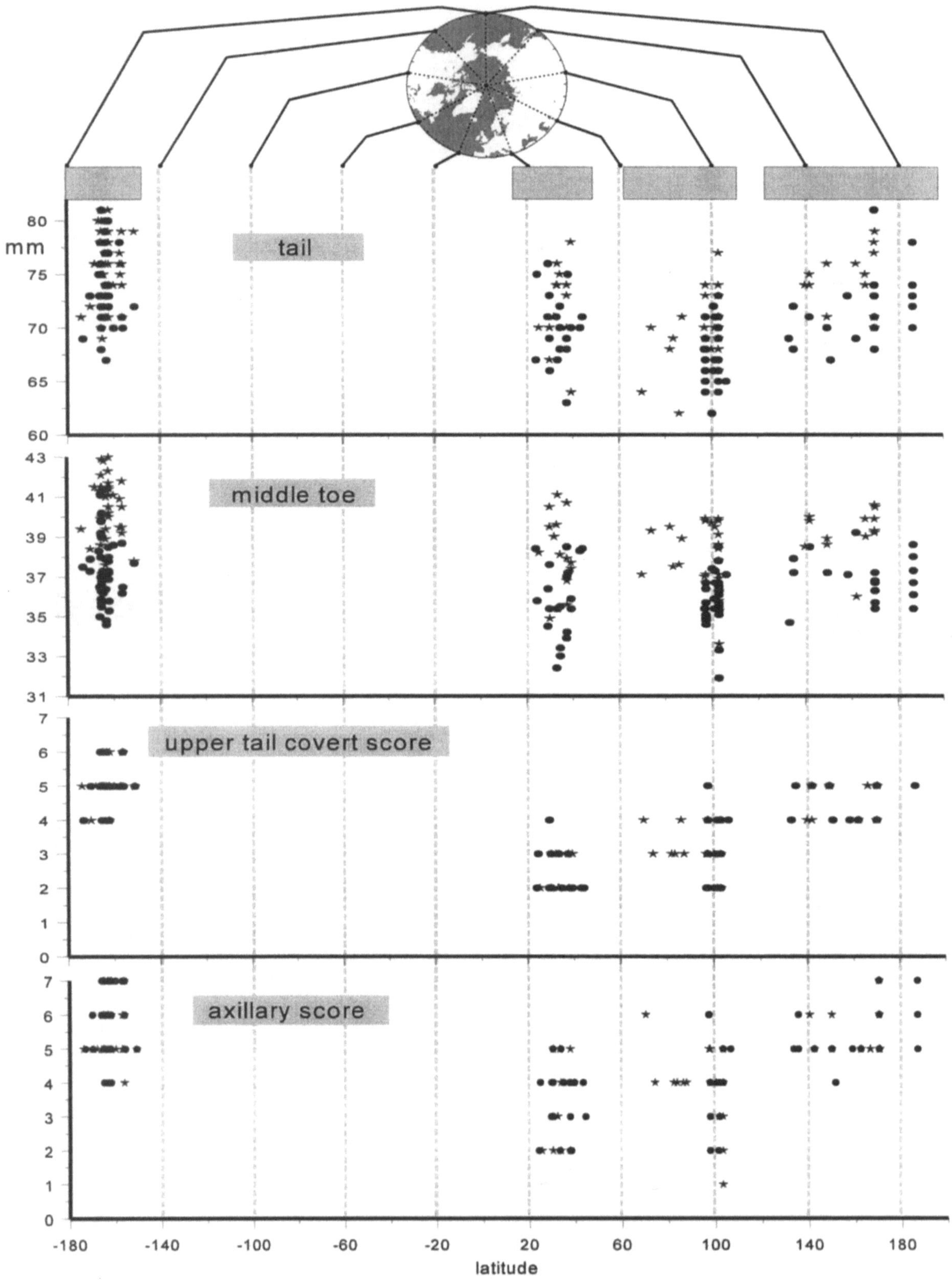

Fig. 88. *Measurements of Bar-tailed Godwits from the breeding grounds. The following measurements are presented from top to bottom: tail length, middle toe length, the upper tail-covert score and the axillary score (description in chapter 2). Explanation of the used symbols: ● ♂♂, ★ ♀♀.*

Table 78. *Sexual size dimorphism in the Bar-tailed Godwit. The mean percentage difference between the measurements of both sexes was calculated for every single variable and breeding population, according to: $(x_{♀}-x_{♂})/x_{♂}*100$. A negative value thus implies smaller measurements for ♀♀. The calculations were based upon the means of Table 76. The obtained values were averaged per variable and summed per breeding population.*

	Wing	Culmen	Tarsus	Secondary	Tail	Middle toe	Σ
N Fennoscandia	5.8	24.8	8.3	8.5	1.4	5.5	54.3
Yamal & Taymyr	5.6	21.8	7.4	6.2	4.6	5.6	51.2
Central Siberia	5.2	27.3	6.1	5.4	6.4	3.5	53.9
E Siberia	4.7	27.3	7.4	7.2	2.9	8.4	57.9
N & W Alaska	9.1	27.3	10.4	6.0	4.1	7.7	64.6
x	6.1	25.7	7.9	6.7	3.9	6.1	
SD	1.7	2.4	1.6	1.2	1.9	2.0	

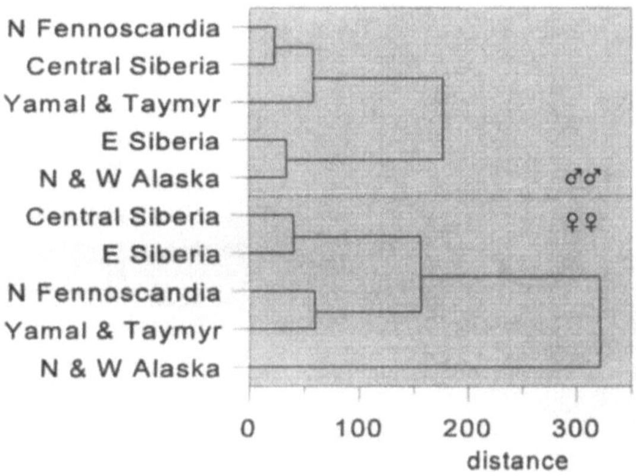

Fig. 89. *Cluster analysis in the Bar-tailed Godwit based upon the 6 measurements presented in Table 76. The samples of both sexes were analysed separately.*

Sexual differences were larger than geographical differences. Sexual dimorphism was smallest in the samples from Yamal and Taymyr and largest in the ones of N and W Alaska (Table 78). The different measurements were highly correlated, especially in ♀♀ (Table 79). Secondary lengths correlated least with other measurements.

The squared euclidean distances as calculated with cluster analysis were large, indicating large distances between samples (Fig. 89). Moreover, the calculated distances between samples were larger in the ♀♀ than in the ♂♂ which contrasts with the results of cluster analysis in most of the other species analysed here. The relationships between samples also differed for ♂♂ and ♀♀: mean measurements of ♂♂ from N Fennoscandia were nearly similar to those of Central Siberia. The E Siberia population was most closely related to the Alaska population. Measurements of the Yamal & Taymyr ♂♂ were close to those of ♂♂ from N Fennoscandia and Central Siberia. The interrelationship of the ♂ samples corresponds more with the published opinion on geographical variation (see introduction) than those of the ♀♀. The ♀♀ from Central Siberia were most closely related with the E Siberia ♀♀ and the ♀♀ of N Fennoscandia with the Yamal & Taymyr ♀♀. The ♀-sample of Alaska was most distinctive.

Table 79. *Pearson's product-moment correlation between measurements of Bar-tailed Godwits. Data of ♂♂ and ♀♀ were analysed separately. Two-tailed probabilities: * $0.01 < P < 0.05$; ** $0.001 < P < 0.01$; *** $P < 0.001$.*

	Wing	Culmen	Tarsus	Secondary	Tail	Middle toe	
Wing length		0.565***	0.718***	0.384***	0.644***	0.679***	
Culmen length	0.493***		0.617***	0.286**	0.374***	0.548***	
Tarsus length	0.567***	0.578***		0.137	0.313**	0.672***	
Secondary length	0.411***	0.277**	0.265**		0.275**	0.289**	♀♀
Tail length	0.523***	0.303**	0.460***	0.271**		0.415***	
Middle toe length	0.477***	0.473***	0.478***	0.197*	0.266**		
			♂♂				

Table 80. *Results of MANOVA analysis concerning the morphometric differentiation of the Bar-tailed Godwit. Pillai's trace values are presented together with the level of significance as based upon F-values (Norušis 1992c). In these series of pairwise comparisons 10 groups and different sets of measurements were involved. There were 6 and 168 df in the '6 variable' analyses; 3 and 265 df in the '3 variable' analyses and 2 and 273 df in the last group of analyses. Significance levels: *0.01 < P < 0.05; **0.001 < P < 0.01; ***P < 0.001.*

Locality	n	N Fennoscandia ♀	Yamal & Taymyr ♀	Central Siberia ♀	E Siberia ♀	N & W Alaska ♀	N Fennoscandia ♂	Yamal & Taymyr ♂	Central Siberia ♂	E Siberia ♂
wing-, culmen-, tarsus-, secondary-, tail- & middle toe										
N Fennoscandia ♀	13									
Yamal & Taymyr ♀	23	0.075*								
Central Siberia ♀	5	0.037	0.107**							
E Siberia ♀	7	0.167***	0.265***	0.099**						
N & W Alaska ♀	34	0.392***	0.536***	0.167***	0.059					
N Fennoscandia ♂	18	**0.439*** **	0.360***	0.366***	0.500***	0.714***				
Yamal & Taymyr ♂	29	0.547***	**0.487*** **	0.447***	0.577***	0.803***	0.094*			
Central Siberia ♂	5	0.234***	0.149***	**0.244*** **	0.324***	0.460***	0.030	0.056		
E Siberia ♂	14	0.348***	0.295***	0.291***	**0.339*** **	0.560***	0.170***	0.293***	0.082*	
N & W Alaska ♂	47	0.464***	0.462***	0.325***	0.369***	**0.657*** **	0.364***	0.531***	0.156***	0.061
wing-, culmen- & tarsus length										
N Fennoscandia ♀	20									
Yamal & Taymyr ♀	29	0.064***								
Central Siberia ♀	16	0.064***	0.184***							
E Siberia ♀	13	0.082***	0.208***	0.102***						
N & W Alaska ♀	43	0.298***	0.490***	0.286***	0.049**					
N Fennoscandia ♂	31	**0.441*** **	0.334***	0.531***	0.514***	0.739***				
Yamal & Taymyr ♂	41	0.530***	**0.442*** **	0.596***	0.585***	0.799***	0.029*			
Central Siberia ♂	11	0.230***	0.124***	**0.336*** **	0.337***	0.518***	0.020	0.061***		
E Siberia ♂	22	0.314***	0.224***	0.443***	**0.367*** **	0.587***	0.135***	0.239***	0.057***	
N & W Alaska ♂	51	0.371***	0.293***	0.494***	0.389***	**0.657*** **	0.294***	0.427***	0.123***	0.046**
wing-, & culmen length										
N Fennoscandia ♀	23									
Yamal & Taymyr ♀	29	0.056***								
Central Siberia ♀	16	0.060***	0.177***							
E Siberia ♀	13	0.083***	0.196***	0.077***						
N & W Alaska ♀	43	0.312***	0.477***	0.257***	0.047***					
N Fennoscandia ♂	33	**0.447*** **	0.324***	0.521***	0.508***	0.738***				
Yamal & Taymyr ♂	41	0.539***	**0.436*** **	0.589***	0.577***	0.794***	0.029*			
Central Siberia ♂	12	0.238***	0.127***	**0.336*** **	0.343***	0.531***	0.018	0.060***		
E Siberia ♂	22	0.303***	0.194***	0.418***	**0.361*** **	0.581***	0.118***	0.214***	0.053***	
N & W Alaska ♂	52	0.381***	0.293***	0.487***	0.387***	**0.656*** **	0.291***	0.424***	0.135***	0.027*

Three MANOVA-analyses were performed, in which samples were contrasted pairwise, and which differed with respect to the number of variables included and thus the sample sizes (Table 80). The following conclusions could be drawn:

1. All ♂ samples were statistically distinct from ♀ samples.
2. The large measurements of Bar-tailed Godwits from E Siberia and Alaska always differed from the western samples.
3. The largest distances were estimated between the samples of Yamal & Taymyr and N & W Alaska.
4. The measurements of the N Fennoscandian and Central Siberian birds hardly differed, except for ♀♀ when using wing-, culmen- and tarsus lengths only.
5. The Yamal & Taymyr birds were the smallest ones examined. The ♀♀ of Yamal & Taymyr differed always from other ♀ samples and the ♂♂ nearly always from other ♂ samples.
6. In both ♂♂ and ♀♀, the Central Siberian samples showed more similarity with the N Fennoscandian samples than with the E Siberian ones.
7. The E Siberian samples always differed from the Central Siberian ones and mostly from the Alaskan ones. This is rather surprising as both the E Siberian and the Alaskan population are generally considered to belong to the subspecies *baueri*. This difference was most convincing when 3 measurements were analysed and was mainly caused by differences in wing length (Fig. 87).
8. When 2 or 3 measurements were used, all ♀ samples differed significantly from each other at the 1 % level.

All selected measurements were important for the distinction between sexes and breeding locations (Table 77). Culmen-, wing-, secondary, tarsus-, middle toe- and tail length (in decreasing order of importance) were important to identify sexes. Wing-, tarsus-, tail-, culmen-, middle toe- and secondary length (in decreasing order of importance) were important to separate between breeding locations.

Both wing- and culmen lengths were important for the analysis of sexual and geographical variation (Table 77). MANOVA analysis with wing- and culmen lengths produced mostly identical results as MANOVA analysis with 3 measurements (compare Table 80 lower and middle block). Our data did not show clear differences in wing- and culmen lengths between ♂♂ of N Fennoscandia and Central Siberia, though ♀♀ of these groups differed strongly.

Next to standard measurements the coloration of the upper tail-coverts and the colour-pattern of the axillaries was quantified (see Introduction under 'The selected variables'). Large differences in plumage scores between the western and eastern samples occurred, western ones having whitish axillaries and upper tail-coverts, while axillaries and upper tail-coverts of samples from Central Siberia, E Siberia and Alaska were much darker (Fig. 90). Males and ♀♀ from the same population had not difference in plumage characters (Table 81). Plumage colors of the Central Siberian birds were closest to the ones of the E Palearctic, though their measurements were more similar to W Palearctic ones. Yamal & Taymyr birds had a generally darker plumage than birds from N Fennoscandia but were not as dark as the Central Siberian ones.These plumage characters thus have an important additional value for separating populations, especially since they improve the distinction between N Fennoscandian and Central Siberian breeding birds.

Table 81. *Testing the difference between both sexes with respect to the scores of axillaries and upper tail-coverts. Chi^2-tests were used to test the differences between sexes within a population.*

	Axillaries				Upper tail-coverts			
Locality	Chi²	*df*	*P*	*n*	Chi²	*df*	*P*	*n*
N Fennoscandia	1.010	3	0.799	34	0.858	2	0.651	32
Yamal & Taymyr	2.074	5	0.839	56	4.664	3	0.198	56
Central Siberia	3.111	2	0.211	12	0.343	1	0.558	12
E Siberia	3.500	2	0.174	21	0.154	1	0.694	21
N & W Alaska	4.100	3	0.251	87	3.977	2	0.137	87

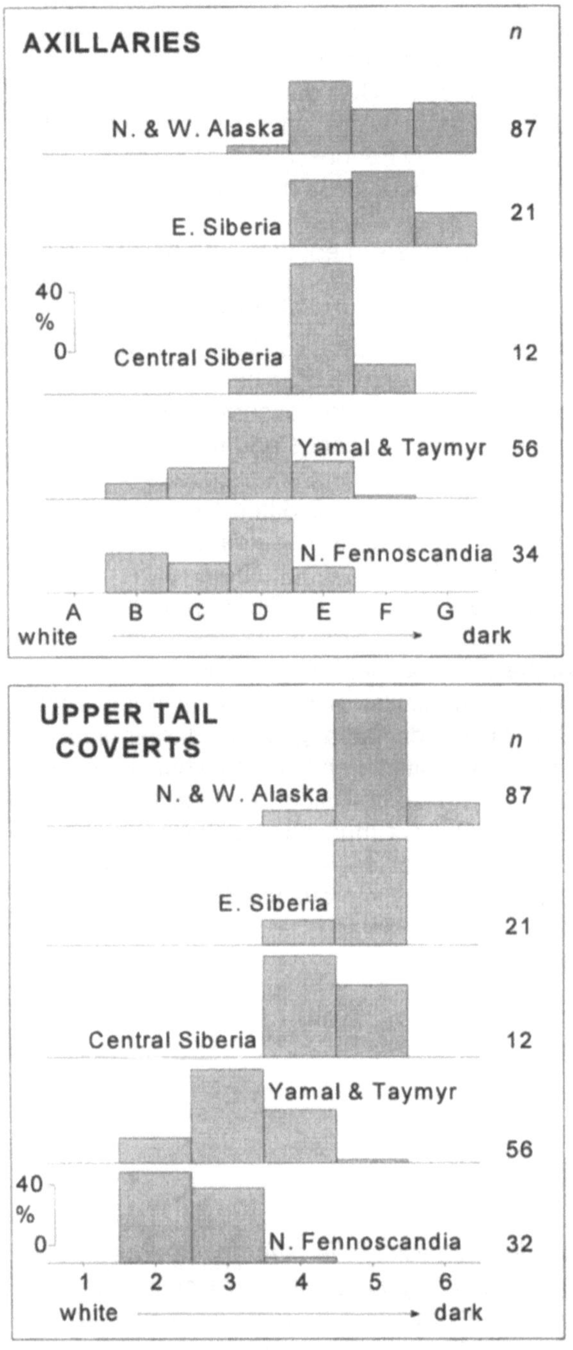

Fig. 90. *Frequency distributions of the scores of axillary patterns and upper tail-coverts in different populations of the Bar-tailed Godwit.*

PREDICTION OF BREEDING ORIGIN

Due to a large size dimorphism, both sexes ought to be treated separately. Moreover, the cluster- and MANOVA analyses identified 5 distinct breeding populations: (1) N Fennoscandia, (2) Yamal & Taymyr, (3) Central Siberia, (4) E Siberia, and (5) Alaska. Thus 10 POSCON sample groups were distinguished, each of which can be best characterized with the use of all 6 measurements and both plumage scores (see the preceding chapters). As plumage scores are categorical data , with up to (6x7 =) 42 possible combinations, such an analysis needs a lot more data than available. Every combination ought to contain at least 16 birds (10 POSCON groups, 6 measurements) which means a sample of at least 6.720 Bar-tailed Godwits from the breeding areas which is far beyond reality. In fact, we had a complete set of data of 5 ♂♂ and 5 ♀♀ from Central Siberia and 14 ♂♂ and 7 ♀♀ for E Siberia. The limited number of full-data samples forced us to assume that the covariance matrices were equal for all samples. When analysing the accuracy- and exclusivity rates, predictions lacking information on a bird's sex were separated from the ones with known sex.

Accuracy

Sexual differences were generally so large that knowledge of a bird's sex will only significantly improve the prediction results of the intermediate POSCON groups: E Siberian ♂♂, Alaska ♂♂ and Yamal & Taymyr ♀♀ (compare tables 82 and 83). It is therefore essential to know the sex of a Bar-tailed Godwit when any of these 3 populations might be involved. But knowledge on a bird's sex improved less the predictional qualities for the small-sized western ♂♂ or the large-sized eastern ♀♀.

The estimated accuracy rates demonstrate how difficult it is to obtain proper predictions. For instance, when wing-, culmen- and tarsus lengths were used the N Fennoscandian ♀♀ clearly differed -at the 0.1 % level- from all other samples (Table 80). But at the same time the estimated accuracy in the N Fennoscandian ♀-group showed that only 45 % of them have the highest resemblance with the own POSCON group (Table 83). Thus 55 % of the birds in this group more closely resembled another sample group. Nevertheless, the estimated accuracy rates were quite high especially when the large number of samples is considered. In general, the accuracy rates are negatively correlated with the number of POSCON groups.

Table 82. *Accuracy of prediction in the Bar-tailed Godwit when sexes are* ***known****. The columns show the predicted proportions in relation to the total number per breeding area (last column).*

	Predicted										
Locality	Fennoscandia ♂	Yamal & Taymyr ♂	Central Siberia ♂	E Siberia ♂	Alaska ♂	Fennoscandia ♀	Yamal & Taymyr ♀	Central Siberia ♀	E Siberia ♀	Alaska ♀	*n*
	Wing-, culmen- , tarsus-, secondary-, tail- and middle toe length										
Fennoscandia ♂[1]	44.4	11.1	38.9	5.6	-	-	-	-	-	-	18
Yamal & Taymyr ♂	10.3	58.6	27.6	3.4	-	-	-	-	-	-	13
Central Siberia ♂	20.0	-	80.0	-	-	-	-	-	-	-	5
E Siberia ♂	-	7.7	7.7	61.5	23.1	-	-	-	-	-	13
Alaska ♂	2.1	2.1	8.5	19.1	68.1	-	-	-	-	-	47
Fennoscandia ♀	-	-	-	-	-	38.5	30.8	23.1	-	7.7	13
Yamal & Taymyr ♀	-	-	-	-	-	26.1	73.9	-	-	-	23
Central Siberia ♀	-	-	-	-	-	20.0	-	80.0	-	-	5
E Siberia ♀	-	-	-	-	-	-	-	14.3	71.4	14.3	7
Alaska ♀	-	-	-	-	-	-	-	10.0	26.7	63.3	30
	Wing-, culmen- and tarsus length										
Fennoscandia ♂[1]	31.3	28.1	34.4	3.1	3.1	-	-	-	-	-	32
Yamal & Taymyr ♂	21.4	66.7	9.5	2.4	-	-	-	-	-	-	42
Central Siberia ♂	18.2	9.1	54.5	18.2	-	-	-	-	-	-	11
E Siberia ♂	4.8	-	19.0	52.4	23.8	-	-	-	-	-	21
Alaska ♂	5.5	-	5.5	21.8	67.3	-	-	-	-	-	55
Fennoscandia ♀	-	-	-	-	-	45.0	25.0	20.0	5.0	5.0	20
Yamal & Taymyr ♀	-	-	-	-	-	30.3	66.7	3.0	-	-	33
Central Siberia ♀	-	-	-	-	-	18.8	-	75.0	-	6.3	16
E Siberia ♀	-	-	-	-	-	7.7	7.7	23.1	23.1	38.5	13
Alaska ♀	-	-	-	-	-	4.5	-	2.3	27.3	65.9	44
	Wing- and culmen length										
Fennoscandia ♂	17.6	38.2	29.4	11.8	2.9	-	-	-	-	-	34
Yamal & Taymyr ♂	19.0	73.8	2.4	4.8	-	-	-	-	-	-	42
Central Siberia ♂	16.7	8.3	58.3	16.7	-	-	-	-	-	-	12
E Siberia ♂	4.8	-	28.6	38.1	28.6	-	-	-	-	-	21
Alaska ♂	3.6	1.8	5.4	26.8	62.5	-	-	-	-	-	56
Fennoscandia ♀	-	-	-	-	-	39.1	26.1	21.7	8.7	4.3	23
Yamal & Taymyr ♀	-	-	-	-	-	21.2	69.7	6.1	3.0	-	33
Central Siberia ♀	-	-	-	-	-	12.5	-	75.0	6.3	6.3	16
E Siberia ♀	-	-	-	-	-	15.4	-	23.1	15.4	46.2	13
Alaska ♀	-	-	-	-	-	4.5	-	-	29.5	65.9	44

The high accuracy rates in the E Siberian and Alaskan POSCON groups were remarkable and clearly demonstrate the morphometrical differences between both. High accuracy rates were also reached in the ♂ POSCON groups. POSCON groups with intermediate measurements had a higher error rate.

When only wing- and culmen lengths were available, the average accuracy rate was 56.4 % with unknown sex and 61.8 % with known sex. Additional measuring of tarsus length resulted in a small increase with resp. 2.8 and 2.2 %.

Table 83. *Accuracy of prediction in the Bar-tailed Godwit when sexes are* ***unknown****. The columns show the predicted proportions in relation to the total number per breeding area (last column).*

	Predicted										
Locality	Fennoscandia ♂	Yamal & Taymyr ♂	Central Siberia ♂	E Siberia ♂	Alaska ♂	Fennoscandia ♀	Yamal & Taymyr ♀	Central Siberia ♀	E Siberia ♀	Alaska ♀	*n*
	Wing-, culmen- , tarsus-, secondary-, tail- and middle toe length										
Fennoscandia ♂	44.4	11.1	38.9	5.6	-	-	-	-	-	-	18
Yamal & Taymyr ♂	10.3	58.6	27.6	3.4	-	-	-	-	-	-	13
Central Siberia ♂	20.0	-	80.0	-	-	-	-	-	-	-	5
E Siberia ♂	-	7.7	-	61.5	23.1	-	7.7	-	-	-	13
Alaska ♂	2.1	2.1	8.5	17.0	63.8	-	-	-	4.3	2.1	47
Fennoscandia ♀	-	-	-	-	-	38.5	30.8	23.1	-	7.7	13
Yamal & Taymyr ♀	-	-	-	4.3	-	26.1	69.6	-	-	-	23
Central Siberia ♀	-	-	-	-	-	20.0	-	80.0	-	-	5
E Siberia ♀	-	-	-	-	-	-	-	14.3	71.4	14.3	7
Alaska ♀	-	-	-	-	3.3	-	-	10.0	26.7	60.0	30
	Wing-, culmen- and tarsus length										
Fennoscandia ♂	31.3	28.1	34.4	3.1	3.1	-	-	-	-	-	32
Yamal & Taymyr ♂	21.4	66.7	9.5	2.4	-	-	-	-	-	-	42
Central Siberia ♂	18.2	9.1	45.5	18.2	-	-	9.1	-	-	-	11
E Siberia ♂	4.8	-	9.5	47.6	23.8	-	14.3	-	-	-	21
Alaska ♂	5.5	-	1.8	18.2	63.6	-	5.5	-	5.5	-	55
Fennoscandia ♀	-	-	-	-	-	45.0	25.0	20.0	5.0	5.0	20
Yamal & Taymyr ♀	-	-	6.1	6.1	12.1	30.3	42.4	3.0	-	-	33
Central Siberia ♀	-	-	-	-	-	18.8	-	75.0	-	6.3	16
E Siberia ♀	-	-	-	-	7.7	7.7	-	23.1	23.1	38.5	13
Alaska ♀	-	-	-	-	-	4.5	-	2.3	27.3	65.9	44
	Wing- and culmen length										
Fennoscandia ♂	17.6	38.2	29.4	11.8	2.9	-	-	-	-	-	34
Yamal & Taymyr ♂	19.0	73.8	2.4	4.8	-	-	-	-	-	-	42
Central Siberia ♂	16.7	8.3	50.0	16.7	-	-	8.3	-	-	-	12
E Siberia ♂	4.8	-	19.0	33.3	28.6	-	14.3	-	-	-	21
Alaska ♂	3.6	1.8	1.8	25.0	53.6	1.8	8.9	-	3.6	-	56
Fennoscandia ♀	-	-		-	-	39.1	26.1	21.7	8.7	4.3	23
Yamal & Taymyr ♀	-	-	6.1	3.0	9.1	21.2	51.5	6.1	3.0	-	33
Central Siberia ♀	-	-	-	-	-	12.5	-	75.0	6.3	6.3	16
E Siberia ♀	-	-	-	-	7.7	15.4	-	23.1	7.7	46.2	13
Alaska ♀	-	-	-	-	-	4.5	-	-	29.5	65.9	44

Exclusivity

When birds were not sexed, it was impossible to reach some degree of exclusivity. The estimated exclusivity rates were 1-2 %, irrespective of the inclusion of 2 to 3 measurements. When birds were sexed, the exclusivity rates were estimated at 2-4 % with wing- and culmen lengths and 2-8 % with wing-, culmen- and tarsus lengths. Exclusivity was only reached in the smallest- and largest-sized samples.

BODY MASS

In the W Palearctic most of the body masses obtained were from ♂♂. Here the average increased from 230 gm in late June towards 300 gm in mid-July, apparently because of premigratory fattening. The daily mass gain as based on these data was estimated at approx. 4 gm for ♂♂.
Some body masses of E Siberian and Alaskan Bar-tailed Godwits were obtained from birds collected rather early in the breeding season (Fig. 91). ♀♀ were heavier than ♂♂ and both sexes decreased in mass during June. There was no sign of premigratory fattening during July and early August with ♀♀ weighing approx. 285 g and ♂♂ 250 gm.

PRIMARY MOULT

No breeding birds in active primary moult were present in the material studied: Bar-tailed Godwits apparently do not moult their primaries on the breeding grounds.

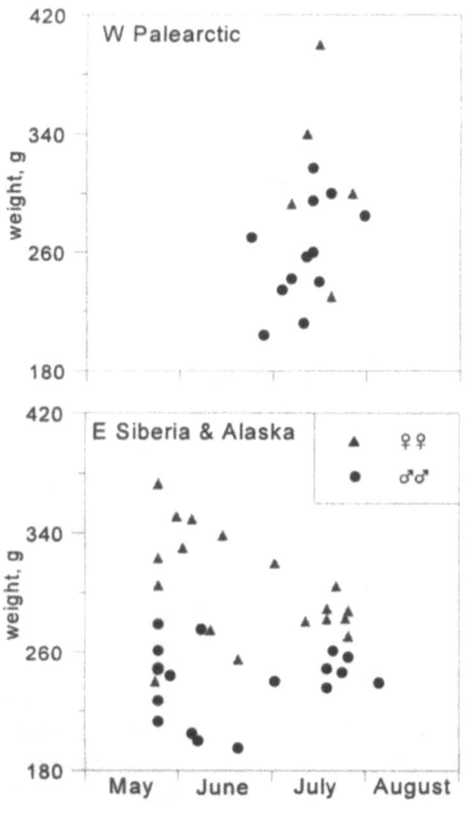

Fig. 91. *Body mass of Bar-tailed Godwits during the breeding season.*

DISCUSSION

The taxonomy of the Bar-tailed Godwit

The data presented above indicate the existence of 5 distinct populations. There are good reasons to distinguish them all 5 on the subspecies level as: (1) all 5 populations have spatially segregated breeding grounds, (2) most can be distinguished with morphometric data, and (3) if measurements do not result in enough separation, plumage data can do. In fact, there are no reasons to suppose heterogeneity or the existence of clinal variation within these populations since the relative standard deviations of the measurements within each population are not extremely large and comparable with the ones found in other species (e.g. Black-tailed Godwit - Table 70).

In the W Palearctic 2 lightly coloured subspecies can be distinguished and the Swedish population ought to be considered as the nominate type. The Taymyr-population ought to receive a new subspecies' name since no other names have been used in the past for this population. In the E Palearctic it is more complicated to apply the nomenclatural rules, due to the presence of several older latin names which are no longer in use nowadays. The Central Siberian population can be recognized as *menzbieri* according to the older literature. However, 4 names might apply to both Bering Sea populations: *Limosa baueri*, *Limosa novaezealandiae*, *Limosa foxii* and *Limosa uropygialis*. The corresponding type-specimen belonging to these taxa are described below:

1. *Limosa baueri* Naumann, 1836, Naturgeschichte Vögel Deutschlands VIII, p. 429, 'Neuholland' [errore].
   Type: NMW 40.990 [Wien; old catalogue nr. 1826.V80], 21 September 1804, Norfolk Island (Tasman Sea), sex unknown (see Schifter 1992).
   Measurements from skin (formerly mounted, supplied by Dr Ernst Bauernfeind d.d. 1 July 1996): wing 235 mm (somewhat curved), culmen ± 110 (tip damaged), tarsus 62 (somewhat unreliable due to iron in leg),

middle toe 37.2.
Plumage (concluded from photo supplied): like adult ♀ breeding, but body a mixture of old winter, rather old breeding, and fresh winter plumage, upper wing-coverts and tertials generally heavily worn and wing appears fairly new. Eventually a summering 1- or 2-year-old; wing certainly not juvenile, though may not have attained a length comparable with fully adult (CSR).
Remarks: according to Hartert & Steinbacher (1938. p. 481) and Hellmayr & Conover (1948) a *nomen nudum*, but name available according to Ridgway (1919, vol. 8), Peters (1934, vol. 2), Portenko (1936), Vaurie (1959), and HANZAB (1996, vol. 3, p. 82)

2. *Limosa novaezealandiae* G.R. Gray, 1845, Zool. Voy. HMS Erebus & Terror 1839-1843, vol. 1, Birds, p. 13, New Zealand. Based on a small series, of which 2 birds are still available.
co-type 1: BMNH 42.12.16.61 [Tring], Bay of Islands, New Zealand, 'July' 1841' [is probably August 1841 or later, as ships 'Erebus' and 'Terror' did not arrive there until 18 August 1841 and departed 3 months later], sex unknown, original collection number 60. Mentioned as 'the type' and said to be collected by P. Earl (Warren 1966), but unknown on what evidence; collector on label reads Surgeon R. McCormick (R. Prŷs-Jones). In card system of Warren: 'of the 2 co-types, this one selected as type'. Also, type of *Limosa brevipes* Gray, 1844, *nomen nudum*.
Measurements from skin (formerly mounted, supplied by R. Prŷs-Jones d.d. 30 August 1996): wing 224 mm, culmen 99.0, tarsus ± 52.0, middle toe with claw ± 34.0 ('don't put too much evidence on tarsus and toe; measurements untrustworthy' - RPJ).

co-type 2: BMNH 44.1.18.141, [Tring], 'New Zealand', also collected during the voyage of HM vessels Erebus and Terror, but no further data; omitted in Warren (1966); said to be ♀, but unlikely to be so (R. Prŷs-Jones).
Measurements from skin (formerly mounted, supplied by R. Prŷs-Jones d.d. 30 August 1996): wing 222 mm, culmen 79.0, tarsus ± 56.0, middle toe with claw ± 35.0.

Plumage (both co-types from photo's supplied): Both generally in rather worn winter plumage, but with some fresh breeding plumage feathers admixed (far fewer than Vienna bird); apparently oversummering 1- or 2-year-olds; wing coverts and tertials old, wing rather old, but highly probably not juvenile, lacking obvious white fringes to unworn inner feathers; thus both 2-year old, as Vienna bird, though with less breeding plumage? Summer plumage feathers of both co-types extend well to belly, as in ♂, but differences in bill length and -shape suggest 1 is ♀ and 2 is ♂ (CSR).

3. *Limosa foxii* Peale, 1848, US Explor. Exped. 8, p. 231, Rose Island, Samoa.
Type: USNM 15367 [Washington], Rose Island (14°31'30" S, 168°08'30" W), Samoa, 'adult', date and sex unknown, collected 1838-1842 (collector J.L. Fox according to Deignan 1961, p. 85), original collection nr. 630. 'Apparently adult, no idea about sex' (J.P. Angle).
Measurements from skin (formerly mounted, supplied by J. Phillip Angle d.d. 11 Oct. 1996): wing 218 mm ('measured in European way'), tail 71.5, culmen 74.7, tarsus 56.7, middle toe ± 34.2 (slightly curved).

4. *Limosa uropygialis* Gould, 1848, Proc. Zool. Soc. London 16, p. 38, 'Australia' [= New South Wales, SE Australia].
Type: ANSP 11395 [Acad. Nat. Sci. Philadelphia], no further data from label supplied, 'probably immature ♀' (D. Agro).
Measurements from skin (supplied by David Agro d.d. 6 Jan. 1997): wing 221 mm ('measured in European way'), tail 64, culmen 78.6, tarsus 69.9, middle toe 42.9.

With the obtained measurements from these type specimen it was tried to locate the geographical breeding origins by using POSCON (described in Annex 3). Before such analyses could be performed, measurements had to be made comparable which means corrections for age and post-mortem shrinkage:

- Apart of the Godwit from Samoa, all type specimen were immatures. This makes correction of the immature wing lengths to 'adult' lengths necessary. A 2% increase was used according to Engelmoer *et al.* (in prep.). It also appeared from this study that it was not necessary to correct culmen lengths for age differences.

- Corrections for post-mortem shrinkage were based upon the conclusions in Chapter 2. The wing length was multiplied with a factor 1.0194 and the culmen length with 1.0172 (corresponds with 1.69% shrinkage) .

The corrected measurements and the outcome of the POSCON-analyses are presented in Table 84. The results are as follows:

- The *baueri*-type might well belong to both the Anadyr- and Alaska population.
- The first co-type of *novaezealandiae* is mostly resembling the Central Siberian population.
- The second co-type of *novaezealandiae* and the *foxii*-type have most resemblance with the Anadyr-population.
- The *uropygialis*-type did not resemble any of the known populations which might be due to the long tarsus.

These results are ambiguous and do not help us to characterize the Alaskan- and Anadyr breeding populations in a taxonomic way. They also strengthen us in the opinion that non-breeding birds ought never be used to characterize subspecies, since in many wader species several breeding populations intermix on the wintering grounds. In a final attempt to seize the taxonomic position of either the Alaskan or the Anadyr population, we analysed the measurements of another 10 birds wintering in New Zealand and E Australia (Table 85). These were all adults and thus wing- and culmen lengths were only corrected for post-mortem shrinkage. These birds might breed in Central Siberia, Ana-

Table 84. *Results of POSCON-analysis used to reveal the geographical breeding origin of some type-specimen of the Bar-tailed Godwit. These non-breeders were collected on the Southern Hemisphere in the Pacific region. [1]The corrected wing- and culmen lengths are presented here (correction-values presented in text). The wing length of the Samoa-bird was only corrected for museum-shrinkage. Two alternatives were analysed: (1) ALL - all available measurements involved; (2) WCT - wing-, culmen- and tarsus lengths involved.*

Coll. number	40990		42121661		44118141		15367		11395	
Subspecies	*baueri*		*novaezealandiae*		*novaezealandiae*		*foxii*		*uropygialis*	
Area	Norfolk Island (Tasman Sea)		Bay of Islands (New Zealand)		'New Zealand'		Rose Island Samoa		New S Wales - SE Australia	
Sex on label	-		-		♀		-		-	
Presumed sex	♀♀						♂♂			
Wing length[1]	244.5		233.0		230.9		222.2		229.9	
Culmen length[1]	111.9		100.7		80.4		76.0		80.0	
Tarsus length	62.0		52.0		56.0		56.7		69.9	
Secondary length	-		-		-		-		-	
Tail length	-		-		-		71.5		64.0	
Middle toe length	37.2		34.0		35.0		34.2		42.9	
Typicality probabilities										
	ALL	WCT	ALL	WCT	ALL	WCT	ALL	WCT	ALL	WCT
Central Siberia ♂	0	0	0.004	0.0404	0.0252	0.1675	0.0171	0.0883	0	0
Central Siberia ♀	0.0284	0.0467	0.0512	0.2485	0.006	0.001	0	0	0	0
Anadyr ♂	0.001	0	0.0382	0.0388	0.4067	0.7337	0.1012	0.183	0	0
Anadyr ♀	0.0523	0.4895	0.007	0.0857	0.005	0.007	0	0	0	0
Alaska ♂	0	0	0.031	0.1335	0.1304	0.447	0.0115	0.0268	0	0
Alaska ♀	0.0233	0.5189	0	0.0215	0	0.001	0	0	0	0

Table 85. *Results of POSCON-analysis used to reveal the geographical breeding origin of 10 wintering birds in New Zealand and SE Australia. [1]The corrected wing- and culmen lengths are presented here (correction-values presented in text). Two alternatives were analysed: (1) all available measurements involved; (2) wing-, culmen- and tarsus lengths involved.*

Museum	Berkeley	Berkeley	Berkeley	Toronto	Tring	Berkeley	Berkeley	Berkeley	Toronto	Berkeley
Collection number	100633	100635	100636	9111930	?	100632	100634	82045	38231	142184
Area	New Zealand					New Zealand				Australia
Sex	♂♂					♀♀				♀
Wing length[1]	231.4	221.2	234.5	228.0	255.0	235.5	246.7	246.7	246.0	226.3
Culmen length[2]	85.5	89.8	82.7	88.9	82.2	112.0	113.4	107.5	111.0	101.8
Tarsus length	52.4	54.6	56.8	54.1	56.4	58.9	61.8	60.0	59.2	54.1
Secondary length	92.0	92.0	90.0	-	-	97.0	104.0	101.0	-	96.0
Tail length	76.0	71.0	71.0	-	-	72.0	85.0	85.0	-	71.0
Middle toe length	37.4	42.0	40.0	-	-	-	-	-	-	38.4
Typicality probabilities based upon all available measurements										
Central Siberia ♂	0.3899	0.1673	0.0488	-	-	0.002	0	0	-	0.2777
Central Siberia ♀	0.079	0.005	0	-	-	0.8063	0.0118	0.0455	-	0.9025
Anadyr ♂	0.8686	0.006	0.1948	-	-	0.004	0	0	-	0.0953
Anadyr ♀	0.0415	0	0.004	-	-	0.3874	0.0218	0.1239	-	0.1418
Alaska ♂	0.9859	0.005	0.2936	-	-	0.002	0	0	-	0.0529
Alaska ♀	0.023	0	0.003	-	-	0.3217	0.008	0.0997	-	0.0438
Typicality probabilities based upon wing-, culmen- and tarsus lengths										
Central Siberia ♂	0.311	0.7614	0.0364	0.6343	0	0	0	0	0	0.0783
Central Siberia ♀	0.002	0.0146	0	0.0194	0	0.5897	0.0195	0.0313	0.0885	0.6654
Anadyr ♂	0.6742	0.4453	0.4666	0.904	0	0	0	0	0	0.0294
Anadyr ♀	0.004	0.004	0.003	0.0209	0	0.5602	0.405	0.6952	0.7542	0.1001
Alaska ♂	0.9711	0.1756	0.3541	0.8579	0	0	0	0	0	0.0328
Alaska ♀	0	0	0	0.002	0	0.179	0.6495	0.9836	0.9694	0.006

dyr or Alaska. Probabilities were calculated 17 times (Table 85). Central Siberia and Alaska had 6 times the highest scores and Anadyr 3 times. In the remaining 2 estimates the bird did not resemble any of the known populations.

The above presented results do not lead to an answer on questions like: 'Is it the Alaskan or the Anadyr population which ought to be considered as *baueri*?' or 'What would be a correct name for the remaining population?'. We therefore propose to follow the recent literature (Vaurie 1965, Dement'ev *et al.* 1969, Prater *et al.* 1977, Cramp & Simmons 1983), in which the Alaskan population is considered as *baueri*: the Alaskan breeding birds were taken to represent *baueri* in its purest form, being the largest-sized and darkest population, whereas the Anadyr population is considered to be influenced by *menzbieri*. The following subspecies are then distinguished:

- *Limosa lapponica lapponica* (Linneaus, 1758), breeding in N Fennoscandia, along the north coast of the White Sea and on the Kanin Peninsula. Nominate *lapponica* was originally described from Swedish Lappland (Vaurie 1965). Therefore, the westernmost breeding population has to be referred to as *L. l. lapponica*. It is one of 2 intermediate-sized subspecies. The other is *L. l. menzbieri*. The whitish axillaries and the paucity of brown marks on the white upper tail-coverts differentiate this subspecies from *L. l. menzbieri*. The ♀♀ of both populations also differ morphometrically, especially in culmen lengths.

- *Limosa lapponica taymyrensis* (Engelmoer & Roselaar, *nov. ssp*) breeds from the Yamal Peninsula east to the delta of the Anabar River. The holotype is at the Zoological Institute (Petersburg) coll. nr. 45805, an adult ♂ collected near Lake Taymyr (74.35°N 103°E) on 8 June 1843 without brood patches, and with measurements wing 215, culmen 80.0, tarsus 49.7, secondary 88, tail 65, middle toe 35.7, axillary score 'D', upper tail-covert score '3'. This subspecies is the smallest measured, and morphometrically distinct from every other population. It has most resemblance with the N Fennoscandian population, but it is smaller and the axillaries and upper tail-coverts are darker.

- *Limosa lapponica menzbieri* (Portenko, 1936) breeds from the Lena Delta east to the Chaunsk Bay. The type locality of this subspecies is Russkoye Ust'ye, Indigirka Delta (71.04°N 149.00°E); This is the other of 2 intermediate-sized subspecies - see *L. l. lapponica* above for the differences. Due to its description *menzbieri* was often considered as an intermediate between nominate *lapponica* and *baueri*, not deserving recognition. Since its characters were nearer to *baueri*, it was usually included in the latter (Vaurie 1965, Glutz *et al.* 1977, Cramp & Simmons 1983). However, it can clearly be separated on morphometrical grounds from the E Siberian and Alaskan breeding populations. Moreover, the breeding areas of *baueri* and *menzbieri* are spatially segregated.

- *Limosa lapponica anadyrensis* (Engelmoer & Roselaar, *nov. ssp.*) breeds at Anadyr and along the south part of the Chukotsk Peninsula. Holotype at the Zoological Institute (St. Petersburg) coll. nr. 45871, an adult ♀ collected near Markovo (Magadan region; 64.40°N 170.24°E) on 3 June 1897 with brood patches, and with measurements wing 244, culmen 106.4, tarsus 57.7, secondary 98, tail 79, middle toe 40.5, axillary score 'E', upper tail-covert score '5'. This is the first of 2 large-sized subspecies. The other is *baueri*. The E Siberian birds have on average smaller measurements than the Alaskan ones, although the secondary lengths are longer, resulting in a more blunt wing shape. Both populations share the dark axillaries and upper tail-coverts.
  Application of the earlier applied names *novaezealandiae*, *foxii* and *uropygialis* is considered impossible, since the breeding origin of their type specimens remained unclear (see above).

- *Limosa lapponica baueri* (Naumann, 1836) is breeding in Alaska. Originally, it was described from Norfolk Island, Tasman Sea (see Schifter 1992). Australian wintering birds are breeding in E Siberia and Alaska. It is the second of the 2 large-sized and darkly coloured subspecies: see above. It is distinguishable from *anadyrensis* by the more pointed wing shape: 6-10 mm longer wing lengths in combination with somewhat smaller secondary lengths. The other measurements average 0.5 mm longer.
  In order to dissolve the confusion with the type specimens (see above), we proclaim as neotype for this subspecies an adult ♂ in the collection of the Museum of Vertebrate Zoology, Berkeley, collection nr. 60467, collected near Cape Etolin, Nunivak (60° N 166° W) on 3 June 1931. This bird had not yet developed brood patches. Its measurements are: wing length 226 mm, culmen 83.4, tarsus 53.8, secondary 92, tail 73, middle toe 35, axillary score 'F', upper tail covert score '6'. The here presented wing- and culmen lengths were not corrected for post-mortem shrinkage.

SUMMARY

Measurements and plumage characters of all breeding populations of the Bar-tailed Godwit were analysed with multivariate statistics. The species showed marked sexual and geographical variation. Sexual differences were larger than geographical differences and ♂♂ were smaller-sized than ♀♀. Sexual dimorphism was smallest in the breeding population of Yamal & Taymyr and largest in the Alaskan population. The species is polytypic and 5 subspecies were recognized: (1) *L. l. lapponica*, breeding in N Fennoscandia, the north part of the White Sea and on the Kanin Peninsula; (2) *L. l. taymyrensis* nov. ssp., breeding from Yamal east to the delta of the Anabar River; (3) *L. l. menzbieri*, breeding from the Lena Delta east to the Chaunsk Bay;

(4) *L. l. anadyrensis* nov. ssp., breeding at Anadyr and along the south part of the Chukotsk Peninsula; and (5) *L. l. baueri*, breeding in Alaska. *Taymyrensis* had small measurements, *lapponica* and *menzbieri* were intermediate-sized and *anadyrensis* and *baueri* were large-sized. Both intermediate-sized subspecies differed from each other in the axillary pattern and darkness of the upper tail. Both large-sized subspecies differed in measurements: Alaskan birds were generally larger and had more pointed wings.

All variables contributed significantly to the group distinctions. The darkness of the upper tail-coverts and the color pattern of the axillaries were important criteria as well. The decrease in body mass after arrival on the breeding grounds was 50-100 gm. The breeding birds of Taymyr and N Fennoscandia increased in body mass during the first half of July and prior to departure from the breeding grounds. No primaries were moulted during breeding.

Chapter 14

Whimbrel

-

Numenius phaeopus

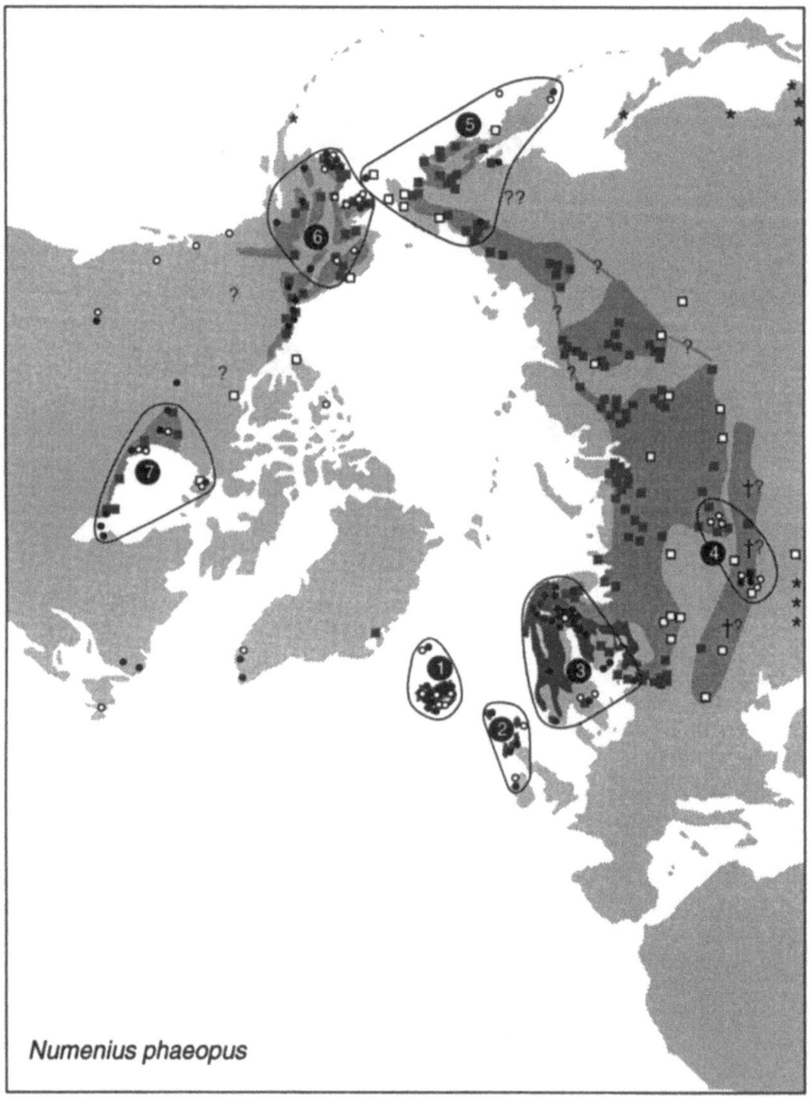

Fig. 92. *Breeding distribution of the Whimbrel.*

INTRODUCTION

The Whimbrel is breeding throughout the northern Holarctic (Fig. 92). The existence of geographical variation in the breeding populations of the Whimbrel is summarized by Vaurie (1965), Glutz *et al.* (1977), Prater *et al.* (1977), Cramp & Simmons (1983), Stepanyan (1990) and Rogacheva (1992). Often, up to 4 races are distinguished in the Whimbrel:

- *N. p. phaeopus*, breeding from Iceland in the west to SW Taymyr, the middle Yenisei and Omsk in the east.
- *N. p. alboaxillaris*, breeding in the steppes east of the lower Volga and south and south-east of the Urals.
- *N. p. hudsonicus*, breeding in N America.
- *N. p. variegatus*, breeding in NE Siberia. The west border of its breeding range stretches to the middle Yana basin and perhaps SE Taymyr.

Salomonsen (1947) separated the Icelandic breeding population as *N. p. islandicus*, based on larger dimensions. The subspecies is generally not accepted nowadays, but was found to be valid. *N. p. phaeopus* and *N. p. variegatus* have more or less the same measurements but differ in plumage (Vaurie 1965, Cramp & Simmons 1983). It is not clear whether the breeding ranges of both subspecies overlap or not. According to Stepanyan (1990) there is overlap on the Taymyr Peninsula and in the Yenisei valley, but Rogacheva (1992) stated that both subspecies inhabit separate ranges in Central Siberia, with the nominate race breeding at least as far east as 90°E (Lake Khantayskoye, incl. NW Putorana) and *variegatus* west to 105°E (east part of Evenkia district). Vaurie (1965) and Stepanyan (1975, 1990) did not accept the distinction of *N. p. alboaxillaris*: they considered these breeding birds as the lightest individuals of the variably coloured nominate subspecies. This steppe-breeding population is near to extinction nowadays (Belik 1994).

Zink *et al.* (1995) recently showed that *variegatus* and *hudsonicus* importantly differed both in mtDNA pattern as well as morphology. They proposed to distinguish them as separate species. And they suggested that possibly even 3 species may be involved: *Numenius phaeopus*, *Numenius variegatus*, *Numenius hudsonicus*.

MORPHOMETRICAL DIFFERENTIATION

In total, measurements of 176 breeding Whimbrels (56.8% ♂♂) were examined from a large part of the breeding range (Fig. 92, Table 86). Since relatively few Central and E Siberian skins were measured, 37 skins of non-breeders from E- and SE Asia were also used in the morphometric analyses.

The breeding grounds are reached from May onwards (Fig. 93). Breeding birds in Fennoscandia were already collected by early May, the ones on Iceland by early June. Birds in Kazakh-

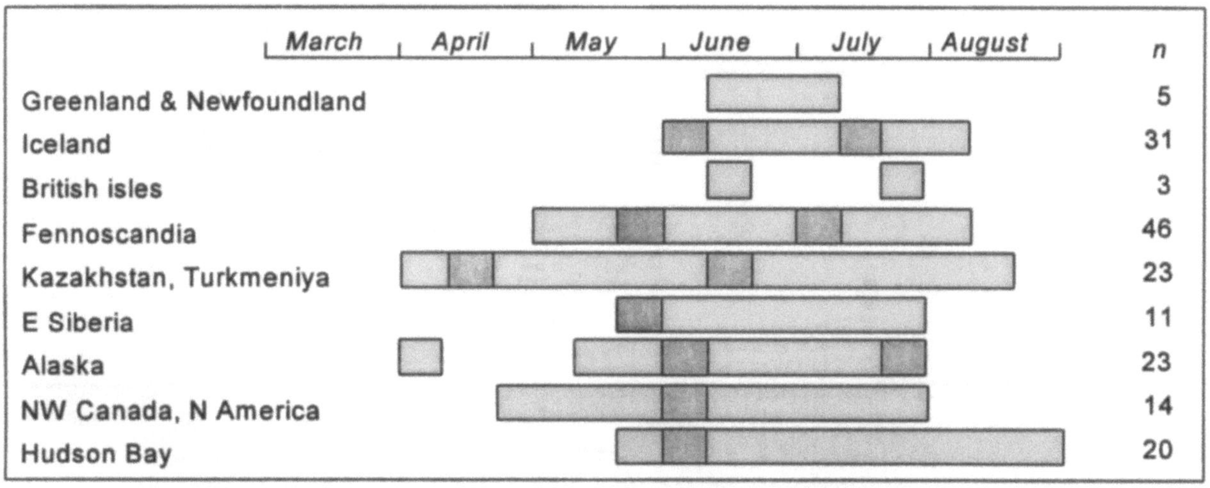

Fig. 93. *Collection dates of Whimbrels. Most birds had been collected during the periods marked by the heavily shaded bars. The sample sizes are presented on the right side.*

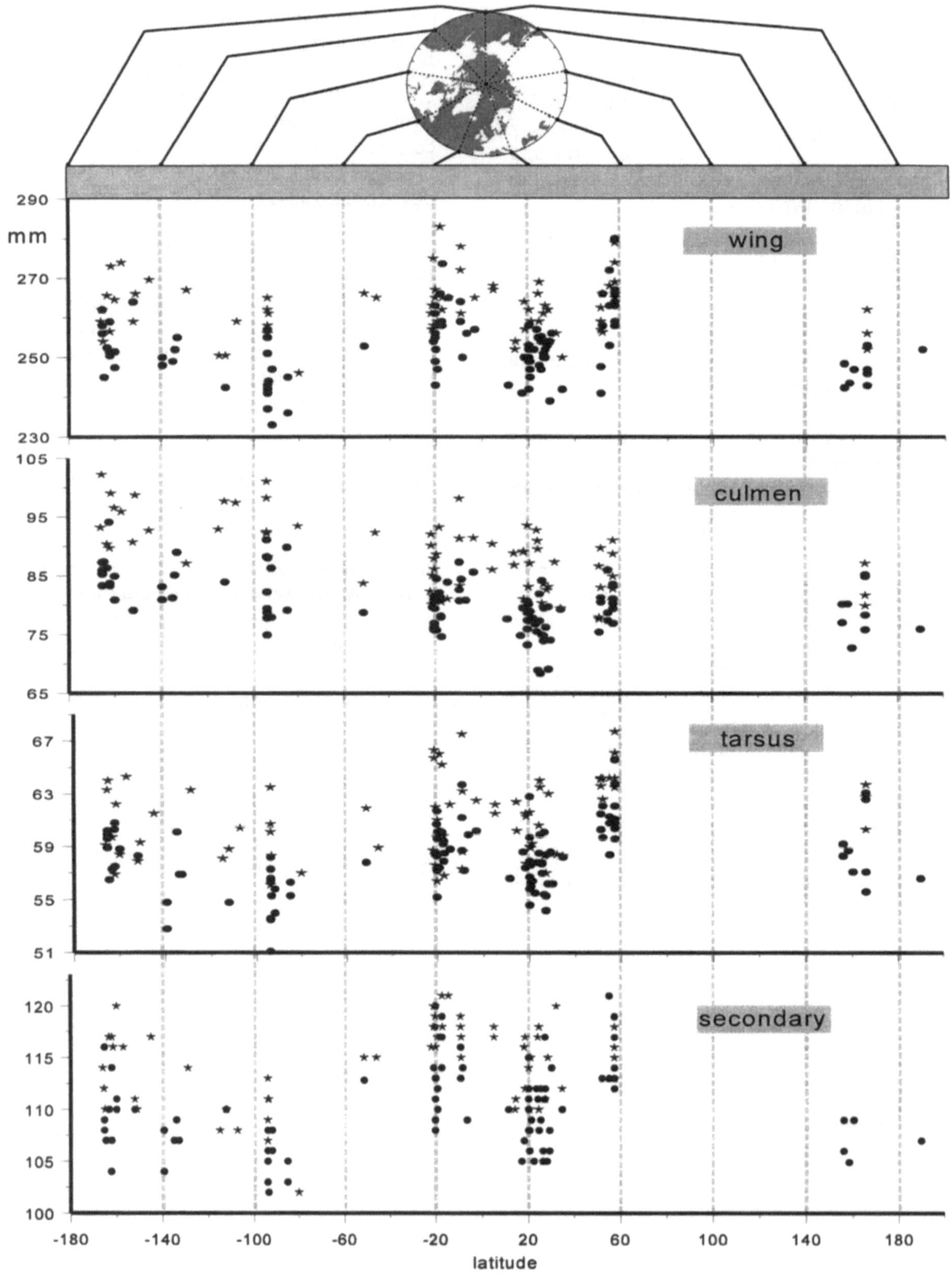

Fig. 94. *Measurements of Whimbrels on the breeding grounds. The following measurements are presented from top to bottom: wing length, culmen length, tarsus length and secondary length (description in chapter 2). Explanation of the used symbols: ● ♂♂, ★ ♀♀.*

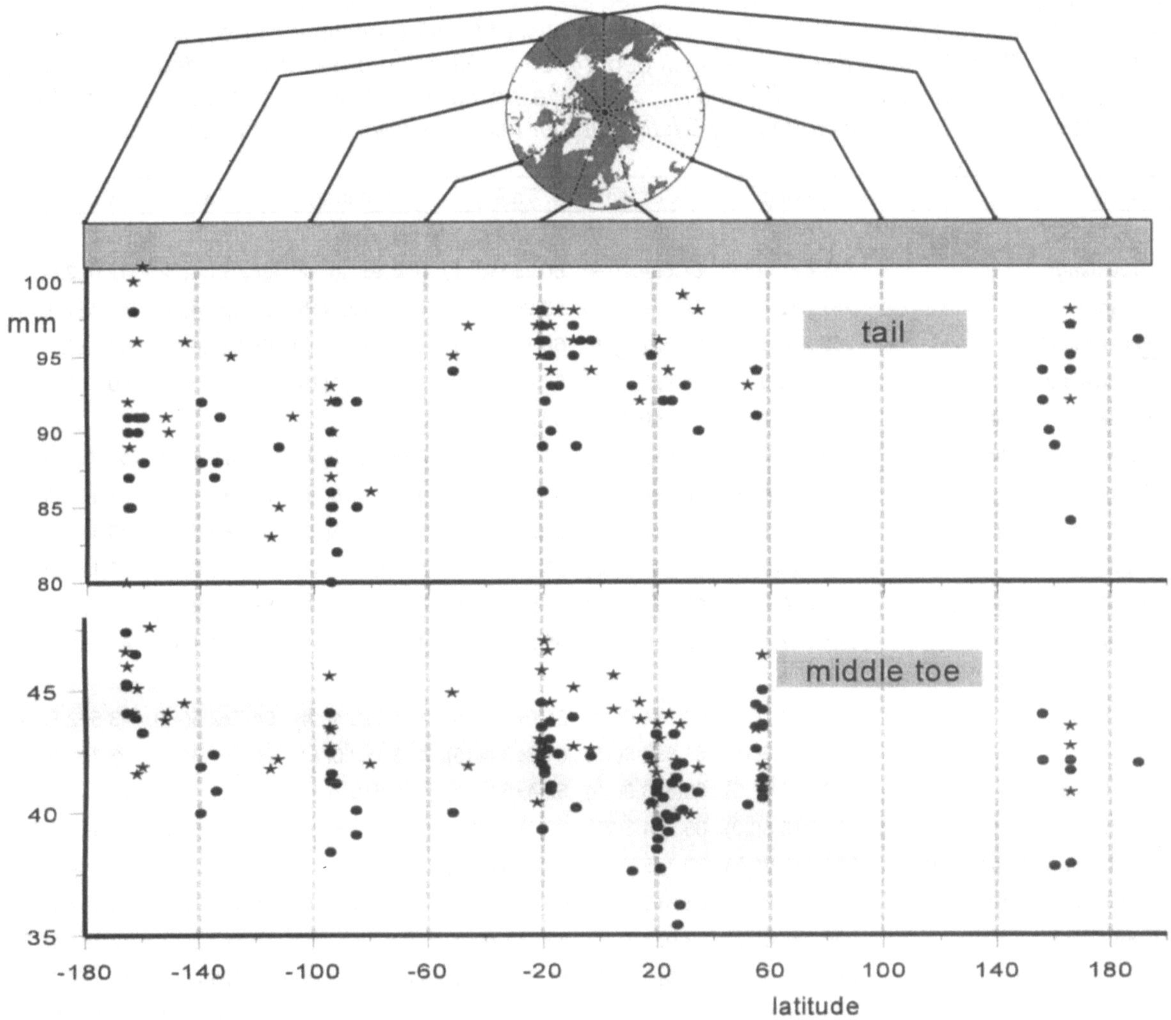

Fig. 95. *Measurements of Whimbrels on the breeding grounds. Both tail- and middle toe lengts are presented. Explanation of the used symbols: ● ♂♂, ★ ♀♀.*

stan were present from April. The breeding birds of Alaska seem to arrive a few weeks earlier than the ones in the Hudson Bay. Most of the skins had been collected during June and July. Only few birds had been collected after the end of July.

Initially, 7 distinct breeding locations were distinguished for the morphometric analyses (Fig. 92, Table 86). However, the skins from Kazakhstan and Turkmeniya were separated in 2 groups. This resulted in the distinction of 8 different sample groups. The skins from Kazakhstan and Turkmeniya were split according to their coloration and axillary pattern.The 2 distinguished types were (1) *alboaxillaris* and (2) nominate *phaeopus*. In *alboaxillaris* the under wing-coverts, axillaries and upper tail-coverts were white and the fringes of the upper wing-coverts were paler than in nominate *phaeopus*.

The collected birds of 2 regions were not used beforehand in the morphometric analyses:

1. Greenland & Newfoundland. This sample was not used since no regular breeding areas are recognized here, even though some birds had brood patches and were collected during the breeding season.

2. NW Canada. The measurements of nearctic breeders showed a gradual decrease from west to east with the Alaskan birds being large-sized and the Hudson Bay

Table 86. *Measurements of Whimbrels. Presented are the means ± 1 SD (n).*
Notes: (1) 3 Whimbrels (2♂♂, 1♀) from the northwest coast of Newfoundland included; (2) 2 Jan Mayen birds included; (3) a skin from the Faeroes included; (4) 3 NW Russian breeders included, (5) 5 spring-migrants from Turkmeniya, Mangyshlak and Bashkirskaya included, (6) 9 spring-migrants on Kamchatka and the Commander Islands included, (7) 2 late-summer migrants from James Bay included.

	Locality	Wing	Culmen	Tarsus	Secondary	Tail	Middle toe
			♂♂				
	Greenland[1]	263.4±15.0(2)	81.6±4.0(2)	60.1±3.2(2)	113.9±1.6(2)	95.0±1.4(2)	42.0±2.8(2)
1	Iceland[2]	257.6±7.5(17)	79.1±3.0(17)	58.9±1.5(17)	113.9±3.7(15)	94.1±3.4(16)	42.3±1.3(14)
2	British isles[3]	258.3±5.3(6)	83.6±2.7(6)	60.1±2.2(6)	113.0±2.5(5)	95.0±3.0(6)	42.1±1.9(3)
3	Fennoscandia[4]	250.0±5.0(30)	76.8±3.7(30)	57.4±1.8(30)	109.2±3.1(29)	92.4±1.5(7)	40.1±1.9(27)
4	Kazakhstan - *phaeopus*[5]	243.4±3.7(3)	75.6±0.3(2)	59.8±2.0(3)	-	-	-
4	Kazakhstan - *alboaxillaris*	264.1±7.1(12)	81.0±2.7(12)	61.3±1.9(12)	114.7±3.2(10)	92.5±2.1(2)	42.7±1.7(11)
5	E Siberia[6]	246.9±3.8(9)	78.9±4.2(9)	58.7±2.6(9)	107.2±1.8(5)	92.3±4.1(9)	41.1±2.3(7)
	E & SE Asia (non-breeding)	242.9±7.1(22)	78.0±7.6(21)	57.2±2.5(22)	105.7±1.5(3)	87.7±4.7(15)	40.1±1.6(15)
6	Alaska	254.6±5.8(12)	85.1±3.8(12)	58.9±1.4(11)	109.6±3.3(11)	89.7±3.6(11)	45.1±1.5(7)
	NW Canada & N America	248.6±6.4(9)	83.5±3.6(9)	55.9±2.1(9)	107.9±2.1(9)	88.2±3.2(9)	41.8±1.5(6)
7	Hudson Bay[7]	244.3±7.4(12)	82.8±5.6(12)	55.3±2.0(12)	105.2±2.0(10)	85.8±4.1(12)	41.2±1.8(9)
			♀♀				
	Greenland[1]	269.0±9.5(4)	90.9±4.9(4)	61.0±2.9(4)	116.5±2.4(4)	94.7±2.5(3)	43.6±2.5(4)
1	Iceland[2]	265.7±7.5(15)	85.6±4.6(15)	61.3±3.4(15)	118.1±2.0(13)	96.7±1.3(11)	43.7±1.9(13)
2	British isles	269.7±4.0(3)	93.6±3.9(3)	62.4±5.1(3)	118.5±0.7(2)	95.0±1.4(2)	42.6 - (1)
3	Fennoscandia[4]	259.2±5.8(20)	86.2±4.3(20)	60.4±2.1(20)	113.7±3.8(19)	95.7±2.6(6)	42.4±1.7(17)
4	Kazakhstan - *phaeopus*[5]	256.9±7.2(7)	83.1±5.7(7)	63.4±0.8(3)	-	93.0 - (1)	-
4	Kazakhstan - *alboaxillaris*	268.0±6.5(8)	86.3±3.5(8)	64.4±1.9(8)	115.2±1.9(6)	94.0 - (1)	42.6±2.1(6)
5	E Siberia[6]	256.7±5.0(3)	82.9±3.7(3)	62.3±1.8(3)	-	95.7±3.2(3)	42.3±1.4(3)
	E & SE Asia (non-breeding)	250.9±5.5(15)	89.0±5.2(15)	61.3±2.9(15)	112.0 - (1)	92.7±5.2(6)	41.1±1.7(6)
6	Alaska	263.9±6.5(11)	94.2±4.7(11)	60.6±2.6(11)	114.5±3.4(11)	92.8±6.4(9)	44.5±1.9(10)
	NW Canada & N America	259.7±8.7(6)	92.6±4.2(6)	60.2±1.8(6)	110.3±3.4(6)	89.5±4.6(6)	43.2±1.8(4)
7	Hudson Bay[7]	259.0±7.0(8)	92.7±4.9(8)	59.4±2.3(8)	108.8±3.9(6)	89.5±2.4(8)	43.0±1.6(6)

birds being small (figs. 94 & 95). For this reason and because there is nearly no evidence of breeding in NW Canada, we did not include the NW Canada & N America sample as a separate sample in the morphometric analyses.

The morphometric differences were analysed and presented in the tables 86 - 90 and in the figures 94 - 96.

MANOVA-analysis: overall results. There was much sexual and geographical variation in the Whimbrel (Table 87). Sexual differences were larger than geographical differences. These differences were evident in all 3 performed types of MANOVA-analyses and thus sexual as well as geographical variation could already be shown to exist with only analysing wing- and culmen lengths. Other variables contributed also significantly to the group distinctions. Sexes differed in culmen-, tarsus-, wing-, secondary- and tail length (in decreasing order of importance). Middle toe lengths were not important to distinguish between sexes. Culmen-, secondary-, tail-, wing-, tarsus- and middle toe length (also in decreasing order of importance) were important to distinguish between breeding locations.

Sexual dimorphism. A large differentiation between both sexes was measured in the samples from Fennoscandia, E Siberia & E Asia and the Hudson Bay (Table 88). Sexual dimorphism was smallest in both Kazakhstan-samples. The largest differences between sexes occurred with respect to the culmen lengths: the mean differ-

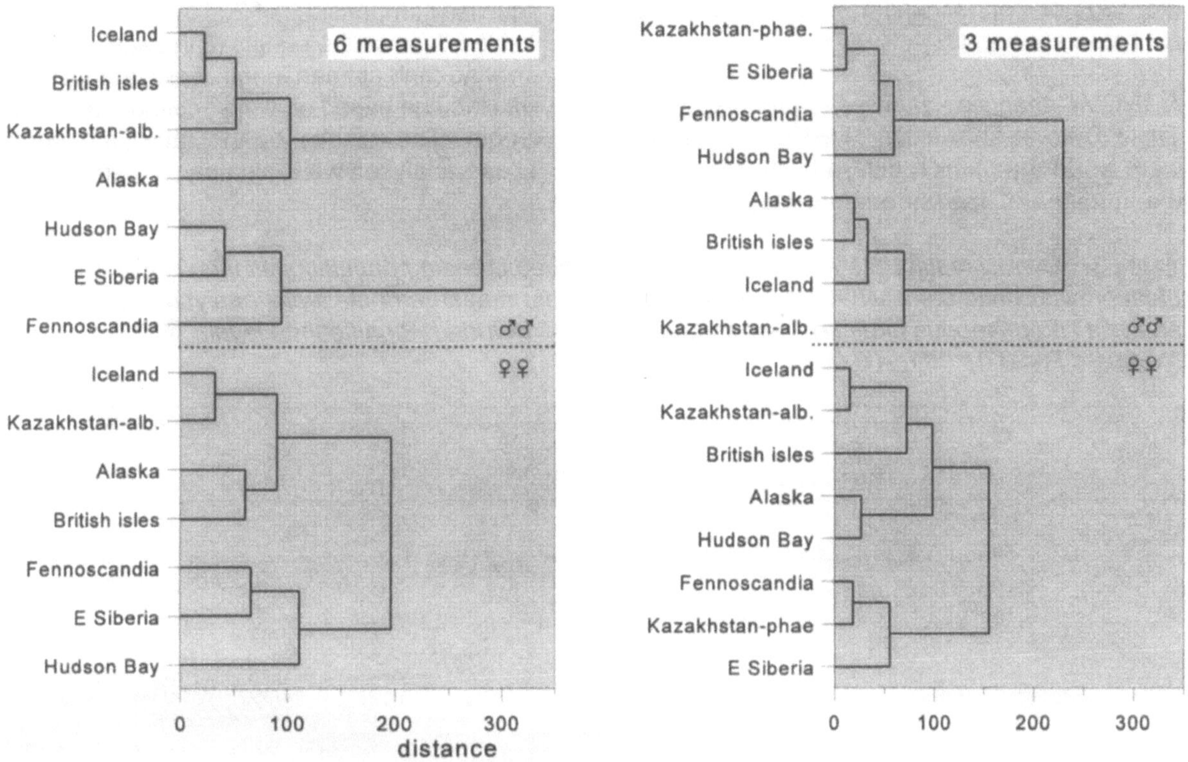

Fig. 96. *Cluster analysis in the Whimbrel based upon the mean measurements presented in Table 86. The samples of both sexes were analysed separately.*

ence between both sexes was 11 %. The sexual differences with respect to tarsus-, wing- and secondary lengths were (with 3 - 5 %) importantly smaller. There were only minor differences with respect to the tail- and middle toe lengths, resp. 2.4 and 2.5 %.

Relatedness of measurements. The different measurements were often correlated with each other (Table 89). Measurements of ♂♂ were more often and stronger correlated than those of ♀♀. In the ♂♂ wing- and tarsus lengths had high correlation coefficients and culmen lengths low ones. The ♀-measurements correlated less good than the ♂-measurements. This is especially true for wing and tarsus.

Cluster analysis. Two analyses were performed: one including the mean wing-, culmen- and tarsus lengths per sample group and the other with additionally also secondary-, middle toe and tail lengths. Since in the Kazakhstan - *phaeopus* sample no secondaries, middle toes and tails were measured, this sample is lacking in the '6 measurements' analysis. Both sexes were treated separately.

Cluster analysis showed the geographical differences to be as large as in the Bar-tailed Godwit (compare Fig. 96 with Fig. 89) but smaller than in the Black-tailed Godwit (Fig. 84) and in the Curlew (Fig. 102). The distances were larger in ♂ than in ♀ samples.

There was an important distinction in 2 separate sets of groups in both types of analyses which appeared to be valid for both the ♂- and the ♀-groups. One set included the samples from Iceland, the British isles, Kazakhstan *alboaxillaris* and Alaska. The other set included always the samples from Fennoscandia and E Siberia and 3 of the 4 times the Hudson Bay birds. When the Kazakhstan *phaeopus*-sample could be included, it was also present in this set. The first set of samples concerned the populations with large-sized individuals and the second group concerned the populations with small-sized individuals. The degree of relatedness within these sets regularly differed between ♂♂ and ♀♀ or between the '6-' and '3-measurements' analyses. Nevertheless, the populations of Iceland and the British isles were 3 of the 4 times closely related and the same applies to the birds of Fennoscandia, E Siberia and Kazakhstan *phaeopus*.

MANOVA-analysis: pairwise comparisons. In Table 87 it was shown that 3 MANOVA-analyses were performed which differed with respect to the number of measurements included. The pairwise comparisons of 2 of them are shown in Table 90, since the results of the analysis based on wing- and culmen length were largely similar with the one based on wing-, culmen- and tarsus length. The results of the latter are shown. These analyses were also performed for pair-

Table 87. *Overall results of MANOVA for all distinguished breeding populations in the Whimbrel in combination with univariate ANOVA's in which the different measurements were analysed separately. In analysis I no phaeopus-birds from Kazakhstan could be included. Significance levels: *0.01 < P < 0.05; **0.001 < P < 0.01; ***P < 0.001.*

	Analysis I			Analysis II			Analysis III		
Variables	**wing, culmen, tarsus, secondary, tail, middle toe**			**wing, culmen, tarsus**			**wing, culmen**		
n	69			211			212		
Multivariate analysis of variance									
Effects	Interaction	Area	Sex	Interaction	Area	Sex	Interaction	Area	Sex
Pillai's trace	0.654	1.659	0.483	0.090	1.009	0.441	0.064	0.789	0.431
F-value	1.122	3.502***	7.787***	0.861	14.120***	50.676***	0.930	18.247***	73.730***
Hypoth. & error *df*	36/330	36/330	6/50	21/585	21/585	3/193	14/392	14/392	2/195
Univariate F tests									
df	6/55	6/55	1/55	7/195	7/195	1/195	7/196	7/196	1/196
wing length	0.229	4.435**	17.231***	1.216	27.264***	73.965***	1.211	27.056***	75.345***
culmen length	0.978	11.272***	29.527***	0.876	11.741***	102.194***	0.882	11.628***	103.811***
tarsus length	2.603*	3.257**	20.977***	0.761	9.813***	47.791***			
secondary length	0.712	9.870***	8.757**						
tail length	0.649	6.680***	8.454**						
middle toe length	1.359	4.741**	1.444						

Table 88. *Sexual size dimorphism of the Whimbrel. The mean percentage difference between the measurements of both sexes was calculated for every single variable and breeding population, according to: $(x_{♀}-x_{♂})/x_{♂}*100$. A negative value thus implies smaller measurements for ♀♀. The calculations were based upon the means of Table 86. The obtained values were averaged per variable and summed per breeding population.*

Locality	Wing	Culmen	Tarsus	Secondary	Tail	Middle toe	Σ_3	Σ_6
Iceland	3.2	8.2	4.1	3.7	2.8	3.5	15.5	25.5
British isles	4.4	12.0	3.8	4.9	0.0	1.1	20.2	26.2
Fennoscandia	3.7	12.3	5.2	4.1	3.5	5.8	21.2	34.6
Kazakhstan -*phaeopus*	5.5	9.9	3.5	-	-	-	18.9	-
Kazakhstan - *alboaxillaris*	1.5	6.6	5.0	0.4	1.6	-0.4	13.1	14.7
E Siberia & E Asia	3.2	12.4	6.6	5.1	4.8	2.6	22.2	34.6
Alaska	3.7	10.6	2.9	4.5	3.4	-1.3	17.2	23.8
NW Canada & N America	4.5	10.9	7.6	2.3	1.4	3.3	23.0	29.9
Hudson Bay	6.0	12.0	7.4	3.5	4.4	4.4	25.4	37.7
x	3.8	10.6	4.8	3.4	2.4	2.5		
SD	1.4	1.9	2.0	1.5	1.8	2.3		

Table 89. *Pearson's product-moment correlations between measurements in the Whimbrel. Both sexes are separately analysed. Two-tailed probabilities: * 0.01 < P < 0.05; ** 0.001 < P < 0.01; *** P < 0.001.*

	Wing	Culmen	Tarsus	Secondary	Tail	Middle toe	
Wing		0.165	0.242*	0.586***	0.379**	0.365**	
Culmen	0.247**		0.129	-0.155	-0.368**	0.262*	
Tarsus	0.525***	0.210*		0.209	0.220	0.381***	
Secondary	0.684***	-0.061	0.424***		0.677***	0.107	♀♀
Tail	0.542***	0.205	0.439***	0.560***		-0.142	
Middle toe	0.480***	0.326**	0.510***	0.288*	0.148		
				♂♂			

wise contrasts with 6 measurements included, but the use of 6 measurements caused a strong reduction in sample sizes. When 3 measurements were used in the analyses, the results seemed to be more reliable to us. Nevertheless, both the '6-' and '3-measurement'-analyses are presented in Table 90, but our conclusions are mostly based upon the MANOVA-analyses with wing-, culmen- and tarsus lengths. The following conclusions were drawn:

1. In 59 out of 64 tests the ♂♂ were significantly distinct from the ♀♀.

2. When the British sample was not considered, nearly all ♀ samples were statistically distinct from each other at the 5 % level. There are 2 exceptions: (1) Alaska ♀♀ did not differ from Hudson Bay ♀♀, and (2) Fennoscandian ♀♀ did not differ from Kazakhstan - *phaeopus* ♀♀.

3. In the ♂♂ the results were largely similar to the ♀♀: 23 of the 28 contrasts were significantly different. Also here the Fennoscandian ♂♂ did not differ from the Kazakhstan - *phaeopus* ♂♂. The latter neither differed from the E Siberia ♂♂. Moreover, 3 pairwise contrasts involving the British ♂♂ were insignificant.

4. The Hudson Bay population was morphometrically distinct from nearly every other sample group. The only exception concerned the ♀♀ being not distinct from the Alaskan ♀♀. Such can be imagined with ♀♀ having a lower breeding site fidelity than ♂♂ (Table 2). Nevertheless, the Hudson Bay birds (both ♂♂ and ♀♀) always had smaller mean measurements than the Alaskan breeding birds (Fig. 94, Table 86).

5. The measurements of the British breeding population had much in common with the ones from Iceland and Alaska (being all large-sized populations). When 3 measurements were considered, they were more similar to the Alaska breeders than to the Iceland ones. But when also secondary-, tail- and middle toe were included, the difference with the Alaskan birds became larger and with the Icelandic smaller: the Alaskan population has systematically shorter secondaries and tails and longer middle toes. For this reason and because the distance to Iceland is much smaller than to Alaska, we consider the British breeding birds to belong to the same population as the Icelandic ones. Nevertheless some differentiation between the British isles and Iceland has already taken place.

6. The Fennoscandian breeding birds were nearly always smaller than the breeders from Iceland and the British isles. These differences were statistically discernable (Table 90) and were obvious in all measurements with the only exception being the culmen length in the ♀♀.

7. The Fennoscandian samples differed from every other sample except the one of nominate *phaeopus* in Kazakhstan. Both sets of samples might well belong to one and same population.

Table 90. *Results of MANOVA analysis concerning the morphometric differentiation in the Whimbrel. The values presented are approximate F-values which are derived from Pillai's trace statistic (Norušis 1992c). In these series of pairwise comparisons 16 groups (2 sexes, 8 geographical groups) were compared and 2 different sets of variables were used. Significance levels: *0.01 < P < 0.05; **0.001 < P < 0.01; ***P < 0.001.*

Locality	*n*	Alaska ♀	Hudson Bay ♀	Iceland ♀	British isles ♀	Fennoscandia ♀	Kazakhstan -albo. ♀	E Siberia ♀	Kazakhstan -phae. ♀	Alaska ♂	Hudson Bay ♂	Iceland ♂	British isles ♂	Fennoscandia ♂	Kazakhstan -albo ♂	E Siberia ♂
wing-, culmen-, tarsus-, secondary-, tail- & middle toe length (*df* 6 and 50)																
Alaska ♀	8															
Hudson Bay ♀	5	0.184														
Iceland ♀	8	0.378***	0.478***													
British isles ♀	1	-	-	-												
Fennoscandia ♀	4	0.276**	0.317**	0.190	-											
Kazakhstan-*albo.* ♀	1	-	-	-	-	-										
E Siberia ♀	1	-	-	-	-	-	-									
Kazakhstan-*phae.* ♀	-	-	-	-	-	-	-	-								
Alaska ♂	6	**0.305****	0.231*	0.362**	-	0.266*	-	-	-	-						
Hudson Bay ♂	8	0.572***	**0.383*****	0.590***	-	0.423***	-	-	-	0.340**						
Iceland ♂	11	0.584***	0.564***	**0.269***	-	0.154	-	-	-	0.417***	0.553***					
British isles ♂	2	0.283**	0.283**	0.145	-	0.042	-	-	-	0.203	0.305**	0.062				
Fennoscandia ♂	5	0.602***	0.538***	0.446***	-	**0.272***	-	-	-	0.459***	0.357**	0.214	0.151			
Kazakhstan-*albo.* ♂	2	0.321**	0.373***	0.106	-	0.149	-	-	-	0.196	0.373***	0.099	0.085	0.257*		
E Siberia ♂	7	0.679***	0.599***	0.564***	-	0.351**	-	-	-	0.487***	0.414***	0.318**	0.202	0.086	0.286**	
Kazakhstan-*phae.* ♂	-	-	-	-	-	-	-	-	-	-	-	-	-	-	-	-
wing-, culmen- & tarsus length (*df* 3 and 193)																
Alaska ♀	11															
Hudson Bay ♀	9	0.043														
Iceland ♀	15	0.129***	0.149***													
British isles ♀	3	0.017	0.045*	0.037												
Fennoscandia ♀	20	0.108***	0.090***	0.054*	0.055*											
Kazakhstan-*albo.* ♀	8	0.170***	0.201***	0.049*	0.048*	0.121***										
E Siberia ♀	18	0.172***	0.116***	0.201***	0.102***	0.095***	0.195***									
Kazakhstan-*phae.* ♀	7	0.155***	0.141***	0.059**	0.082**	0.038	0.066**	0.057*								
Alaska ♂	11	**0.111*****	0.069**	0.104***	0.082**	0.024	0.172***	0.064**	0.055*							
Hudson Bay ♂	12	0.299***	**0.199*****	0.350***	0.215***	0.250***	0.385***	0.218***	0.206***	0.118***						
Iceland ♂	17	0.266***	0.227***	**0.115*****	0.129***	0.098***	0.171***	0.210***	0.051*	0.077***	0.234***					
British isles ♂	6	0.102***	0.089***	0.031	**0.063****	0.008	0.077**	0.059**	0.014	0.018	0.147***	0.022				
Fennoscandia ♂	30	0.397***	0.315***	0.311***	0.212***	**0.251*****	0.325***	0.286***	0.119***	0.132***	0.168***	0.083**	0.083**			
Kazakhstan-*albo.* ♂	12	0.223***	0.226***	0.036	0.085**	0.095***	**0.055***	0.219***	0.042*	0.128***	0.340***	0.060**	0.034	0.223***		
E Siberia ♂	30	0.431***	0.331***	0.410***	0.248***	0.324***	0.388***	**0.253*****	0.157***	0.170***	0.117***	0.221***	0.136***	0.098***	0.338***	
Kazakhstan-*phae.* ♂	2	0.164***	0.138***	0.110***	0.137***	0.080**	0.122***	0.064**	**0.041***	0.063**	0.070**	0.045*	0.050*	0.021	0.080***	0.014

8. The *alboaxillaris* breeding birds in Kazakhstan were large-sized. They differed distinctly from nearly every other population. The lack of distinction between the *alboaxillaris* ♂♂ and the British ♂♂ was the only exception. More data might well reveal significant differences between both. The differences with the Kazakhstan *phaeopus* population are evident. *Alboaxillaris* has importantly larger measurements.
9. The E Siberian population was morphometrically distinct from nearly every other sample group. The only exception concerned the♂♂ being not distinct from the Kazakhstan *phaeopus* ♂♂.
10. No single sample was smallest in all measurements, in contrast to most of the other species. For instance, the Fennoscandian sample had the smallest culmen length, but the Hudson Bay sample had the smallest secondary length.

Plumage differences. Nearctic and Palearctic samples clearly differed in the basic colour of the axillaries. Nearctic birds had axillaries with a black-brown pattern on an orange-coloured basis, whereas Palearctic birds had axillaries with a whitish basis. Moreover, patterns of the axillaries and color of the rump in the Palearctic samples differed, but these differences were not quantified. Quantificatition might well have an important additional value for the distinction between populations.

The pairwise comparisons with MANOVA and the plumage differences result in the recognition of the following distinct breeding populations: (1) Iceland & British isles, (2) Fennoscandia & Kazakstan *phaeopus*, (3) Kazakhstan *alboaxillaris*, (4) E Siberia, (5) Alaska, (6) Hudson Bay. This ought to be reflected in a subspecies distinction.

PREDICTION OF BREEDING ORIGIN

Due to much sexual dimorphism, both sexes should be treated separately. Moreover, cluster and MANOVA analyses resulted in the recognition of 6 distinct groups (see above). Thus 12 POSCON groups (6 per sex) were constructed and especially the sample groups with >3 measurements were rather incomplete. The sample groups with 3 measurements had to contain at least 15 birds per sample as long as covariance matrices were not assumed to be equal. We therefore had to assume equal covariance matrices in this species (NEC-model POSCON - annex 3). When investigating the accuracy rates some populations were left out of consideration due to the small sample sizes (British isles, Kazakhstan *phaeopus*) or questionable status (Greenland, NW Canada). The accuracy in some samples was rather high when sexes were known (Table 91): Alaska ♂♂, *alboaxillaris* ♂♂ and *alboaxillaris* ♀♀. In other samples reasonable results were obtained (Hudson Bay ♂♂, Iceland ♂♂, Fennoscandia ♂♂, Iceland ♀♀, Fennoscandia ♀♀ and E Siberia ♀♀). In some samples there was relatively much confusion with other samples: (1) E Siberian ♂♂ with Fennoscandian ♂♂, (2) Alaska ♀♀ with Hudson Bay ♀♀. The accuracy rates became worse when sexes are unknown. Decreases of more than 10 % were estimated in the Alaska ♂♂, *alboaxillaris* ♂♂, Iceland ♀♀ and Fennoscandia ♀♀. These results can importantly be improved by including additional measurements (secondary-, tail- and middle toe lengths) and with additional information on plumages as is shown in the other species.

BODY MASS

Mass data of 29 Whimbrels collected on the breeding grounds were obtained: 16 from Fennoscandia, 5 from Iceland (♂♂ 15 June 350 gm, 18 July 338 gm, 18 July 375 gm; ♀♀ 15 June 400 gm, 18 July 439 gm), 5 from Alaska (♂♂ 1 June 350.1 gm, 27 June 382 gm; ♀♀ 1 June 381.8 gm, 16 June 424 gm, 27 June 440 gm) and 3 from N Canada (♂♂ 4 June 363.6 gm, 9 June 311.6 gm; migrant ♀ 27 April 566.8 gm). On average, ♀♀ were 65 gm heavier than ♂♂ during breeding. The Fennoscandian birds

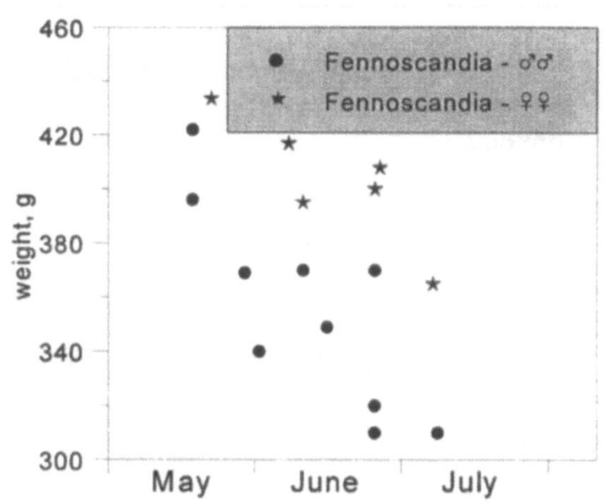

Fig. 97. *The body mass of breeding Whimbrel in Fennoscandia.*

Table 91. *Accuracy of prediction in the Whimbrel for both situations that a-priori sexes are known as well as unknown. These predictions have been based upon an analysis including 3 measurements: wing-, culmen- and tarsus lengths. The columns show the predicted proportions in relation to the total number per breeding area (last column).*

	Predicted												
Locality	Alaska ♂	Hudson Bay ♂	Iceland ♂	Fennoscandia ♂	*alboaxillaris* ♂	E Siberia ♂	Alaska ♀	Hudson Bay ♀	Iceland ♀	Fennoscandia ♀	*alboaxillaris* ♀	E Siberia ♀	*n*
					sex known								
Alaska ♂	72.7	9.1	18.2	-	-	-							11
Hudson Bay ♂	16.7	58.3	8.3	-	-	16.7							12
Iceland ♂	-	-	58.3	17.6	23.5	-							17
Fennoscandia ♂	3.3	3.3	23.3	53.3	-	16.7							30
alboaxillaris ♂	-	-	-	8.3	91.7	-							12
E Siberia ♂	3.3	16.7	3.3	40.0	-	36.7							30
Alaska ♀							36.4	54.5	-	-	-	9.1	11
Hudson Bay ♀							25.0	50.0	12.5	12.5	-	-	8
Iceland ♀							-	6.7	40.0	20.0	26.7	6.7	15
Fennoscandia ♀							5.0	5.0	20.0	45.0	5.0	20.0	20
alboaxillaris ♀							-	-	12.5	-	75.0	12.5	8
E Siberia ♀							5.6	11.1	-	16.7	16.7	50.0	18
					sex unknown								
Alaska ♂	27.3	9.1	9.1	-	-	-	-	-	-	45.5	-	9.1	11
Hudson Bay ♂	8.3	58.3	8.3	-	-	16.7	-	8.3	-	-	-	-	12
Iceland ♂	-	-	52.9	17.6	11.8	-	-	-	11.8	5.9	-	-	17
Fennoscandia ♂	3.3	3.3	23.3	53.3	-	13.3	-	-	-	-	-	3.3	30
alboaxillaris ♂	-	-	-	8.3	50.0	-	-	-	16.7	8.3	16.7	-	12
E Siberia ♂	3.3	6.7	-	40.0	-	26.7	-	-	-	-	-	13.3	30
Alaska ♀	-	-	-	-	-	-	36.4	54.5	-	-	-	9.1	11
Hudson Bay ♀	-	-	-	-	-	-	25.0	50.0	12.5	12.5	-	-	8
Iceland ♀	-	-	20.0	-	20.0	-	-	6.7	20.0	6.7	20.0	6.7	15
Fennoscandia ♀	15.0	-	15.0	5.0	5.0	-	5.0	5.0	15.0	15.0	-	20.0	20
alboaxillaris ♀	-	-	-	-	12.5	-	-	-	-	-	75.0	12.5	8
E Siberia ♀	5.6	-	5.6	5.6	11.1	-	5.6	11.1	-	5.6	-	50.0	18

seemed to loose weight continuously during breeding with an average rate of about 1.7 gm.day^{-1}.

PRIMARY MOULT

No breeding birds in active primary moult were present in the material studied: Whimbrels apparently do not moult their primaries on the breeding grounds.

DISCUSSION

The taxonomy of the Whimbrel

Our data confirmed the existence of -at least- six clearly identifiable populations. Zink *et al.* (1995) proposed to distinguish *hudsonicus* as a separate species and they suggested to distinguish *variegatus* also at the species level. We follow their proposition concerning *hudsonicus* and consider them as allospecies belonging to one superspecies *Numenius [phaeopus]*: *Numenius [phaeopus] phaeopus* and *Numenius [phaeopus] hudsonicus*. Also in our study E Siberian *variegatus* appeared to be clearly distinct from the W

Palearctic populations, but the data set of the E Palearctic populations was as yet too small to allow separation on the species level. Thus we continue to distinguish *variegatus* at the subspecies' level.

Within *Numenius [p.] phaeopus* 4 distinct breeding populations can be recognized as different subspecies:

- *N. [p.] p. phaeopus* Linnaeus (1758) breeds from Scandinavia and the Baltic republics in the west to SW Taymyr, the Yenisei valley and the Omsk area in the east. The nominate form was originally described from Sweden. Birds present on this type locality are most likely to be local breeding birds. It has a whitish lower back and rump. It has short wing lengths and the ♂♂ have the shortest culmen lengths measured. On the axillaries the white parts are mostly clearly wider than the brown bars like type 'D' in the Bar-tailed Godwit (Fig. 13 - own obs.).

- *N. [p.] p. islandicus* Brehm (1831) breeds on Iceland, the Færoes, the Orkneys, the Shetlands and in Scotland. The type locality is Iceland. *Islandicus* averages larger than nominate *phaeopus*. It has a whitish lower back and rump and has especially longer wing- and secondary lengths, but also the culmen-, tarsus-, tail- and middle toe average larger. Salomonsen (1947) was correct with the distinction of this subspecies *islandicus*, since there are clear morphometric differences with nominate *phaeopus*. These differences are not in line with the expectations when it were considered as clinal variation. Moreover, the mean measurements of British Whimbrel are generally similar to or larger than the Icelandic ones. Therefore, we consider the British breeding population as *islandicus*. On the axillaries the white parts are mostly not wider than the brown bars like type 'E' in the Bar-tailed Godwit (Fig. 13 - own obs.).

- *N. [p.] p. alboaxillaris* Lowe (1921) breeds in the steppes east of the lower Volga and south and south-east of the Urals. The type locality is Inhambane (SE Mozambique) in winter. This subspecies is rather pale above and below, due to larger whitish fringes of feathers and paler feather-centres. The breast feathers to the under tail-coverts are white as are the under wing-coverts and axillaries. It has the longest wing-, tarsus- and secondary lengths of the identified subspecies. This subspecies has white axillaries like type 'A' of the Bar-tailed Godwit, or it has brown bars along one side with one or two brown bars on top of the other side (combination of type 'B' and 'C' in the Bar-tailed Godwit - Fig. 13).

- *N. [p.] p. variegatus* Scopoli (1786) breeds in NE Siberia. The type locality is Luzon (Philippines) in winter. It has a whitish lower back spotted with brown and a barred white rump. It still appears as a contrastingly pale spot in flight which is not so with both Nearctic subspecies (Paulson 1993). Nominate *phaeopus* has a nearly pure white lower back. It has axillaries with broad brown bars and small white parts in between like type 'G' in the Bar-tailed Godwit (Fig. 13 - own obs.). *Variegatus* averages smaller than the American Whimbrels (Paulson 1993 - see also Annex 1). According to Stepanyan (1990) the breeding ranges of nominate *phaeopus* and *variegatus* overlap on the Taymyr Peninsula and in the Yenisei valley. Rogacheva (1992) showed that the breeding range of the species in Central Siberia is not disjunct, but she had too limited collected material to trace a zone of intergradation between both subspecies. We also had too few data.

Within *Numenius [p.] hudsonicus* 2 distinct breeding populations can be recognized as different subspecies:

- *N. [p.[h. hudsonicus* Latham (1790) breeds along the south and west coast of the Hudson Bay. Its type locality is the Hudson Bay (Canada). It has a buffish-brown back and rump. The birds from this breeding population have systematically smaller measurements than the Alaskan subspecies. The breeding locations of both these populations are also spatially separated. It has the shortest wing-, tarsus-, secondary- and tail lengths of the identified subspecies. It has axillaries with the same ground patterns and colours as in *rufiventris* (ZMA).

- *N. [p.[h. rufiventris* Vigors (1829) breeds from W Alaska east to -at least- the Melville Hills in the NW Territories. The type locality is NW North America. This subspecies is characterized by its brownish appearance and its large dimensions. It has the longest culmen lengths of the identified subspecies. It has axillaries with dark-brown bars on both sides of the shaft which are smaller than the reddish brown parts in between (pattern like type 'D' in the Bar-tailed Godwit with white being reddish-brown - Fig. 13. Own obs.). Gibson & Kessel (1997) consider this subspecies as a widespread breeder throughout Alaska.

SUMMARY

There was much sexual and geographical variation in the Whimbrel. Sexual differences were larger than geographical ones and ♂♂ were smaller-sized than ♀♀. Sexual dimorphism was smallest in the population breeding in Kazakhstan and largest in the one of Hudson Bay. This species complex ought to be considered as a superspecies with at least 2 allospecies: *Numenius [phaeopus] phaeopus* (Eurasian Whimbrel) and *Numenius [phaeopus] hudsonicus* (Hudsonian Whimbrel). Possibly the E Palearctic populations can also be recognized as allospecies. Both the Eurasian and Hudsonian Whimbrel are polytypic. In the Eurasian Whimbrel 4 subspecies were recognized: (1) *N. [p.] p. phaeopus* breeding from Fennoscandia in the west to SW Taymyr, the Yenisei valley and the Omsk area in the east; (2) *N. [p.] p. islandicus* breeding on Iceland, the Færoes, and the British isles; (3) *N. [p.] p. alboaxillaris* breeding east of the lower Volga and south and southeast of the Urals; (4) *N. [p.] p. variegatus* breeding in NE Siberia. In the Hudsonian Whimbrel 2 subspecies were recognized: (1) *N. p. [h.] hudsonicus* breeding along the south and west coast of the Hudson Bay, and (2) *N. p. rufiventris* breeding in Alaska west to -at least- the Melville Hills in the NW Territories. These taxa can be recognized by combining measurements and plumage characters. The Hudsonian Whimbrel has a buffish-brown lower back and rump with *hudsonicus* having smaller dimensions than *rufiventris*. The Eurasian Whimbrel has a whitish lower back and rump with *alboaxillaris* being the palest of the 4. Nominate *phaeopus* and *islandicus* have a nearly pure white back and rump, but *islandicus* has larger measurements than *phaeopus*. Wing-, culmen-, and tarsus lengths were important characters for the distinction between sexes as well as breeding populations. This might also be true for secondary- and tail lengths. Whimbrels did not moult primaries on the breeding grounds. Nearctic birds had less broad wings than Icelandic and Fennoscandian ones. There were no differences in wing structure between the sexes.

Chapter 15

Eurasian Curlew

-

Numenius arquata

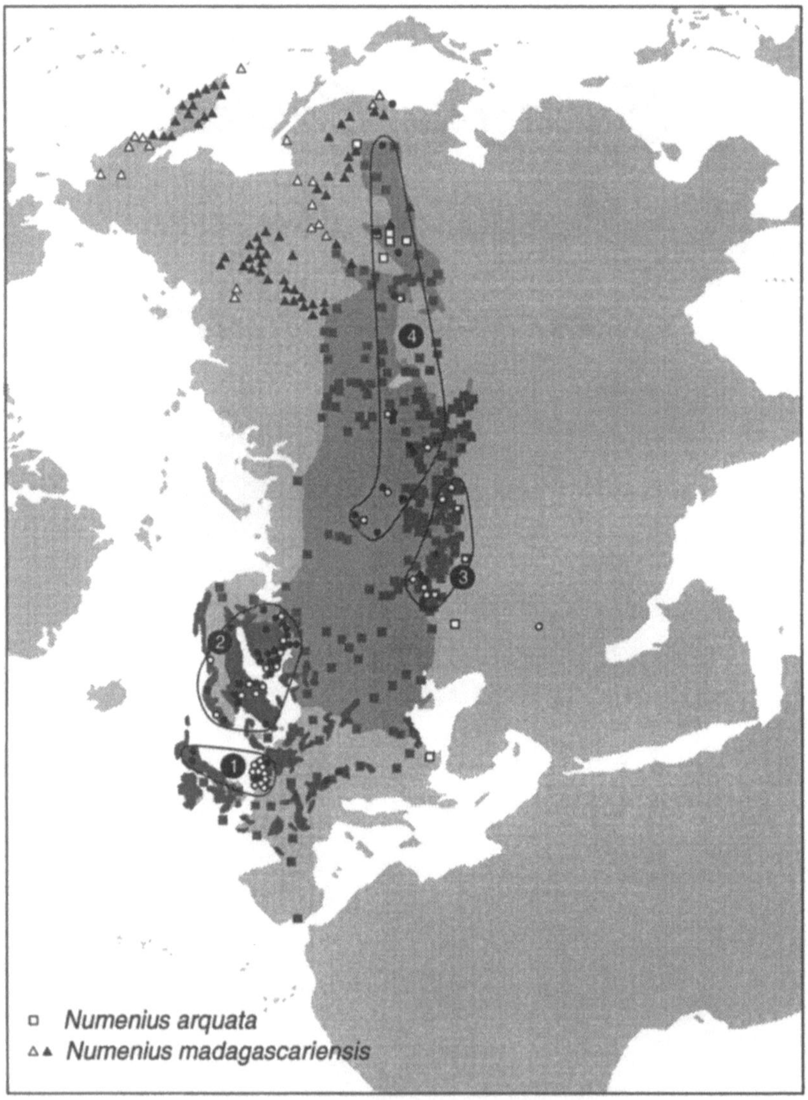

Fig. 98. *Breeding distribution of the Curlew.*

INTRODUCTION

The Eurasian Curlew is breeding in temperate, boreal and subarctic habitats scattered over the Palearctic (Fig. 98). It is replaced by the Far-eastern Curlew *Numenius madagascariensis* in the E Palearctic. Mostly 2 distinct subspecies are recognized (Vaurie 1965, Glutz *et al.* 1977, Cramp & Simmons 1983, Stepanyan 1990):

- *N. a. arquata*, breeding in W and Central Europe.
- *N. a. orientalis*, breeding in Central Asia.

There is an extended zone of intergradation between both subspecies covering SE Europe and W Siberia, including clinal changes in both measurements and plumage characters. Birds from the southern Ural region and Kazakhstan are sometimes separated as *suschkini*, because they combine the smaller size of *arquata* with the pale axillaries of *orientalis* (Neumann 1929, Johansen 1960, 1961a). As only a few authors have examined birds of this race, it is generally ignored.

MORPHOMETRICAL DIFFERENTIATION

In total, measurements of 128 Curlews (57.8% ♂♂) from throughout the breeding range had been collected. The breeding grounds were reached from early April onwards (Fig. 99). Most skins had been collected during May and June.

Initially, 4 different breeding populations were distinguished (Fig. 98). The mean measurements per sample are presented in Table 92. The morphometric variation in this species was analysed and presented in the tables 93 - 97 and the figures 100 - 102.

Table 92. *Measurements of Eurasian Curlews, collected on the breeding grounds. Numbers in the first column refer to the different locations as used in Fig. 98. Presented are the means ± 1 SD (n).*

Locality	Wing	Culmen	Tarsus	Secondary	Tail	Middle toe
			♂♂			
1 W Europe	300.3±7.7(23)	123.6±6.8(23)	79.2±4.0(23)	137.4±2.9(5)	98.0 - (1)	49.8±2.8(8)
2 N Europe	298.1±6.9(33)	121.0±8.2(34)	77.5±3.7(34)	138.4±6.8(32)	97.5±0.7(2)	49.3±2.1(30)
3 S Ural & Kazakhstan	301.3±5.5(8)	126.5±6.0(8)	83.8±5.5(8)	137.7±4.5(8)	92.0 - (1)	51.0±4.1(8)
4 Central Siberia	301.4±6.7(15)	139.4±6.4(14)	84.3±4.5(15)	137.3±4.3(15)	100.0±2.8(2)	50.1±1.8(15)
			♀♀			
1 W Europe	319.9±9.8(9)	148.7±8.0(9)	83.2±1.9(9)	140.0 - (1)	-	49.8±4.6(2)
2 N Europe	317.3±8.6(25)	148.6±9.8(25)	82.3±4.7(25)	145.5±5.8(23)	96.0 - (1)	50.8±2.3(18)
3 S Ural & Kazakhstan	310.0±6.8(6)	137.6±5.7(6)	83.5±1.3(6)	136.3±9.6(6)	95.0±1.4(2)	49.6±1.9(5)
4 Central Siberia	319.5±5.6(9)	173.9±5.2(9)	86.6±2.7(9)	145.4±3.6(8)	116.0 - (1)	52.4±2.4(7)

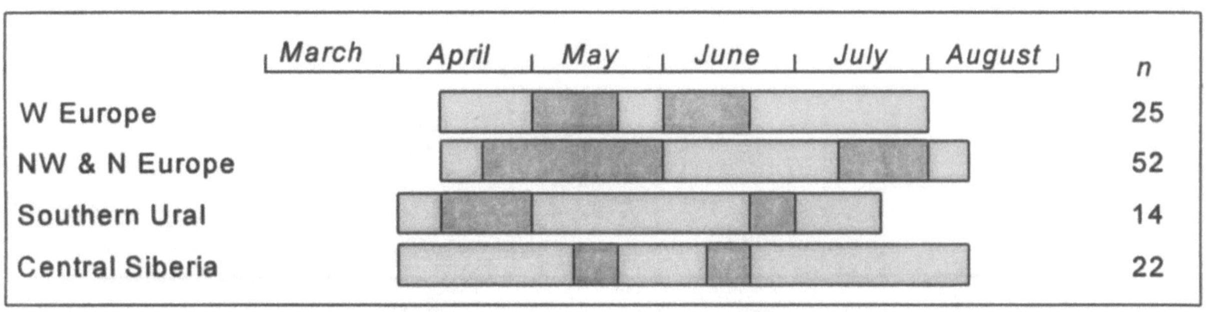

Fig. 99. *Collection dates of Curlews. Most birds had been collected during the periods marked by the heavily shaded bars. The sample sizes are presented on the right side.*

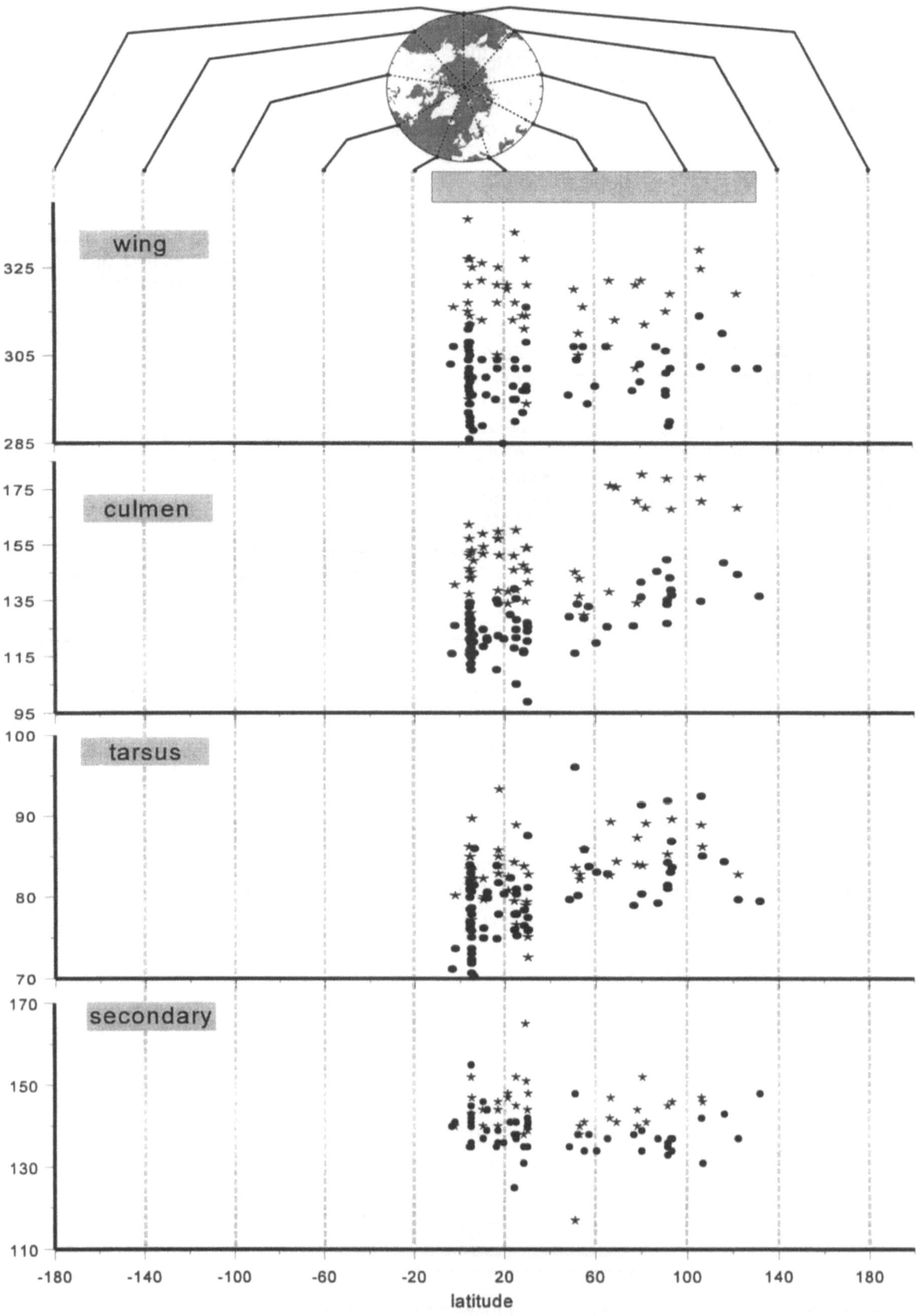

Fig. 100. *Measurements of Curlews on the breeding grounds. The following measurements are presented from top to bottom: wing length, culmen length, tarsus length and secondary length (description in chapter 2). Explanation of the used symbols: ● ♂♂, ★ ♀♀.*

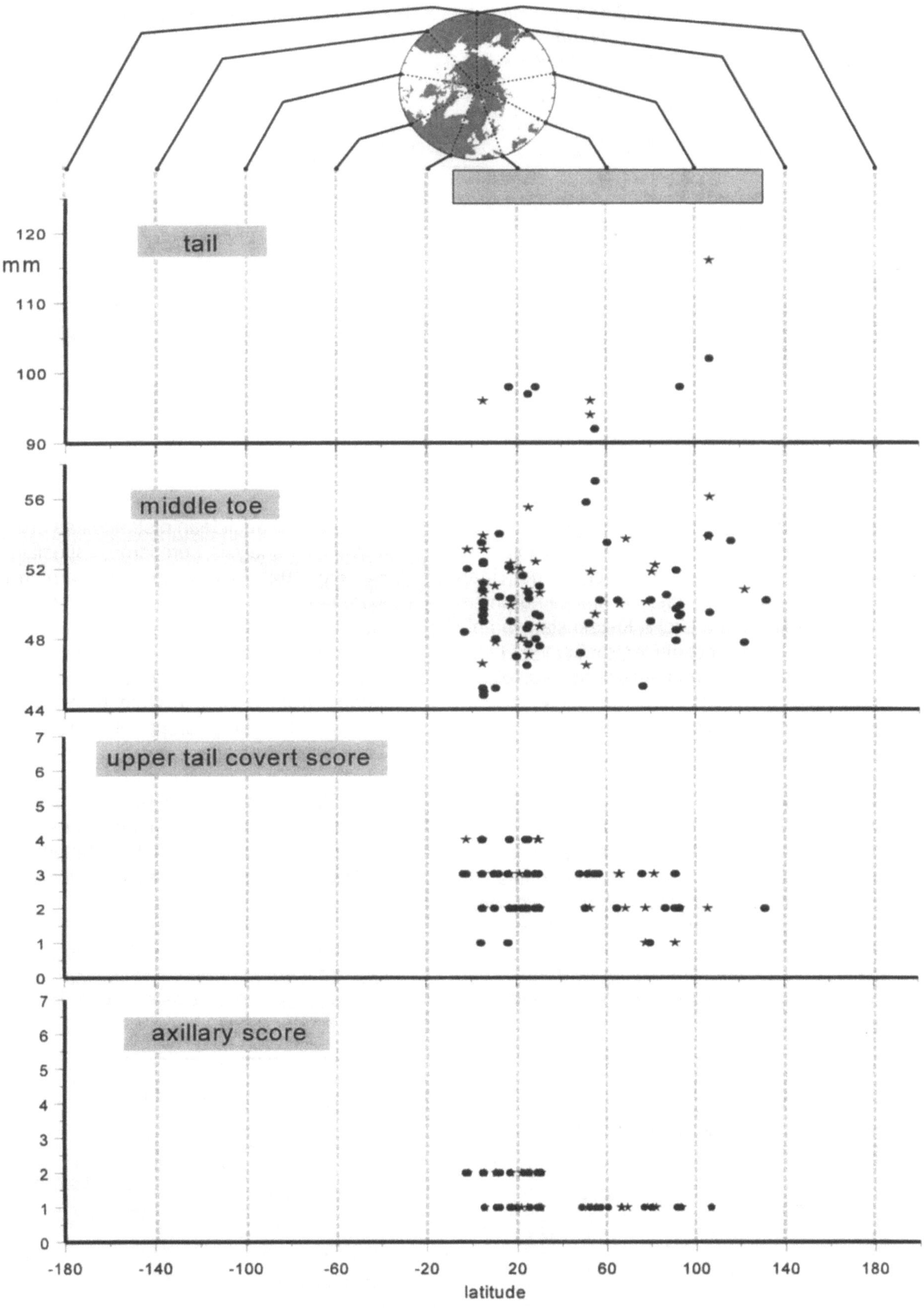

Fig. 101. *Measurements of Curlews on the breeding grounds. The following measurements are presented from top to bottom: tail length, middle toe length, upper-tail covert score, and axillary score (description in chapter 2). Explanation of the used symbols: ● ♂♂, ★ ♀♀.*

Table 93. *Overall results of MANOVA for all distinguished breeding populations in the Curlew in combination with univariate ANOVA's in which the different measurements were analysed separately. Significance levels: *0.01 < P < 0.05; **0.001 < P < 0.01; ***P < 0.001.*

	Analysis I			Analysis II			Analysis III		
Variables	**wing, culmen, tarsus, secondary, middle toe**			**wing, culmen, tarsus**			**wing, culmen**		
n	85			127			127		
Multivariate analysis of variance									
Effects	Interaction	Area	Sex	Interaction	Area	Sex	Interaction	Area	Sex
Pillai's trace	0.312	0.893	0.597	0.143	0.682	0.726	0.119	0.553	0.726
F-value	1.744*	6.355***	21.606***	1.986*	11.678***	103.245***	2.505*	15.161***	156.146***
Hypoth. & error *df*	15/ 225	15/ 225	5/ 73	9/ 357	9/ 357	3/ 117	6/ 238	6/ 238	2/118
Univariate F tests									
df	3/ 77	3/ 77	1/ 77	3/ 119	3/ 119	1/ 119	3/ 119	3/ 119	1/ 119
wing length	1.276	1.201	43.601***	1.418	1.422	112.575***	1.418	1.422	112.575***
culmen length	6.554***	41.618***	74.816***	4.151**	45.339***	245.699***	4.151**	45.339***	245.699***
tarsus length	0.432	10.325***	3.197	1.511	12.864***	12.009***			
secondary length	0.134	2.179	3.536						
middle toe length	0.150	1.766	2.506						

All measurements. The Curlew showed marked sexual and geographical variation (Table 93). Sexual differences were much larger than geographical differences. Sexual dimorphism was smallest in the S Ural & Kazakhstan sample and largest in the one of Central Siberia (Table 94). The different measurements correlated badly which is in contrast to the measurements in other wader species (Table 95).

Table 94. *Sexual size dimorphism of the Curlew. The mean percentage difference between the measurements of both sexes was calculated for every single variable and breeding population, according to: $(x_{♀}-x_{♂})/x_{♂}*100$. A negative value thus implies smaller measurements for ♀♀. The calculations were based upon the means of Table 92. The obtained values were averaged per variable and summed per breeding population.*

Locality	Wing	Culmen	Tarsus	Secondary	Middle toe	Σ
W Europe	6.6	20.8	5.3	0.9	-0.2	33.4
N Europe	6.5	22.1	5.9	5.1	3.0	42.6
S Ural & Kazakhstan	3.2	9.6	-2.8	-2.1	-5.9	2.0
Central Siberia	5.9	25.8	3.5	5.3	4.0	44.5
x	5.6	19.6	3.0	2.3	0.2	
SD	1.6	7.0	4.0	3.6	4.5	

The squared euclidean distances as calculated with cluster analysis were large, especially in the ♀♀ (Fig. 102). This was mainly caused by the extremely long culmen lengths of the Central Siberian ♀♀ which was much more extreme as in the ♂ samples. The W European and the N European samples were largely similar. The sample from S Ural & Kazakhstan was more similar to the European populations than to the Central Siberian one.

MANOVA-analysis was initially restricted to analysis with 5 measurements (wing-, culmen-, tarsus-, secondary- and middle toe length) since too few tail lengths had been measured.

Table 95. *Pearson's product-moment correlation between measurements of Curlews. Data of ♂♂ and ♀♀ were separately analysed. Two-tailed probabilities: * 0.01 < P < 0.05; ** 0.001 < P < 0.01; *** P < 0.001.*

	Wing	Culmen	Tarsus	Secondary	Middle toe	
Wing		0.240	0.275	0.318*	0.355*	
Culmen	0.098		0.470**	0.302	0.569***	
Tarsus	0.273*	0.366**		0.095	0.435*	♀♀
Secondary	0.238	-0.040	-0.040		0.517**	
Middle toe	0.205	0.165	0.438***	0.100		
			♂♂			

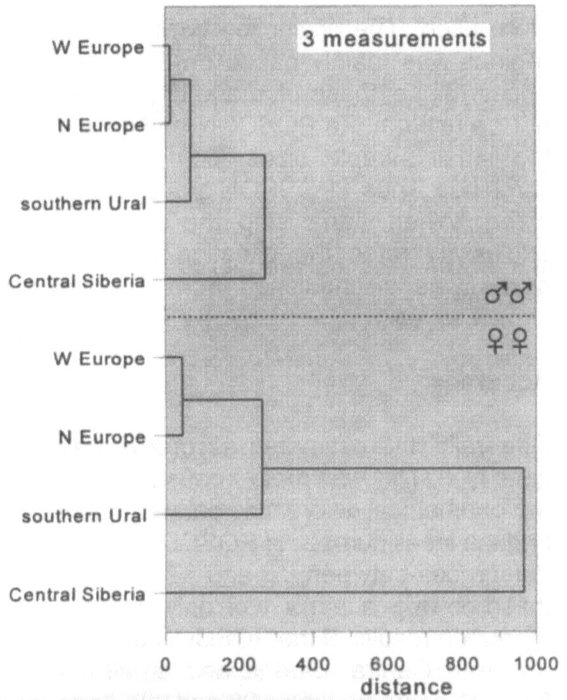

Fig. 102. *Cluster analysis in the Curlew based on the mean measurements presented in Table 92. The samples of both sexes were analysed separately.*

Moreover, overall analysis clearly revealed the insignificant role of secondary- and middle toe lengths (Table 93). Pairwise comparisons were therefore based upon an analysis just including wing-, culmen en tarsus lengths (Table 96). The results from this analysis closely corresponded with the results from cluster analysis. The following conclusions were drawn:

(1) Nearly all ♂ samples were statistically distinct from the ♀ samples (Table 96). There was no difference between Central Siberian ♂♂ and S Ural & Kazakhstan ♀♀.
(2) Central Siberian Curlews always strongly differed from western birds. This was due to a difference in culmen- and tarsus length, not to wing length.
(3) The measurements of the W and N European samples did not differ significantly.
(4) The S Ural & Kazakhstan sample differed from the European and Central Siberian samples, but this was not because of intermediacy in measurements: ♀♀ from S Ural & Kazakhstan had the smallest average wing- and culmen lengths measured amongst the ♀♀. The mean culmen length of ♂♂ was somewhat larger than the one of the W and N European samples, but importantly smaller than the one of Central Siberia.

Selection of variables. Wing-, culmen- and tarsus lengths contributed significantly to the group distinctions and secondary- and middle toe lengths did not play a role of importance (Table 93). Wing lengths were not important for the distinction between geographical regions. Generally, the following order of importance was found: culmen-, wing-, tarsus length.

Table 96. *Results of MANOVA analysis on morphometric differentiation of Curlew. Shown are approximate F-values which are derived from Pillai's trace statistic (Norušis 1992c). In these series of pairwise comparisons 10 groups (2 sexes, 5 geographical groups) were compared and 3 measurements (wing-, culmen- and tarsus length) were used. There were 3 and 114 df. Significance levels: *0.01 < P < 0.05; **0.001 < P < 0.01; ***P < 0.001.*

Locality	*n*	W Europe ♀	N Europe ♀	S Ural & Kazachstan ♀	Central Siberia ♀	W Europe ♂	N Europe ♂	S Ural & Kazachstan ♂
W Europe ♀	9							
N Europe ♀	25	0.0070						
S Ural & Kazach. ♀	6	0.099**	0.112**					
Central Siberia ♀	9	0.297***	0.385***	0.405**				
W Europe ♂	23	0.455***	0.583***	0.162**	0.714***			
N Europe ♂	33	0.533***	0.676***	0.243***	0.757***	0.0380		
S Ural & Kazach. ♂	8	0.329***	0.392***	0.086*	0.591***	0.067*	0.139**	
Central Siberia ♂	14	0.262***	0.329***	0.0540	0.520***	0.282***	0.398***	0.108**

Wing- and culmen length. Both wing- and culmen lengths were important for the analysis of sexual and geographical variation (Table 93). A MANOVA analysis restricted to wing- and culmen lengths produced nearly identical results as one including 3 measurements. Therefore, pairwise contrasts based on the use of wing- and culmen lengths were not presented in Table 96. The only difference with the '3 measurements' analysis was that the measurements of the S Ural & Kazakhstan ♂♂ were not significantly different from the ones of both European ♂ samples.

Plumage characters. Apart from the standard measurements we quantified the coloration of the upper tail-coverts and the colour pattern of the axillaries (Fig. 101).

Western and eastern samples differed greatly in plumage. W and N European birds mostly had striped axillaries (56.6 %; N-total = 53) and the S Ural & Kazakhstan and Central Siberian birds all had white ones (*n*-total = 22; Chi^2 = 25.895; *df* = 3; *P* = 0.000). The upper tail-coverts of western birds were also darker than the ones of eastern birds (Fig. 101; Chi^2 = 27.750; *df* = 9; *P* = 0.001). Within populations, sexes did not differ in plumage characters (Table 97). The axillaries and upper tail-coverts thus have an important additional value when predicting a bird's breeding origin.

Table 97. *Results of testing sexual differences in axillary scores and upper tail-covert scores. Chi^2-tests were performed per breeding population. [1] Birds from Kazakhstan included.*

Locality	Axillaries				Upper tail-coverts			
	Chi^2	*df*	*P*	*n*	Chi^2	*df*	*P*	*n*
W Europe	♂♂ & ♀♀ striped			4	2.000	2	0.368	4
N Europe	0.1	1	0.758	53	1.592	2	0.451	54
S Ural[1]	♂♂ & ♀♀ white			13	1.887	2	0.170	13
Central Siberia	♂♂ & ♀♀ white			9	1.078	2	0.583	17

PREDICTION OF BREEDING ORIGIN

Because of the strong sexual dimorphism, both sexes were considered separately. Both the cluster- and MANOVA analyses resulted in the identification of 3 geographical regions: (1) W and N Europe, (2) S Ural & Kazakhstan, and (3) Central Siberia. Since the plumage data -having a discrete character- were too scarce to be included, only wing-, culmen- and tarsus lengths were used. The distinction between both first regions was mainly based on plumage differences. Since, these plumage characters could not be used in the POSCON analyses, due to too small sample sizes, both regions were combined. Thus 4 POSCON groups were recognized. When analysing the accuracy- and exclusivity rates the situation with *unknown* sexes was distinguished from the one with *known* sexes.

Accuracy

Generally, the estimated accuracy rates were quite high. This was mainly caused by the large morphometrical differences between both sexes and the small number of POSCON groups. The differences between ♂♂ and ♀♀ were so large that knowing a bird's sex only improved the prediction results of the POSCON groups (large ♂♂ from Central Siberia and small ♀♀ from Europe - compare tables 98 and 99). As much of the confusion between sexes arises from both these groups, it is essential to know the sex of a

Table 98. *Accuracy of prediction in Curlew (sex* ***unknown*** *a priori). These predictions are based on morphometric analysis and include (A) wing-, culmen- and tarsus length and (B) wing- and culmen length. The columns show the predicted proportion in relation to the total per breeding area (last column).*

Locality	Predicted				
	Europe ♂	Central Siberia ♂	Europe ♀	Central Siberia ♀	*n*
(A) Wing-, culmen- and tarsus length					
Europe ♂	84.6	15.4	-	-	52
C. Siberia ♂	6.7	80.0	13.3	-	15
Europe ♀	5.7	11.4	80.0	2.9	35
C. Siberia ♀	-	-	-	100.0	7
(B) Wing- and culmen length					
Europe ♂	86.5	13.5	-	-	52
C. Siberia ♂	13.3	73.3	13.3	-	15
Europe ♀	5.7	14.3	77.1	2.9	35
C. Siberia ♀	-	-	-	100.0	7

Table 99. *Accuracy of prediction in Curlew (sex **known** a priori). These predictions are based on morphometric analysis and include (A) wing-, culmen- and tarsus length and (B) wing- and culmen length. The columns show the predicted proportion in relation to the total per breeding area (last column).*

Locality	Predicted				
	Europe ♂	Central Siberia ♂	Europe ♀	Central Siberia ♀	*n*
(A) Wing-, culmen- and tarsus length					
Europe ♂	84.6	15.4	-	-	52
C. Siberia ♂	6.7	93.3	-	-	15
Europe ♀	-	-	97.1	2.9	35
C. Siberia ♀	-	-	-	100.0	7
(B) Wing-, and culmen length					
Europe ♂	86.5	13.5	-	-	52
C. Siberia ♂	13.3	86.7	-	-	15
Europe ♀	-	-	97.1	2.9	35
C. Siberia ♀	-	-	-	100.0	7

Curlew. Knowledge of a bird's sex did not improve the quality of prediction of the small-sized western ♂♂ or the large-sized eastern ♀♀.

The average accuracy rates as based upon predictions with wing- and culmen lengths were somewhat lower than the ones also including tarsus lengths (tables 98 and 99). However, these lower accuracy rates were restricted to the POSCON groups with birds of intermediate size. Additional inclusion of tarsus lengths increased the accuracy rates with maximally 7 %.

Exclusivity

The estimated exclusivity rates were extremely high since (1) both sexes differed markedly in measurements and (2) only 2 morphometrically distinct breeding locations were recognized. The estimated exclusivity rates show that (1) the highest exclusivity rates were reached in the POSCON groups of European ♀♀ and Central Siberian ♀♀, (2) the average rates were also high and varied between 47 and 59 %, (3) the additional inclusion of tarsus lengths in the analyses did not result in systematically larger exclusivity rates, and (4) knowing a bird's sex improved the average outcome with about 12 % and in case of the European ♀♀ with 37 % (Table 100).

Table 100. *Estimated exclusivity rates when predicting the breeding origin and/or sex in Curlews. These predictions were based upon 2 or 3 measurements.*

Sample	Wing-, culmen- & tarsus- length		Wing- & culmen length		*n*
	Sex known	Sex unknown	Sex known	Sex unknown	
Europe ♂	46.2	46.2	44.2	46.2	52
C. Siberia ♂	37.5	25.0	37.5	25.0	24
Europe ♀	84.2	47.4	84.2	47.4	38
C. Siberia ♀	70.0	70.0	70.0	70.0	10
x	59.5	47.2	59.0	47.2	
SD	21.5	18.4	21.9	18.4	

CONTACT ZONES

Mostly, the morphometric variation in this species is considered as clinal variation with western birds having shorter measurements and being darker coloured than the eastern ones (Glutz *et al.* 1977; Cramp & Simmons 1983). This view mostly concerned the northern populations, since data from S Ural & Kazakhstan were lacking or scarce. The data presented here show it to be a too generalized picture. Though some characters changed clinally (e.g. culmen length in Curlew ♂♂; scores upper tail-coverts, figs. 100 & 101), others did not (e.g. wing dimensions and middle toe lengths) or changed abruptly at about 40°E: (1) culmen length of ♀♀, (2) tarsus length of both sexes, and (3) color pattern on axillaries (no birds with brown-striped axillaries occurred east of 40°E). Therefore, this variation cannot be interpreted in terms of clinal variation, but show the existence of 3 clearly different breeding populations. Since few birds of the S Ural & Kazakhstan population were sampled, it was impossible to draw exact subspecies' limits, but at least the Central Siberian population seems to contact the S Ural & Kazakhstan population in the south part of the W Siberian lowlands and in the Altay region.

BODY MASS

The body masses of 20 Scandinavian breeding Curlews were traced (Fig. 103). These data showed that ♀♀ were heavier than ♂♂ and that ♀♀ lost about 200 gm during May and June. Females were about 100 gm heavier than ♂♂ during June.

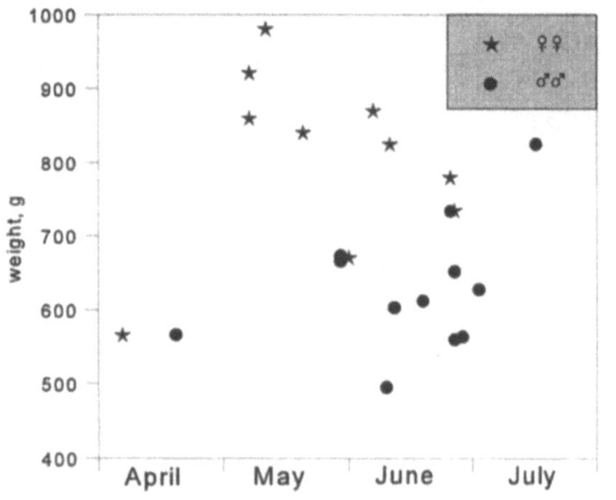

Fig. 103. *Body mass of Fennoscandian Curlews during the breeding season.*

PRIMARY MOULT

No breeding birds in active primary moult were present in the material studied: Curlews apparently do not moult primaries on the breeding grounds.

DISCUSSION

The taxonomy of the Curlew

The above presented information clearly showed the existence of 3 distinct populations:

- *N. a. arquata* Linnaeus (1758) is breeding in W-, N- and Central Europe east to -at least- 40°E. The type locality of this subspecies is Sweden.This subspecies has small measurements and barred plumage characters. The measurements are rather comparable with *suschkini*, but this latter subspecies has whitish axillaries and underwings. The ♂♂ have the shortest culmen- and tarsus lengths measured. It has rather broad streaks on upperparts and chest. The belly and upper tail-coverts are often marked with black blobs. The axillaries are barred and mostly with black subterminal blobs.

- *N. a. orientalis* Brehm (1831) is breeding in the European and Asiatic part of Russia from about 40°E to the east limit of the species' range, where it overlaps the breeding range of the Far Eastern Curlew *N. madagascariensis*. The type locality of this subspecies are the East Indies. It is characterized by long culmen- and tarsus lengths and white axillaries and underwing. It has narrow and brownish-black streaks on head, neck and upperparts and it is mostly lacking black blobs on the flanks and upper tail-coverts.

- *N. a. suschkini* (Neumann 1929) is breeding in the S Ural region and in Kazakhstan. The type locality of this subspecies is Dagana (Senegal) outside the breeding range. Neumann (1929) also had 3 specimen collected within the breeding range (Orenburg and the Kurgan steppe) to his disposal, but none of these was allocated as type specimen. *Suschkini* is characterized by its short measurements. The ♀♀ even have shorter wing-, culmen-, and secondary lengths than nominate *arquata*. *Suschkini* ♂♂ have a bit longer or equal the measurements of nominate *arquata*. *Suschkini* has pale axillaries and upper tail-coverts and sharp banded tail feathers with alternating bands of nearly pure white and black (Neumann 1929). *Suschkini* contacts *orientalis* near the Volga River, in the south part of the W Siberian lowlands and in the Altay region. There are no indications for geographical variation within the subspecies.

These 3 subspecies contact in several situations. However, our data were not adequate enough to analyse the character of contact in detail.

Size differences in the Curlew

It is intriguing to find such huge differences in size dimorphism between both sexes in the different subspecies: (1) the differences in *suschkini* are small as compared with *orientalis*, and (2) within *arquata* there are also differences in sexual size dimorphism between the W- and N European population. It does not seem to be accidental that the largest subspecies of the Eurasian Curlew -*orientalis*- is neighbouring or even overlapping the breeding range of the Far Eastern Curlew *N. madagascariensis*. They have their large measurements in common. Moreover, the smallest subspecies of the Eurasian Curlew -*suschkini*- has (or 'had' due to extinction) an overlapping breeding range with the small-sized Slender-billed Curlew *N. tenuirostris* and the Whimbrel subsp. *alboaxillaris*. These species have their small dimensions and white underparts in common. These similarities might be due to inter-species interactions: ♂♂ or ♀♀ from one species mating with the opposite sex of the other species. However, there is no evidence from field-data. It might also be the result of selectional forces, driving the easterly Curlew species (especially ♀♀) to huge dimensions and the S Russian species to pale plumages and smaller dimensions. This selection then results in large size differences between the sexes in the E Palearctic species and to small size differences in the S Russian species. The driving force behind the evolutionary selection of such huge culmen lengths is then very puzzling. During migration and winter selection in favour of long-beaked Curlews might be caused by a restricted availability of deeply burrowed preys. During breeding it might also be caused by differences in vegetation heights: the bogs in Central Asia reaching higher than the arid steppes of Kazakhstan. However, these are mere speculations. In the same way, it might be possible that selectional forces favour white underwings in the steppes of the S Ural and Kazakhstan. It is rather well imaginable that a better heat reflection of white underparts favours the reproductional output of these individuals. But a factor like heat reflection is not a logical one to explain reduced sexual size dimorphism in the S Russian Curlew populations.

SUMMARY

The pattern of geographical variation in the Eurasian Curlew was studied. There was marked sexual and geographical variation. Sexual differences were larger than geographical differences and ♂♂ were smaller than ♀♀. Sexual dimorphism was smallest in the S Ural & Kazakhstan population and largest in the Central Siberian population. The species is polytypic and 3 subspecies were recognized: (1) *N. a. arquata*, breeding in Europe, (2) *N. a. orientalis*, breeding in Central Siberia, and (3) *N. a. suschkini*, breeding in the S Ural & Kazakhstan. These 3 subspecies can easily be recognized by a combination of measurements and plumage characters. *Orientalis* has on average long culmen- and tarsus lengths in combination with white axillaries and light upper tail-coverts. Nominate *arquata* and *suschkini* have smaller measurements, but nominate *arquata* nearly always has striped axillaries and dark upper tail-coverts while *suschkini* generally has white axillaries and light upper tail-coverts.

Wing-, culmen- and tarsus lengths were important to be measured with wing- and culmen lengths being of major importance for the distinction between sexes, and culmen- and tarsus lengths being important for the distinction between locations. Breeding ♀♀ in Scandinavia decreased in body mass with about 200 gm in the course of the breeding season. Curlews did not moult their primaries on the breeding grounds.

Chapter 16

Redshank

-

Tringa totanus

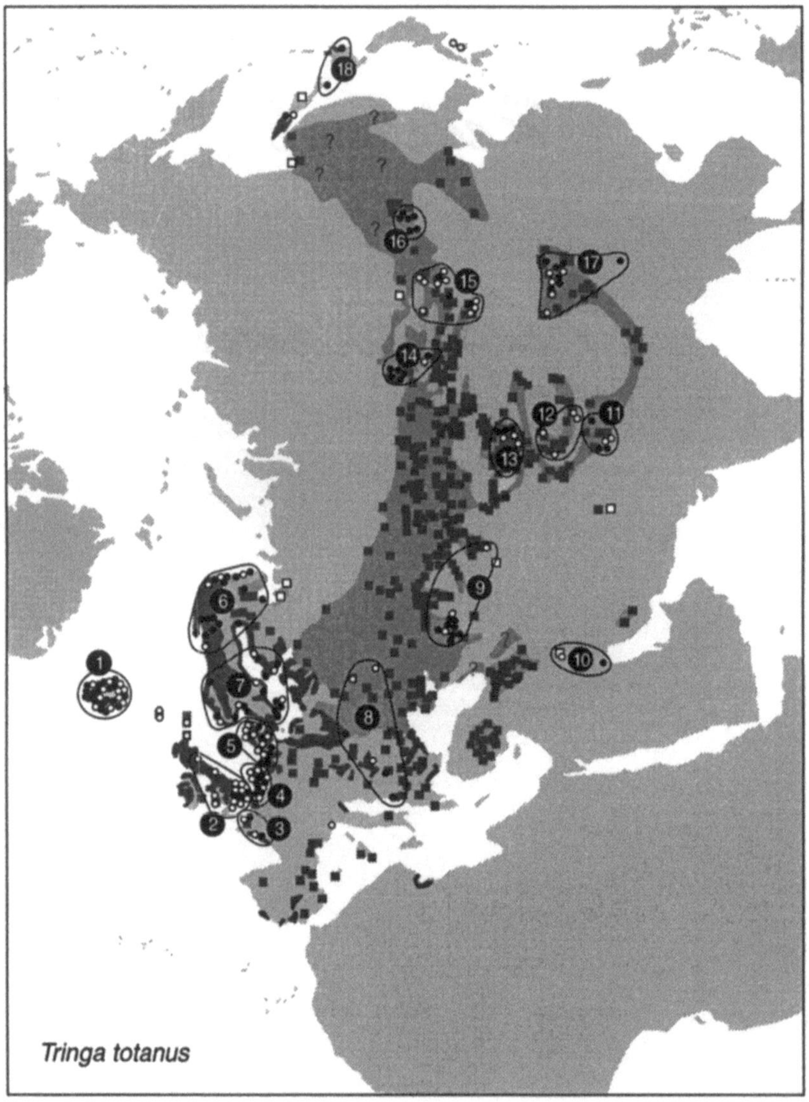

Fig. 104. *Breeding distribution of the Redshank.*

INTRODUCTION

The Redshank has a wide breeding distribution throughout the Palearctic, where it is breeding in moist and wet grasslands in different climate zones, ranging from the Mediterranean to the subalpine and subarctic zones (Fig. 104).

The taxonomy of the Redshank is complicated. Six subspecies are recognized nowadays (Hale 1971, Glutz *et al.* 1977, Cramp & Simmons 1983, Stepanyan 1990). The different subspecies are in contact with each other, resulting in processes of introgression and intergradation. The various subspecies are characterized by (1) measurements, (2) colouration and plumage pattern, and (3) amount of winter feathers retained in breeding plumage. Basically 2 morphs are distinguished throughout the various subspecies. In the first morph the basic colour of the upper parts is brownish-grey and the spots in the central parts of the feathers are tinged olive and rather small. The basic colour of the upper parts of the second morph is buff with larger and darker spots. The proportional occurrence of these morphs varies between the subspecies.

The following subspecies were recognized by Hale (1971):

- *T. t. robusta*, breeding on Iceland and the Faeroes. Polymorphic. Hale considered these birds as belonging to a stabilized hybrid population.
- *T. t. totanus*, breeding in N Fennoscandia and Spain. Polymorphic.
- *T. t. eurhinus*, breeding in N India, S Tibet and the Pamirs. Monomorphic.
- *T. t. terrignotae*, breeding in S Manchuria and E China. Monomorphic.
- *T. t. ussuriensis*, breeding in central and E Russia (east of the Urals and the Emba river and south to the Aral Sea and the Tyan-Shan), N Manchuria, and Mongolia. Monomorphic.
- *T. t. craggi*, breeding in China, Mongolia and the east part of Russia. Monomorphic.

Hale (1971, 1980) recognized several contact zones between the different subspecies. These were: (1) W Europe and (2) the Aral Sea region, both between *totanus* and *ussuriensis*, and (3) N Tibet and Tien-Shan, between *eurhinus* and *ussuriensis*. *Robusta* was considered to be related directly to *ussuriensis*. *Terrignotae*,

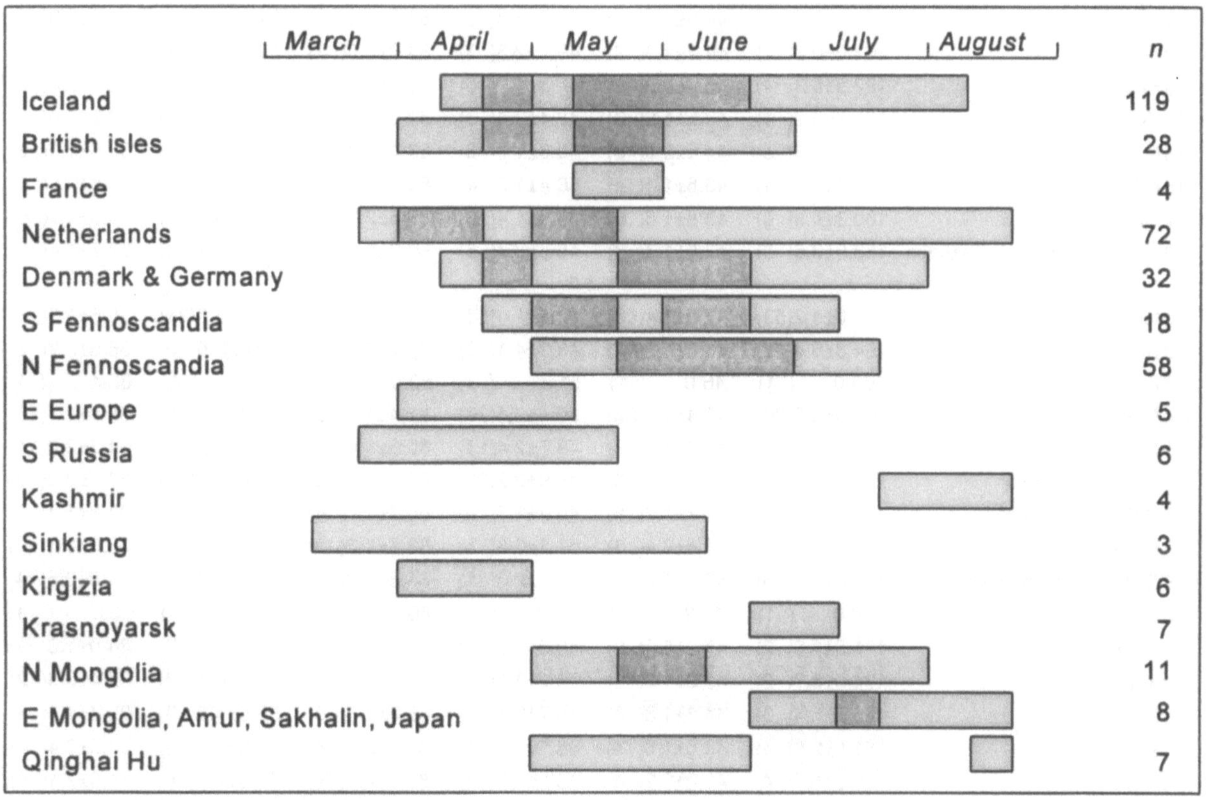

Fig. 105. *Collection dates of Redshanks. Most birds had been collected during the periods marked by the heavily shaded bars. The sample sizes are presented on the right side.*

craggi and *totanus* had more or less the same measurements. The measurements of *ussuriensis* decreased clinally from west to east.

Glutz *et al.* (1977) and Cramp & Simmons (1983) mostly supported Hale's conclusions. However, Cramp & Simmons (1983) did not consider the European breeding population as a hybrid one, but recognized it as nominate *totanus*, thus being a polymorphic subspecies with a breeding range covering Europe east to S Russia. Summers *et al.* (1988b) were able to separate the British and Icelandic Redshanks by using morphometric data.

Table 101. *Measurements of breeding populations of Redshanks. Notes: (1) some birds of Greenland and the Faeroes included; (2) above 60°N; (3). below 60°N; (4) is the Xinjiang Uygur Zizhiqu autonomous region in China; (4) also known under the older name of Koko Nor. Presented are the means ± 1 SD (n).*

Locality	Wing	Culmen	Tarsus	Secondary	Tail	Middle toe
			♂♂			
1 Iceland[1]	168.7±4.5(67)	40.0±2.0(67)	47.0±2.0(67)	86.1±3.0(23)	64.1±2.9(26)	35.1±1.2(12)
2 British isles	160.3±2.6(16)	42.1±1.9(16)	47.1±2.9(15)	81.7±3.1(3)	60.3±4.0(3)	33.7±1.4(2)
3 France	160.3±2.3(3)	40.3±0.6(3)	46.5±1.3(3)	81.3±2.5(3)	59.7±1.5(3)	33.5±1.1(3)
4 Netherlands	160.0±3.5(46)	42.8±1.7(45)	49.4±1.7(45)	80.4±1.9(29)	59.2±2.7(29)	34.2±1.7(26)
5 Denmark & Germany	159.6±4.3(20)	42.8±1.8(20)	47.9±1.9(20)	80.8±1.7(13)	60.2±2.9(13)	34.4±1.4(13)
6 N Fennoscandia[2]	156.7±2.6(32)	40.6±1.7(32)	45.7±2.1(32)	78.6±1.7(23)	58.0±2.5(23)	33.5±1.2(23)
7 S Fennoscandia[3]	157.2±4.0(16)	41.1±2.0(16)	46.2±2.3(16)	78.8±3.1(12)	60.7±3.7(12)	34.2±1.5(10)
8 E Europe	155.0 - (1)	43.5 - (1)	48.8 - (1)	82.0 - (1)	65.0 - (1)	-
9 S Russia & Kazakhstan	162.3±4.2(7)	43.0±1.7(5)	52.5±2.0(5)	83.2±2.9(5)	60.4±2.9(5)	35.0±1.6(3)
10 SW Iran	161.0 - (1)	42.2 - (1)	48.2 - (1)	82.0 - (1)	59.0 - (1)	31.6 - (1)
11 Kashmir	164.0±6.0(3)	44.9±2.6(3)	49.2±1.8(3)	87.3±2.1(3)	61.3±2.1(3)	34.8±0.9(3)
12 Xinjiang[4]	162.0±0.0(2)	44.2±1.3(2)	49.7±2.3(2)	84.0±0.0(2)	59.0±1.4(2)	31.9 - (1)
13 Kirgizia	162.5±5.3(4)	45.2±0.9(5)	49.8±1.1(5)	83.2±2.8(5)	60.2±1.6(5)	33.6±2.4(4)
14 Krasnoyarsk	161.7±5.9(3)	42.5±1.9(3)	48.4±2.4(3)	82.3±2.5(3)	60.0±6.2(3)	32.9±0.9(3)
15 N Mongolia	161.5±5.8(8)	44.5±2.3(8)	48.3±1.9(8)	83.4±3.7(8)	59.8±3.7(8)	32.6±1.9(8)
16 E Mongolia	158.3±5.1(4)	43.8±1.4(4)	50.4±1.3(4)	82.3±3.2(4)	59.0±2.2(4)	33.8±0.8(4)
17 Qinghai Hu[5]	160.3±4.8(6)	43.5±1.6(6)	47.3±1.4(6)	83.5±2.3(6)	57.5±3.0(6)	34.9±1.7(6)
18 Japan, NE China, Sakhalin	153.9±3.6(3)	42.5±1.3(3)	45.3±2.7(3)	80.0±1.0(3)	55.7±2.1(3)	32.7±1.4(3)
			♀♀			
1 Iceland[1]	172.0±4.2(53)	42.5±1.4(53)	48.1±1.9(53)	88.3±2.2(19)	63.7±2.8(21)	36.6±1.2(13)
2 British isles	164.8±5.4(12)	43.6±1.2(12)	48.2±4.1(12)	82.7±2.1(3)	60.0±2.0(3)	35.9±3.7(2)
3 France	165.0 - (1)	46.0 - (1)	51.3 - (1)	82.0 - (1)	57.0 - (1)	35.4 - (1)
4 Netherlands	164.5±3.2(26)	45.4±1.6(26)	50.4±2.5(26)	82.4±1.2(15)	60.1±2.2(15)	35.1±1.7(13)
5 Denmark & Germany	160.8±3.8(13)	43.5±1.6(13)	48.7±2.4(13)	80.8±1.0(4)	57.8±1.0(4)	34.3±2.0(4)
6 N Fennoscandia[2]	161.4±2.8(29)	41.8±1.7(29)	46.9±2.2(29)	80.6±2.2(19)	58.8±3.8(19)	33.7±0.9(19)
7 S Fennoscandia[3]	161.0±3.9(5)	41.7±1.2(5)	46.8±1.7(5)	80.7±1.7(4)	58.7±1.7(4)	32.8±1.4(3)
8 E Europe	165.3±1.2(3)	44.8±2.4(3)	51.3±0.8(3)	83.7±1.2(3)	60.3±1.5(3)	35.0±1.1(2)
9 S Russia & Kazakhstan	173.6±3.5(5)	46.0±1.3(5)	52.2±2.3(5)	86.8±2.2(5)	62.8±2.5(5)	34.8±0.5(4)
10 SW Iran	170.0 - (1)	44.2 - (1)	50.8 - (1)	86.0 - (1)	58.0 - (1)	33.1 - (1)
11 Kashmir	171.5±7.8(2)	47.2±5.3(2)	49.9±4.2(2)	88.5±9.2(2)	62.0±1.4(2)	34.0±1.3(2)
12 Xinjiang[4]	164.0±5.7(2)	45.6±1.3(2)	48.2±3.3(2)	84.0±2.8(2)	59.5±6.4(2)	32.4±0.3(2)
13 Kirgizia	166.7±6.4(3)	44.3±1.8(3)	48.7±2.8(3)	85.3±2.5(3)	59.0±4.6(3)	33.3±1.8(3)
14 Krasnoyarsk	161.0±2.9(4)	43.6±1.6(4)	48.5±3.8(4)	84.3±2.9(4)	57.0±3.5(4)	33.3±0.4(4)
15 N Mongolia	165.8±5.9(4)	43.2±3.2(4)	48.1±2.2(4)	86.0±4.5(4)	56.3±7.1(4)	34.0±2.4(4)
16 E Mongolia	150.0±4.2(2)	42.7±0.5(2)	47.5±1.6(2)	79.0±1.4(2)	57.0±1.4(2)	32.2 - (1)
17 Qinghai Hu[5]	164.4±5.1(5)	44.7±0.9(5)	47.4±2.1(5)	86.8±3.6(5)	58.0±2.1(5)	34.0±2.1(4)
18 Japan, NE China, Sakhalin	159.1±6.5(5)	40.6±2.0(5)	44.2±1.2(5)	82.4±3.4(5)	58.2±3.8(5)	32.4±1.8(5)

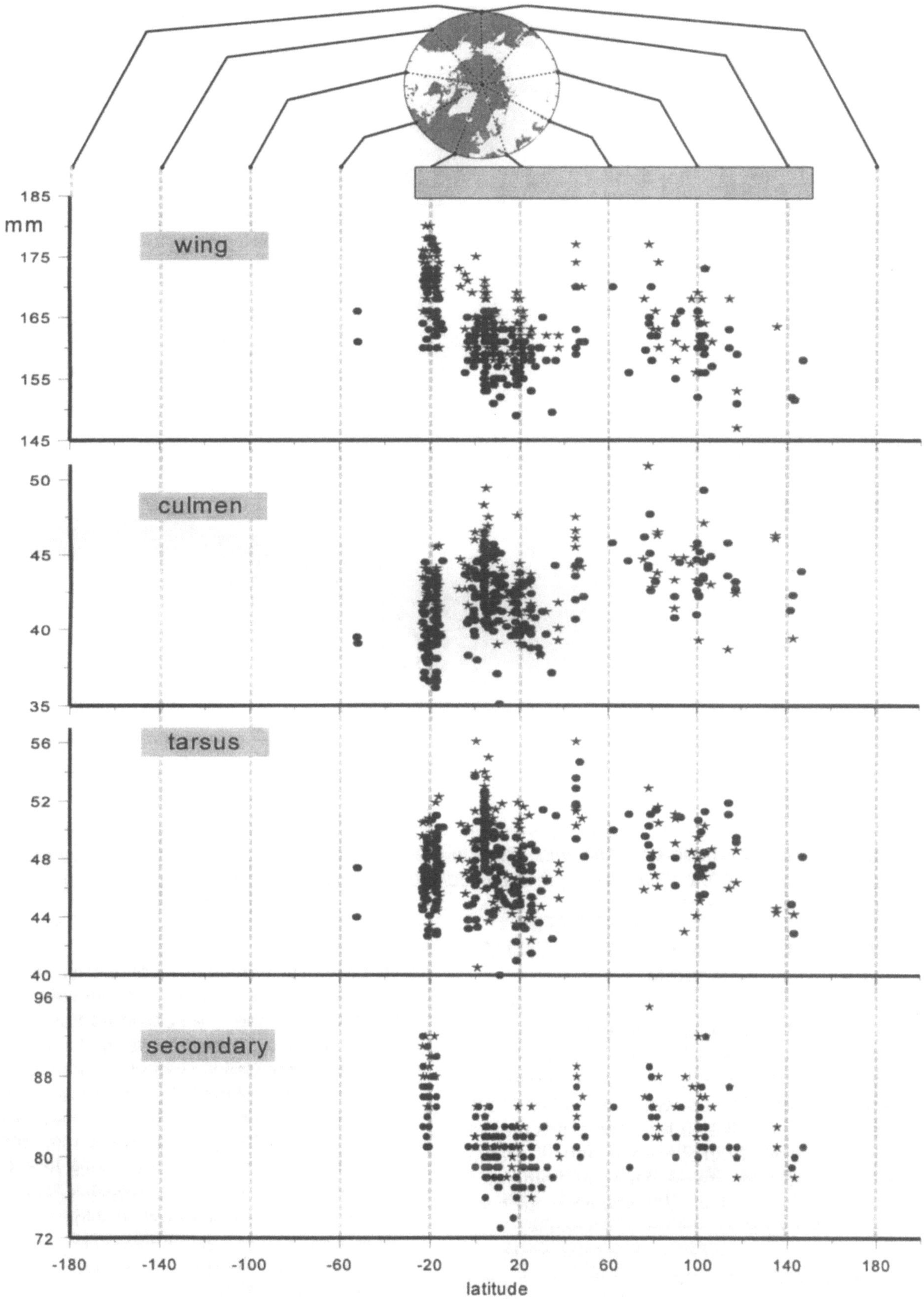

Fig. 106. *Measurements of Redshanks from the breeding grounds. The following measurements are presented from top to bottom: wing length, culmen length, tarsus length and secondary length (described in chapter 2). Explanation of the used symbols: ● ♂♂, ★ ♀♀.*

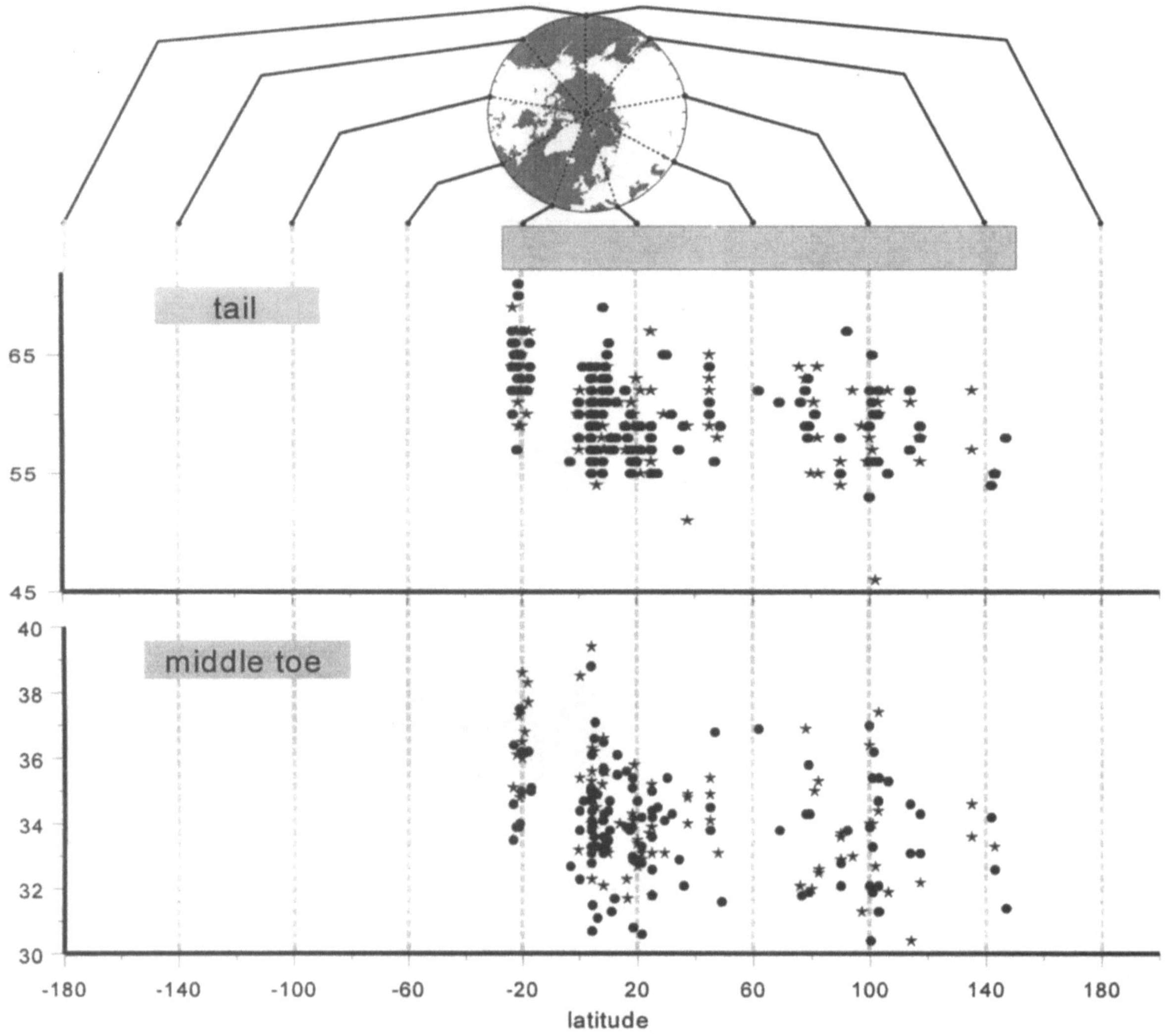

Fig. 107. *Measurements of Redshanks from the breeding grounds. The following measurements are presented from top to bottom: tail length and middle toe length (described in chapter 2). Explanation of the used symbols: ● ♂♂, ★ ♀♀.*

MORPHOMETRICAL DIFFERENTIATION

In total, 418 Redshanks from nearly all over the breeding range were sampled (Fig. 104). Few birds from the E Palearctic were examined, since our prime interest was focussed on the W Palearctic during the early stages of this project. Also, few skins from E Europe were available. We failed to measure any breeders from the W Mediterranean region. Males were more common (58.1%) than ♀♀ in the samples. The breeding grounds were reached from late March onwards (Fig. 105). Northern breeding birds arrived later at the breeding grounds than southern ones: Icelandic breeders late April and N Fennoscandian ones mid-May. Most of the skins examined had been collected during May and June.

Initially, 18 breeding locations were distinguished (Fig. 104, Table 101). However, since too few birds had been collected from south of the Black and Caspian Seas, and from Iran or Irak, these populations were left out of consideration. The morphometric differences between the samples were analysed and presented in the tables 101 - 107 and in the figures 106 - 108. No attention was paid to the existence of different colour morphs in the Redshank, since this is rather hazardous with museum skins: the history of preservation of a skin plays an important role in detecting different colour morphs. Moreover, different light conditions in various museum collections might easily lead to incomparable results. Unlike Hale (1971) we could not compare all skins under standard light conditions on a single spot. It was first tried to reduce the number of initially distinguished breeding locations in both the E and W Palearctic with MANOVA-analysis. The measurements in the W Palearctic were influenced by both breeding location and sex (Table 102). Much morphometric variation occurred around the North Sea basin (Table 103) which made it impossible to combine the data from the British isles, the Netherlands, Denmark, Germany and S Fennoscandia. The preliminary analyses in the E Palearctic gave less meaningful results, whilst a complicated pattern of geographical variation in measurements was expected because of (1) the complicated taxonomy of the Redshank and (2) the described variation of different 'mountain'- and 'desert' morphs in this region (Hale 1980). However, surprisingly little morphometric variation was found. Every analysis again showed that the eastern samples could be split into 3 categories: (1) large-sized birds breeding in S Russia east to the Aral Sea and southeast to Kashmir, (2) intermediate-sized birds breeding in Xinjiang, Kirgizia, N & W Mongolia, and Qinghai Hu, and (3) small-sized birds breeding in the Krasnoyarsk region, E Mongolia, Amur, Sakhalin and on the Kuriles (Table 102 and Fig. 107).

Table 102. *Overall results of MANOVA analysis involving different breeding populations in different compositions. Analyses with 6 (wing-, culmen-, tarsus-, secondary-, tail- and middle toe lengths) as well as 3 (wing-, culmen- and tarsus lengths) measurements were performed. The numbers in the first column correspond with the numbered breeding sites from Fig. 104. The involved degrees of freedom are presented between brackets. [1]These concern the birds breeding in the regions numbered 1, 2, 4, 5, 6, and 7 in Table 101. [2]These concern the birds breeding in the regions numbered 9, 11, 12, 13, 14, 15, 16 and 17 in Table 101.*

Geographical groups	Interaction		Locality		Sex		
	F	*df*	F	*df*	F	*df*	
			W Palearctic - 6 measurements				
all[1]	1.056	42/714	4.838***	42/714	2.176*	6/114	135
2, 3, 4, 5, 7	1.292	24/316	1.929**	24/316	1.896	6/76	91
			W Palearctic - 3 measurements				
all[1]	1.613*	21/933	22.727***	21/933	13.525***	3/309	327
2, 3, 4, 5, 7	1.926*	12/513	6.617***	12/513	8.502***	3/169	181
			E Palearctic - 6 measurements				
all[2]	0.964	42/240	1.480*	42/240	2.942*	6/35	56
8, 9, 13	1.553	6/6	1.821	12/14	9.841**	6/6	16
11, 12, 13	0.472	12/4	1.825	12/4	5.901	6/1	12
11, 12, 17	0.515	12/16	1.035	12/16	0.484	6/7	18
14, 15, 16	0.775	12/28	0.601	12/28	1.076	6/13	24
11.12.13/ 14.15	1.400	6/23	0.899	6/23	1.470	6/23	32
9/ 11.12.13/ 14.15	1.401	12/62	2.185*	12/62	2.903*	6/30	41
			E Palearctic - 3 measurements				
all[2]	1.087	21/141	2.834***	21/141	3.054*	3/45	63
8, 9, 13	0.770	6/30	1.862	6/30	4.157*	3/14	22
11, 12, 13	0.331	6/14	0.546	6/14	1.650	3/6	14
11, 12, 17	0.193	6/26	0.893	6/26	1.844	3/12	20
14, 15, 16	1.582	6/36	2.391*	6/36	0.326	3/17	25
11.12.13/ 14.15	0.757	3/28	1.092	3/28	1.507	3/28	34
9/ 11.12.13/ 14.15	1.131	6/78	4.031***	6/78	5.072**	3/38	46

Table 103. *Results of MANOVA analysis concerning the size differentiation in the W European populations of the Redshank. Significance levels: *0.01 < P < 0.05; **0.001 < P < 0.01; ***P < 0.001. The F-values presented in the upper right corner of this table concern ♂♂ and the ones in the lower left corner ♀♀.*

Locality	British isles	Netherlands	Denmark/ Germany	S Fennoscandia	
British isles		4.163**	1.790	9.542***	
Netherlands	6.768***		3.092*	15.045***	♂♂
Denmark/ Germany	3.207*	5.821***		4.626**	
S Fennoscandia	1.478	14.617***	2.352		
		♀♀			

Either, we had too few data to find the expected differences in the E Palearctic, or most differentiation between these samples are plumage differences rather than differences in body dimensions. These preliminary analyses resulted in the distinction of 9 breeding locations in the Redshank: (1) Iceland and the Faeroes, (2) the British isles, (3) the Netherlands, (4) Denmark & Germany, (5) Fennoscandia north of 60°N, (6) Fennoscandia south of 60°N, (7) SE Russia & SW Asia, (8) Xinjiang, Kirgizia, N Mongolia, and Qinghai Hu, and (9) the Krasnoyarsk region, E Mongolia, NE China, Amur and Sakhalin.

The Redshank showed a fair degree of sexual variation and strong geographical variation (Table 104). Nearly always ♀♀ were larger than ♂♂ (Table 105). The populations with large body dimensions also had a large degree of dimorphism: Iceland, Netherlands, SE Russia/ SW Asia. The differences between the sexes in the populations with small body dimensions were also smaller: N Fennoscandia, S Fennoscandia and the E Asian population.

The squared euclidean distances as calculated with cluster analysis were large, indicating large distances between samples (Fig. 108). Both ♂- and ♀-samples can generally be divided in 3 size categories with cluster analysis:

- groups with small-sized birds (N Fennoscandia, S Fennoscandia, NE Asia);
- groups with intermediate-sized birds (British isles, Netherlands, Central Asia);
- groups with large-sized birds (Iceland, SE Russia/ SW Asia).

Some other samples wandered between 2 of these groups, e.g. the sample of the ♂♂ from SE Russia/ SW Asia had closet resemblance with the 'intermediate-size' group and the sample of ♀♀ from this population with the 'large-size' group. The same applies to the sample of Denmark/ Germany.

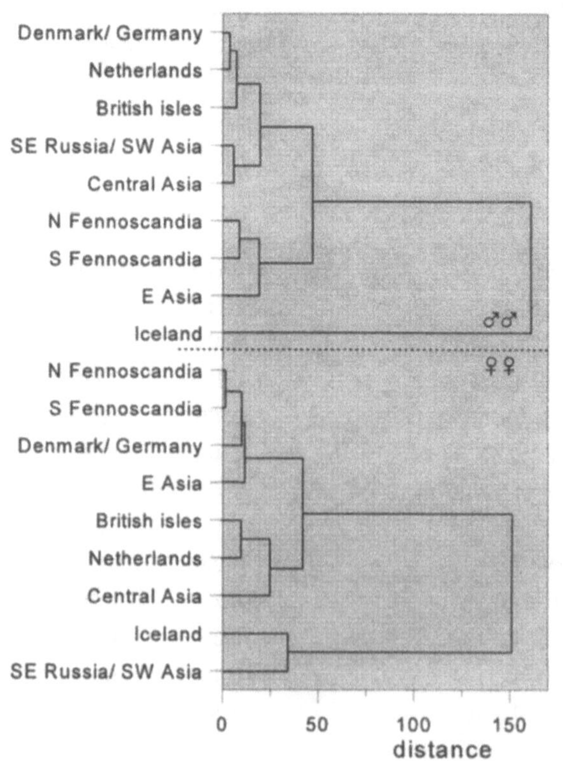

Fig. 108. *Cluster analysis in the Redshank. The analysis was based upon 6 measurements. The samples of both sexes were analysed separately.*

Table 104. *Overall results of MANOVA for all distinguished breeding populations in the Redshank in combination with univariate ANOVA's in which the different measurements were analysed separately. Significance levels: *0.01 < P < 0.05; **0.001 < P < 0.01; ***P < 0.001.*

	Analysis I			Analysis II			Analysis III		
Variables	wing, culmen, tarsus, secondary, tail, middle toe length			wing, culmen, tarsus			wing, culmen		
n	210			410			412		
Multivariate analysis of variance									
Effects	Interaction	Area	Sex	Interaction	Area	Sex	Interaction	Area	Sex
Pillai's trace	0.309	1.496	0.172	0.114	1.026	0.192	0.097	0.918	0.183
F-value	1.302	7.969***	6.474***	1.939**	25.472***	30.807***	2.518**	41.776***	44.061***
Hypoth. & error *df*	48/1152	48/1152	6/187	24/1176	24/1176	3/390	16/788	16/788	2/393
Univariate F tests									
df	8/192	7/192	1/192	8/392	8/392	1/392	8/394	8/394	1/394
wing length	1.370	27.343***	24.486***	1.883	62.696***	63.987***	1.906	62.651***	65.336***
culmen length	2.582*	17.388***	11.782**	3.185**	28.707***	32.601***	3.218**	28.838***	33.149***
tarsus length	1.801	13.428***	5.404*	1.218	19.605***	3.649			
secondary length	0.788	33.657***	9.580**						
tail length	1.209	10.260***	1.169						
middle toe length	1.214	7.011***	1.835						

Table 105. *Sexual size dimorphism in the Redshank. The mean percentage difference between the measurements of both sexes was calculated for every single variable and breeding population, according to: $(x_{♀}-x_{♂})/x_{♂}*100$. A negative value thus implies smaller measurements for ♀♀. The calculations were based upon the means of Table 101. The obtained values were averaged per variable and summed per breeding population.*

Locality	Wing	Culmen	Tarsus	Secondary	Tail length	Middle toe	Σ_3	Σ_6
Iceland	2.0	6.2	2.3	2.6	-0.6	4.3	10.5	16.8
British isles	2.8	3.6	2.3	(1.2)	(-0.5)	(6.5)	8.7	-
Netherlands	2.8	6.0	2.0	2.5	1.5	2.6	10.8	17.4
Denmark/ Germany	0.8	1.6	1.7	(0.0)	(-4.0)	(-0.3)	4.1	-
N Fennoscandia	3.0	3.0	2.6	2.5	1.4	0.6	8.6	13.1
S Fennoscandia	2.4	1.5	1.3	(2.4)	(-3.2)	(-4.1)	5.2	-
SE Russia/ SW Asia	7.0	7.0	-0.6	4.3	4.0	(-0.6)	13.4	-
Qinghai Hu	2.6	2.8	0.2	4.0	0.9	(-2.6)	5.6	-
E Asia	3.4	-4.5	-2.4	3.0	4.5	-0.9	-3.5	3.1
x	2.7	2.4	0.4	2.3	-0.5	0.7		
SD	1.7	3.5	1.9	1.3	3.1	3.0		

Table 106. *Pearson's product-moment correlation between measurements of Redshanks. Data of ♂♂ and ♀♀ were separately analysed. Two-tailed probabilities: * $0.01 < P < 0.05$; ** $0.001 < P < 0.01$; *** $P < 0.001$.*

	Wing	Culmen	Tarsus	Secondary	Tail length	Middle toe	
Wing		0.065	0.344***	0.779***	0.571***	0.533***	
Culmen	-0.157*		0.496***	0.164	0.104	0.253*	
Tarsus	0.131*	0.576***		0.259**	0.172	0.458***	
Secondary	0.767***	0.170*	0.215**		0.521***	0.379***	♀♀
Tail	0.620***	-0.070	0.158	0.535***		0.311**	
Middle toe	0.349***	0.126	0.218*	0.246**	0.222*		
				♂♂			

Secondaries, tails and/or middle toes had been measured of 55 % of the birds. Therefore, 3 different MANOVA-analyses were performed in which samples were pairwise contrasted while using varying sets of variables: (1) all 6 measurements, (2) wing-, culmen- and tarsus lengths, and (3) wing- and culmen lengths. The following conclusions were drawn from these analyses (tables 107 & 108, Fig. 109):

(1) Measurements of ♂♂ mostly differed from those of ♀♀, especially when 3 variables were involved. Within regions, ♂♂ differed from ♀♀ in the samples of Iceland, the British isles, the Netherlands, N Fennoscandia, SE Russia/ SW Asia and Central Asia. No significant differences between sexes were found in the samples from Denmark & Germany, S Fennoscandia and E Asia.

(2) The Icelandic samples were clearly distinct from every other sample, irrespective of a bird's sex. Icelandic birds were characterized by long wing-, secondary-, tail-, and middle toe lengths in combination with short culmen- and tarsus lengths. The samples of both sexes in this population were statistically different from each other with ♂♂ having smaller measurements.

(3) In general, N and W European birds were considerably smaller than Icelandic Redshank. Especially N Fennoscandian birds were small-sized. W European birds had smaller wing-, secondary-, tail-, and middle toe lengths and longer culmen- and tarsus lengths than the Icelandic birds. The N Fennoscandian population was (together with the E Asian) the smallest measured.

(4) The Dutch sample differed slightly from other W European samples in (1) longer culmen lengths (♀ only) and (2) longer tarsus lengths (both sexes).

(5) The British sample had intermediate measurements between the Icelandic and Dutch population. British Redshank might thus be influenced by both the Icelandic and Dutch populations (see below).

(6) Birds from Germany and Denmark had somewhat smaller measurements than those from the Netherlands.

(7) Redshanks of S Fennoscandia more closely resembled the ones of N Fennoscandia than the ones of Germany and Denmark.

(8) Redshanks from SE Russia/ SW Asia were large in size. These birds were separable from nearly all other populations. The only other large-sized population, on Iceland, differed clearly in characteristically short culmen- and tarsus lengths.

(9) Central Asian Redshanks were intermediate between large-sized birds from SE Russia/ SW Asia and small-sized ones

Table 107. *Results of MANOVA analysis with 6 variables concerning the morphometric differentiation in Redshank. The values presented are Pillai's trace statistic (Norušis 1992c). In these series of pairwise comparisons 18 groups (2 sexes, 9 populations) were compared on the basis of 6 measurements (wing-, culmen-, tarsus-, secondary-, tail- and middle toe length) with 6 and 187 df. Significance levels: *0.01 < P < 0.05; **0.001 < P < 0.01; ***P < 0.001.*

Locality	*n*	Iceland ♀	British isles ♀	Netherlands ♀	Denmark/ Germany ♀	N Fennoscandia ♀	S Fennoscandia ♀	SE Russia/ SW Asia ♀	Central Asia ♀	NE Asia ♀	Iceland ♂	British isles ♂	Netherlands ♂	Denmark/ Germany ♂	N Fennoscandia ♂	S Fennoscandia ♂	SE Russia/ SW Asia ♂	Central Asia ♂
Iceland ♀	13																	
British isles ♀	2	0.177***																
Netherlands ♀	13	0.401***	0.014															
Denmark/ Germany ♀	4	0.240***	0.018	0.036														
N Fennoscandia ♀	19	0.392***	0.069*	0.223***	0.048													
S Fennoscandia ♀	3	0.176***	0.071*	0.113***	0.039	0.010												
SE Russia/ SW Asia ♀	11	0.267***	0.085**	0.140***	0.113***	0.301***	0.133***											
Central Asia ♀	12	0.303***	0.082*	0.202***	0.095**	0.227***	0.095**	0.103**										
NE Asia ♀	11	0.372***	0.065*	0.183***	0.048	0.077*	0.035	0.212***	0.083*									
Iceland ♂	12	**0.103****	0.172***	0.396***	0.223***	0.311***	0.132***	0.320***	0.300***	0.308***								
British isles ♂	2	0.108**	**0.049**	0.069*	0.037	0.020	0.015	0.074*	0.036	0.014	0.079*							
Netherlands ♂	25	0.486***	0.026	**0.137*****	0.023	0.155***	0.079*	0.326***	0.298***	0.169***	0.415***	0.065*						
Denmark/ Germany ♂	13	0.382***	0.026	0.087**	**0.019**	0.058	0.045	0.235***	0.204***	0.092**	0.320***	0.030	0.051					
N Fennoscandia ♂	23	0.527***	0.079*	0.316***	0.073	**0.082***	0.043	0.442***	0.353***	0.142***	0.426***	0.048	0.184***	0.106**				
S Fennoscandia ♂	10	0.397***	0.061	0.198***	0.068*	0.074*	**0.051**	0.337***	0.293***	0.147***	0.293***	0.046	0.094**	0.049	0.047			
SE Russia/ SW Asia ♂	13	0.374***	0.027	0.058	0.038	0.188***	0.094**	**0.145*****	0.145***	0.125***	0.329***	0.056	0.089**	0.085*	0.274***	0.174***		
Central Asia ♂	16	0.406***	0.039	0.103**	0.046	0.185***	0.081*	0.151***	**0.085****	0.066*	0.353***	0.035	0.159***	0.101**	0.281***	0.198***	0.033	
NE Asia ♂	8	0.420***	0.041	0.150***	0.038	0.107**	0.067*	0.259***	0.153***	**0.044**	0.357***	0.044	0.089**	0.075*	0.092**	0.112***	0.103**	0.075*

Table 108. *Results of MANOVA analysis with 3 variables concerning the morphometric differentiation in Redshank. The values presented are Pillai's trace statistic (Norušis 1992c). In these series of pairwise comparisons 18 groups (2 sexes, 9 populations) were compared on the basis of 3 measurements (wing-, culmen- and tarsus length) with 3 and 390 df. Significance levels: *0.01 < P < 0.05; **0.001 < P < 0.01; ***P < 0.001.*

Locality	*n*	Iceland ♀	British isles ♀	Netherlands ♀	Denmark/ Germany ♀	N Fennoscandia ♀	S Fennoscandia ♀	SE Russia/ SW Asia ♀	Central Asia ♀	NE Asia ♀	Iceland ♂	British isles ♂	Netherlands ♂	Denmark/ Germany ♂	N Fennoscandia ♂	S Fennoscandia ♂	SE Russia/ SW Asia ♂	Central Asia ♂
Iceland ♀	53																	
British isles ♀	12	0.088***																
Netherlands ♀	26	0.263***	0.034**															
Denmark/ Germany ♀	13	0.195***	0.020	0.037**														
N Fennoscandia ♀	29	0.245***	0.032**	0.148***	0.030**													
S Fennoscandia ♀	5	0.079***	0.015	0.053***	0.012	0.000												
SE Russia/ SW Asia ♀	13	0.104***	0.050***	0.046***	0.096***	0.170***	0.081***											
Central Asia ♀	13	0.121***	0.007	0.043***	0.030**	0.050***	0.023*	0.063***										
NE Asia ♀	12	0.209***	0.035**	0.096***	0.018	0.015	0.006	0.148***	0.036**									
Iceland ♂	67	0.157***	0.126***	0.364***	0.204***	0.198***	0.057***	0.228***	0.179***	0.189***								
British isles ♂	16	0.195***	0.029**	0.111***	0.021*	0.001	0.000	0.148***	0.044***	0.009	0.155***							
Netherlands ♂	42	0.399***	0.062***	0.110***	0.014	0.078***	0.021*	0.176***	0.096***	0.049***	0.390***	0.046***						
Denmark/ Germany ♂	20	0.272***	0.036**	0.088***	0.005	0.023*	0.007	0.151***	0.050***	0.010	0.256***	0.012	0.018					
N Fennoscandia ♂	31	0.420***	0.118***	0.273***	0.080***	0.059***	0.015	0.294***	0.139***	0.030**	0.333***	0.035**	0.137***	0.059***				
S Fennoscandia ♂	16	0.296***	0.078***	0.176***	0.045***	0.029**	0.009	0.221***	0.092***	0.014	0.229***	0.017*	0.068***	0.026*	0.003			
SE Russia/ SW Asia ♂	18	0.251***	0.040***	0.033**	0.014	0.084***	0.032**	0.092***	0.065***	0.063***	0.275***	0.062***	0.018	0.035**	0.160***	0.098***		
Central Asia ♂	16	0.209***	0.015	0.027*	0.004	0.047***	0.018	0.086***	0.018	0.025*	0.245***	0.034**	0.038**	0.017	0.119***	0.070***	0.023*	
NE Asia ♂	8	0.224***	0.057***	0.090***	0.020*	0.036**	0.016	0.154***	0.061***	0.010	0.199***	0.024*	0.029**	0.012	0.025*	0.013	0.049***	0.032**

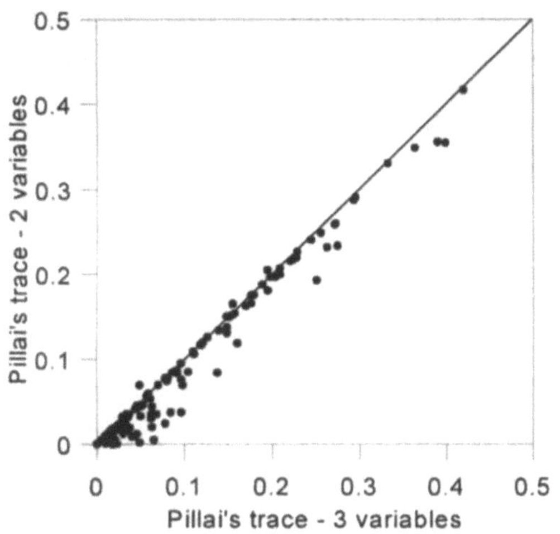

Fig. 109. *Relationship between the results of 2 different MANOVA-analyses. On the X-axis the Pillai's trace values resulting from the analyses with 3 variables (wing-, culmen- and tarsus lengths) are presented and on the Y-axis the ones from '2-variable' analysis (wing- and culmen length).*

from E Asia; tarsus lengths corresponded more with the ones from E Asia, and secondary lengths more with the ones of SE Russia/ SW Asia.

(10) E Asian populations were the smallest measured in the E Palearctic. The small size was matched by N Fennoscandian birds in the west which differed in slightly longer wing and slightly shorter culmen and tarsus.

All 6 selected measurements were important for the distinction between samples (Table 104). Secondary-, wing-, culmen- and tarsus lengths were most important to separate breeding populations and wing-, secondary-, and culmen lengths were most important to identify sexes. Tail- and middle toe lengths were marginally important for identifying both geographical location and sex.

MANOVA-analysis including only wing- and culmen lengths produced nearly identical results as with the MANOVA-analysis including wing-, culmen- and tarsus lengths (Fig. 109). Therefore an additional third table with the MANOVA-results based upon wing- and culmen lengths is not presented. There were some differences. The major one concerns the calculated distances between the samples which were generally smaller in the '2-variable' calculations. There was no significant distinction between samples anymore in the following pairs (only ♂- and ♀-groups among each other): (1) Dutch ♀♀ versus Central Asian ♀♀; (2) Denmark/ Germany ♀♀ versus Central Asia ♀♀, (3) British isles ♂♂ versus Netherlands ♂♂; (4) SE Russian ♂♂ versus Central Asian ♂♂.

PREDICTION OF BREEDING ORIGIN

Measurements of ♂♂ and ♀♀ from the same breeding locations usually differed, and hence both could not be combined in one POSCON group. The cluster- and MANOVA analyses presented above resulted in a prudent identification of 6 POSCON groups per sex: (1) Iceland, (2) The Netherlands, (3) N Fennoscandia, (4) S Russia/ SE Asia, (5) Central Asia, and (6) E Asia. The samples from the British isles, Germany, Denmark, S Fennoscandia and E Europe were not used, due to their mixed character.

Prediction of breeding origin was restricted to the W Palearctic populations because of the uncertain geographical limits of the E Palearctic populations. Six POSCON groups were constructed: (1) Icelandic ♂♂; (2) Icelandic ♀♀; (3) ♂♂ from the Netherlands; (4) ♀♀ from the Netherlands; (5) Fennoscandian ♂♂ north of 60°N; and (6) Fennoscandian ♀♀ north of 60°N. Predictions of breeding origin were based upon: (1) wing-, culmen-, tarsus-, secondary- and tail lengths, (2) wing-, culmen- and tarsus lengths, and (3) wing- and culmen lengths. With these POSCON groups accuracy- and exclusivity rates could be estimated for the W Palearctic, if excluding probability of occurrence of E Palearctic breeding birds in the W Palearctic. Two types of analyses were performed: (1) predictions based upon *known* sexes, and (2) predictions based upon *unknown* sexes.

Accuracy

High accuracy rates were reached when 5 measurements were used and the birds' sexes were known (Table 109). Then, all 3 populations could easily be distinguished from each other. On average, 92.6 % had the highest resemblance with the own POSCON group in this situation. This high percentage decreased to 86.2% with 2 measurements and to 82.0% with 2 measurements being involved which is a relatively small decrease in comparison with the results in other species. It clearly demonstrates

Table 109. *Accuracy of prediction in W Palearctic Redshank (sex* **known** *a priori). These predictions have been based upon morphometric analysis and included (A) wing-, culmen-, tarsus-, secondary- and tail length; (B) wing-, culmen- and tarsus length; and (C) wing- and culmen length. The columns show the predicted proportions in relation to the total number per breeding area (last column).*

	Predicted						
Locality	Iceland ♂	Netherlands ♂	N Fennoscandia ♂	Iceland ♀	Netherlands ♀	N Fennoscandia ♀	*n*
			(A) All measurements				
Iceland ♂	100.0	-	-	-	-	-	15
Netherlands ♂	6.7	90.0	3.3	-	-	-	30
N Fennoscandia ♂	4.5	13.6	81.8	-	-	-	22
Iceland ♀	-	-	-	100.0	-	-	14
Netherlands ♀	-	-	-	-	88.9	11.1	18
N Fennoscandia ♀	-	-	-	5.3	-	94.7	19
			(B) Wing-, culmen- and tarsus length				
Iceland ♂	91.7	3.3	5.0	-	-	-	60
Netherlands ♂	1.9	90.6	7.5	-	-	-	53
N Fennoscandia ♂	-	13.3	86.7	-	-	-	30
Iceland ♀	-	-	-	89.1	4.3	6.5	46
Netherlands ♀	-	-	-	2.9	80.0	17.1	35
N Fennoscandia ♀	-	-	-	6.9	13.8	79.3	29
			(C) Wing- and culmen length				
Iceland ♂	88.3	6.7	5.0	-	-	-	60
Netherlands ♂	3.6	76.4	20.0	-	-	-	55
N Fennoscandia ♂	-	13.3	86.7	-	-	-	30
Iceland ♀	-	-	-	89.4	4.3	6.4	47
Netherlands ♀	-	-	-	11.4	68.6	20.0	35
N Fennoscandia ♀	-	-	-	6.9	10.3	82.8	29

that these birds can be predicted to occur in mixed samples, if the birds' sexes are known. When only wing- and culmen lengths were used, the resulting predictions for the Dutch population were poor, but better results were reached with the inclusion of tarsus lengths. Then the outcome was better than for the N Fennoscandian POSCON group. The percentage of correctly predicted birds decreased with about 20 % when the birds' sexes were unknown (compare Table 110 with 109). The percentages of correctly predicted birds became 72.4, 66.4 and 60.0 % respectively when 5, 3 and 2 measurements were included. Most of the confusion arose between small-sized N Fennoscandian ♀♀ and intermediate-sized Dutch ♂♂. When the birds' sexes were unknown and only 2 or 3 measurements were taken, the prediction results for both these POSCON groups were meagre, and it might be better to combine these groups then. Predicting both sexes of the Iceland population, of N Fennoscandian ♂♂, and of Dutch ♀♀ was much better then even with so few measurements available.

Exclusivity

The POSCON groups described above concerned breeding populations of Iceland, the Netherlands and N Fennoscandia only. In reality, these populations might well occur rather mixed in several locations of the W Palearctic. Estimation of the exclusivity rates should therefore not include measurements of birds from these mixed zones. Thus the measurements of all sampled W Palearctic Redshanks were used to estimate the exclusivitiy rates (Table 111). When the birds' sexes were known, the estimated exclusivity rates were as high as 31.5%. This high percentage was reduced when more measurements were used: the predictions become more accurate but less certain (compare tables 109 and 111). When the birds' sexes were unknown the estimated exclusivity rates were about 20% lower than with known sexes.

Table 110. *Accuracy of prediction in W Palearctic Redshank (sex* **unknown** *a priori). These predictions have been based upon morphometric analysis and included (A) wing-, culmen-, tarsus-, secondary- and tail length; (B) wing-, culmen- and tarsus length; and (C) wing- and culmen length. The columns show the predicted proportions in relation to the total number per breeding area (last column).*

Locality	Predicted: Iceland ♂	Netherlands ♂	N Fennoscandia ♂	Iceland ♀	Netherlands ♀	N Fennoscandia ♀	*n*
			(A) All measurements				
Iceland ♂	80.0	-	-	13.3	-	6.7	15
Netherlands ♂	3.3	63.3	3.3	-	-	30.0	30
N Fennoscandia ♂	-	9.1	72.7	-	-	18.2	22
Iceland ♀	28.6	-	-	71.4	-	-	14
Netherlands ♀	-	-	-	-	88.9	11.1	18
N Fennoscandia ♀	10.5	15.8	15.8	-	-	57.9	19
			(B) Wing-, culmen- and tarsus length				
Iceland ♂	70.0	-	1.7	11.7	5.0	11.7	60
Netherlands ♂	-	52.8	3.8	-	24.5	18.9	53
N Fennoscandia ♂	-	6.7	80.0	-	-	13.3	30
Iceland ♀	15.2	-	-	73.9	4.3	6.5	46
Netherlands ♀	-	8.6	-	2.9	77.1	11.4	35
N Fennoscandia ♀	3.4	17.2	17.2	3.4	13.8	44.8	29
			(C) Wing- and culmen length				
Iceland ♂	70.0	-	1.7	16.7	3.3	8.3	60
Netherlands ♂	1.8	30.9	18.2	-	20.0	29.1	55
N Fennoscandia ♂	-	13.3	80.0	-	-	6.7	30
Iceland ♀	14.9	-	-	74.5	4.3	6.4	47
Netherlands ♀	-	14.3	-	11.4	62.9	11.4	35
N Fennoscandia ♀	3.4	24.1	17.2	6.9	6.9	41.4	29

Table 111. *Estimated exclusivity rates in the predictions concerning the breeding origin and sex of Redshank breeding in the W Palearctic. ¹Wing- and culmen lengths; ²Wing-, culmen- and tarsus lengths; ³Wing-, culmen-, tarsus-, secondary- and middle toe lengths.*

Locality	Sex known			Sex unknown		
n of variables	2¹	3²	5³	2¹	3²	5³
n	325	323	157	325	323	157
Iceland ♂	11.4	10.5	5.1	4.0	2.8	1.3
Netherlands ♂	6.8	7.4	3.8	-	0.9	1.3
N Fennoscandia ♂	2.2	3.1	0.6	1.5	0.9	-
Iceland ♀	4.6	3.1	6.4	0.6	0.3	1.9
Netherlands ♀	3.1	4.6	0.6	2.5	4.3	0.6
N Fennoscandia ♀	3.4	2.8	5.1	-	-	0.6
Total %	31.5	31.5	21.6	8.6	9.2	5.7

As with known sexes more measurements resulted in lower exclusivity rates and higher accuracy rates.

CONTACT ZONES

Measurement-data from two areas in W Europe were subjected to a more detailed POSCON analysis concerning the intermixing of populations. For these analyses, wing-, culmen- and tarsus lengths of the following 3 sample groups were used: (1) Iceland, (2) The Netherlands, and (3) N Fennoscandia (Fig. 110). A priori a warning is needed: the more sample groups are used for prediction, the smaller is the chance that a bird can be predicted to resemble exclusively one particular group with the exclusion of all other sample groups. In this situation, the differences between the three sample groups are significant, though small. It then rarely happens that a peculiar bird can be assigned definitely to only one sample group. In most cases, individu-

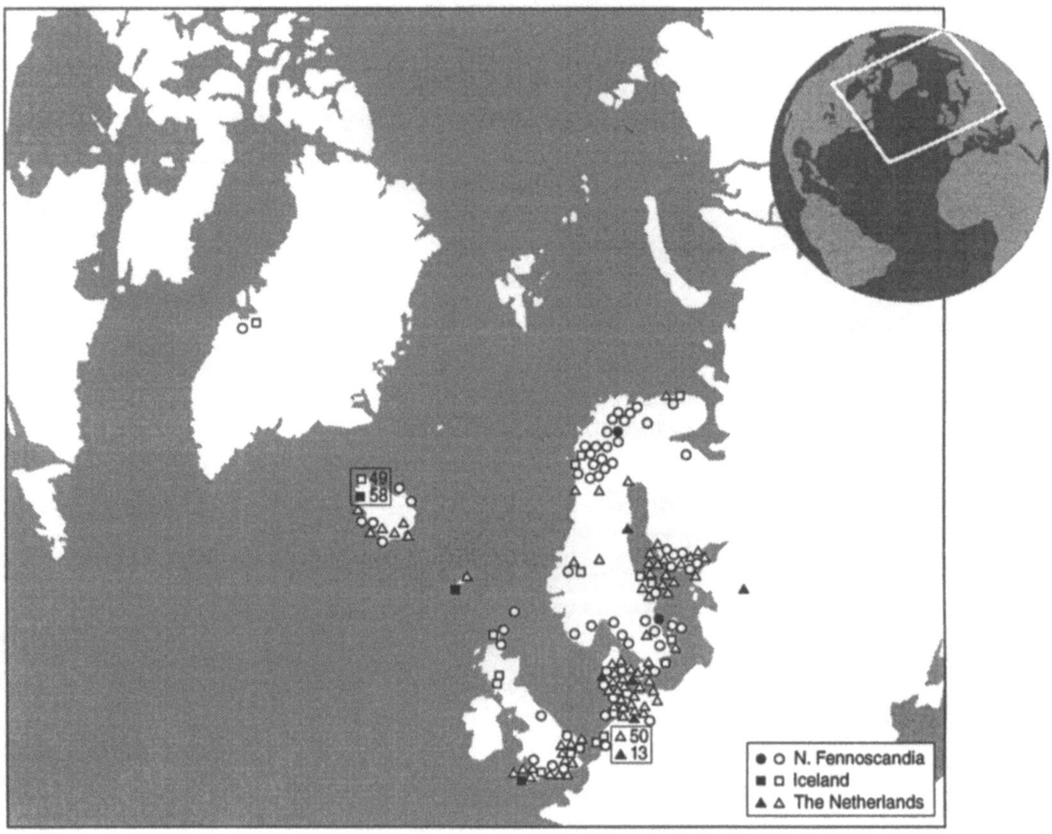

Fig. 110. *The results of predicting the breeding origin of Redshank in W- and NW Europe. Circles indicate the predicted occurrence of the N Fennoscandian population, squares of the Icelandic population, and triangles of the North Sea population. Open symbols are less certain than closed symbols. This is due to the fact that although one population was predicted to have the highest probability, both other could not be excluded: one or both typicality probabilities might still be > 0.05%. When closed symbols are presented, these remaining probabilities are < 0.05%.*

als have a high resemblance with one population, whilst they cannot be excluded with 95 % certainty from any of the other populations. The investigated areas were:

(1) *The British isles.* Hale (1971) considered the British breeding population of Redshank as a hybrid swarm between the Icelandic and W European breeding populations, with higher frequencies of interbreeding in the north and west parts of the British isles. This view was based on differences in the completeness of summer plumage in both populations. Summers *et al.* (1988b) rejected this hypothesis since their British samples were not different from each other. However, they used 2 samples of breeding birds from (1) NW England and (2) Scotland. They did not have measurements of breeding birds from S and E England which were essential to test Hale's (1971) hypothesis: these SE English birds represented most purely the continental breeding population. Here Hale's ideas were tested with POSCON-analysis. If Hale's (1971) hypothesis is true, one would expect (1) a clinal change in measurements towards the north and west, (2) SE English birds would resemble the breeding birds from Belgium, the Netherlands and Germany; and (3) Scottish birds would resemble Icelandic breeding birds.
Our POSCON-analyses clearly show that the British breeding population is not homo-

geneous (Fig. 110). The birds from S and E England indeed had most resemblance with the Dutch sample. Some W Scottish Redshanks also had measurements which mostly corresponded with the ones of Iceland. Moreover Fig. 110 also shows that some birds along the NE coast and on the Orkneys and Shetlands have most resemblance with the type present in N Scandinavia. And thus Hale's (1971) hypothesis seems to be true, since W European-type birds are mainly present in S and E England, and Icelandic-type birds in W Scotland. However, it is necessary to analyse more morphometrical data of British breeding birds, before firm conclusions can be reached on the exact nature of intermixing of these three subspecies on the British isles.

(2) *Germany, Denmark and S Fennoscandia.* It was already evident that N Fennoscandian Redshank differ from W European ones. It was also evident that S Fennoscandian ♂♂ were a bit larger-sized than N Fennoscandian ones (Table 101). Moreover, the breeding birds of Denmark and Germany were somewhat smaller than the Netherlands' ones. But it does not become evident from this table whether there are abrupt or gradual changes or whether the drawn boundaries were drawn correctly beforehand. Therefore, membership probabilities were calculated with POSCON-analysis as was indicated above. These calculations revealed that Fennoscandian Redshanks breeding south of 60°N resembled mostly the N Fennoscandian population (Fig. 110). Even the S Swedish and S Norwegian birds more closely resembled the N Fennoscandian breeding birds. Oppositely, Danish & German Redshanks had most resemblance with W European breeding birds. The breeding birds of Åland and S Finland are somewhat exceptional: they regularly resembled the W European population and not as expected the N Fennoscandian ones.

BODY MASS

Mass data of 35 breeding Redshanks were collected. The average body masses of the Dutch and Fennoscandian ♂♂ were 111.3 (n = 7) and 103.1 gm (n = 9) respectively, whilst this was 126.5 (n = 4) and 128.6 gm (n = 7) for ♀♀. On Iceland, ♂♂ weighed on average 133.3 (n = 3) and ♀♀ 158.3 gm (n = 3). Dutch ♂♂ tended to loose mass during breeding from 120 to 95 gm between mid-May and early July which is about 0.56 gm per day.

PRIMARY MOULT

No breeding birds in active primary moult were present in the material studied: Redshanks apparently do not moult primaries on the breeding grounds.

DISCUSSION

The taxonomy of the Redshank

The morphometric information presented above indicated the existence of -at least- 6 different subspecies in this species. Since our data of the W Palearctic were much better than the ones of the E Palearctic, the existence of other taxa in the E Palearctic cannot be excluded. Three subspecies can be distinguished in the W Palearctic since (1) all three differ morphometrically, (2) they can be predicted to occur with a high accuracy with help of these measurements, (3) the breeding grounds are spatially segregated. However, broad intergradation zones occur in between. In the E Palearctic, up to 4 subspecies are currently recognized. Our morphometric data revealed the existence of 3 different taxa: (1) large-sized birds in S Russia & Kazakhstan; (2) intermediate-sized birds in Xinjiang, Kirgizia, the Krasnoyarsk region, N Mongolia and Qinghai Hu, and (3) small-sized birds from E Mongolia eastwards. However, we did not analyse the geographical variation in the E Palearctic further since (1) the currently used subspecies were not only characterized by their measurements but also by plumage characters (of which we had virtually no data), and (2) the sample sizes were as yet too small.

The proposed W Palearctic subspecies are:

(1) *T. t. totanus* (Linnaeus, 1758), breeding throughout Fennoscandia with the most typical individuals breeding in Scandinavia north of the Arctic circle. Nearly all measurements were small. Nominate *totanus* was originally described from Sweden by Linneaus (1758 after Vaurie 1965). It is one of the 2 smallest-sized subspecies. The

other is the population breeding from E Mongolia eastwards.

(2) *T. t. robusta* (Schiøler 1919), breeding on Iceland, the Faeroes, the Orkneys (and the Shetlands?). It is characterized by long wing-, secondary, tail-, and middle toe lengths, in combination with short culmen- and tarsus lengths.

(3) *T. t. britannica* (Mathews 1935), breeding in the Netherlands, Germany, Denmark, the S Baltic, S- and E England, and probably as far east as the Black Sea. This subspecies has intermediate wing-, secondary-, and tail lengths but long culmen- and tarsus lengths. *Britannica* was distinguished by Mathews (1935) with the type locality being Yarmouth (Clancey 1947) or Hampshire (Meinertzhagen 1947). The recognition of this subspecies was not accepted by Hale (1971), since he considered it to be almost impossible to distinguish from *robusta*, except in a few cases. Instead, all W European birds were considered as hybrids. Wattel (in Cramp & Simmons 1983) considered all W European populations as nominate *totanus*. However, our data showed large size differences from both Icelandic and N Fennoscandian populations. Since we did not agree with the hybridization criteria postulated by Hale (see below), it was impossible to follow his classification. We therefore re-introduce *britannica* for the breeding populations of the Netherlands, Germany, Denmark, the S Baltic and S and E England.

The amount of breeding plumage as a criterion for subspeciation

Hale (1971) noticed that breeding Redshanks from W Central Europe and the Tien Shan had the most incomplete breeding plumage of all the populations studied. He interpreted this phenomenon as evidence for the existence of hybridization in these populations. The degree of completeness of the summer plumage was also used as a criterion in his other taxonomical studies on Golden Plover, Ringed Plover, Dunlin and Black-tailed Godwit (Hale 1980). His argument was based on an analogy with the moulting process in hybridized, captive pheasants, where the normal moulting process was also disrupted.

There are 2 major arguments against the use of such a criterion for the recognition of hybridization: (1) many birds in captivity show disrupted moulting patterns whether hybrids or not; and (2) the onset of the moulting process and egg-laying are both photoperiodically influenced. In temperate bird species, moult and egg-laying seem to exclude each other (Payne 1972, Murton & Westwood 1977, Meyer 1988). The administration of androgens and oestrogens leads to disruption of the normal moulting process (Meyer 1988). In nature, these hormones are on a high level at the onset of breeding. The onset of post-breeding moult is influenced by the day length (Meyer 1988), and photo-refractoriness might also be the trigger to start pre-breeding moult. This has special consequences for the Redshank. Populations breeding in continental W Europe and in Fennoscandia winter largely in W Africa (Salomonsen 1954, Hale 1973, Dick 1975, Wymenga *et al.* 1990, Meltofte 1993). Although the pre-breeding moult will start at about the same time in both populations, the departure to the breeding grounds is differently timed: W European populations depart mainly by late March and N Fennoscandian ones from the second half of April (Piersma *et al.* 1990). The W European populations have started to breed (and stopped body moult) when the Fennoscandian breeding birds are still migrating with moulting body feathers (Meltofte 1993, own observations). The same phenomenon seems to hold true for the Icelandic and British breeding populations: these winter mainly in W Europe (Hale 1973, Glutz von Blotzheim 1977, Cramp & Simmons 1983) and both start the pre-breeding moult there. But the breeding season on the British isles starts about a month earlier than the one on Iceland. Such a difference in timing will have consequences for the continuation of the pre-breeding moult. Hale (1971) presented evidence that the British breeding birds retain more winter feathers than the Icelandic ones in their breeding plumage. He also showed the occurrence of clinal differentiation in the amount of breeding plumage over the British isles. He suggested that such was an indication of a process of hybridization with *robusta*. However, it might also be explained by a later start of the breeding season (and thus a longer period available for moult) further north. Such would inevitably result in different plumages. The occurrence of such different breeding and migration schedules must have a profound effect on the moulting cycle even when no genetical differences were included. In our view, differences in the retention of winter feathers might not necessarily have a single genetical basis, but might just as well be influenced by physiological processes. Moreover, part of the so-called 'retained winter feathers' in summer plumage are new summer-plumage feathers which were moulted during spring but resemble the old winter feathers (own observation). Therefore we do not support Hale's theory that the occurrence of an incomplete summer plumage in these wader populations is the result of large-scale hybridization processes.

Hale (1980) used plumage characters to hypothesize on the occurrence of hybrids in Europe. His theory was that (1) Asian *ussuriensis* had followed a westerly colonization route after the last glaciations and (2) *totanus* had followed a northerly route from the SE European refugia during the last glaciation. Both these colonization processes had resulted in hybridization throughout Europe. These ideas were based upon similarities in plumage characters, since Spanish Redshank resembled N Fennoscandian ones and Redshank from west of the Urals had more or less the same breeding plumage as the Icelandic birds. Since, we lack measurement data from the Mediterranean breeding populations, we cannot verify the degree of resemblance between the Mediterranean and N Fennoscandian populations. But our results showed no resemblance between the SE Russia/ SW Asia sample and the Icelandic one. The Iceland ♂♂ had long and broad wings and long tails whereas the SE Russia/ SW Asia ♂♂ were characterized by their long culmen- and tarsus lengths. The differentiation in wing- and tail size was less evident in the ♀ populations, but the differences in culmen- and tarsus lengths were also evident. Correspondingly we cannot agree with such a far reaching interpretation that Iceland was colonized by the Asian *ussuriensis* after the last glaciation. On the other side it would be very interesting to establish the evolutionary age of the different breeding populations by using DNA-analysis. DNA-analysis is needed to find evolutionary explanations for questions concerning such long-term colonization processes. Also because then short-term factors might be considered as explanatory factors causing morphometric variation. For instance, migration distances (influencing wing dimensions), breeding conditions (influencing breeding plumage patterns) and feeding circumstances (influencing culmen lengths) might also have selective influence on morphometrics and plumage characters. For instance, it is well imaginable that selection favours greyish breeding birds both in Kazakhstan and on Iceland. Also winter conditions might well select towards heavy wintering birds. Such heavy birds necessarily need a large wing- and tail area to transport such weights.

SUMMARY

The pattern of geographical variation in the Redshank was studied. Measurements were obtained from specimens all over the breeding range and were analysed with multivariate statistics. The species showed a fair degree of sexual variation and a high degree of geographical variation. In general, ♂♂ had smaller measurements than ♀♀. Sexual dimorphism was small in the breeding populations with small body dimensions (Fennoscandia and E Asia) and large in the ones with large body dimensions (Iceland, the Netherlands, SE Russia/ SW Asia). The species is polytypic and 3 subspecies were identified in the W Palearctic: *totanus*, *robusta* and *britannica*. We had too few data to consider the complicated subspecific status of the E Palearctic populations, although basically 3 morphometrically different types could be recognized: (1) large-sized birds in SE Russia/ SW Asia; (2) intermediate-sized birds in Xinjiang, Kirgizia, N Mongolia, and Qinghai Hu; and (3) small-sized birds in the Krasnoyarsk area and from E Mongolia eastwards. Morphometric differences between populations were described. Wing-, culmen- and secondary lengths were most important for separation of the sexes, and secondary-, wing-, culmen-, tarsus- and tail lengths for geographical distinction. Predicting a bird's sex and breeding origin had to be restricted to W Palearctic populations and concerned each sex of the following regions: (1) Iceland, (2) Netherlands, and (3) Fennoscandia north of 60°E. Prediction results were good when a bird's sex was known.
Dutch ♂♂ decreased on average 25 gm during the breeding season. Redshanks did not moult primaries on the breeding grounds.
Our morphometric results were discussed in relation to results of earlier studies. In general, the pattern of geographical variation as recognized by Hale (1971) was supported, although the breeding ranges of the various subspecies differed markedly. We also re-introduced the subspecies *britannica* for a better understanding of the geographical variation in W Europe. By doing so it was not needed to hypothesize on the occurrence of large-scale hybridisation in Europe. We did not agree with Hale's interpretation that retention of winter feathers in the breeding plumage indicates hybridization.

Chapter 17

Ruddy Turnstone

-

Arenaria interpres

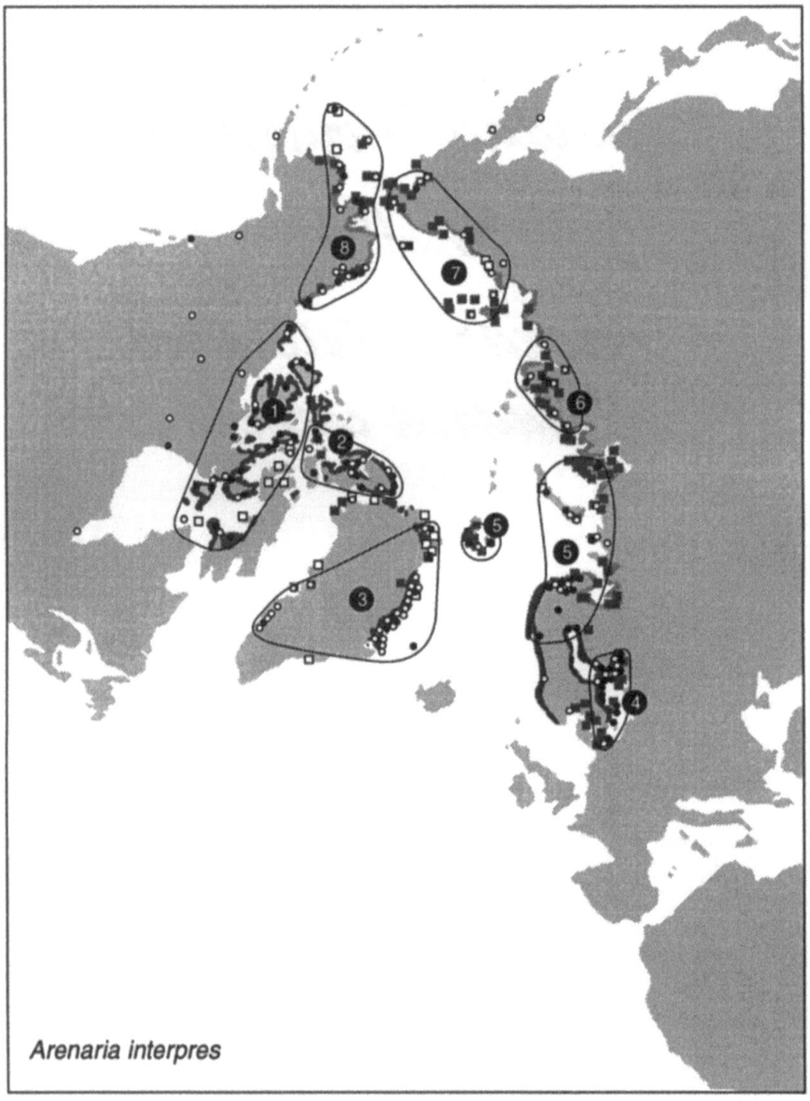

Fig. 111. *Breeding distribution of the Turnstone.*

INTRODUCTION

The Turnstone breeds in coastal areas throughout the Holarctic, in climate zones ranging from the temperate regions in the Baltic to the high arctic of Canada (Fig. 111). It is a polytypic species, in which often 2 and sometimes 3 subspecies are distinguished (Dement'ev *et al.* 1969, Glutz *et al.* 1977, Prater *et al.* 1977; Cramp & Simmons 1983; Godfrey 1986; Stepanyan 1990; Paulson 1993, Gibson & Kessel 1997):

- *A. i. interpres*, breeding on Axel Heiberg Island, Ellesmere Island (high arctic Canada), Greenland, and in N Eurasia east to the Taymyr Peninsula;
- *A. i. oahuensis*, breeding in the E Palearctic from Taymyr east to Alaska (Point Barrow). Nominate *interpres* and *oahuensis* intergrade on Taymyr.
- *A. i. morinella* is breeding in the Canadian low Arctic below 74°N and along the Alaskan coast of the Beaufort Sea.

The characters of the subspecies are mostly based on differences in plumage. *Morinella* has more brightly coloured upper parts in summer plumage than *interpres*. The latter shows more black above. Both subspecies also have slightly different measurements, *interpres* showing slightly larger wing length but smaller culmen- and tarsus length. The rufous upper part colours of *oahuensis* are more dark rusty-chestnut, whilst those of *interpres* are more light-buffish (Stepanyan 1990). Hayman *et al.* (1986) reported upon a cline in the colouration of Palearctic birds, from a pale, orangey colour on scapulars and wing-coverts in the west to deep chestnut red in the east. *Interpres* and *oahuensis* do not differ in measurements. Glutz *et al.* (1977), Cramp & Simmons (1983), Godfrey (1986) and Paulson (1993) did not recognize *oahuensis* as a separate subspecies because of the large similarities with *interpres*. Summers *et al.* (1989) as well as Paulson (1993) considered the Alaskan breeders as *interpres* rather than *morinella* as was claimed by Hayman *et al.* (1986). According to Paulson (1993) it is difficult in the field to recognize the Alaskan breeding birds as any of both subspecies, since their plumage is intermediate between *morinella* and the typical Eurasian *interpres*. Alaskan breeding birds in the hand are distinctly different (Paulson 1993).

Table 112. *Measurements of Turnstones, collected on the breeding grounds. Presented are the means ± 1 SD (n). Notes: [1]breeding south of 74°N; [2]breeding north of 74°N.*

	Locality	Wing	Culmen	Tarsus	Secondary	Tail	Middle toe
				♂♂			
1	Low arctic Canada[1]	152.0±3.4(67)	23.5±1.1(67)	26.0±0.9(66)	66.6±1.8(58)	55.7±2.7(58)	27.1±1.3(46)
2	High arctic Canada[2]	154.5±3.7(39)	22.2±1.0(39)	25.7±0.9(39)	68.0±1.8(37)	58.4±2.4(37)	27.0±1.1(30)
3	Greenland	154.1±4.0(65)	23.0±0.9(64)	25.7±0.8(21)	67.3±2.9(6)	57.8±3.8(8)	27.1±1.6(5)
4	Baltic coast	156.1±5.7(28)	23.1±1.3(28)	25.7±1.3(26)	67.6±1.9(25)	58.3±2.7(25)	27.1±1.1(23)
5	N Europe	156.0±3.9(30)	22.7±1.3(30)	25.5±0.8(28)	66.4±2.0(14)	59.2±2.8(13)	27.1±1.1(11)
6	Taymyr	155.5±3.0(13)	21.8±1.2(13)	25.4±1.0(13)	64.3±1.6(10)	58.0 - (1)	25.8±1.4(10)
7	E Siberia	154.9±2.5(11)	22.5±1.4(11)	25.3±0.8(11)	65.8±1.5(5)	57.6±2.1(5)	26.0±0.8(3)
8	Alaska	154.4±4.4(23)	22.6±1.0(23)	25.7±1.1(23)	65.6±1.9(23)	56.0±2.2(22)	27.3±0.9(20)
				♀♀			
1	Low arctic Canada[1]	154.1±4.2(40)	24.2±1.1(40)	26.3±0.8(40)	67.6±1.7(35)	57.1±2.4(35)	27.3±0.9(31)
2	High arctic Canada[2]	156.4±3.4(19)	23.1±1.1(19)	25.9±1.3(19)	68.8±1.4(18)	59.3±2.1(18)	27.0±1.1(13)
3	Greenland	157.1±5.1(49)	23.3±1.1(49)	26.1±0.9(16)	68.5±2.6(4)	59.4±3.6(7)	26.4±1.1(7)
4	Baltic coast	160.8±1.9(11)	23.9±1.0(11)	25.9±1.3(11)	68.1±2.1(11)	60.1±2.8(11)	27.3±1.1(11)
5	N Europe	157.6±2.9(17)	23.5±1.1(17)	25.8±1.1(16)	66.8±1.9(12)	61.0±4.3(6)	27.5±1.1(12)
6	Taymyr	157.5±2.1(2)	23.5±0.3(2)	25.3±0.7(2)	65.0 - (1)	56.0 - (1)	24.6 - (1)
7	E Siberia	157.7±4.0(10)	23.9±0.9(9)	25.9±0.7(10)	65.0 - (1)	55.0 - (1)	-
8	Alaska	157.3±5.0(16)	23.2±1.0(16)	26.0±0.7(16)	67.5±1.7(16)	57.3±1.1(16)	26.9±1.0(10)

Summers *et al.* (1989) split up the breeding range of *interpres* into 3 parts: (1) NE Canada and Greenland; (2) Fennoscandia and W Russia; and (3) Siberia from the Kara Sea eastwards. They showed that these populations clearly differed in migration schedules and routes, though only minor morphometric differences existed between these populations.

MORPHOMETRICAL DIFFERENTIATION

In total, 440 Turnstones were examined from all over the breeding range (Table 112, Fig. 111). Data from Siberia were rather scarce. Males predominated (62.7 %) in the samples. In most areas arrival took place from the 10th of May onwards (Fig. 112), although Turnstones on the Baltic coast were already present by late March. Most birds had been collected during June and early July.

Initially, 16 samples (8 breeding locations and both sexes) were distinguished (Fig. 111). Female data from Taymyr and E Siberia were scarce respectively incomplete. The analyses on the morphometric differences between samples are presented in the tables 112 - 116 and the figures 113 - 115.

The Turnstone showed slight sexual and geographical variation (Table 113) and the degree of differentiation strongly depended upon the number of variables included (and thus upon the sample size). A bird's sex and breeding location appeared to be important for explaining the variation in morphometrics. For instance, when all 5 measurements were used, it appeared that the variation in tarsus lengths was not significantly influenced by sex or breeding location, but when only wing-, culmen- and tarsus lengths were included and thus larger sample sizes were reached, sex differences appeared to be significant.

On average, ♂♂ were smaller than ♀♀ with culmen- and tail lengths showing the largest differences between sexes (Table 114). Sexual dimorphism was possibly somewhat more pronounced in the E Siberian and Baltic samples than in others, but the differences were small. The 6 measured variables correlated badly with each other (Table 115). Nevertheless, 2 sets of related measurements were evident: (1) wing-, secondary- and tail lengths; and (2) culmen-, tarsus- and middle toe lengths.

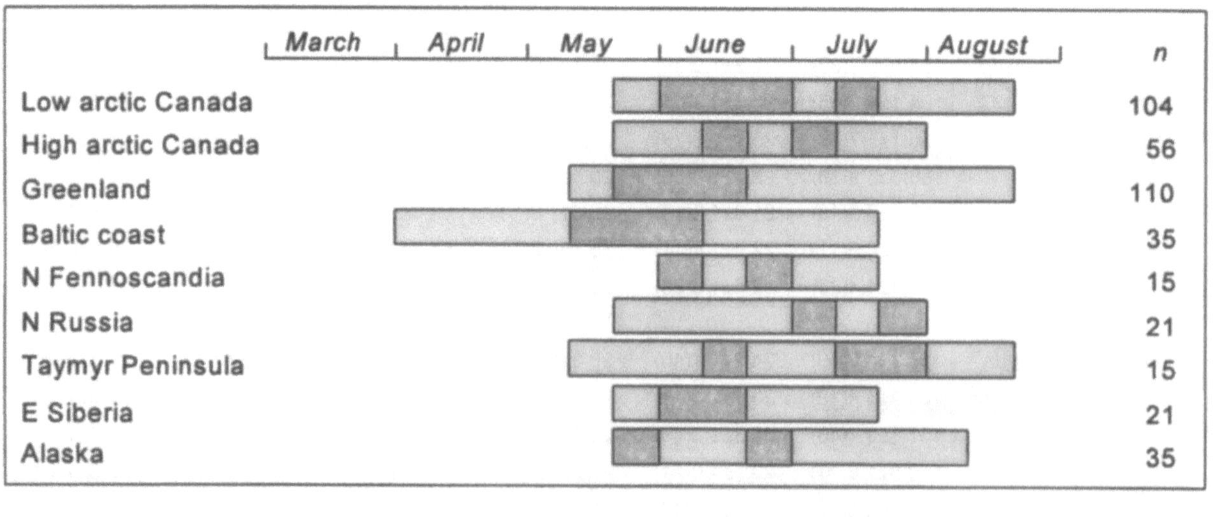

Fig. 112. *Collection dates of Turnstones. Most birds had been collected during the periods marked by the heavily shaded bars. The sample sizes are presented on the right side.*

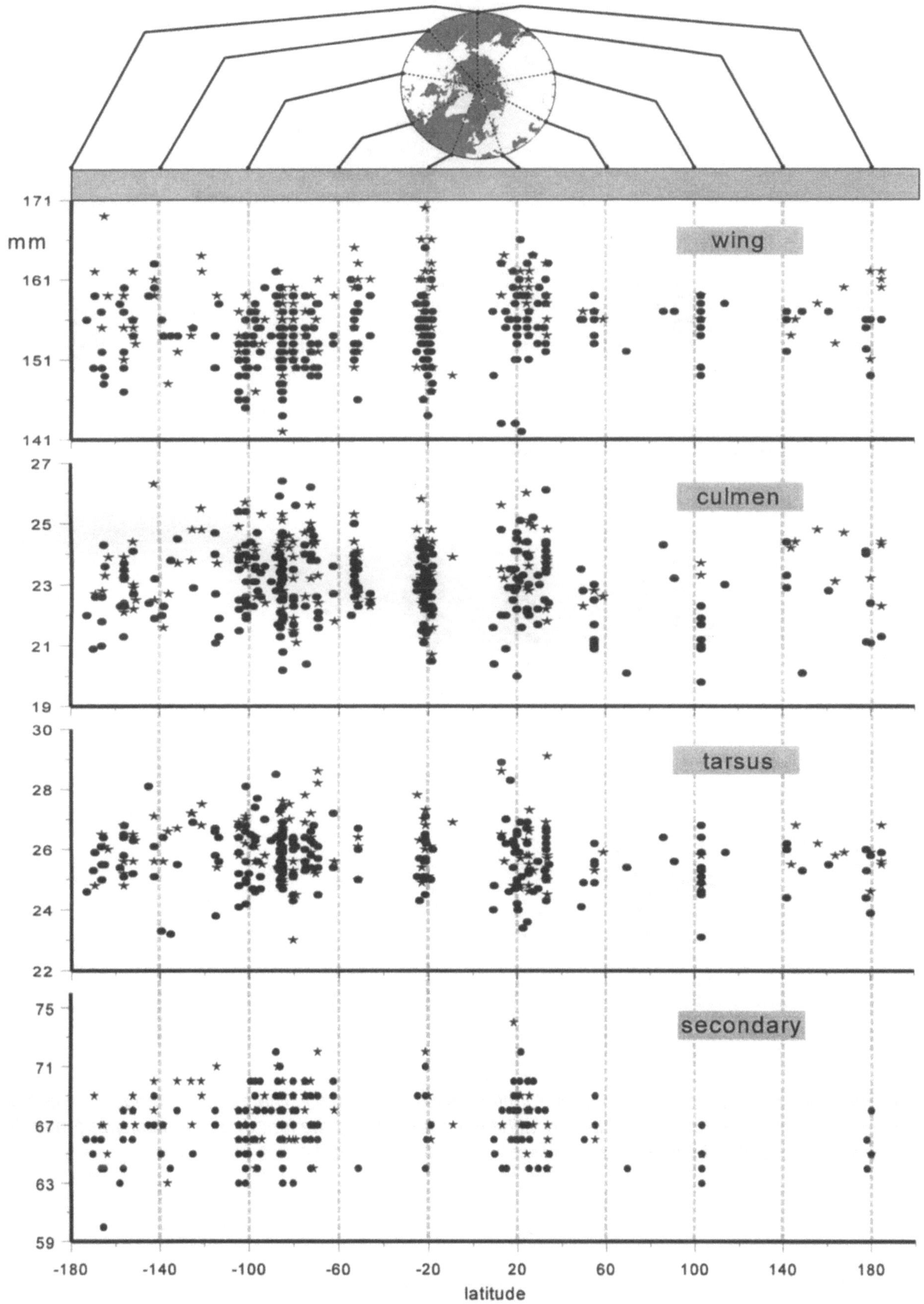

Fig. 113. *Measurements of Turnstones from the breeding grounds. The following measurements are presented from top to bottom: wing length, culmen length, tarsus length and secondary length (described in chapter 2). Explanation of the used symbols: ● ♂♂, ★ ♀♀.*

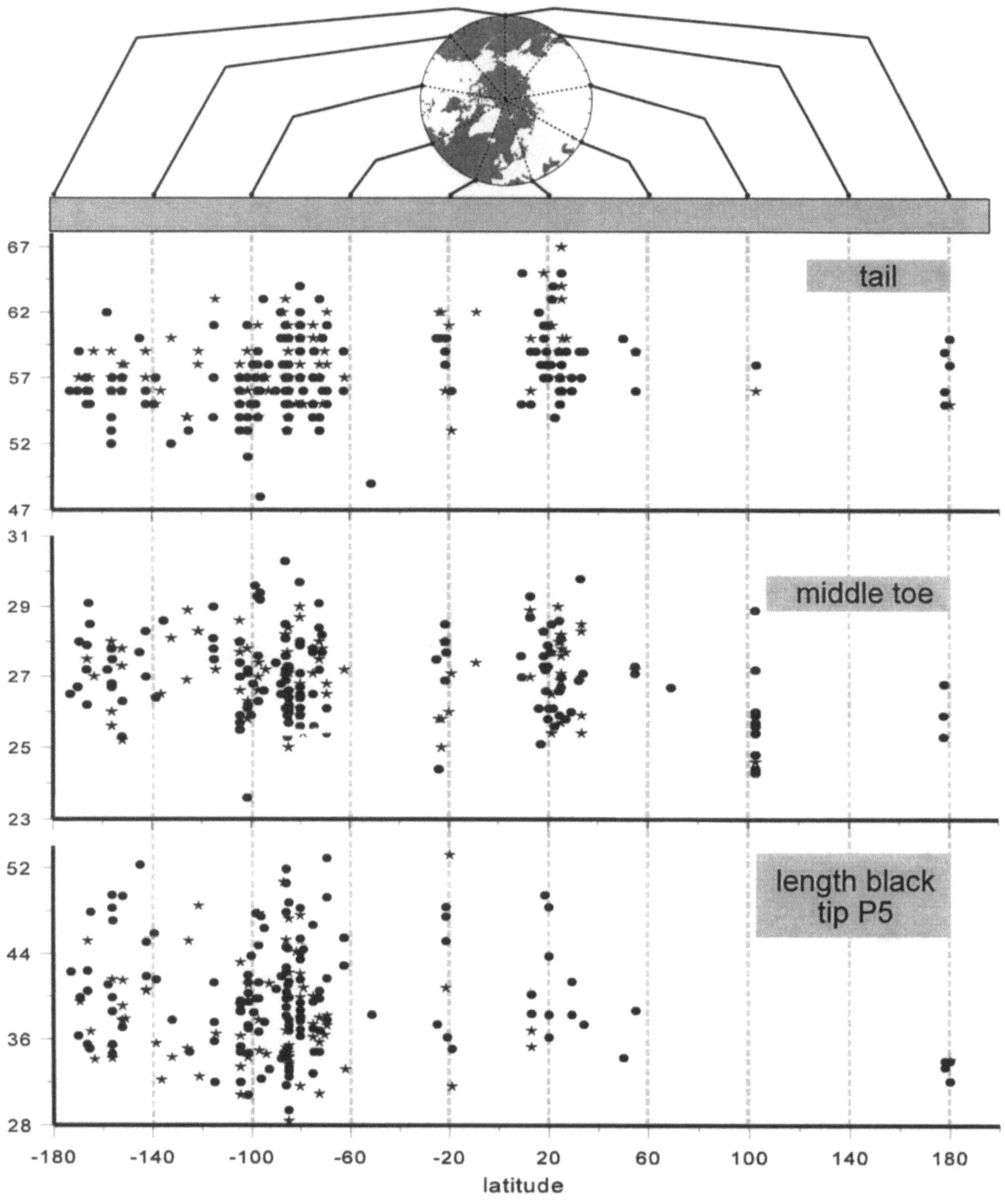

Fig. 114. *Measurements of Turnstones from the breeding grounds. The following measurements are presented from top to bottom: tail length, middle toe length and the length of the black pattern on the 5th primary (described in chapter 2). Explanation of the used symbols: ● ♂♂, ★ ♀♀.*

Table 113. *Overall results of MANOVA for all distinguished breeding populations in the Turnstone in combination with univariate ANOVA's in which the different measurements were analysed separately. Significance levels: *0.01 < P < 0.05; **0.001 < P < 0.01; ***P < 0.001.*

	Analysis I			Analysis II			Analysis III		
Variables	wing, culmen, tarsus, secondary, tail			wing, culmen, tarsus			wing, culmen		
n	257			356			438		
Multivariate analysis of variance									
Effects	Interaction	Area	Sex	Interaction	Area	Sex	Interaction	Area	Sex
Pillai's trace	0.088	0.661	0.076	0.027	0.330	0.117	0.031	0.255	0.100
F-value	0.724	5.261***	3.899**	0.438	5.998***	14.937***	0.951	8.799***	23.446***
Hypoth. & error *df*	30/1210	35/1210	5/238	21/1020	21/1020	3/338	14/844	14/844	2/421
Univariate F tests									
df	6/242	7/242	1/242	7/340	7/340	1/340	7/422	7/422	1/422
wing length	0.865	7.950***	2.766	0.458	8.456***	20.134***	0.612	8.589***	22.796***
culmen length	0.810	6.694***	17.485***	0.549	7.103***	35.424***	1.132	7.382***	34.221***
tarsus length	0.170	1.893	0.579	0.206	2.042*	4.816*			
secondary length	0.684	4.252***	1.989						
tail length	0.267	9.219***	1.989						

Table 114. *Sexual size dimorphism of the Turnstone. The mean percentage difference between the measurements of both sexes was calculated for every single variable and breeding population, according to: $(x_♀-x_♂)/x_♂*100$. The calculations were based upon the means presented in Table 112. The obtained values were averaged per variable and summed per breeding population.*

Locality	Wing	Culmen	Tarsus	Secondary	Tail	Middle toe	Σ_3	Σ_6
Low arctic Canada	1.3	3.0	1.1	1.4	2.6	0.9	5.4	10.2
High arctic Canada	1.2	4.1	0.7	1.1	1.6	-0.3	6.1	8.6
Greenland	2.0	1.3	1.7	-	2.9	-2.3	5.0	-
Baltic coast	3.0	3.1	0.9	0.8	3.1	0.5	7.1	11.4
N Europe	1.0	3.5	1.2	0.7	3.0	1.7	5.8	11.2
E Siberia	1.8	6.3	2.1	-	-	-	10.2	-
Alaska	1.9	2.6	1.4	2.9	2.4	-1.6	5.9	9.6
x	1.8	3.4	1.3	1.4	2.6	-0.2		
SD	0.7	1.5	0.5	0.9	0.6	1.5		

Table 115. *Pearson's product-moment correlations between measurements in Turnstones. Both sexes are separately analysed. Two-tailed probabilities: * 0.01 < P < 0.05; ** 0.001 < P < 0.01; *** P < 0.001.*

	Wing	Culmen	Tarsus	Secondary	Tail length	Middle toe	
Wing		0.146	0.062	0.316**	0.416**	0.067	
Culmen	0.095		0.352**	-0.056	-0.066	0.410**	
Tarsus	0.161*	0.316**		-0.026	-0.117	0.343**	
Secondary	0.438**	0.178*	0.146		0.488**	0.098	♀♀
Tail	0.420**	-0.106	-0.043	0.338**		-0.043	
Middle toe	0.115	0.245*	0.348**	0.163	0.023		
			♂♂				

Cluster analysis clearly demonstrated the lack of difference between various samples: the calculated cluster distances between geographical regions were small (Fig. 115). The samples from low arctic Canada were most distinct. The samples of high arctic Canada and Greenland were closely similar. The same was true for the samples of E Siberia and Taymyr.

The results of MANOVA-analysis, in which samples were pairwise contrasted, mostly supported the findings of cluster analysis. Only the wing-, culmen- and tarsus lengths were used in the MANOVA analyses due to the scarcity of data in both ♀ groups of Taymyr and E Siberia. The following conclusions were derived from this analysis (Table 116):

(1) Males differed significantly in 4 of the 8 tests from ♀♀ of the same breeding location and in 34 of the 56 tests (=61%) from ♀♀ of other locations.
(2) The measurements of Turnstones from low arctic Canada nearly always differed from the ones of other samples. If not it was caused by too small sample sizes (Taymyr ♀♀).
(3) The distinct position of the low arctic Canada samples was more pronounced in the ♂- than in the ♀ samples.
(4) There were nearly no morphometric differences between the various Palearctic samples.
(5) Within this scope of little variation the ♂♂ from Taymyr seemed to be as yet the most distinct sample group of the Palearctic ones. These ♂♂ had relatively small culmen-, tarsus- and secondary lengths.

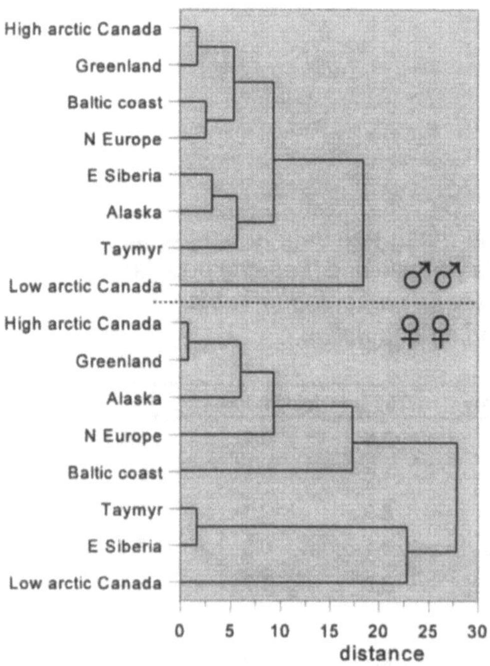

Fig. 115. *Cluster analysis in the Turnstone. The analysis was based upon 5 measurements: wing-, culmen-, tarsus-, secondary- and tail length. The samples of both sexes were analysed separately.*

Differences between sexes and geographical regions played a small but significant role when explaining the observed variation in measurements (Table 113). Both sexes differed in culmen-, wing- and tarsus length (in decreasing order of importance). Nearly all measurements differed significantly when distinguishing between locations. These were in decreasing order of importance: tail-, wing-, culmen-, secondary- and tarsus lengths.

When exclusively wing- and culmen lengths were analysed, it was only possible to distinguish the low arctic Canadian birds from all others. They always differed from the other samples at the 0.1% level. The Turnstones from low arctic Canada had a somewhat shorter and

Table 116. *Results of MANOVA analysis concerning the morphometric differentiation in the Turnstone. Pillai's trace values are presented together with the level of significance as based upon the derived F-values (Norušis 1992c). In this series of analyses 3 measurements and 8 populations were included (3 and 338 df): wing-, culmen- and tarsus lengths. Significance levels: *0.01 < P < 0.05; **0.001 < P < 0.01; ***P < 0.001.*

Locality	*n*	Low arctic Canada ♀	High arctic Canada ♀	Greenland ♀	Baltic ♀	N Europe ♀	Taymyr ♀	E Siberia ♀	Alaska ♀	Low arctic Canada ♂	High arctic Canada ♂	Greenland ♂	Baltic ♂	N Europe ♂	Taymyr ♂	E Siberia ♂
Low arctic Canada ♀	40															
High arctic Canada ♀	19	0.062***														
Greenland ♀	16	0.038**	0.006													
Baltic ♀	11	0.082***	0.029*	0.018												
N Europe ♀	16	0.053***	0.006	0.003	0.012											
Taymyr ♀	2	0.014	0.004	0.005	0.004	0.002										
E Siberia ♀	9	0.033*	0.012	0.004	0.006	0.003	0.003									
Alaska ♀	16	0.058***	0.002	0.003	0.020	0.004	0.004	0.008								
Low arctic Canada ♂	66	**0.038****	0.065***	0.061***	0.120***	0.075***	0.016	0.057***	0.070***							
High arctic Canada ♂	39	0.176***	**0.029***	0.058***	0.091***	0.058***	0.013	0.060***	0.037**	0.141***						
Greenland ♂	21	0.059***	0.004	**0.016**	0.048**	0.016	0.005	0.022	0.011	0.044**	0.023					
Baltic ♂	26	0.074***	0.002	0.010	**0.031***	0.006	0.003	0.013	0.005	0.079***	0.041**	0.005				
N Europe ♂	27	0.114***	0.006	0.025*	0.045**	**0.018**	0.004	0.027*	0.013	0.105***	0.020	0.007	0.006			
Taymyr ♂	14	0.156***	0.038**	0.065***	0.080***	0.060***	**0.016**	0.065***	0.044**	0.127***	0.010	0.037**	0.046**	0.024*		
E Siberia ♂	11	0.073***	0.010	0.027*	0.045**	0.021	0.006	**0.029***	0.017	0.054***	0.008	0.006	0.009	0.002	0.012	
Alaska ♂	23	0.087***	0.010	0.029*	0.063***	0.029*	0.008	0.035**	**0.018**	0.060***	0.009	0.003	0.013	0.007	0.024*	0.004

smaller wing combined with longer culmen lengths (Fig. 113, Table 112). No systematic differences between other samples were found.

Some additional information on the length of the black tip of primary 5 was collected. Also, the coloration of scapulars and coverts in the different samples received some attention. However, these data appeared to be rather unreliable, since there were larger differences between collections than between locations. Fading because of wear as well as the history of preservation in a collection seem to play an important role in detecting different colours: chestnut may bleach towards paler yellowish-ochre colours.

The length of the black tip on primary 5 was measured from the distal feather end along the outer edge of the feather. Information on this plumage character was mainly sampled from the Alaskan and Canadian breeding populations (Fig. 114). Almost no contrasts between samples produced significant differences, except for the contrasts between the ♀♀ from low arctic Canada with the ♂ samples of resp. high arctic Canada (5 % level) and Alaska (1 % level).

Table 117. *Accuracy of prediction in Turnstone (sex* ***known*** *a priori). These predictions have been based upon morphometric analysis and included (A) wing-, culmen-, secondary- and tail length; (B) wing-, culmen- and tarsus length; and (C) wing- and culmen length. The columns show the predicted proportions in relation to the total number per breeding area (last column).*

	Predicted				
Locality	*interpres* ♀	*morinella* ♀	*interpres* ♂	*morinella* ♂	*n*
	(A) Wing, culmen, secondary & tail				
interpres ♀	82.8	17.2	-	-	58
morinella ♀	20.0	80.0	-	-	35
interpres ♂	-	-	76.3	23.7	114
morinella ♂	-	-	31.6	68.4	57
	(B) Wing, culmen & tarsus				
interpres ♀	78.0	22.0	-	-	91
morinella ♀	25.0	75.0	-	-	40
interpres ♂	-	-	81.3	18.7	166
morinella ♂	-	-	30.3	69.7	66
	(C) Wing & culmen				
interpres ♀	71.2	28.8	-	-	125
morinella ♀	20.0	80.0	-	-	40
interpres ♂	-	-	74.2	25.8	213
morinella ♂	-	-	37.3	62.7	67

PREDICTION OF BREEDING ORIGIN

Since the measurements of ♂♂ and ♀♀ within breeding locations differed, both could not be combined in one POSCON group. Moreover, the cluster- and MANOVA analyses showed justified the recognition of 2 breeding regions: (1) low arctic Canada; and (2) Alaska, high arctic Canada, Greenland and the Palearctic. Thus 4 POSCON groups were constructed. The predictions were based upon: (1) wing-, culmen-, secondary-, and tail lengths; (2) wing-, culmen- and tarsus lengths; and (3) wing- and culmen lengths. When analysing the accuracy- and exclusivity rates the situation with known sexes was distinguished from the one with unknown sexes.

Accuracy

When 4 measurements were used and the bird's sexes were known, the accuracy rates in both ♀ POSCON groups were estimated at > 80 % (Table 117). These figures decreased to about 75 % when fewer measurements were used.

Table 118. *Accuracy of prediction in Turnstone (sex* ***unknown*** *a priori). These predictions have been based upon morphometric analysis and included (A) wing-, culmen-, secondary- and tail length; (B) wing-, culmen- and tarsus length; and (C) wing- and culmen length. The columns show the predicted proportions in relation to the total number per breeding area (last column).*

	Predicted				
Locality	*interpres* ♀	*morinella* ♀	*interpres* ♂	*morinella* ♂	*n*
	(A) Wing, culmen, secondary & tail				
interpres ♀	53.4	12.1	25.9	8.6	58
morinella ♀	14.3	54.3	5.7	25.7	35
interpres ♂	23.7	3.5	54.4	18.4	114
morinella ♂	13.8	12.1	22.4	51.7	58
	(B) Wing, culmen & tarsus				
interpres ♀	51.6	11.0	25.3	12.1	91
morinella ♀	15.0	50.0	10.0	25.0	40
interpres ♂	25.3	4.2	56.6	13.9	166
morinella ♂	10.6	18.2	22.7	48.5	66
	(C) Wing & culmen				
interpres ♀	44.0	16.8	28.0	11.2	125
morinella ♀	10.0	57.5	10.0	22.5	40
interpres ♂	21.1	9.9	52.1	16.9	213
morinella ♂	9.0	17.9	26.9	46.3	67

The estimated accuracy rates of ♂♂ were lower and ranged between 63 % (2 measurements) and 81 % (3 measurements). When the birds' sexes were unknown, the quality of the predictions decreased drastically (Table 118). The average accuracy rates were 53.5 % with 4 measurements, 51.7 with 2 measurements and 50.0 with 2 measurements. These low accuracy rates were because of confusion between both sexes within the same region. It may thus be better not to rely heavily on morphometric information when sexing Turnstones.

Exclusivity

The estimated exclusivity rates were highest when the bird's sexes were known (Table 119), since then only 2 sample groups had to be distinguished. The exclusivity rates decrease when more variables are included. These results also show that increasing accuracy rates coincide with decreasing exclusivity rates. In other words: when the prediction results become better, we also become less certain on the outcome.

Table 119. *Estimated exclusivity rates with predicting the breeding origin and sex of Turnstones as derived from morphometric data. Remarks: [1]wing- and culmen lengths; [2]wing-, culmen- and tarsus lengths; [3]wing-, culmen-, secondary- and tail lengths.*

	Sex known			Sex unknown		
n of variables	2[1]	3[2]	4[3]	2[1]	3[2]	4[3]
n	440	363	264	440	363	265
interpres ♀	5.4	4.1	2.3	0.9	2.2	0.4
morinella ♀	1.8	1.1	0.8	0.9	0.3	-
interpres ♂	7.9	5.2	4.2	2.9	1.7	1.1
morinella ♂	2.5	2.5	2.3	-	-	-
Total %	17.6	12.9	9.6	4.7	4.2	1.5

BODY MASS

Body mass was notified in 25 % of the labels examined. On average, ♂♂ had the same body mass as ♀♀ (*t*-test; difference between ♂♂ and ♀♀ 2.1 gm ± 5.7 95% confidence interval; Table 120). The mean body mass fluctuated between 98 and 127 gm. Most mass data concerned Canadian Turnstones. These birds tended to decrease in mass towards mid-July, though the variation between individuals was large (Fig. 116). Some Turnstones were below 90 gm (mainly *morinella*).

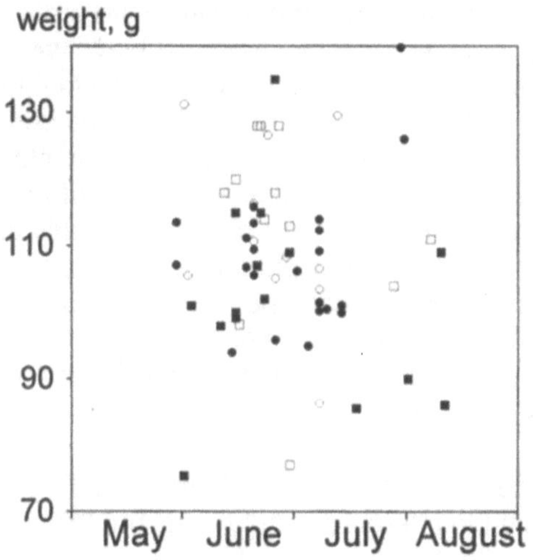

Fig. 116. *Body masses of Turnstones collected in arctic Canada: morinella in low arctic Canada (■ - ♂♂, □ - ♀♀) and interpres in high arctic Canada (● - ♂♂, ○ - ♀♀*

Table 120. *Mean body mass of Turnstones, collected on the breeding grounds during the breeding season.*

Locality	♂♂	♀♀
Low arctic Canada	101.8±14.5(1 5)	113.1±14.7(12)
High arctic Canada	108.1±10.5(22)	110.9±13.0(12)
Greenland	122.7± 2.1(3)	108.0±15.5(4)
Baltic	127.5± 2.1(2)	-
N Europe	98.3± 7.5(6)	106.7± 5.1(4)
Siberia	113.0± 9.1(4)	105.7± 6.1(5)
Alaska	106.2± 1.6(2)	108.2±20.7(4)

PRIMARY MOULT

None of the birds showed primary moult on the breeding grounds.

DISCUSSION

The taxonomy of the Ruddy Turnstone

From the above, it became evident that 2 subspecies can be distinguished morphometrically. Separating both subspecies with POSCON analysis performed better than separating the sexes, though slight sexual dimorphism was found. To some extent, it appeared to be possible to predict a bird's breeding origin with morphometrics. But the quality of the predictions was only acceptable when a bird's sex was known, or when it was not necessary to know its sex. During summer, both subspecies can best be distinguished by their mantle colour, Nearctic *morinella* showing more extensive rufous than *interpres*. However, this character is only useful for birds in summer plumage. In other periods, measurements can be used to distinguish between both: *morinella* has shorter wing dimensions, shorter tails and longer culmen- and tarsus lengths.

Some minor geographical differentiation exists within the *interpres* populations, but this was far too little to allow the distinction of separate subspecies or to make a distinction between the W and E Palearctic populations. We therefore support the conclusions of Cramp & Simmons (1983) and Godfrey (1986) (contra Stepanyan 1990) that the Alaskan and E Siberian populations should all be considered as *interpres*.

In the introduction, we mentioned the splitting of the breeding range of *interpres* in 3 regions by Summers *et al.* (1989): (1) NE Canada and Greenland, (2) Fennoscandia and W Russia, and (3) Siberia from the Kara Sea eastwards. They showed the existence of minor morphometric differences between these regions. However, our data did not support their findings, since these three were morphometrically not as homogeneous as suggested.

SUMMARY

Measurements and plumage characters of Ruddy Turnstones from nearly all breeding areas had been collected and analysed with multivariate statistics. Sexual and geographical variation was small. ♂♂ were somewhat smaller than ♀♀. Sexual dimorphism was slightly more pronounced in the Baltic and N European samples than in the other ones. The species is polytypic and the recognition of 2 subspecies was supported: *interpres* and *morinella*. The amount of rufous on the mantle feathers in summer plumage and the different measurements are the best characters. *Morinella* can be distinguished from *interpres* by short wing- and tail length in combination with long culmen- and tarsus length. Small morphometric differentiation existed within the breeding range of *interpres*, but this was too small to allow recognition of different subspecies. Culmen- and wing lengths were necessary to distinguish sexes, and tail-, wing-, culmen- and secondary lengths to distinguish between locations. Two POSCON groups per sex were constructed for predictional purposes: (1) *morinella* - low arctic Canada and (2) *interpres* - Alaska, high arctic Canada, Greenland and the Palearctic. The prediction results were good as long as birds were sexed. Sexing birds with the use of morphometrics is not advised. Turnstones did not moult primaries on the breeding grounds.

Chapter 18

General discussion

In the course of this study morphometric data on 68 distinct populations divided over 24 species were collected. In this chapter the observed patterns in geographical variation, size dimorphism and sex ratio amongst the various species are discussed. In the first subchapter the morphometric (dis)similarities between species and/or populations are presented by using cluster analysis. In the second subchapter it is tried to show that migration has a profound impact on the shape of waders. There is much scientific research conducted to the size differences between sexes. Our study on waders produced a lot of material, which can be used to compare various species and/or populations with respect to size differences between sexes. This comparison is presented in the third subchapter. In the fourth subchapter the observed skewed sex ratios on the breeding grounds are discussed.

The final three subchapters are directed towards the future. The size of wader populations is an important item in the international wetland conservation. Since most wader species are boreal and (sub)arctic breeders it is often difficult to establish the size of a breeding population. Nevertheless it is possible to conduct counts on the wintering grounds, since most wader species gather in restricted number of sites. Regular counting programs are carried out all over the world in order to estimate the size of these wintering populations. Once it is possible to estimate the population composition of a wintering population in terms of morphometrically distinct populations, it becomes also possible to produce estimates on the size of breeding populations. This theme is discussed in the fifth subchapter. In the chapters on Dunlin and Red Knot the results from morphometric research and DNA-studies were already parallelled. This theme returns in the sixth subchapter. In the final subchapter the attention is focused on the quality of morphometric variables for their use in future trapping studies: 'Which measurements ought to be collected during trapping in order to obtain the best possible results?'

Variation among and within species

In the species' chapters our approach was mainly focused on the smallest possible morphometric distinction between samples. If distinctions were evident samples were recognized as distinct subspecies. In this discussion cluster analysis was performed over 66 ♂- and 64 ♀ samples in order to facilitate comparisons between species. In total, 6 measurements were included: wing-, culmen-, tarsus-, secondary-, middle toe- and tail length. The ♂- and ♀ samples were separately analyzed while using the mean measurements presented in Annex 1 (Figs. 117 and 118).

This cluster analysis show that the earlier distinguished allospecies justify their position with having large distances between the related taxa:

- The Ringed Plover and the Semipalmated Plover were considered as allospecies belonging to the superspecies *Charadrius hiaticula*. The morphometric difference between both is evident.

- The Purple Sandpiper and the Rock Sandpiper are closely related. However, the cluster analyses of both ♂♂ and ♀♀ show larger morphometric differences between the Icelandic Purple Sandpipers and the others than between the Rock Sandpiper and the other Purple Sandpipers. There are thus no morphometric arguments to separate the Rock Sandpiper as an allospecies. The different plumage characters are decisive.

- The easterly Black-tailed Godwit differ importantly from the western. The morphometric differences are large enough to separate them as allospecies. The arguments are based upon morphometrics as well as plumage data. It is necessary to include the characteristics of the Hudsonian Godwit as well.

- The Nearctic and Palearctic populations of the Whimbrel are generally considered as different subspecies. The Alaskan ♂♂ show most resemblance with *islandicus* and *alboaxillaris* and the Canadian ♂♂ with *phaeopus* and *variegatus*. The Nearctic ♀♀ form a cluster on their own with large differences to the other subspecies. This large morphometric differences together with large plumage differences justify separation as allospecies: the Hudsonian and the Eurasian Whimbrel.

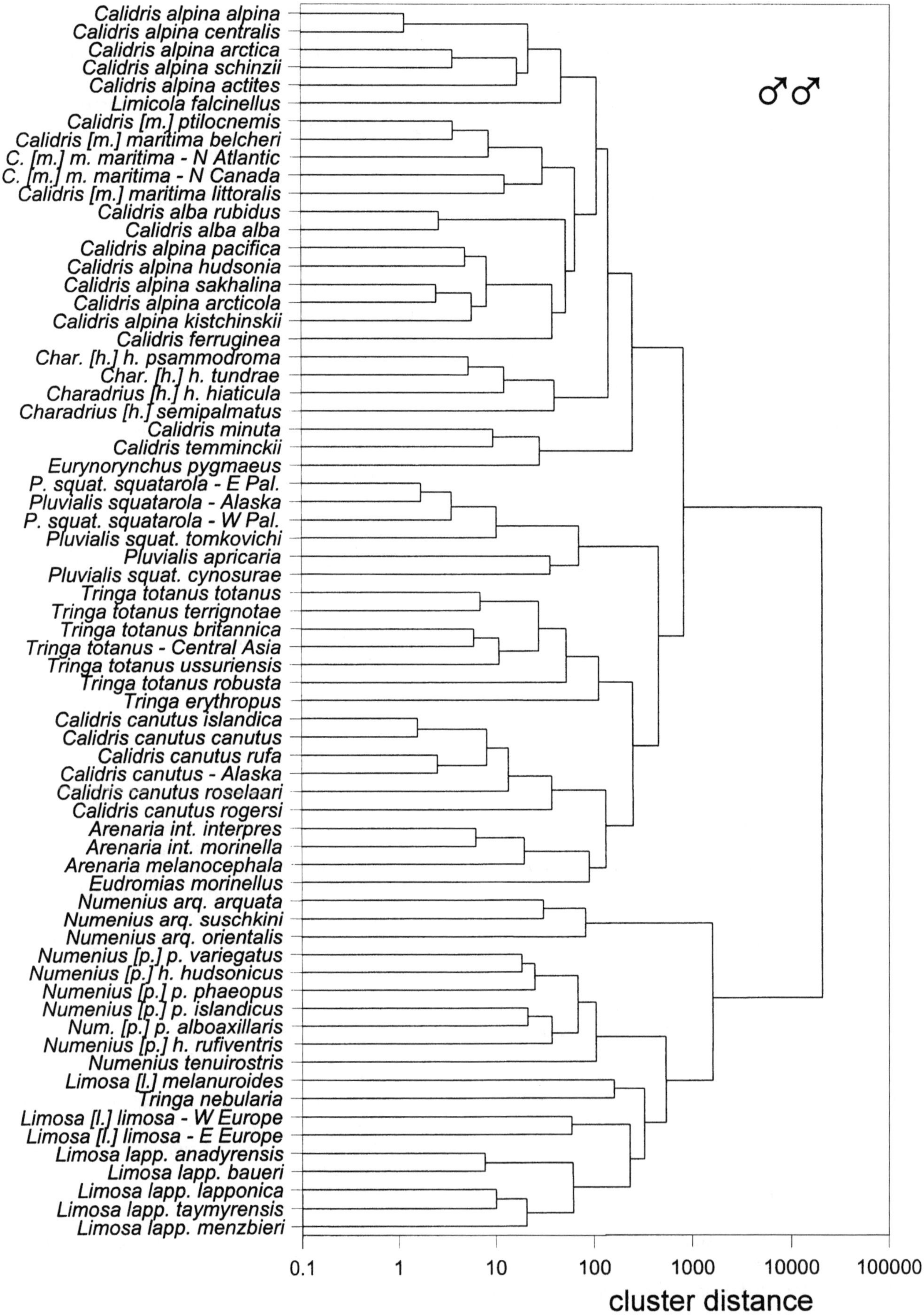

Fig. 117. *The results of a cluster analysis in which the ♂♂ of all distinguished species and subspecies were included.*

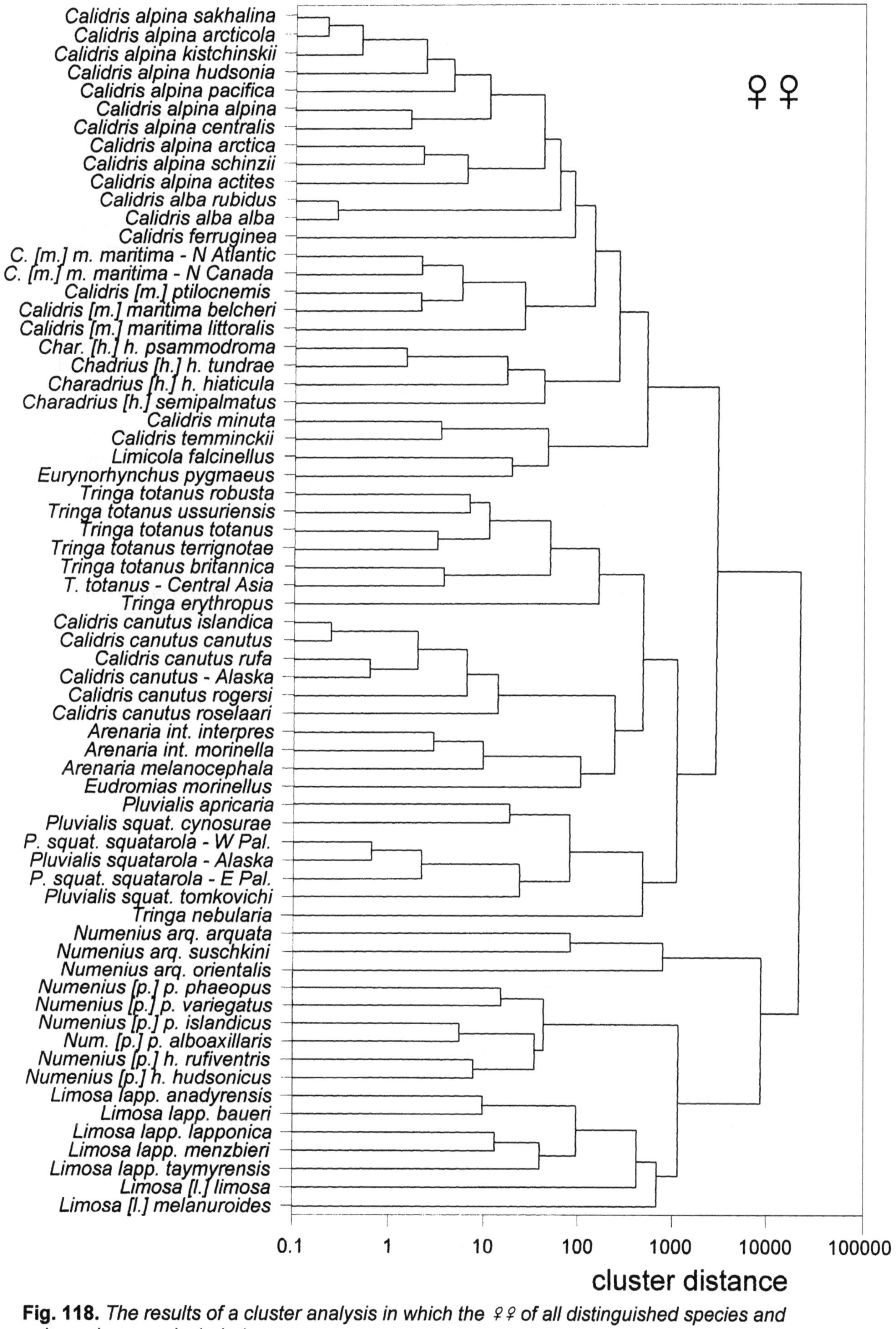

Fig. 118. *The results of a cluster analysis in which the ♀♀ of all distinguished species and subspecies were included.*

The results from the figs. 117 and 118 lead to the taxonomical order as presented in Table 121. Some samples were morphometrically rather similar, whilst they were distinguished as separate subspecies. In these situations additional criteria like plumage characters were used to identify the different subspecies:

- The E Siberian Dunlin *C. a. sakhalina* and *C. a. kistchinski*.
- The Atlantic subspecies of the Dunlin *C. a. arctica* and *C. a. schinzii*.
- Both subspecies of the Sanderling: *C. a. rubidus* and *C. a. alba*.
- The Atlantic subspecies of the Knot: *C. c. rufa* and *C. c. canutus*.
- Both subspecies of the Ruddy Turnstone: *A. i. interpres* and *A. i. morinella*.
- In the Bar-tailed Godwit: *L. l. lapponica* and *L. l. menzbieri*.
- The Whimbrel from Iceland and the British isles *N. p. islandica* and the S Russian *N. p. alboaxillaris*.

In Table 121 no attention was paid to those breeding areas where different subspecies intermix like for instance the Red Knot in Alaska or the Ringed Plover in S Scandinavia.

In 5 of the 14 analyzed species the degree of diffferentiation between the ♂ samples was higher than between the ♀ samples. These species were the Red Knot, Purple Sandpiper, Dunlin, Whimbrel and Turnstone. In the Bar-tailed Godwit and the Redshank showed the ♀♀ more differentiation than the ♂♂. In the other 7 species there was not a clear difference between the ♂- and ♀-groups. This difference is most likely caused by the larger sample sizes in the ♂♂ possibly resulting in more evident differences between groups.

Table 121. *The taxonomic listing of the (super)species analyzed in this study. The taxa are ranked from superspecies (bold face) and species (bold face italics) to subspecies (italics). The breeding ranges for the lowest levels are given.*

Nomenclature	Breeding range
Charadrius [hiaticula]	
Charadrius [h.] hiaticula (Ringed Plover)	
hiaticula	W Europe
psammodroma	NE Canada, Greenland, Jan Mayen, Iceland
tundrae	N Europe, Siberia
Charadrius [h.] semipalmatus (Semipalmated Plover)	
	Alaska, N Canada
Pluvialis apricaria (Eurasian Golden Plover)	
apricaria?	W Europe
altifrons?	Greenland, Iceland, NW Europe, W Siberia
Pluvialis squatarola (Grey Plover)	
squatarola	N Europe - E Siberia (excl. Wrangel Island), Alaska
cynosurae	N Canada
tomkovichi	Wrangel Island
Calidris canutus (Red Knot)	
canutus	N Taymyr
islandica	NE Canada, N & E Greenland, Svalbard
rufa	N Canada
rogersi	New Siberian islands
roselaari	Wrangel Island
ssp. ?	mainland E Siberia

Table 121 - continued

Nomenclature	Breeding range
Calidris alba (Sanderling)	
alba	NE & E Greenland, Jan Mayen, Svalbard, Franz Josephland, Taymyr
rubidus	Alaska, N Canada, E & Central Siberia?
Calidris ferruginea (Curlew Sandpiper)	
	High arctic Siberia, NW Alaska
Calidris [maritima]	
Calidris [m.] maritima (Purple Sandpiper)	
maritima	N Canada, Greenland, Jan Mayen, Bear Island, Svalbard, N Fennoscandia, Franz Josephland, Novaya Zemlya, W Siberia
littoralis	Iceland
belcheri	East part Hudson Bay, James Bay
Calidris [m.] ptilocnemis (Rock Sandpiper)	
ptilocnemis	Pribiloffs, St. Matthew, Hall islands
tschuktschorum	St. Lawrence, Nunivak islands, W Alaska, E part Chukotsk Peninsula
couesi	Aleutian islands, Alaska Peninsula
quarta	Commander islands, Kurile islands, S Kamchatka
Calidris alpina (Dunlin)	
alpina	N Fennoscandia - Yenisei Delta
arctica	E & NE Greenland
schinzii	S & SE Greenland, Iceland, Faeroes, British isles, W Europe, Baltic, S Scandinavia
centralis	Taymyr Peninsula - Indigirka Delta
actites	N Sakhalin Island
kistchinski	SW Koryak Highland, NE coast Ochotsk Sea, Kamchatka, N Kuril islands
sakhalina	Wrangel Island, Chukotsk Peninsula, Anadyr
arcticola	N Alaska
pacifica	W & SW Alaska
hudsonia	N Canada
Limosa [limosa]	
Limosa [l.] limosa (Western Black-tailed Godwit)	
limosa	W & E Europe
islandica	Iceland
Limosa [l.] melanuroides (Eastern Black-tailed Godwit)	
Limosa lapponica (Bar-tailed Godwit)	
lapponica	N Fennoscandia, N coast White Sea, Kanin Peninsula
taymyrensis	Yamal Peninsula east to delta Anabar River
menzbieri	Lena Delta east to Chaunsk Bay
anadyrensis	Anadyr, S part Chukotsk Peninsula
baueri	Alaska
Numenius [phaeopus]	
Numenius [p.] phaeopus (Eurasian Whimbrel)	
phaeopus	Fennoscandia east to SW Taymyr, Yenisei valley and Omsk area
islandicus	Iceland, Faeroes, N part British isles
variegatus	NE Siberia
alboaxillaris	Steppes east of the lower Volga, south and south-east of the Urals

Table 121 - continued

Nomenclature	Breeding range
Numenius [p.] hudsonicus (Hudsonian Whimbrel)	
hudsonicus	Hudson Bay
rufiventris	W Alaska east to Melville Hills, NWT
Numenius arquata (Eurasian Curlew)	
arquata	W, N & Central Europe
sushkini	S Ural, Kazakhstan
orientalis	Central Asia
Tringa totanus (Redshank)	
totanus	Fennoscandia
robusta	Iceland, Faeroes, Orkneys
britannica	North Sea countries east to the Black Sea
ussuriensis	SE Russia, SW Asia
ssp. ?	Central Asia
terrignotae	NE Asia
Arenaria interpres (Ruddy Turnstone)	
interpres	NE Canada, Greenland, Palearctic
morinella	low arctic Canada

Have long-distance migrants short and small wings?

In the E Atlantic flyway several populations are known to cover larger distances during migration than their conspecifics, since most species are confronted with the choice whether they will stay and winter in W European coastal areas or whether they will travel 8.000-10.000 km more to reach the W African wintering grounds. The populations covering the largest distances to the wintering grounds appear to have the smallest wings. We will give some examples to substantiate these propositions. The information on the migration distances and wintering grounds of the various breeding populations is based on Cramp & Simmons (1983), Smit & Piersma (1989), Drent & Piersma (1990), Piersma & Jukema (1990), Wymenga *et al.* (1990), Davidson & Wilson (1992), Piersma *et al.* (1992), Meltofte (1993), Meltofte *et al.* (1994), Tomkovich & Soloviev (1996). The involved populations are:

- In the Ringed Plover both *psammodroma* and *tundrae* have smaller wings than *hiaticula*. Nominate *hiaticula* is breeding in temperate W Europe and the other two in the boreal and subarctic regions of the Palearctic. All 3 subspecies are wintering mainly along the coast of W Africa, but *hiaticula* is covering smaller distances than both others. Nominate *hiaticula* also has the largest and broadest wings of the three subspecies.

- The spatial distribution of Grey Plover breeding populations over the wintering grounds is poorly understood. However the westernmost breeding population in N Europe has the smallest wings.

- Two major populations of Red Knot are thought to occur in the E Atlantic flyway: (1) *islandica* is breeding in NE Canada and Greenland and wintering in W Europe; (2) *canutus* is breeding in Siberia (Taymyr?) and wintering in W Africa. The Taymyr birds might well have smaller wings than the ones from NE Canada and Greenland. However this picture is not evident since also methodological differences in measuring might have influenced the results.

- The Greenland and Iceland populations of Dunlin winter in W Africa, whilst the N European and Taymyr populations stay to winter in W Europe. The Greenland and Iceland populations have to travel larger distances than the N European Dunlin and have correspondingly smaller wing- and secondary lengths.

- The N Fennoscandian population of the

Bar-tailed Godwit (nominate *lapponica*) is probably wintering in W Europe and the Yamal & Taymyr population (ssp. *taymyrensis*) in W Africa. Nominate *lapponica* is thus migrating over smaller distances and has -as the preceding species- larger wings.

- Nearly all Whimbrels winter in W Africa, but the Iceland population is relatively most abundant in Mauritania and the Fennoscandian population most in Guinea-Bissau. The Fennoscandian birds are thus travelling longer distances and again have correspondingly smaller wings.

- In the Redshank 3 populations are distinguished: (1) Icelandic *robusta* is wintering in W Europe, (2) W European *britannica* are predominantly wintering in Mauritania, and (3) Fennoscandian *totanus* are mainly wintering in Guinea-Bissau. Thus *totanus* is covering the largest distances, followed by respectively *britannica* and *robusta*. The wing lengths are negatively related to the traveled distances, since *robusta* has the largest wings and *totanus* the smallest.

There is also an exception on this rule: Black-tailed Godwits from Iceland winter in W & SW Europe, whilst the W European Black-tailed Godwits winter in the Sahel countries. Nominate *limosa* thus covers larger distances but also has longer wings.

It might be that either (1) long migrations in the E Atlantic flyway have resulted in small wing dimensions or (2) short flight distances have evolved to large wing dimensions. Both hypotheses can be justified with energetic arguments. It would be interesting to look in more detail to this inverse relationship between flight distances and wing dimensions and link it to the hypotheses presented by Meltofte (1996). It is also interesting to know whether such an inverse relationship also exists in the other major flyways. For the moment it must be enough to settle the pattern in the E Atlantic Flyway.

Morphometrics in relation to sexual size dimorphism

Sexual size dimorphism is mostly related to body mass (Fairbair & Shine 1993, Abouheif & Fairbairn 1997) or wing length (Payne 1984) as the expression of size. However, Teather & Nol (1997) recently showed that the sexual dimorphism in the Semipalmated Plover was of a mixed character: ♂♂ had the longest toes and bills and were most colorful, whilst ♀♀ were heaviest and had the longest wings. They addressed to the problem that in a sexually dimorphic species one sex is not necessarily consistently larger than the other. They also stated that mixed dimorphism occurred in Ringed Plovers *C. hiaticula*, Mongolian Plovers *C. mongolius* and Killdeers *C. vociferus* according to the data presented by Jehl & Murray (1986). We calculated our data in the same way, but also distinguished between morphometrically distinct populations within species (Table 122). Teather & Nol's (1997) assumption was adopted, stating that '*a male to female ratio of a particular character equaling 1.00 ± 0.01 or less reflects monomorphism*'.

These calculations show that the ♀♀ of nearly all populations had longer wing lengths, secondary lengths and weights than the ♂♂ had except for some monomorphic populations of the Grey Plover and the Ringed Plover. This is in agreement with the idea that wing dimensions and size express the same sort of information. These calculations also show that no strictly monomorphic species were analysed in this study. The best example is the Grey Plover. When the geographical variation in this species was analysed with MANOVA-analysis, it appeared to be useless to distinguish between both sexes (Table 20). However, the W Palearctic *squatarola* population is monomorphic in 5 of the 7 studied aspects, according to the criteria of Teather & Nol (1997). In the same way the E Palearctic *squatarola* population, the subspecies *tomkovichi* and the N Canadian *cynosurae* population are monomorphic in 3 of the 7 aspects. Finally the Alaska population is monomorphic in 4 of the 7 aspects. Differences between both sexes were estimated to occur in all measured variables: wing length (1x), culmen length (2x), tarsus length (3x), secondary length (1x), middle toe length (3x), tail length (5x) and culmen height (3x). Part of these so-called 'differences' might not be classified like this when proper tests were carried out and are based upon the arbitrarily chosen border value of 1.00±0.01 (see above). We were not aware of a ratio-test including the confidence interval of that ratio. On the other side the ♂♂ in 4 of 5 Grey Plover populations have on average longer tails than the ♀♀, which might well have a sound biological reason as for instance a function in the courtship display of ♂♂. In the same way it is remarkable that in most populations of the Charadrinae the ♂♂ have longer toes and tarsi and heavier bills than the ♀♀, whilst they mostly have smaller wing dimensions and lower weights. With these findings in mind we agree with Teather & Nol (1997) that the Charadrinae have a pattern of mixed dimorphism. It can be added that this pattern is different for the various species.

Mixed dimorphism occurred rarely in the Calidrinae (Table 122). It was not observed in the Red Knot, Sanderling, Little Stint, Rock Sandpiper, Purple Sandpiper and the Spoonbill Sandpiper. It occurred in the Temminck's Stint, Curlew Sandpiper, Dunlin ssp. *arcticola*, Broad-billed Sandpiper and possibly also in the Dunlin ssp. *hudsonia*. It concerned 5 times longer tail lengths in the ♂♂ and 1 time longer tarsi. In the Tringinae it was just as rare a phenomenon as in the Calidrinae. Mixed dimorphism was not recorded in the Western Black-tailed Godwit, Eastern Black-tailed Godwit, Bar-tailed Godwit, most races of Whimbrel, Slender-billed Curlew, Curlew ssp. *arquata* and *orientalis* and Greenshank. It occurred in the Whimbrel ssp. *rufiventris*, Curlew ssp. *sushkini*, 4 of the 6 Redshank populations and possibly in the Spotted Redshank. Mixed dimorphism did not occur in both Turnstone subspecies.

Table 122. *Male to female ratio in morphometrically distinct breeding populations. Only ratio's calculated from mean values based upon a sample size of 5 or more birds are presented.*

Species	Subspecies/ population	Wing	Culmen	Tarsus	Secondary	Tail	Middle toe	Culmen height	Weight
Ringed Plover	*hiaticula*	0.9754	0.9849	1.0006	0.9919	0.992	1.0195	0.9808	-
	psammodroma	1.0018	1.004	1.011	1.0029	0.9913	0.9987	0.9952	-
	tundrae	0.9949	1.0172	1.0211	0.9913	0.9775	1.0245	1.0302	0.9371
Semipalmated Plover		0.9901	1.0187	1.0135	0.9968	1.0035	1.0352	1.0151	0.9565
Dotterel		0.9541	0.967	0.9747	0.9621	0.9931	1.0089	-	-
Golden Plover		0.9837	0.9921	1.0072	0.9892	0.9965	1.0143	0.993	0.8784
Grey Plover	*squatarola* - W Palearctic	1.0015	1.0004	1.0046	0.9972	1.0422	0.9982	1.0226	-
	squatarola - E Palearctic	0.9906	1.0048	1.0368	1.0025	1.0152	1.0472	1.0103	-
	tomkovichi	0.9763	1.0174	0.9864	0.9959	0.9730	1.0070	1.0048	-
	Alaska - ssp. ?	1.001	1.0293	1.0134	0.9939	1.0316	0.9956	1.0038	-
	cynosurae	1.0006	1.0039	1.0100	1.0158	1.0296	1.0177	1.0106	-
Red Knot	*canutus*	0.9691	0.8970	0.9587	0.9683	0.9712	0.9879	-	-
	rufa	0.9761	0.9498	0.9960	0.9746	0.9843	0.9913	-	-
	islandica	0.9755	0.9330	0.9816	0.9670	0.9764	0.9895	-	0.8113
	rogersi	0.9564	0.9020	0.9494	0.9332	0.9438	0.9753	-	-
	roselaari	0.9714	0.9558	0.9931	0.9663	0.9371	0.9880	-	-
	Alaska - ssp. ?	0.989	0.9361	0.9594	0.9932	0.9881	0.9835	-	0.9224
Sanderling	*alba*	0.9737	0.9291	0.9668	0.9634	0.9789	0.9847	-	0.9841
	rubidus	0.975	0.9263	0.9661	0.9833	0.9818	0.9983	-	0.8515
Little Stint		0.9607	0.9726	0.9371	-	-	-	-	-
Temminck's Stint		0.9822	0.9593	1.0118	0.9906	1.0096	0.9793	-	-
Curlew Sandpiper		0.9816	0.9008	0.9462	0.9860	1.0153	0.9590	-	0.956
Rock Sandpiper		0.9491	0.8688	0.9698	0.9406	0.9602	0.9649	-	-
Purple Sandpiper	*maritima* - W Palearctic	0.9586	0.8599	0.9691	0.9446	0.9730	0.9695	-	0.8676
	maritima - N Canada	0.9717	0.8717	0.9708	0.9695	0.9794	0.9806	-	0.8608
	belcheri	0.968	0.8695	0.9401	0.9546	0.9824	0.9690	-	-
	littoralis	0.9654	0.8853	0.9607	0.9561	0.9808	0.9618	-	-
Dunlin	*alpina*	0.9764	0.9125	0.9602	0.9724	0.9633	0.9732	-	0.9368
	arctica	0.9626	0.8939	0.9569	0.9469	0.9845	0.9756	-	-
	schinzii	0.9648	0.8711	0.9453	0.9673	0.9798	0.9582	-	0.9643
	centralis	0.9686	0.8706	0.9417	0.9697	0.9816	0.9417	-	0.9129
	kistchinski	0.976	0.8850	0.9534	0.9697	0.9806	0.9663	-	-
	sakhalina	0.9832	0.8868	0.9604	0.9850	0.9909	0.9956	-	0.9058
	arcticola	0.9879	0.9087	0.9683	0.9801	1.0127	0.9601	-	0.8968

Species	Subspecies/ population	Wing	Culmen	Tarsus	Secon-dary	Tail	Middle toe	Culmen height	Weight
	pacifica	0.9666	0.9210	0.9634	0.9655	0.9854	0.9704	-	0.9086
	hudsonia	0.9901	0.9351	0.9797	0.9852	1.0051	0.9954	-	0.8476
Broad-billed Sandpiper		0.9744	0.9499	0.9691	0.9865	1.0141	1.0099	-	-
Spoonbill Sandpiper		0.9668	0.9673	0.9913	0.9773	-	-	-	-
Western Black-tailed Godwit	*limosa* - W Europe	0.9491	0.8436	0.9044	-	-	-	-	0.8616
	limosa - E Europe	0.9551	0.8990	0.9420	-	0.9576	0.9713	-	-
	islandica	0.9408	0.8268	0.8905	-	-	-	-	-
Eastern Black-tailed Godwit		0.9821	0.8681	0.9074	0.9589	0.9448	0.9768	-	-
Bar-tailed Godit	*lapponica*	0.9422	0.8000	0.9182	0.9215	0.9808	0.9389	-	-
	taymyrensis	0.9485	0.8252	0.9298	0.9416	0.9562	0.9465	-	-
	menzbieri	0.9506	0.7856	0.9426	0.9485	0.9395	0.9665	-	-
	anadyrensis	0.9487	0.7949	0.9318	0.9323	0.9731	0.9224	-	0.7472
	baueri	0.9409	0.7953	0.9045	0.9436	0.9637	0.9208	-	0.8741
Whimbrel	*phaeopus* - N Europe	0.9634	0.8958	0.9603	0.9604	0.9642	0.9447	-	0.8822
	phaeopus - Iceland	0.9649	0.9267	0.9639	0.9620	0.9789	0.9695	-	-
	variegatus	0.9658	0.9083	0.9458	-	0.9655	0.9825	-	-
	alboaxillaris	0.9849	0.9383	0.9520	0.9959	-	1.0042	-	-
	hudsonia - Alaska	0.9647	0.9040	0.9722	0.9571	0.9671	1.0128	-	-
	hudsonia - N Canada	0.9481	0.8959	0.9317	0.9649	0.9728	0.9634	-	-
Slender-billed Curlew		-	0.9453	0.9923	-	0.9648	-	-	-
Curlew	*arquata*	0.9406	0.8248	0.9489	0.9514	-	0.9744	-	0.8042
	suschkini	0.9718	0.9190	1.0036	1.0104	-	1.0293	-	-
	orientalis	0.9434	0.7939	0.9716	0.9325	-	0.9486	-	-
Spotted Redshank		0.9768	0.9756	0.9884	0.9719	1.0001	0.9931	-	-
Redshank	*totanus*	0.9708	0.9713	0.9743	0.9744	0.9857	0.9931	-	0.8796
	robusta	0.9811	0.9409	0.9766	0.9748	1.0063	0.9592	-	-
	brittannica	0.9773	0.9536	0.9840	0.9813	1.0020	0.9779	-	-
	ussuriensus	0.9576	0.9537	0.9937	0.9776	0.9869	0.9923	-	-
	Central Asia - ssp. ?	0.9821	0.9974	1.0106	0.9758	1.0244	0.9951	-	-
	terrignotae	0.9764	1.0079	1.0434	0.9834	0.9756	1.0090	-	-
Greenshank		0.9813	0.9974	0.9909	0.9781	0.9765	0.9926	-	-
Turnstone	*interpres*	0.9833	0.9699	0.9874	0.9853	0.9816	0.9979	-	1.0197
	morinella	0.9868	0.9709	0.9891	0.9862	0.9749	0.9916	-	0.9001

Table123. *Pearson product-moment correlations between the mean measurements of ♂♂ and the corresponding differences (in percentages) between ♂♂ and ♀♀. Significant linear regressions are presented in Fig. 119. Correlations were calculated for (1) all data, (2) the plovers - Charadrinae, (3) the small sandpipers - Calidrinae (4) the godwits, curlews, large sandpipers and turnstones - Tringinae & Arenariinae.*

	all		Charadrinae		Calidrinae		Tringinae & Arenariinae	
	r	*n*	*r*	*n*	*r*	*n*	*r*	*n*
wing length	0.366	65	0.195	11	0.032	26	0.503**	28
culmen length	0.595***	66	0.293	11	0.123	26	0.660***	29
tarsus length	0.478	66	0.030	11	0.138	26	0.499	29
secondary length	0.430	61	0.270	11	0.247	26	0.155	24
tail length	0.164	58	0.464	11	0.434*	24	0.391	23
middle toe length	0.078	61	0.110	11	0.055	24	0.032	26
weight	0.516**	24	0.987	2	0.359	14	0.512	8
$\sum$-diff. - weight	0.293	34	0.403	7	0.071	16	0.786**	11

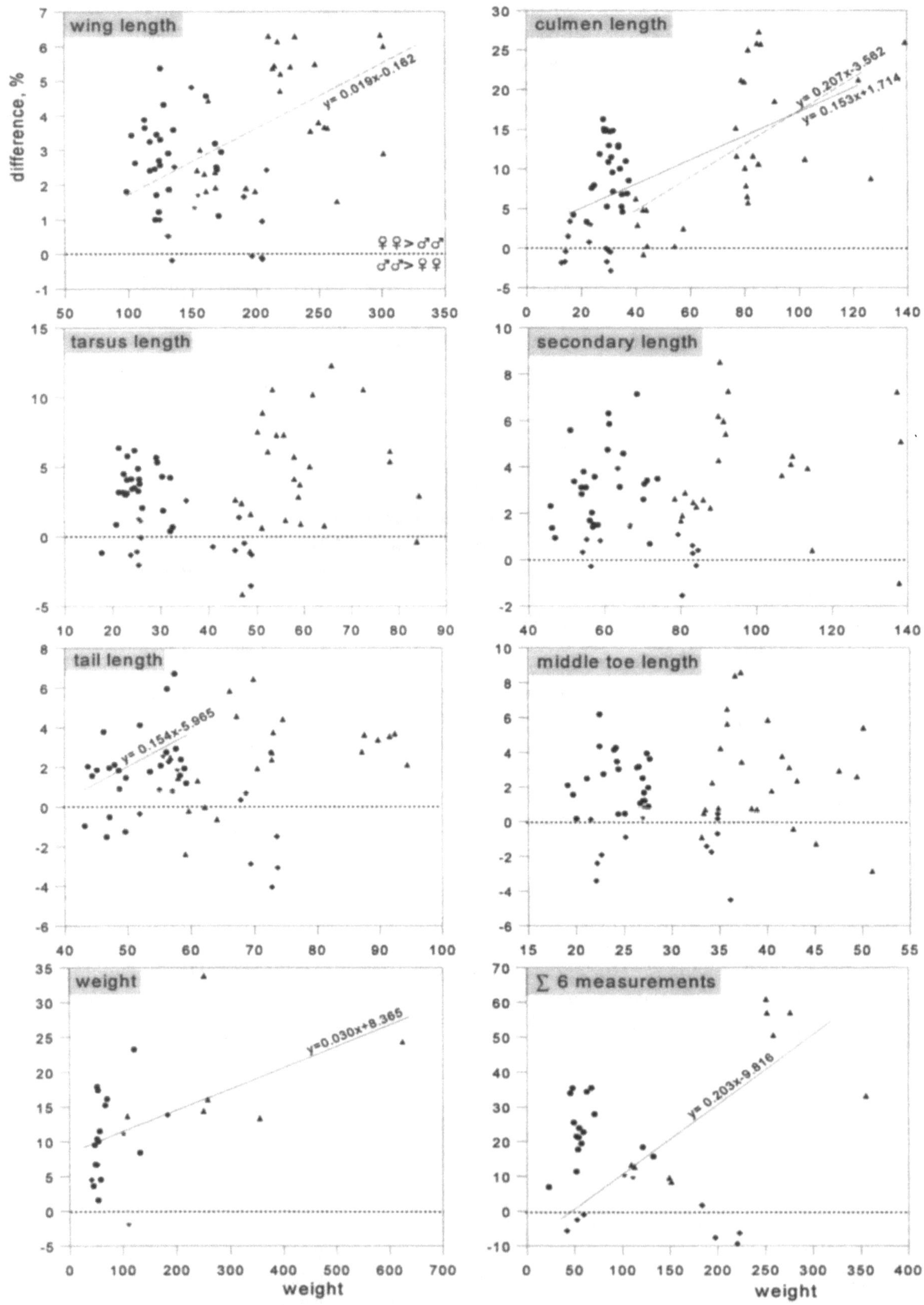

Figure 119. *Size dimorphism in relation to the mean measurements of the ♂♂. The distinguished populations and their measurements are presented in Annex 1. The Charadrinae ('◆' symbol), Calidrinae ('●'), Tringinae ('▲') and Arenariinae ('★') were distinguished.*

The differences between both sexes were also related to the average size of the ♂♂ (Fig. 119). In this figure 4 families were distinguished: the Charadrinae, Calidrinae, Tringinae and the Arenariinae. The correlations between the percentage differences and the mean ♂ sizes are presented in Table 122. Six analyses produced significant correlations. The relations based upon these correlations are given in Fig. 119. All 6 were positive relationships: increasing measurements lead to larger differences between ♂♂ and ♀♀.

There were no significant correlations in the Charadrinae and nearly no in the Calidrinae. This virtual lack of significance in the latter group was mainly due to the low degree of dimorphism in the heaviest Calidris-species the Red Knot. In the Tringinae (& Arenariinae) positive relations between percentage difference and the wing- and culmen lengths. This significancy was especially due to the large degree of dimorphism in the Curlew.

Skewed sex-ratios on the breeding grounds

In all species, more ♂♂ than ♀♀ were present in the museum collections: the ♂ percentages ranged between 55.3 in the Knot to 70.5 % in Eurasian Golden Plover. Such might be logical when ♂♂ occur more frequent and longer on the breeding grounds as is the case in most species. But it not logical for Curlew Sandpiper, since here nearly exclusively ♀♀ take care for the reproduction. Even in this species 62 % of the birds examined were ♂♂. Moreover, species with a strong male-biased parental care, e.g. Spotted Redshank, were not collected with higher male proportions than species with shared parental efforts (e.g. Eurasian Golden Plover).

High male percentages occur especially during arrival- and departure times (Table 124).

Table 139. *Male-percentages in different shorebird species throughout the breeding season.*

		March		April		May		June		July		August	
Species	Subspecies	%	*N*	%	*N*	%	*N*	%	*N*	%	*N*	%	*N*
		boreal / arctic (sub)species											
Ringed Plover	*tundrae*	-	0	-	0	75	24	67	106	63	100	88.9	9
Semipalmated Plover		-	0	66.7	3	68.2	22	57.4	44	58	69	80	50
Eurasian Golden Plover	*altifrons*	-	0	-	0	80	15	86.7	15	70	10	-	0
Grey Plover		-	0	-	0	75	24	63.9	119	62.4	85	75.9	29
Red Knot		-	0	-	0	75	4	51.2	127	59.5	79	100	4
Sanderling		-	0	-	0	71.4	14	56.9	123	56.9	72	47.4	19
Curlew Sandpiper		-	0	-	0	81.8	11	70.5	44	9.1	11	50	2
Purple Sandpiper		-	0	55.6	9	55.6	27	55	140	67	88	48.3	58
Dunlin		-	0	-	0	59.9	117	57.5	273	63.3	147	61.1	18
Bar-tailed Godwit		-	0	-	0	51.1	47	58.9	112	58.9	95	0	2
Redshank	*totanus*	-	0	71.4	7	50	26	53.5	58	63.2	19	-	0
Redshank	*robusta*	-	0	50	14	57.1	35	58.5	41	50	14	20	5
Ruddy Turnstone		-	0	100	1	56.8	74	55.6	216	72.3	119	87	23
		temperate / boreal (sub)species											
Ringed Plover	*hiaticula*	68.2	22	33.3	6	73	37	64	86	62.3	69	75	12
Eurasian Golden Plover	*apricaria*	33.3	3	80	5	28.6	7	75	12	66.7	6	-	0
Dunlin	*schinzii*	50	6	68.2	22	57.5	80	56.1	82	60.6	66	-	0
Black-tailed Godwit		40	10	64.2	53	63.5	52	57.7	26	25	4	100	1
Eurasian Curlew		-	2	66.7	21	34.1	41	72	50	73.3	15	71.4	7
Spotted Redshank		-	0	-	0	78.9	19	65.7	35	71.4	7	100	1
Redshank	*brittannica*	-	0	54.8	31	77.5	40	62.5	16	100	6	-	0
Greenshank		-	0	40	5	68.6	35	44.8	58	71.1	38	40	5

In all plover- and most sandpiper species the ♂ percentages exceed 60 % at arrival. This is not so for Purple Sandpiper, Dunlin and several larger-sized species (Bar-tailed Godwit, Icelandic- and W European Redshank, Ruddy Turnstone and Greenshank). In several species, ♀♀ not only arrive later, but also depart earlier as is known from literature. Such early departure might well occur in Ringed Plover, Semipalmated Plover, Grey Plover, Red Knot, Dunlin, Scandinavian Redshank (ssp. *totanus*), Spotted Redshank and Ruddy Turnstone according to the high ♂ percentages found during July and Augustus. Therefore, we think that the longer presence of ♂♂ on the breeding grounds results in a predominance of ♂♂ collected.

Nevertheless, the June-data also show a majority of ♂♂, but less extreme as with all data. Some species had male percentages larger than 60 %: Ringed Plover, Eurasian Golden Plover, Grey Plover, Curlew Sandpiper, Eurasian Curlew, Spotted Redshank and W European Redshank. The high June male-percentages in the Curlew Sandpiper might reflect the departure of ♂♂ from the breeding grounds, since the ♂ percentage in July is very low in this species. In the other Sandpiper species the ♂ percentages range between 51 and 58 % during June. Possibly, male plovers have more a defensive task making them more vulnerable for being shot. It might be interesting to know whether the male predominance reflects the overall sex ratios in these species, or whether it reflects behavioural differences between the sexes on the breeding grounds.

Estimating the composition of wader populations

The morphometric research presented in this book results from a study on the population composition of waders in the Dutch part of the Wadden Sea. Such estimates can now be made by comparing morphometric data from the breeding grounds with morphometric data from birds in the Wadden Sea. Such can also be done with other migrant or wintering populations. The population composition can be estimated by using multivariate, predictive statistics (e.g. POSCON). Such estimates are important when interpreting the population sizes of waders as collected by Wetlands International. Their population figures are used for the identification of internationally important wetlands throughout the world (Rose & Scott 1994). Quantitative estimates of the composition of migrant and wintering populations will provide us with insight on:

- *the population size of breeding populations.* Generally waders congregate during migration and winter, whilst they have a very dispersed distribution during breeding. Most of these species also breed in regions from which it is difficult to obtain population estimates. It would thus be very profitable to have quantitative estimates on the population composition of non-breeding populations especially for the numerically most important sites within a flyway. In recent years

- *the migration and wintering habits of waders.* Quantitave estimates of the population composition can now be made as long as morphometric differences were evident. These morphometrically different populations can be traced during migration and wintering.

Our morphometric studies necessitated adjustments in the geographical limits of various breeding populations e.g. Ringed Plover, Purple Sandpiper, Curlew, Whimbrel. We hope that it will have consequences for the distinctions as used for the population delineations and estimates produced by Wetlands International (Rose & Scott 1994). Quantitative information on the population composition of wintering and migrating populations is regularly lacking or questionable (explained in the introduction)., Morphometric analysis might fill this gap of knowledge.

Morphometrics and DNA-studies

It was acknowledged in the introduction that measurement characters are strongly inherited, but that they are also influenced by the environment. Nearly always more genes are involved in the expression of a character which (1) might result in a relative complex pattern of inheritance, and (2) requires DNA-research in various genes. With the results of DNA-analysis becoming visible, we have reached the state of science to explain the morphometric variation in terms of genetics and/or environmental conditions. It is also possible to consider questions concerning the evolutionary age of distinct populations which might help to identify the isolational mechanisms leading to a process of subspeciation in waders.

It is necessary to make some basic distinctions when considering results from both DNA- and morphometric research:

- *populations and individuals.* Breeding populations of waders ought to be considered as all populations in the wild: being genetically and phenotypically variable. Most individuals in a population must possess geno- and phenotypes which permit them to survive under the existing conditions and habits during breeding, migration as well as wintering. Such a collection of individuals is shaped through generations and characterization might be based upon variables which might be shaped long ago or just quite recently. And thus heterogeneity in a breeding population is the rule and not the exception. In this respect there is no difference between morphometric data or DNA-type of data. With both types overlap occurs. The definition of breeding populations, based upon the characters of sampled individuals, is not necessarily better with one type of data or the other. We certainly do not agree with the statement of Wenink (1994) that morphological characterization (of breeding populations) has a higher chance to be based upon '*poor criteria*' than characterization in terms of DNA types.

- *Two different stages: description and prediction.* In our study, morhometric data from breeding populations were collected in order to predict the geographical breeding origin of migrants or wintering birds. And thus a *description*-stage in which breeding populations are characterized (in terms of morphometrics) is followed by a *prediction*-stage. The same stages can be distinguished with DNA-analysis. In the description phase a breeding population is characterized by a frequency distribution of haplotypes. In the prediction phase firstly a non-breeding individual is charactized by its haplotype or (in case of a wintering population) a frequency distribution of haplotypes. Secondly the geographical breeding origin of these non-breeders is predicted with a chosen set of analytical procedures. When comparing the strengths and weaknesses of analysis with resp. morphometrics and DNA it is necessary to distinguish both stages. What appears to be a weakness in one stage might well turn out to be a strength in the other.

- *Continuous and discontinuous data.* In the descriptive phase of research breeding populations are characterized in morphometric studies by continuous, normally distributed data (e.g. culmen length 38.9 mm) or categorical data (e.g. axillary score). DNA studies produce discontinuous, categorical data (e.g. haplotype 'SIB6'). The characters of all sampled birds are resp. averaged or categorized. A breeding population is then characterized by mean values, standard deviations and sample sizes of that sample and/or by frequency distributions.

We present an example derived from the Dunlin-studies to show the (dis)advantages of the various methods (Table 125). In this example 4 breeding populations are characterized in terms of morphometrics and DNA-analysis. In both types of analysis breeding populations include individuals with different characters and the mean values as well as the frequency distributions characterizing these populations are overlapping. Only 13 of the 24 haplotypes were unique for any of the 4 presented populations, whilst less than 20 birds were sampled in 3 of the 4 populations. The other 11 haplotypes occurred in 2 or more populations, resulting in overlap of frequency distributions between these 4 populations. There is one character (EUR1) occurring in all 4 populations. Larger sample sizes will inevitably result in more overlap. It becomes even more difficult when ♂♂ and ♀♀ will be separated as in morphometric analysis. Because the breeding site fidelity in ♀♀ is generally smaller than in ♂♂ (Table 2), it is expected that the ♀♀ in a breeding population have more heterogeneous mtDNA-characters than ♂♂. Just the ♀♀ are essential in mtDNA analysis, since the mtDNA-characters are inherited from mother to offspring (Wenink 1994).

Table 125. *The geographic distribution of haplotypes in Dunlin (after Wenink 1994) in combination with the measurements of the same breeding populations according to Table 57 from our study.*

Variable	S. Fennoscandia	N. Fennoscandia	Kanin-Yamal	Taymyr
		mtDNA-analysis		
SIB1	-	-	2	2
SIB2	-	1	-	5
SIB3	-	-	-	1
SIB4	-	-	1	-
SIB5	-	1	1	4
SIB6	2	-	-	3
SIB7	-	-	1	-
EUR1	22	6	6	1
EUR2	1	-	-	-
EUR3	1	-	-	-
EUR5	-	-	-	1
EUR6	2	1	-	-
EUR7	1	-	-	-
EUR8	-	1	-	-
EUR9	-	1	-	-
EUR10	1	1	-	-
EUR11	2	-	1	-
EUR12	1	1	-	-
EUR13	1	-	-	-
EUR16	2	1	-	-
EUR17	-	1	-	-
EUR18	-	1	-	-
EUR19	6	-	-	-
EUR20	1	-	-	-
N	43	16	12	17
		Morphometrics ♂♂ (this study)		
wing length	115.2±3.7(22)	117.3±2.6(22)	117.1±2.1(18)	117.6±1.9(24)
culmen length	28.8±1.9(22)	30.6±1.7(22)	32.4±2.1(18)	31.8±1.5(24)
tarsus length	23.6±0.9(22)	23.6±1.0(22)	24.3±0.8(17)	24.6±1.0(23)
secondary length	52.1±1.8(22)	54.1±2.4(17)	53.3±1.5(3)	53.9±1.1(11)
tail length	44.4±2.9(22)	46.1±2.1(17)	44.7±1.5(3)	45.4±3.0(11)
middle toe length	22.8±1.0(22)	22.7±1.0(17)	24.0±1.2(3)	22.1±1.0(11)
		Morphometrics ♀♀ (this study)		
wing length	118.3±3.5(7)	120.5±2.1(20)	120.2±2.8(20)	120.8±2.0(13)
culmen length	31.9±1.8(7)	34.3±1.7(20)	34.6±2.2(19)	36.4±1.3(13)
tarsus length	24.6±0.9(7)	25.0±0.8(20)	25.1±0.8(20)	26.1±1.1(13)
secondary length	54.1±1.8(7)	55.5±1.5(10)	55.0 - (1)	55.6±1.9(5)
tail length	45.1±2.5(7)	47.9±2.1(10)	49.0 - (1)	46.0±1.9(5)
middle toe length	23.7±0.9(6)	23.7±0.8(10)	23.0 - (1)	23.9±1.2(5)

These results show that Wenink's remark (*'the ranges of morphological characters often overlap among populations'* - Wenink 1994, page 1) also applies to mtDNA data, and that it is not an exclusive 'problem' for morphometric analysis. This incomplete remark is the result of not making a proper distinction between the characters of individuals (e.g. having a culmen length of 38.9 mm or a haplotype 'EUR1') and the characters of populations (mean values or frequency distributions).

In the prediction stage there are always 2 types of predictions. One is dealing with individuals and the other with populations:

1. It is tried to calculate the probability of an *individual* migrant or wintering bird to

belong to some predefined breeding populations (here called 'sample groups'). A migrant can be characterized by some continuous variables (wing- and culmen length) and/or by some categorical data (plumage-class, DNA-haplotype). If that migrant is characterized as haplotype 'EUR1', it will be a logical question to ask what probability this bird has to belong to any of the 4 presented breeding populations. When it is the only measured character, the answer will be something like: a probability of resp. 51.2 % (22/43*100) for S Fennoscandia, 37.5 % for N Fennoscandia, 50.0 % for Kanin-Yamal and 5.9 % for Taymyr. It is then assumed that the obtained frequency distributions completely reflect reality (which is not true with these small sample sizes). These are all probabilities of > 5 % and it must thus be concluded that this bird is breeding somewhere in the region between S Fennoscandia and Taymyr which is not very accurate. Also the problem of estimating the confidence limits around these percentage figures has to be solved then: how accurate are these estimates. Moreover, assessing the DNA-category is like assessing the plumage-category. Both are categorical data and suffer from the same difficulties when predicting a bird's breeding origin (Annex 3 - POSCON analysis).

2. It is tried to estimate the composition of a migrant or wintering *population*. If it is based upon categorical data like plumage-characters or DNA-haplotypes we even do not see a start of a solution without the inclusion of unrealistically large sample sizes from the breeding populations. The small sample sizes in this study regularly force us to obtain estimates of the population composition under the assumption of and as long as they are based on normally distributed data (Engelmoer *et al.* in prep.). The assumption of equal covariance matrices is not necessarily true. However, such an assumption is not needed anymore when sample sizes are larger than about 50 individuals. Such estimates of the population composition which are based upon normal distributions also include confidence limits. We are not aware of operational methods to calculate the confidence limits of breeding origin estimates, based upon categorical data. One of the major difficulties will be the huge sample sizes needed for such operations.

It was not our intention to contrast both methods (morphometrics and DNA-analysis) in order to score the 'best' one. We have tried to show that prediction of breeding origin involves more than measuring a wing length or assessing a haplotype. We conclude that it is not an 'or'- but an 'and'-type of question: doing one thing and don't forget the other. When both types of data become available in the description stage, interpretation of results becomes more exciting. We will give 3 examples derived from the Dunlin-studies:

1. **The distinction between *arctica*, *schinzii* and *alpina***. There are clear morphometrical and plumage differences between the 3 subspecies which were undetectable with mtDNA-analysis. We agree with Wenink *et al.* (1996), who already cautioned against the lumping of all 3 in one *alpina* group. However, their argument was based upon future DNA-work which might provide the distinctions and not upon the clear morphometric and plumage differences. The latter already produce so much argument that DNA-analysis is not essential to distinguish them. Nevertheless, it is most useful to know that these 3 subspecies are relatively recent (< 10.000 years). Speculations on the causation of this recent process of subspeciation might thus include present-day ecological circumstances.

2. **The distinction between *alpina* and *centralis***. Morphometric research did never give clear evidence to the distinction of both. The mtDNA-results clearly showed the existence of both lineages with a broad zone of intermixing in between. Since the results from mtDNA-studies it is also possible to recognize them with the proper phenotypical characters and to study the hybrid zone in N Europe and W Siberia. See also p. 166-168.

3. **The Alaskan breeding populations**. MtDNA-analysis clearly showed the existence of one lineage in Alaska be-

ing assigned to *C. a. pacifica* without a sign of mixing with the N Canadian or the E Siberian populations. This apparent lack of mixing might partly be due to the small sample sizes (Anadyr 4, W Alaska 7, N Alaska 15 and N Canada 16 birds). But apart of that the morphometric characters in the N and W Alaskan populations clearly differed (tables 63 & 64, Fig. 72). They differed so much that subspecies recognition solely based on morphometric data is justifiable. However, the measurements of the N Alaskan populations differed badly from the ones of the E Siberian populations and it would be more logical to consider *sakhalina* and *arcticola* as one and the same subspecies. The differentiation in measurements between the N and W Alaskan populations is puzzling. It must have taken place in the last 70.000 years (Wenink *et al.* 1996). If the N Alaskan population has developed a migration route with crossing the Bering Strait in order to winter along the Asian coastline, it is strange that not any E Siberian ♀ crossed the Bering Strait in the opposite direction, whilst (1) Dunlin ♀♀ are least site faithful and (2) other species are regularly successful in crossing the Bering Strait towards Alaska (Bar-tailed Godwit, Curlew Sandpiper, Grey Plover).

We suggest some other items, which might be interesting enough to be clarified with DNA analysis:

1. Palearctic Grey Plover clearly show differentiation: the westernmost population clearly differ from the easternmost ones. It would be interesting to learn more about the variation in haplotypes in these breeding populations.
2. The Alaskan Knot population was considered as a population with mixed characters. Is it a clearly identifiable population or is influenced by both *roselaari* and *rufa*?
3. There were morphometric differences between the Purple Sandpiper populations of N Canada, Greenland and the N Atlantic. However the differences were too slight to settle distinct subspecies. Such might be possible with DNA analysis. Especially the position of the Greenland population might be considered carefully.
4. The distinction between the subspecies *kistchinski* and *sakhalina* in the Dunlin is strongly based upon plumage characters. Morphometrics are not decisive. DNA analysis might settle the dispute.
5. There seems to be differentiation between the W- and E European populations of the Western Black-tailed Godwit. These might also include genetical differences.
6. The Bar-tailed Godwit populations at Anadyr and in W Alaska were separated as distinct subspecies. Is it supported by the results from DNA analysis?
7. It would be interesting to learn more about the geographical variation in the Hudsonian Whimbrel.
8. The E Palearctic subspecies on the Whimbrel *variegatus* is still considered as a subspecies. Is this correct or is it a distinct species?
9. The distinctions between the Central Asian populations of Redshank are strongly plumage based. Morphometrics are not decisive: the measurements of these populations suggest the presence of 2-3 subspecies. DNA analysis might result in a clear picture concerning the distinction of *ussuriensis*, *eurhinus*, *terrignotae* and *craggi* and might clarify their geographical boundaries.
10. The S European breeding populations of the Redshank are considered to be related to the Fennoscandian ones (Hale 1971). Does DNA analysis support this view?

Geographical isolation in low arctic Canada

In the Dunlin *C.a. hudsonia* appeared to be the eldest lineage being about 223.000 years old (Wenink *et al.* 1996). The evolutionary age of the Knot *C. c. rufa*, occupying breeding ranges northwards of the north part of the Hudson Bay, was estimated less than 20.000 years (Baker *et al.*. 1994). Most other studied species also have clearly definable taxa breeding in N Canada or the Hudson Bay, being distinguished as allospecies or subspecies:

1. Low arctic Semipalmated Plovers and Ringed Plovers might have been isolated from each other in this region.
2. The Grey Plover ssp. *cynosurae* is

breeding in N Canada and distinctive from the Palearctic and Alaskan populations. *Cynosurae* is relatively small-sized and has also distinctive plumage characters.

3. The Sanderling ssp. *rubidus* is breeding in N Canada and N Alaska and is distinct from the W Palearctic populations. It is mainly characterized by its plumage characters.
4. The Purple Sandpiper ssp. *belcheri* is breeding in the east part of the Hudson Bay and James Bay and is the smallest-sized of the studied populations.
5. The Hudsonian Godwit and the Black-tailed Godwit are considered as allo-species with the Hudsonian Godwit breeding around the Hudson Bay and in Alaska. The latter was not included in this study, but it might well be possible that geographical variation exists in this species. Nominate *hudsonicus* is the smallest-sized of both.
6. The Hudson Bay population of the Whimbrel ssp. *hudsonicus* differs from every other population including the closest related Alaskan population subsp. *rufiventris*.
7. The Turnstone ssp. *morinella* is breeding in the north part and north of the Hudson Bay. It is mainly characterized by its plumage characters.

There appear to be 2 groups of species in this region:

- In the first group the distinctive characters are importantly based on measurements. This group includes the Semipalmated Plover, Grey Plover, Purple Sandpiper, Dunlin, Hudsonian Godwit and Whimbrel. All except the Dunlin are characterized by their relatively small size. The Dunlin ssp. *hudsonia* is larger-sized than most other taxa in the Dunlin. In the Semipalmated Plover and the Grey Plover also plumage differences play a role.
- The second group includes taxa which are mainly separated on the basis of plumage differences: Red Knot, Sanderling and Turnstone. All 3 taxa have a somewhat larger size than their relatives.

These analogies produce several questions. Some might be solved with analysing DNA samples. Are these analogous developments and can they be related to the same timescale? Is the splitting in the first group of older date than in the second one? Are there any ecological factors somewhere in the year's cycle resulting in a analogous size-reduction of the involved populations?

Curlew species in Kazakhstan and S Russia

There is a remarkable resemblance in phenology and distribution between 3 Curlew species as was already mentioned on p. 225. The involved (sub)species are the Slender-billed Curlew *Numenius tenuirostris*, the Eurasian Curlew ssp. *suskini* and the Whimbrel ssp. *alboaxillaris*. All 3 are rather palish with white or whitish axillaries and upper-tail coverts. The Slender-billed Curlew and *alboaxillaris* have their endangered status in common (Gretton 1991, Belik 1994, Danilenko *et al.* 1996). There is no information on the breeding status of *suschkini*. The Slender-billed Curlew and *suschkini* are also characterized by their small dimensions which is not the case with *alboaxillaris*. Both *alboaxillaris* and *suschkini* are breeding in more or less the same region: the southern Ural and Kazakhstan. It is not known where the Slender-billed Curlew is breeding. It would be interesting to know the timescale of this process of (sub)speciation.

Which measurements ought to be included in morphometric studies?

The MANOVA-analyses performed in the species chapters have also provided insight in the relative importance of the morphometric variables. This information was presented in the second table of each chapter and is summarized here (Table 126). Ranks went from '1' to '7' with decreasing priorities.

Both culmen- and wing lengths were often important variables. The tarsus- and (surprisingly enough) secondary lengths mostly had an intermediate ranking. The tail- and middle toe lengths did not produce as well. It is suggested to measure the secondary lengths also more often during trapping studies.

Table 126. *The priority of measuring in the various wader species. This priority is scaled from '1' (important) to maximally '7'. When a measurement did not show significant differentiation in the overall MANOVA analyses this is shown with a '-'sign.*

Species	Table nr.	Wing	Culmen	Tarsus	Secondary	Tail	Middle toe	Culmen height
		distinction between populations						
Ringed- & Semipalmated Plover	8	1	2	6	3	5	7	4
Eurasian Golden Plover	15	-	-	1	-	-	3	2
Grey Plover	20	1	6	3	2	4	7	5
Red Knot	30	4	1	3	2	-	-	
Sanderling	40	-	1	4	2	3	-	
Curlew Sandpiper	46	-	-	-	-	-	-	
Purple- & Rock Sandpiper	50	1	3	4	2	?	5	
Dunlin	60	2	1	4	3	6	5	
Black-tailed Godwit	71	2	1	3	?	?	?	not measured
Bar-tailed Godwit	77	1	4	6	2	3	5	
Whimbrel	87	5	1	6	2	3	4	
Eurasian Curlew	93	-	1	2	-	?	-	
Redshank	104	2	3	4	1	5	6	
Ruddy Turnstone	113	2	3	-	4	1	?	
		distinction between sexes						
Ringed- & Semipalmated Plover	8	-	-	2	-	4	1	3
Eurasian Golden Plover	15	1	-	-	-	-	2	-
Grey Plover	20	-	-	-	-	1	-	-
Red Knot	30	2	5	4	3	4	-	
Sanderling	40	2	1	4	3	-	-	
Curlew Sandpiper	46	4	1	2	-	-	3	
Purple- & Rock Sandpiper	50	3	1	4	2	?	5	
Dunlin	60	2	1	3	4	-	5	
Black-tailed Godwit	71	2	1	3	?	?	?	not measured
Bar-tailed Godwit	77	2	1	4	3	6	5	
Whimbrel	87	3	1	2	4	5	-	
Eurasian Curlew	93	2	1	-	-	?	-	
Redshank	104	1	2	4	3	-	-	
Ruddy Turnstone	113	-	1	-	-	-	?	

Chapter 19

References

The '✶'-marked references were (also) used to compile the breeding distribution maps.

Abouheif, E. & D.J. Fairbairn. 1997. A comparative analysis of allometry for sexual size dimorphism: assessing Rensch's rule. Am. Nat. 149: 540-562.

Alberico, J.A.R., J.M. Reed & L.W. Oring. 1992. Non-random philopatry of sibling Spotted Sandpipers. Ornis Scand. 23: 504-508.

Allen, J.A. 1877. The influence of physical conditions in the genesis of species. Radical Review 1: 108-40.

Allen, J.R.L., B.J. Hoskins, B.W. Sellwood, R.A. Spicer & P.J. Valdes (eds.). 1994. *Palaeoclimates and their modelling, with special reference to the Mesozoic era.* Chapman and Hall, London, U.K.

Altenburg, W., E. Wymenga & L. Zwarts (eds.). 1992. Ornithological importance of the coastal wetlands of Guinea-Bissau. WIWO report 26. WIWO Foundation, Zeist.

Amadon, D. & L.L. Short. 1992. Taxonomy of lower categories - suggested guidelines. Bull. B.O.C. Centenary Suppl. 112A: 11-38.

Ambergen, A.W. 1989. Statistical uncertainties in posterior probabilities. Ph. D. thesis, Univ. of Groningen, The Netherlands.

✶ **American Ornithologists' Union**, 1957. *Check-list of North American birds.* Baltimore, U.S.A.

American Ornithologists' Union. 1983. *Check-list of North American birds.* 6th edition A.O.U., Lawrence, Kansas, U.S.A.

Atkinson, N.K., R.W. Summers, M. Nicoll & J.J.D. Greenwood. 1981. Population, movements and biometrics of the Purple Sandpiper *Calidris maritima* in eastern Scotland. Ornis Scand. 12: 18-27.

✶ **Bailey, A.M.** 1948. *Birds of arctic Alaska.* Colorado Mus. Nat. Hist. Pop. Ser. Nr. 8. Denver, Colorado, U.S.A.

Baker, A.J. 1992. Molecular genetics of *Calidris*, with special reference to the Knots. W.S.G. Bull. 64, Suppl.: 29-35.

Baker, A.J., T. Piersma & L. Rosenmeier. 1994. Unraveling the intraspecific phylogeography of Knots *Calidris canutus*: a progress report on the search for genetic markers. Pages 39-45 *in* T. Piersma. Close to the edge: energetic bottlenecks and the evolution of migratory pathways in Knots. Ph. D. thesis, Univ. of Groningen, The Netherlands.

Barlow, M. 1972. The establishment, dispersal and distribution of the Spur-winged Plover in New Zealand. Notornis 19: 201-211.

Barter, M. 1992. Distribution, abundance, migration and moult of the Red Knot *Calidris canutus rogersi.* W.S.G. Bull. 64, Suppl.: 64-70.

✶ **Bekhuis, J.** 1992. Breeding bird atlas of Europe. Working report. Part 1: Non-passeriformes. EOAC/SOVON, Beek-Ubbergen, The Netherlands.

Belik, V.P. 1994. Where on earth does the Slender-billed Curlew breed? W.S.G. Bull. 75: 37-38.

✶ **Belopolski, L**. 1933. Zur Avifauna des Anadyr-Gebietes. J. Orn. 81, 416-432.

✶ **Bergman, S**. 1935. *Zur Kenntnis nordost-asiatischen Vögel.* Albert Bonnier, Stockholm, Sweden.

Berthold, P. 1993. *Bird Migration.* Oxford Univ. Press, Oxford, U.K.

Berthold, P. & W. Friedrich. 1979. Die Federlänge: ein neues nützliches Flügelmass. Vogelwarte 30: 11-21.

Berthold, P. & U. Querner. 1982. Genetic basis of moult, wing length, and body weight in a migratory bird species, *Sylvia atricapilla.* Experientia 38: 801-802.

Boag, P.T. & A.J. van Noordwijk. 1987. Quantitative genetics. Pages 45-78 *in*: F. Cooke & P.A. Buckley (eds.), *Avian genetics: a population and ecological approach.* Academic Press, London, U.K.

Bock, W.J. 1959. The status of the Semipalmated Plover. Auk 76: 98-100.

Boere, G.C. 1976. The significance of the Dutch Wadden Sea in the annual life-cycle of arctic, subarctic and boreal waders. Part 1. The function as a moulting area. Ardea 64: 210-291.

Boere, G. C., C. S. Roselaar & M. Engelmoer. 1984. The breeding origins of Purple Sandpipers *Calidris maritima* present in The Netherlands. Ardea 72: 101-109.

∗ **Boertman, D., H. Meltofte & M. Forchhammer.** 1991. Population densities of birds in central Northeast Greenland. Dansk Orn. Foren. Tidsskr. 85: 151-160.

∗ **Bogorodskii, Y.V.** 1989. *Ptitsy yuzhnego Predbaikaliya*. Univ. of Irkutsk, Irkutsk, Russia.

∗ **Brazil, M.A.** 1991. *The Birds of Japan*. Croom Helm, London, U.K.

Breiehagen, T. 1989. Nesting biology and mating system in an alpine population of Temminck's Stint *Calidris temminckii*. Ibis 131: 389-402.

Brooks, D.R. & D.A. Mclennan. 1991. *Phylogeny, Ecology, and Behavior*. Univ. of Chicago Press, Chicago, U.S.A.

Browning, M.R. 1977. Geographical variation in Dunlins *Calidris alpina* of North America. Can. Field-Nat. 91: 391-393.

Browning, M.R. 1991. Taxonomic comments on the Dunlin *Calidris alpina* from northern Alaska and eastern Siberia. Bull. B.O.C. 111: 140-145.

Buckley, P.A. 1987. Mendelian genes. Pages 1-44 *in* F. Cooke & P.A. Buckley (eds.), *Avian Genetics: a population and ecological approach*. Academic Press, London, U.K.

Bub, H. 1958. Untersuchungen an einer Population des Flußregenpfeifers. Beitr. Vogelkunde 5: 268-283.

Bub, H. 1962. Planberingungen am Sandregenpfeifer. J. Orn. 103: 243-249.

Byrkjedal, I. 1978. Variation and secondary intergradation in SW Norwegian Golden Plover *Pluvialis apricaria* populations. Ornis Scand. 9: 101-110.

Cadbury, C.J. & P.J.S. Olney. 1978. Avocet population dynamics in England. Brit. Birds 71: 102-121.

Cairns, W.E. 1982. Biology and behaviour of breeding Piping Plovers. Wilson Bull. 94: 531-545.

Carroll, R.L. 1988. *Vertebrate Paleontology and Evolution*. Freeman & Co, New York, U.S.A.

Castro, G. & J.P. Myers. 1989. Flight range estimates for shorebirds. Auk 106: 474-476.

∗ **Cheng, Tso-hsin.** 1987. *A synopsis of the avifauna of China*. Paul Parey Sci. Verlag, Hamburg, Germany / Science Press, Beijing, China.

Clark, N.A. 1984. Ageing criteria for Dunlins. W.S.G. Bull. 42: 39.

Colwell, M.A. & L.W. Oring. 1989. Return rates of prairie shorebirds: sex and species differences. W.S.G. Bull. 55: 21-24.

Connors, P.G. 1983. Taxonomy, distribution, and evolution of Golden Plovers (*Pluvialis dominica* and *Pluvialis fulva*). Auk 100: 607-620.

Connors, P.G., B.J. Mccaffery & J.L. Maron. 1993. Speciation in Golden Plovers, *Pluvialis dominica* and *P. fulva*: Evidence from the breeding grounds. Auk 110: 9-20.

Cramp, S. & K.E.L. Simmons (eds.). 1977. *The birds of the western Palearctic*. Vol. I. Oxford Univ. Press, Oxford.

Cramp, S. & K.E.L. Simmons (eds.). 1983. *The birds of the western Palearctic*. Vol. III. Oxford Univ. Press, Oxford.

Danilenko, A.K., G.C. Boere & E.A. Lebedeva. 1996. Looking for the recent breeding grounds of Slender-billed Curlew: a habitat-based approach. WSG-Bull. 81: 71-78.

∗ **Danilov, N.N.,V.N. Ryzhanovskii & V.K. Ryabitsev.** 1984. *Ptitsy Yamala*. Nauka, Moscow, Russia.

Davidson, N.C. & J.R. Wilson. 1992. The migration system of European-wintering Knots *Calidris canutus islandica*. W.S.G. Bull. 64 Suppl.: 39-52.

Deignan, H. G. 1961. Type specimens of birds in the United States National Museum. Bull. US Natl. Mus. 221: 1-718.

Dement'ev, G. 1939. On the Palæarctic Forms of the Grey Plover (*Squatarola squatarola* (L.)). Ibis 14: 352-354.

* **Dement'ev, G.P. & N.A. Gladkov (eds.)**. 1951. *Ptitsy Sovetskogo Soyuza 3*. Sovetskaya Nauka, Moscow, Russia.

Dement'ev, G.P., N.A. Gladkov & E.P. Spangenberg. 1969. *Birds of the Soviet Union*. Vol. III. Isr. Prog. for Scient. Transl., Jerusalem, Israel.

Denton, G.H. & T.J. Hughes. 1981. *The last great ice sheets*. Wiley, New York, U.S.A.

Dick W.J.A. (ed.). 1975. Oxford and Cambridge Mauritanian Expedition 1973-report. Cambridge, U.K.

Dick, W.J.A., M.W. Pienkowski, M. Waltner & C.D.T. Minton. 1976. Distribution and geographical origins of Knots *Calidris canutus*, wintering in Europe and Africa. Ardea 64: 22-47.

Dijk, A.J. van, K. van Dijk, L.J. Dijksen, T.M. van Spanje & E. Wymenga. 1986. Wintering waders and waterfowl in the Gulf of Gabes, Tunisia, January-March 1984. WIWO report 11. WIWO foundation, Zeist, The Netherlands.

* **Dolgushin, I.A. (ed.)**. 1962. *Ptitsy Kazakhstana 2*. Akad. Nauk Kazakhskoi SSR, Alma-Ata, Kazakhstan.

* **Dybbro, T**. 1976. *De Danske ynglefugles udbredelse*. Dansk Orn. Forening, Copenhagen, Denmark.

Dyke A.S. & T.F. Morris. 1990. Postglacial history of the Bowhead Whale and of driftwood penetration; implications for Paleoclimate, Central Canadian Arctic. Geol. Survey Canada paper 89-24, Canada.

Drent, R.H. & T. Piersma. 1990. An exploration of the energetics of leap-frog migration in arctic breeding waders. P. 399-412 *in*: E. Gwinner (ed.). *Bird migration: physiology and ecophysiology*. Springer Verlag, Berlin, Germany.

Ebbinge, B.S. & G.C. Boere. 1991. Verslag van de Nederlandse deelname aan de internationale Taimyr-expeditie 1990. RIN-report, Arnhem, The Netherlands.

Eldredge, N. 1989. *Macroevolutionary Dynamics: species, niches and adaptive peaks*. McGraw-Hill, New York, U.S.A.

Elliott, C.C.H., M. Waltner, L.G. Underhill, J.S. Pringle & W.J.A. Dick. 1976. The migration system of the Curlew Sandpiper *Calidris ferruginea* in Africa. Ostrich 47: 191-213.

Engelmoer, M. 1993. Maten en (on)zekerheden: om tureluurs van te worden? Op het Vinketouw 71: 16-24.

Engelmoer, M. 1995. The application of discriminant analysis to the morphometrics of waders. Pages 475-490 *in* E.J.M. Hagemeyer & T.J. Verstrael (eds.). *Bird Numbers 1992. Distribution, monitoring and ecological aspects*. Proc. 12th Int. Conf. IBCC & EOAC, Noordwijkerhout, The Netherlands. Statistics Netherlands, Voorburg/ Heerlen & SOVON, Beek-Ubbergen, The Netherlands.

Engelmoer, M., C.S. Roselaar, G.C. Boere & E. Nieboer. 1983. Post-mortem changes in measurements of some waders. Ring. & Migr. 4: 245-249.

Engelmoer, M., E. Nieboer & G.C. Boere. in prep. The composition of wader populations in the Dutch part of the Wadden Sea. Ardea.

Ens, B.J., T. Piersma, W.J. Wolff & L. Zwarts (eds.). 1988. Report of the Dutch-Mauritanian project Banc d'Arguin 1985-1986. Joint report PNBA/WIWO/RIN/NIOZ, Texel, The Netherlands.

Ens, B.J., T. Piersma, W.J. Wolff & L. Zwarts (eds.). 1990. Homeward bound: problems waders face when migrating from the Banc d'Arguin to their northern breeding grounds in spring. Ardea 78: 1-364.

Ens, B.J., T. Piersma & J.M. Tinbergen. 1994. Towards predictive models of bird migration schedules: theoretical and empirical bottlenecks. NIOZ-Report 1994-5, Texel, The Netherlands.

Ens, B.J., F.J. Weissing & R.H. Drent. 1995. The despotic distribution and deferred maturity: two sides of the same coin. Am. Natur. 146: 625-650.

Evans, P.R. 1991. Seasonal and annual patterns of mortality in migratory waders: some conservation implications. Pages 346-359 *in* C. M. Perrins, J.D. Lebreton & G.J.M. Hirons (eds.). *Bird population studies: relevance to conservation and management*. Oxford Univ. Press, Oxford, U.K.

Fabricius, O. & P. Hald-mortensen. 1969. The Golden Plover as a breeding bird in Denmark 1963-66, with some notes on the races, distribution and number. Dansk Orn. Foren. Tidsskr. 63: 137-160.

Fairbairn J. & R. Shine. 1993. Patterns of sexual size dimorhism in seabirds of the Southern Hemisphere. Oikos 68: 139-145.

Ferns, P.N. & G.H. Green. 1979. Observations on the breeding plumage and prenuptial moult of Dunlins, *Calidris alpina*, captured in Britain. Gerfaut 69: 286-303.

Fjeldså, J. 1983. Ecological character displacement and character release in grebes Podicipedidae. Ibis 125: 463-481.

Flint, R.F. 1947. *Glacial geology*. Wiley, New York, U.S.A.

* **Flint, V.E. (ed.).** 1980. Novoe v izuchenii biologii i rasprostranenii Kulikov. Proc. Conf. 'Faune i ekologii Kulikov, 5-7 February 1979'. Nauka, Moscow, Russia.

* **Flint, V.E. & P.S. Tomkovich (eds.)**. 1988. Kuliki v SSSR: rasprostranenie, biologiya i okhrana. Proc. Conf. 'Rasprostranenie, biologiya i okhrana Kulikov, 29-30 October 1987'. Nauka, Moscow, Russia.

Flint, V.E., R.L. Böhme, Y.V. Kostin & A.A. Kuznetsov. 1984. *A field guide to birds of the USSR.* Princeton Univ. Press, Princeton, U.S.A.

Frenzel, B. 1968. *Grundzüge der Pleistozänen Vegetationsgeschichte Nord- Eurasiens.* Steiner Verlag, Wiesbaden, Germany.

Frenzel, B. & C. Troll. 1952. Die Vegetationszonen Nord-Eurasiens während der letzten Eiszeit. Eiszeitalter Gegen. 2: 154-167.

Frenzel, B., M. Pecsi & A.A. Velichko. 1992. *Atlas of Paleoclimates and Paleoenvironments of the northern Hemisphere: Late Pleistocene - Holocene.* Gustav Fisher Verlag, Stuttgart, Germany.

* **Frimer, O**. 1993. Breeding and summering of North American waders on Qaamassoq, Disko, West Greenland. Dansk Orn. Foren. Tidsskr. 87: 255-257.

* **Gabrielson, I.N. & F.C. Lincoln**. 1959. *The birds of Alaska.* Stackpole Comp., Harrisburg, Pennsylvania, U.S.A.

* **Gibbons, D.W., J.B. Reid & R.A. Chapman** (eds.). 1993. *The new atlas of breeding birds in Britain and Ireland: 1988-1991.* B.T.O., London, U.K.

Gibson, D.D. & B. Kessel. 1997. Inventory of the species and subspecies of Alaska birds. Western Birds 28: 1-32.

Ginn, H.B. & D.S. Melville. 1983. *Moult in birds.* BTO Guide 19, B.T.O., Tring, U.K.

* **Gladkov, N.A.** 1941. Beitrag zur Studium der Vögel der Timan-Tundra. J. Orn. 89: 124-156.

Glutz von Blotzheim, U.N., K.M. Bauer & E. Bezzel (ed.). 1975. *Handbuch der Vögel Mitteleuropas. Band 6 - Charadriiformes (1. Teil).* Akademisches Verlagsgesellschaft, Wiesbaden, Germany.

Glutz von Blotzheim, U.N., K.M. Bauer & E. Bezzel (ed.). 1977. *Handbuch der Vögel Mitteleuropas. Band 7 - Charadriiformes (2. Teil).* Akademisches Verlagsgesellschaft, Wiesbaden, Germany.

* **Godfrey, W.E.** 1986. *The birds of Canada.* National Museums of Canada, Ottawa, Canada.

Godfrey, W.E. 1992. Subspecies of the Red Knot *Calidris canutus* in the extreme northwestern Canadian arctic islands. W.S.G. Bull. 64, Suppl.: 24-25.

Goede, A.A., E. Nieboer & P.M. Zegers. 1990. Body mass increase, migration pattern and breeding grounds of Dunlins, *Calidris a. alpina*, staging in the Dutch Wadden Sea in spring. Ardea 78: 135-144.

* **Gorbunov, G.B.** 1929. Materialy po faune miekopitayushchikh i ptits Novoi Zemli. Trud. Inst. Izucheniya Severa ntu vsnkh SSSR 40: 169-239.

Goss-Custard, J.D., S.E.A. le V. dit Durell, R.T. Clarke, A.J. Beintema, R.G.W. Caldow, P.L. Meininger & C.J. Smit. 1996. Population dynamics: predicting the consequences of habitat change at the continental scale. Pages 352-383 *in*: J.D. Goss-Custard (ed.). *The Oystercatcher: from individuals to populations.* Oxford Univ. Press, Oxford, U.K.

Grant, M.C. 1991. Nesting densities, productivity and survival of breeding Whimbrel *Numenius phaeopus* in Shetland. Bird Study 38: 160-179.

Grant, P.R. 1992. Systematics and micro-evolution. Bull. B.O.C. Centenary Suppl. 112A: 97-106.

Grant, P.R. & B.R. Grant. 1992. Hybridization of bird species. Science 256: 193-197.

Gratto, C.L. 1988. Natal philopatry, site tenacity, and age of first breeding of the Semipalmated Sandpiper. Wilson Bull. 100: 660-663.

Gratto, C.L., R.I.G. Morrison & F. Cooke. 1985. Philopatry, site tenacity, and mate fidelity in the Semipalmated Sandpiper. Auk 102: 16-24.

Graul, W.D. 1975. Breeding biology of the Mountain Plover. Wilson Bull. 87: 6-31.

Green, G.H. 1978. Measurements of waders. Pages 56-60 *in* G.H. Green & J.J.D. Greenwood (eds.). Joint biological expedition to North East Greenland 1974 Univ. of Dundee, Dundee, U.K.

Greenwood, J.G. 1986. Geographical variation and taxonomy of the Dunlin *Calidris alpina* (L.). Bull. B.O.C. 106: 43-56.

Gretton, A. 1991. *Conservation of the Slender-billed Curlew.* ICBP Monograph 6. I.C.B.P., Cambridge, U.K.

Groen, N.M. 1993. Breeding site tenacity and natal philopatry in the Black-tailed Godwit *Limosa l. limosa.* Ardea 81: 107-114.

Gromadzka, J. & B. Przystupa. 1984. Problems with the ageing of Dunlins in autumn. W.S.G. Bull. 41: 19-20.

Gromadzka, J. 1989. Breeding and wintering areas of Dunlin migrating through southern Baltic. Ornis Scand. 20: 132-145.

* **Haftorn, S**. 1971. *Norges Fugler.* Scandinavian Univ. Books, Oslo, Norway.

Haig, S.M. & L.W. Oring. 1988a. Mate, site and territory fidelity in Piping Plovers. Auk 105: 268-277.

Haig, S.M. & L.W. Oring. 1988b. Genetic differentiation of Piping Plovers across North America. Auk 105: 260-267.

Haig, S.M. & L.W. Oring. 1988c. Distribution and dispersal in the Piping Plover. Auk 105: 630-638.

Hailman, J.P. 1986. The heritability concept applied to wild birds. Pages 71-95 *in* R.F. Johnston (ed.). *Current Ornithology*, vol. 4. Plenum Press, New York, U.S.A.

Hale, W.G. 1971. A revision of the taxonomy of the Redshank *Tringa totanus.* Zool. J. Linn. Soc. 50: 199-268.

Hale, W.G. 1973. The distribution of the Redshank, *Tringa totanus*, in the winter range. Zool. J. Linn. Soc. 53: 177-236.

Hale, W.G. 1980. *Waders.* The New Naturalist's Series. Collins, London, U.K.

Hale, W. G. 1984. The changing face of European wintering areas. Pages 311-323 *in* P.R. Evans, J.D. Goss-Custard & W.G. Hale (eds.). *Coastal waders and wildfowl in winter.* Cambridge Univ. Press, Cambridge, U.K.

Hartert, E. 1912-21. *Die Vögel der paläarktischen Fauna.* Berlin, Germany.

Hartert E. & F. Steinbacher. 1932-38. Die Vögel der Paläarktischen Fauna. Ergänzungsband. Friedländer & Sohn, Berlin, Germany.

Harrington, B.A., J.M. Hagan & L.E. Leddy. 1988. Site fidelity and survival differences between two groups of New World Red Knots (*Calidris canutus*). Auk 105: 439-445.

Hayman, P., J. Marchant & T. Prater. 1986. *Shorebirds. An identification guide to the shorebirds of the world.* Croom Helm, London, U.K.

Heldt, R. 1966. Zur Brutbiologie des Alpenstrandläufers, *Calidris alpina schinzii*. Corax 1: 173-188.

Hellmayr, C. E. & B. Conover. 1948. Catalogue of birds of the Americas and the adjacent islands. Part 1, nr. 3. Zool. Ser. Field Mus. Nat. Hist. 13: 1-383.

Higgins, P.J. & S.J.J.F. Davies (eds). 1996. Handbook of Australian, New Zealand & Antarctic Birds. Vol. 3, Snipe to Pigeons. Oxford University Press, Melbourne, Australia.

Hildén, O. 1975. Breeding system of Temminck's Stint *Calidris temminckii*. Ornis Fenn. 52: 117-146.

Hildén, O. 1978. Population dynamics in Temminck's Stint *Calidris temminckii*. Oikos 30: 17-28.

Hildén, O. 1979. Territoriality and site tenacity of Temminck's Stint *Calidris temminckii*. Ornis Fenn. 56: 56-74.

Holland, P.K. & D.W. Yalden. 1991. Population dynamics of Common Sandpipers *Actitis hypoleucos* breeding along an upland river system. Bird Study 38: 151-159.

Holmes, R.T. 1966. Breeding ecology and annual cycle adaptations of the Red-backed Sandpiper *Calidris alpina* in northern Alaska. Condor 68: 3-46.

Holmes, R.T. 1971. Density, habitat and the mating system of the Western Sandpiper (*Calidris mauri*). Oecologia (Berlin) 7: 191-208.

Hoyo, J. del, A. Elliott, J. Sargatal (eds.). 1996. *Handbook of the Birds of the World*. Vol. 3: Hoatzin to Auks. Lynx Edicions, Barcelona.

Horvath, L. & A. Keve. 1956. The subspecies of Dunlin in Hungary. Ann. Hist. Nat. Mus. Hungary N.S. 7: 469-476.

Howe, M.A. 1982. Social organization in a nesting population of eastern Willets (*Catoptrophorus semipalmatus*). Auk 99: 88-102.

Hulscher, J.B. 1985. Growth and abrasion of the Oystercatcher bill in relation to dietary switches. Neth. J. Zool. 35: 124-154.

Hulscher, J.B. 1989. Sterfte en overleving van Scholeksters *Haematopus ostralegus* bij strenge vorst. Limosa 62: 177-183.

∗ **Hyytiä, K., E. Kellomäki & J. Koistinen** (eds.). 1983. *Suomen Lintuatlas*. SLY:n Lintutieto Oy, Helsinki, Finland.

Impe, J. van. 1991. Overleving, sterfte en trek van in België geringde jonge kluten (*Recurvirostra avosetta*). Giervalk 81: 217-243.

Innes, J.L. (ed.) 1980. Cambridge Norwegian Expedition 1978 Report. Univ. of Cambridge, Cambridge, U.K.

Jackson, D.B. Breeding dispersal and site-fidelity in three monogamous wader species in the Western Isles, U.K. Ibis 136: 463-473.

Jehl, J.R. 1973. Breeding biology and systematic relationships of the Stilt Sandpiper. Wilson Bull. 85: 115-147.

Jenni, L. & R. Winkler. 1989. The feather length of small passerines: a measurement for wing length in live birds and museum skins. Bird Study 36: 1-15.

Johansen, H. 1956. Revision und Entstehung der arktischen Vogelfauna. Vol. 1. Soc. Arctica Scan. - Acta arctica 8.

∗ **Johansen, H**. 1960. Die Vogelfauna Westsiberiens 10, Limicolae 1. J. Orn. 101: 472-495.

∗ **Johansen, H**. 1961a. Die Vogelfauna Westsiberiens 11, Limicolae 2. J. Orn. 102: 41-67.

∗ **Johansen, H**. 1961b. Revised list of the birds of the Commander Islands. Auk 78: 44-56.

Johnson, R.A. & D.W. Wichern. 1988. *Applied multivariate statistical analysis*. Prentice Hall, New Jersey, U.S.A.

Jonas, R. 1979. Brutbiologische Untersuchungen an einer Population der Uferschnepfe *Limosa limosa*. Vogelwelt 100: 125-136.

Jönsson, P.E. 1988. Ecology of the southern Dunlin *Calidris alpina schinzii*. Ph. D. thesis Univ. of Lund, Sweden.

Jönsson, P.E. 1991. Reproduction and survival in a declining population of the southern Dunlin *Calidris alpina schinzii*. W.S.G. Bull. 61 Supplement: 56-68.

Jukema, J. 1982. Rui en biometrie van de Goudplevier *Pluvialis apricaria*. Limosa 55: 79-84.

Jukema, J. 1989. Verenkleedkenmerken van noordelijke en zuidelijke Gouplevieren *Pluvialis apricaria* op doortrek door Nederland. Limosa 62: 147-152.

Jukema, J. & T. Piersma. 1987. Special moult of breast and belly feathers during breeding in Golden Plovers *Pluvialis apricaria*. Ornis Scand. 18: 157-162.

Jukema, J. & T. Piersma. 1992. Mannelijke en vrouwelijke Goudplevieren *Pluvialis apricaria* zijn wat betreft hun uitwendige en inwendige afmetingen volkomen identiek. Limosa 65: 147-154.

Kersten, M. 1997. Living leisurely should last longer: energetic aspects of reproduction in the Oystercatcher. Ph. D. thesis, Univ. Of Groningen, The Netherlands.

∗ **Kessel, B. & D.D. Gibson**. 1978. *Status and distribution of Alaska birds*. Studies in Avian Biology no. 1. Cooper Orn. Soc., Los Angeles, U.S.A.

∗ **Kishchinskii, A.A.** 1980. *Ptitsy Koryakskogo Nagor'ya*. Nauka, Moscow, Russia.

∗ **Kishchinskii, A.A.** 1988. *Ornitofauna severo-vostoka Azii*. Nauka, Moscow, Russia.

∗ **Kishchinskii, A.A., P.S. Tomkovich & V.E. Flint**. 1983. Ptitsy basseina Kanchalan (Chukotskii Natsional'nii Okrug). Pages 3-76 *in* V. E. Flint & P. S. Tomkovich (eds.). *Rasprostranenie i sistematika ptits*. Univ. of Moscow, Moscow, Russia.

Kleinschmidt, O. 1900. Arten oder 'Formenkreise'? J. Orn. 48: 134-139.

Knox, A. 1987. Taxonomic status of 'Lesser Golden Plovers'. Brit. Birds 80: 482-487.

∗ **Kondratiev, A.Y.** 1982. *Biologiya Kulikov v tundrakh severno-vostoka Azii*. Akad. Nauk SSSR, Moscow, Russia.

Korte, J. de. 1975. Golden Plover *Pluvialis apricaria* breeding in Jameson Land, East Greenland. Dansk Orn. Foren. Tidsskr. 69: 129-134.

Korte, J. de, C.A. Bosman & H. Meltofte. 1981. Observations on waders (Charadriidae) at Scoresby Sund, East Greenland. Medd. Grønland, Bioscience 7: 4-21.

∗ **Kozlova, E.V.** 1932. The birds of South-West Transbaicalia, northern Mongolia and central Gobi. Ibis 13: 567-596.

Kozlova, E.V. 1962. *Charadriiformes, suborder Charadrii. Fauna SSSR, Pticy, Vol. 2,* Nauka Press, Moscow, Russia.

∗ **Krechmar, A.V., A.V. Andreev & A.Y. Kondratiev**. 1991. *Ptitsy severnykh ravnin*. Nauka, Leningrad, Russia.

∗ **Kumari, E**. 1977. *Der Regenbrachvogel*. Neue Brehm-Bücherei. Ziemsen Verlag, Wittenberg Lutherstadt, Germany.

∗ **Larionov, G.P., V.G. Degtyarev & A.G. Larionov**. 1991. *Ptitsy Leno-Amginskogo Mezhdyrech'ya*. Nauka Press, Novosibirsk, Russia.

Larson, S. 1957. The suborder Charadrii in arctic and boreal areas during the Tertiary and Pleistocene. Acta Vert. 1: 1-84.

Laven, H. 1940. Beiträge zur Biologie des Sandregenpfeifers. J. Orn. 88: 183-287.

Lenington, S. & T. Mace. 1975. Mate fidelity and nesting site tenacity in the Killdeer. Auk 92: 149-151.

Low, G.C. 1938. The supposed races of the Grey- or Black-bellied Plover (*Squatarola squatarola*). Ibis 14: 154-158.

Løvenskiold, H.L. 1950. Den geografiske variasjon hos Fjæreplytten. Dansk. Orn. Foren. Tidsskr. 44: 161-167.

∗ **Løvenskiold, H.L.** 1964. Avifauna Svalbardensis. Norsk Polarinstitut, Oslo, Norway.

∗ **Mal'chevski, A.V. & Y.B. Pukinski.** 1983. Ptitsy Leningradskoy oblasti i sopredel'nykh territoriy 1. Leningrad, Russia.

Macivor, L.H., C.R. Griffin & S. Melvin. 1987. Management, habitat selection and population dynamics of Piping Plovers on outer Cape Cod, Massachusetts. Massachusetts, U.S.A.

Manning, T.H. 1981. Birds of the Twin Islands, James Bay, N.W.T., Canada. Syllogeus 30: 1-50. National Museum of Natural Sciences, Ottawa, Canada.

Mathews, G.M. 1935. A new name for the British Redshank. Brit. Birds 29: 152.

Maron, J.L. & J.P. Myers. 1984. A description and evaluation of two techniques for sexing wintering Sanderlings. J. Field Ornithol. 55: 336-342.

Mayer, P.M. & M.R. Ryan. 1986. Population ecology of the Piping Plover in the northern Great Plains. Cyclostyled report.

Mayr, E. 1942. *Systematics and the origin of species*. Columbia Univ. Press, New York, U.S.A.

Mayr, E. 1951. Speciation in birds. Pages 91-131 *in* Acta X Congressus Internationalis Ornithologici, Uppsala, Sweden.

Mayr, E. 1963. *Animal Species and Evolution*. Harvard Univ. Press, Cambridge, U.S.A.

Mayr, E. 1982. Of what use are subspecies? Auk 99: 593-595.

Meinertzhagen, R. 1947. Some notes and comments on sixteen British Species. Bull. B.O.C. LXVIII: 33.

Meininger, P.L. & N.D. van Swelm. 1989. Biometrisch en ringonderzoek aan steltlopers in de Oosterschelde in het voorjaar van 1984 en 1985. Rijkswaterstaat, Middelburg, The Netherlands.

* **Meltofte, H**. 1976. Ornithological observations in southern Peary Land, North Greenland. Medd. Grønland 205: 1-57.

* **Meltofte, H.** 1979. The population of waders, Charadriidae, at Danmarks Havn, North-East Greenland, 1975. Dansk Orn. Foren. Tidsskr. 73: 69-94.

Meltofte, H. 1985. Populations and breeding schedules of waders Charadriidae in high arctic Greenland. Medd. Grønland, Bioscience 16: 2-43.

Meltofte, H. 1993. Vadefugletrækket gennem Danmark. Dansk Orn. Foren. Tidsskr. 87: 1-180.

Meltofte, H. 1996. Are African wintering waders really forced south by competition from northerly wintering conspecifics? Benefits and constraints of northern versus southern wintering and breeding in waders. Ardea 84: 31-44.

* **Meltofte, H., M. Elander & C. Hjort**. 1981. Ornithological observations in North-East Greenland between 74°30' and 76°00' N.lat., 1976. Medd. Grønland, Bioscience: 1-56.

Meltofte, H., J. Blew, J. Frikke, H.-U. Rösner & C.J. Smit. 1994. Numbers and distribution of waterbirds in the Wadden Sea. IWRB-Publ. 34, / W.S.G. Bull. 74, Special Issue.

* **Meschini, E. & S. Frugis (eds.).** 1993. Atlante degli uccelli nidificante in Italia. Suppl. Ric. Biol. Selvaggina 20: 1-344.

Meyer, T.H.M. 1988. Reproductive decisions in the Kestrel *Falco tinnunculus*. Ph. D. thesis. Univ. of Groningen, Groningen, The Netherlands.

Miller, E. 1977. Breeding biology of the Least Sandpiper, *Calidris minutilla* (Vieill.), on Sable Island, Nova Scotia. Ph. D. thesis, Dahlousie Univ. , Nova Scotia, U.S.A.

Miltenburg, H. 1990. De aanwezigheid van een witte vaan op de vierde slagpen van Bonte Strandlopers in de Waddenzee en de bruikbaarheid voor het traceren van de geografische herkomst. Free Univ., Amsterdam, The Netherlands.

Moreau, R.E. 1972. *The Palaearctic-African bird migration systems*. Academic Press, London, U.K.

Morrison, R.I.G. & B.A. Harrington. 1992. The migration system of the Red Knot *Calidris canutus rufa* in the New World. W.S.G. Bull. 64 Suppl.: 71-84.

Murton, R.K. & N.G. Westwood. 1977. *Avian breeding cycles*. Claredon Press, Oxford, U.K.

NAG. 1991. The NAG Fortran Library Introductory Guide. Mark 15. NAG, Oxford, U.K..

∗ **Nechaev, V.F.** 1969. *Ptitsy yunznykh Kuril'skikh Ostrovov*. Nauka Press, Leningrad, Russia.

∗ **Nechaev, V.F.** 1992. *Ptitsy Ostrov Sakhalin*. Nauka Press, Leningrad, Russia.

Nechaev, V.A. & P.S. Tomkovich. 1987. Noviy podvid Chyernozobika - *Calidris alpina litoralis* ssp. n. (Aves, Charadriidae) s Sachalina. Zool. Zhurn. 66: 1110-1113.

Nechaev, V.A. & P.S. Tomkovich. 1988. [A new name for Sakhalin Dunlin (Aves, Charadriidae).] In Russian. Zool. Zhurn. 67: 1596.

Neumann, O. 1929. Ueber Rassen des Großen Brachvogels. Ornith. Monatsber. 37: 76-78.

∗ **Nicolai, L**. 1990. *Atlas der Brutvögel Ostdeutschlands*. Gustav Fisher Verlag, Jena, Germany.

Nicoll, M., R.W. Summers, L.G. Underhill, K. Brockie & R. Rae. 1988. Regional, seasonal and annual variations in the structure of Purple Sandpiper *Calidris maritima* populations in Britain. Ibis 130: 221-234.

Nicoll, M., R. Rae, R.W. Summers, K.-B. Strann & K. Brockie. 1991. The biometrics of Norwegian and Svalbard Purple Sandpipers *Calidris maritima*. Ring. & Migr. 12: 67-71.

Nieboer, E. 1972. Preliminary notes on the primary moult in Dunlins *Calidris alpina*. Ardea 60: 112-119.

Nieboer, E., J. Cronau, R. de Goede, J. Letschert & T. van der Have. 1985. Axillary feathers colour patterns as indicators of the breeding origin of Bar-tailed Godwits. W.S.G. Bull. 45: 34.

Nisbet, I.C.T., J. Baird, D.V. Howard & K.S. Anderson. 1970. Statistical comparison of wing lengths measured by four observers. Bird-banding, 41: 307-308.

Norušis, M.J. 1992a. *SPSS/PC+™ Base System User's Guide*. Version 5.0. SPSS Inc. Chicago, U.S.A.

Norušis, M. J. 1992b. *SPSS/PC+™ Professional Statistics™*. Version 5.0. SPSS Inc. Chicago, U.S.A.

Norušis, M.J. 1992c. *SPSS/PC+™ Advanced Statistics™*. Version 5.0. SPSS Inc. Chicago, U.S.A.

Olson, S.L. 1988. The fossil record of birds. Pages 79-238 *in* D.S. Farner, J.R. King & K.C. Parkes (eds.). *Avian Biology*, vol. 8. Academic Press, New York, U.S.A.

Parish, D., P. Thompson & J. Coulson. 1996. The breeding biology of the Lapwing *Vanellus vanellus*: the incidence of polygamy, double-brooding and polyterritoriality. W.S.G. Bull. 79: 27.

Parr, R. 1980. Population study of Golden Plover *Pluvialis apricaria* using marked birds. Ornis Scand. 11: 179-189.

Paulson, D. 1993. *Shorebirds of the Pacific Northwest*. UBC Press, Vancouver, Canada.

Payne, R. 1972. Mechanisms and control of molt. Pages 104-157 *in* D.S. Farner & J.R. King (eds.). *Avian Biology*, vol. 2. Academic Press, New York, U.S.A.

Payne, R.B. 1984. Sexual selection, lek and arena behavior, and sexual size dimorphism in birds. Ornithological Monographs 33, A.O.U., Washington D.C., U.S.A.

Payne, R.B. 1990. Natal dispersal, area effects and effective population size. J. Field Orn. 61: 396-403.

Peters, J.L. 1934. *Check-list of birds of the world 2*. Cambridge, Massachusetts, U.S.A.

∗ **Piechocki, R**. 1968. Beiträge zur Avifauna der Mongolei. Teil 1, Non-Passeriformes. Mitteil. Zool. Mus. Berlin 44, 149-292.

Pielou, E.C. 1991.*After the Ice Age: the return of life to to glaciated North America*. Univ. Chicago Press, Chicago, U.S.A.

Pienkowski, M.W. 1984. Breeding biology and population dynamics of Ringed Plovers *Charadrius hiaticula* in Britain and Greenland: nest predation as a possible factor limiting distribution and timing of breeding. J. Zool. Lond. 202: 83-114.

Piersma, T. 1994. Close to the edge: energetic bottlenecks and the evolution of migratory pathways in Knots. Ph. D. thesis, Univ. of Groningen, Groningen, The Netherlands.

Piersma, T. & J. Jukema. 1990. Budgeting the flight of a long-distance migrant: change in the nutrient reserve levels of Bar-tailed Godwits at successive spring staging sites. Ardea 78: 315-337.

Piersma, T., M. Klaassen, J.H. Bruggeman, A.-M. Blomert, A. Gueye, Y. Ntiamoa-baidu & N.E. van Brederode. 1990. Seasonal timing of the spring departure of waders from the Banc d' Arguin, Mauritania. Ardea 78: 123-135.

Piersma, T., P. Prokosch & D. Bredin. 1992. The migration system of Afro-Siberian Knots *Calidris canutus canutus*. W.S.G. Bull. 64 Suppl.: 52-64.

Pike, E. 1985. The Piping Plover at Waugoshance Point. Jack Pine Warbler 63: 36-41.

Ploeger, P.L. 1968. Geographical differentiation in arctic Anatidae as a result of isolation during the last Glacial. Ardea 56: 1-159.

Portenko, L. 1936. The Bar-tailed Godwit and its races. Auk 53: 194-197.

* **Portenko, L.A.** 1972. Ptitsy Chukotskogo Poluostrova i Ostrova Vrangelya 1. Nauka, Leningrad, Russia.

Prager, E.M., A.H. Brush, R.A. Nolan, M. Nakanishi & A.C. Wilson. 1974. Slow evolution of transferrin and albumin in birds according to micro-complement fixation analysis. J. Mol. Evol. 3: 243-262.

Prager, E.M. & A.C. Wilson. 1975. Slow evolutionary loss of the potential for interspecific hybridization in birds: a manifestation of slow regulatory evolution. Proc. Natl. Acad. Sci. USA 72: 200-204.

Prater, A.J. 1974. The population and migration of Knot in Europe. Pages 99-113, Proc. I.W.R.B. symposium 1973, Warsaw, Poland. I.W.R.B., Slimbridge, U.K.

Prater, A.J. 1981. *Estuary birds of Britain and Ireland*. Poyser, Calton, U.K.

Prater, A.J., J.H. Marchant & J. Vuorinen. 1977. *Guide to the identification and ageing of Holarctic Waders*. B.T.O. guide 17. B.T.O., Thetford, U.K.

Prokosch, P. 1988. Das Schleswig-Holsteinische Wattenmeer als Frühjahrs-Aufenthaltsgebiet arktischer Watvogel-Populationen am Beispiel von Kiebitzregenpfeifer (*Pluvialis squatarola*, L. 1758), Knutt (*Calidris canutus*, L. 1758) und Pfuhlschnepfe (*Limosa lapponica*, L. 1758). Corax 12: 274-442.

Prokosch, P. & H. Hötker (eds.) 1995. Faunistik und Naturschutz auf Taimyr-Expeditionen 1989-1991. Corax 16, Sonderheft.

Quinn, T.W. 1992. The genetic legacy of mother goose - phylogeographic patterns of Lesser Snow Goose *Chen caerulescens caerulescens* maternal lineages. Mol. Ecol. 1: 105-117.

Rae, R., M. Nicoll & R.W. Summers. 1986. The distribution of Hardangervidda Purple Sandpipers outwith the breeding season. Scottish Birds 14: 68-73.

Redmond, R.L. & D.A. Jenni. 1982. Natal philopatry and breeding area fidelity of Long-billed Curlews (*Numenius americanus*): patterns and evolutionary consequences. Behav. Ecol. Sociobiol. 10: 277-279.

Redmond, R.L. & D.A. Jenni. 1986. Population ecology of the Long-billed Curlew (*Numenius americanus*) in western Idaho. Auk 103: 755-767.

Reed, J.M. & L.W. Oring. 1993. Philopatry, site fidelity, dispersal and survival of Spotted Sandpipers. Auk 110: 541-551.

* **Rheinwald, G**. 1982. *Brutvogelatlas der Bundesrepublik Deutschland - Kartierung 1980.* Publ. Dachverband Deutschen Avifaunisten Nr. 6. Bonn, Germany.

Rhijn, J.G. van. 1991. *The Ruff.* T. & A.D. Poyser, London, U.K.

* **Risberg, L**. 1990. *Sveriges Fåglar.* Sveriges Ornitologiska Förening, Stockholm, Sweden.

Ridgway, R. 1919. The birds of North and Middle America 8. Bull. US. Natl. Mus. 50(8). Washington DC, U.S.A.

Rittinghaus, H. 1956. Untersuchungen am Seeregenpfeifer auf der Insel Oldeoog. J. Orn. 97: 117-155.

Robinson, J.A. & L.W. Oring. 1997. Natal and breeding dispersal in American Avocets. Auk 114: 416-430.

* **Rogacheva, H**. 1992. *The birds of Central Siberia*. Husum Druck- und Verlagsgesellschaft, Husum, Germany.

Rose, P.M. & D.A. Scott. 1994. *Waterfowl population estimates*. IWRB Publication 29. IWRB, Slimbridge, U.K.

Roselaar, C.S. 1983. Subspecies recognition in Knot *Calidris canutus* and occurrence of races in Western Europe. Beaufortia 33 (7): 97-109.

Ruddiman, W.F. & H.E. Wright (eds.). 1987. *North America and Adjacent Oceans During the Last Deglaciation*. The Geology of North America, vol. K-3, Boulder, U.S.A..

Safriel, U.N., M.P. Harris, M. de L. Brooke & C.K. Britton. 1984. Survival of breeding Oystercatchers *Haematopus ostralegus*. J. Anim. Ecol. 53: 867-877.

Salomonsen, F. 1930. Bemerkungen über die geographische Variation von *Charadrius hiaticula* (L.). J. Orn. 78: 65-72.

* **Salomonsen, F**. 1935. Aves, *in* Zoology of the Faroes 64: 1-269. Copenhagen, Denmark.

Salomonsen, F. 1947. Den geografiske variation hos Lille Regnspove (*Numenius phaeopus* (L.)) i Europa. Dansk Orn. Foren. Tidsskr. 41: 143-145.

Salomonsen, F. 1950. *The Birds of Greenland*. Munksgaard, Copenhagen, Denmark.

Salomonsen, F. 1954. The migration of the European Redshanks (*Tringa totanus* L.). Dansk Orn. Foren. Tidsskr. 48: 94-122.

* **Salomonsen, F**. 1967. *Fuglene på Grønland*. Rhodos, Copenhagen, Denmark.

* **Salomonsen, F.** 1981. *Grønlands Fauna*. Gyldendal, Copenhagen, Denmark.

Schamel, D. & D. Tracy. 1977. Polyandry, replacement clutches, and site tenacity in the Red Phalarope (*Phalaropus fulicarius*) at Barrow, Alaska. Bird-Banding 48: 314-324.

Schifter, H. 1992. Die Vögel aus dem Nachlass Ferdinand Lucas BAUER im Naturhistorischen Museum Wien und die darunter befindlichen Typen. Ann. Naturhist. Mus. Wien 93B, 41-57.

Schnakenwinkel, G. 1970. Studien an der Population des Austernfischers auf Mellum. Vogelwarte 25: 336-355.

* **Scott, D.A., H.M. Hamadani & A.A. Mirhosseyni**. 1975. [*The birds of Iran*]. In Iranese. Dept. of Environment, Tehran, Iran.

* **Semenov-tyan-shanskii, O.I. & A.S. Gilyazov**. 1991. *Ptitsy Laplandii*. Nauka Press, Moscow, Russia.

Shackleton, N.J. 1994. The climate system in the recent geological past. Pages 1-6 *in* J.R.L. Allen, B.J. Hoskins, B.W. Sellwood, R.A. Spicer & P.J. Valdes (eds.). *Palæoclimates and their modelling, with special reference to the Mesozoic era* Chapman & Hall, London, U.K.

Sheldon, F.H. & A.H. Bledsoe. 1993. Avian molecular systematics, 1970's to 1990's. Ann. Rev. Ecol. Syst. 24: 243-278.

Shields, G.F. & A.C. Wilson. 1987. Calibration of mitochondrial DNA evolution in geese. J. Mol. Evol. 24: 212-217.

Skeel, M.A. 1983. Nesting success, density, philopatry, and nest-site selection of the Whimbrel (*Numenius phaeopus*) in different habitats. Can. J. Zool. 61: 218-225.

Sluis, D.M. van der, W. Schaafsma & A.W. Ambergen. 1991. POSCON - User Manual: a decision support system in Diagnosis and Prognosis. Univ. of Groningen, Groningen, The Netherlands.

Smit, C.J. & T. Piersma. 1989. Numbers, midwinter distribution, and migration of wader populations using the East Atlantic Flyway. p.24-63 *in*: H. Boyd & J.-Y. Pirot (eds.). Flyways and reserve networks for water birds. IWRB Spec. Publ. 9, Slimbridge, U.K..

Smith, N.G. 1969. Polymorphism in Ringed Plovers. Ibis 111: 177-188.

Snow, D.W. 1970. *A guide to the moult in British birds.* BTO guide 11. B.T.O., Tring, U.K.

* **Snyder, L.L.** 1957. *Arctic birds of Canada.* Univ. of Toronto Press, Toronto, Canada.

Soikkeli, M. 1970a. Dispersal of Dunlin *Calidris alpina* in relation to sites of birth and breeding. Ornis Fenn. 47: 1-9.

Soikkeli, M. 1970b. Mortality and reproductive rates in a Finnish population of Dunlin *Calidris alpina.* Ornis Fenn. 47: 149-158.

Sordahl, T.A. 1984. Observations on breeding site fidelity and pair formation in American Avocets and Black-necked Stilts. N. Am. Bird Bander 9: 8-11.

Spiekman, H.W., G.O. Keijl & P.S. Ruiters. 1993. Waterbirds in the Kneiss area and other wetlands, Tunisia. WIWO-report 38. WIWO foundation, Zeist, The Netherlands.

* **Stastny, K., A. Randik & K. Hudec.** 1987. *Atlas huizdniho rozsireni ptaku v CSSR 1973/77.* Prague, Czechia.

* **Stegmann, B.** 1930. Die Vögel des dauro-mandschurischen Übergangsgebietes. J. Orn. 78: 389-471.

* **Stegmann, B.** 1936. Die Vögel des nördlichen Baikal. J. Orn. 84: 58-139.

Stepanyan, L.S. 1990. Konspekt Ornitologicheskoy Fauni CCCP. Nauka Press, Moscow, Russia.

* **Stishov, M.S., V.I. Pridatko & V.V. Baranyuk.** 1991. *Ptitsy Ostrova Vrangelya.* Nauka Press, Novosibirsk, Russia.

Stravinsky, S. 1974. An attempt to determine duck populations in transit by a biometrical method. Mater. Conf. Study Conserv. Migratory Birds Baltic Basin, Tallinn: 75-76.

Stravinsky, S. 1976. An attempt to determine duck populations in transit by a biometrical method. Pages 73-75 *in* E. Kumari (ed.). *Bird Migration.* Valgus, Tallinn, Estonia.

* **Sudilovskaya, A.M.** 1973. Ptitsy Tarimskogo basseina i okhruznyushchikh ego gor. Sbornik Trudy Zool. Mus. MGU 14: 50-154.

Summers, R. W. & P. A. Hockey. 1980. Breeding biology of the White-fronted Plover (*Charadrius marginatus*) in the south-western Cape, South Africa. J. Nat. Hist. 14: 433-445.

Summers, R.W., C.J. Corse & D.P. Whitfield. 1987. Purple Sandpiper Studies in North Iceland 1986. Report, Norfolk, U.K.

Summers, R.W., C.J. Corse, M. Nicoll, R. Smith & D.P. Whitfield. 1988a. The biometrics and wintering area of Icelandic Purple Sandpipers. Ring. & Migr. 9: 133-138.

Summers, R. W. , M. Nicoll, L. G. Underhill & A. Petersen. 1988b. Methods for estimating the proportions of Icelandic and British Redshanks *Tringa totanus* in mixed populations wintering on British coasts. Bird Study 35: 169-180.

Summers, R.W., L.G. Underhill, C.F. Clinning & M. Nicoll. 1989. Populations, migrations, biometrics and moult of the Turnstone *Arenaria i. interpres* on the East Atlantic coastline, with special reference to the Siberian population. Ardea 77: 145-168.

* **Suschkin, P.P.**, 1925. *Spisok i raspredelenie ptits russkogo Altaya.* Akad. Nauk SSSR, Moscow, Russia.

* **Suschkin, P.P.** 1938. *Ptitsy Sovetskogo Altaya i prilezhachshikh chastei severo-zapadnoi Mongolii.* Akad. Nauk SSSR, Moscow & Leningrad, Russia.

Sutton, G.M. & D.F. Parmelee. 1955. Breeding of the Semipalmated Plover on Baffin Island. Bird-Banding 26: 137-47.

Svensson, L. 1992. *Identification guide to European passerines.* Sweden.

Taylor, R.C. 1979. Geographical variation in the Ringed Plover *Charadrius hiaticula* L. Ph.D. thesis, Liverpool Polytechnic, Liverpool, U.K.

Teather K.L. & E. Nol. 1997. Mixed sexual dimorphism in Semipalmated Plovers. Condor 99: 803-806.

* **Texeira, R.M. (ed.).** 1979. *Atlas van de Nederlandse broedvogels.* Natuurmonumenten 's-Graveland, The Netherlands.

Thayer, H. & O. Bangs. 1914. Notes on the birds of the arctic coast of East Siberia. Proc. N. Engl. Zool. Club 9: 1-66.

Thomas, D.G. 1969. Breeding biology of the Australian Spur-winged Plover. Emu 69: 81-102.

Thomas, D.G. & A.J. Dartnall. 1970. Differences in size between the sexes of the Curlew Sandpiper. Emu 70: 89.

Thompson, P.S. & W.G. Hale. 1989. Breeding site fidelity and natal philopatry in the Redshank *Tringa totanus*. Ibis 131: 214-224.

Thompson, D.B.A., P.S. Thompson & D. Nethersole-Thompson. 1988. Fidelity and philopatry in breeding Redshanks (*Tringa totanus*) and Greenshanks (*Tringa nebularia*). Pages 563-574 *in* Acta XIX Congressus Internationalis Ornithologici (H. Ouellet, ed.). Ottawa Press, Ottawa, Canada.

Thompson, P.S., D. Baines, J.C. Coulson & G. Longrigg. 1994. Age at first breeding, philopatry and breeding site-fidelity in the Lapwing *Vanellus vanellus*. Ibis 136: 474-484.

Times. 1981. *The Times Atlas of the World. - Comprehensive 6th edition*. Bartholomew & Son, Edinburgh & Times Books, London, U.K.

∗ **Tomialojc, L**. 1990. Ptaki Polski. Panstwowe Wydawnictwo Naukowe, Warsaw, Poland.

∗ **Tomkovich, P.S.** 1985. K biologii morskogo pesochniyka na Zemlya Frantsa-Iosifa. Ornitologiya 20: 3-17.

∗ **Tomkovich, P.S.** 1986. Geograficheskai izmyenchivoste chernozobikov dalenyego vostoka. Biul. Moskv. 91: 3-15.

Tomkovich, P.S. 1990. analys Geografichesko izmenchivostii Yslandskogo Pesochnika *Calidris canutus* (L.). Bull. Mosk. Biol. 95/5: 59-72.

Tomkovich, P.S. 1992a. An analysis of the geographic variability in Knots *Calidris canutus* based on museum skins. W.S.G. Bull. 64, Suppl.: 17-23.

∗ **Tomkovich, P.S.** 1992b. Breeding range and population changes of waders in the former Soviet Union. Brit. Birds 85: 344-365.

∗ **Tomkovich, P.S. & A.G. Sorokin**. 1983. Fauna ptits vostochnoi Chukotki. Pages 77-159 *in* V.E. Flint & P.S. Tomkovich (eds.). *Rasprostranenie i sistematica ptits*. Univ. of Moscow, Moscow, Russia.

Tomkovich, P.S. & M.Y. Soloviev. 1996. Distribution, migrations and biometrics of Knots *Calidris canutus* on Taimyr, Siberia. Ardea 84: 85-98.

Underhill, L.G., M. Waltner & R.W. Summers. 1989. Three-year cycles in breeding productivity of Knots *Calidris canutus* wintering in southern Africa suggest Taimyr Peninsula provenance. Bird Study 36: 83-87.

Underhill, L.G., R.P. Prŷs-jones, E.E. Syroechkovski Jr., N.M. Groen, V. Karpov, H.G. Lappo, M.W.J. van Roomen, A. Rybkin, H. Schekkerman, H. Spiekman & R.W. Summers. 1993. Breeding of waders (Charadrii) and Brent Geese *Branta bernicla bernicla* at Pronchishcheva Lake, northeastern Taymyr, Russia, in a peak and a decreasing lemming year. Ibis 135: 277-292.

∗ **Uspenski, S.M.** 1969. *Die Strandläufer Eurasiens*. Neue Brehm-Bücherei 420, Wittenberg Lutherstadt, Germany.

Uspenski, S.M. 1986. *Life in High Latitudes: a study of bird life*. Balkema, Rotterdam, The Netherlands.

Väisänen, R.A. 1969. Evolution of the Ringed Plover (*Charadrius hiaticula* L.) during the last hundred years in Europe. Ann. Ac. Scient. Fenn. Series A. IV. Biol. 149.

Vark, G.N. van & W. Schaafsma. 1992. Advances in the Quantitative Analysis of skeletal Morphology. Pages 225-257 *in Skeletal Biology of Past Peoples*. Wiley-Liss, New York, U.S.A.

∗ **Vasil'chenko, A.A.** 1987. *Ptitsy Khamar-Dabana*. Nauka, Novosibirsk, Russia.

∗ **Vaughan, R**. 1988. Birds of the Thule District, North-West Greenland. Arctic 41: 53-58.

Vaughan, R. 1992. *In search of arctic birds*. Poyser, London, U.K.

Vaurie, C. 1959. *The birds of the Palearctic fauna*. Witherby, London, U.K.

Vaurie, C. 1964. Systematic notes in Palearctic birds. No. 53. Charadriidae: the genera *Charadrius* and *Pluvialis*. Am. Mus. Novit. 2177: 1-22.

Vaurie, C. 1965. *The birds of the Palaearctic fauna; non-passeriformes*. Witherby, London, U.K.

* **Vaurie, C**. 1972. *Tibet and its birds*. Witherby, London, U.K.

* **Viksne, J. (ed.)**. 1989. *Latvijas Ligzdojoso putnu Atlants 1980-1984*. Zinatne, Riga, Latvia.

Voous, K.H. 1960. *Atlas van de Europese vogels*. Elsevier, Amsterdam, The Netherlands.

Voous, K.H. 1973. List of recent Holarctic bird species. Non-passerines. Ibis 115: 612-638.

Voous, K.H. 1992. Reflections on the genus in ornithology. Bull. B.O.C. Centenary Suppl. 112A: 261-266.

* **Vorobiev, K.A.** 1963. *Ptitsy Yakutii*. Akad. Nauk SSSR, Moscow, Russia.

* **Walasz, K. & P. Mielczarec (eds.).** 1992. *The Atlas of breeding birds in Malopolska 1985-1991*. Biologica Silesiae, Wroclaw, Poland.

Warren, R.L.M. 1966. Type-specimens of birds in the British Museum (Natural History). Vol. 1, Non-passeriformes. Trustees Brit. Mus. Nat. Hist., London, UK.

Warriner, J.S., J.C. Warriner, G.W. Page & L.E. Stenzel. 1986. Mating system and reproductive success of a small population of polygamous Snowy Plovers. Wilson Bull. 98: 15-37.

Wenink, P.W. 1994. Mitochondrial DNA sequence evolution in shorebird populations. Ph. D. thesis, Univ. of Wageningen, The Netherlands.

Wenink, P.W., A.J. Baker & M.G.J. Tilanus. 1993. Hypervariable control region sequences reveal global population structuring in a long-distance migrant shorebird, the Dunlin (*Calidris alpina*). Proc. Natl. Acad. Sci. USA. 90: 94-98.

Wenink, P.W., A.J. Baker & M.G.J. Tilanus. 1994. Mitochondrial control region sequences in two shorebird species, the Turnstone and the Dunlin, and their utility in population genetic studies. Mol. Biol. Evol. 11: 22-31.

Wenink, P.W., A.J. Baker & M.G.J. Tilanus. 1996. Global mitochondrial DNA phylogeography of holarctic breeding Dunlins (*Calidris alpina*). Evolution 50: 318-330.

Whyte, A. 1985. Breeding ecology of the Piping Plover (*Charadrius melodus*) in central Saskatchewan. M. Sc. thesis, Univ. of Saskatchewan, U.S.A.

Wiens, T.P. 1986. Nest-site tenacity, and mate retention in the Piping Plover (*Charadrius melodus*). M. Sc. thesis. Univ. of Minnesota, Duluth, U.S.A.

Wilcox, L.R. 1959. A twenty year banding study of the Piping Plover. Auk 76: 129-152.

Wilson, E.O. & W.L. Brown. 1953. The subspecies concept and its taxonomic application. Syst. Zool. 2: 97-111.

Witherby, H.F., F.C.R. Jourdain, N.F. Ticehurst & B.W. Tucker. 1940. *Handbook of British Birds, vol. IV*. Witherby, London, U.K.

Witts, B. F.1985. Waders. Section 4D *in* The Report of the Joint Services Expedition to Princess Marie Bay, Ellesmere Island, 1980. Emsworth, Hants., U.K.

Wood, A.G. 1987. Discriminating the sex of Sanderling *Calidris alba*: some results and their implications. Bird Study 34: 200-204.

Wymenga E., M. Engelmoer, C.J. Smit & T.M. van Spanje. 1990. Geographical breeding origin and migration of waders wintering in West Africa. Ardea 78: 83-113.

Wynne-Edwards, V.C. 1952. Zoology of the Baird Expedition (1950) I. The birds observed in Central and South-East Baffin Island. Auk 69: 353-391.

Wynne-Edwards, V.C. 1957. The so-called 'Northern Golden Plover'. Scott. Nat. 69: 89-93.

Yalden, D.W. 1986. The habitat and activity of Common Sandpipers *Actitis hypoleucos* breeding by upland rivers. Bird Study 33: 214-222.

* **Yamashina, Y**., 1931. Die Vögel der Kurilen. J. Orn. 79: 491-541.

* **Yanushevich, A.I., P.S. Tyurin, I.D. Yakovleva,**

A. Kydyraliev & N. I. Semenova. 1959. Ptitsy Kirgizii 1. Akad. Nauk Kirgizskoi SSR, Frunze, Kirgizia.

∗ **Yeatman, L**., 1976. *Atlas des oiseaux nicheurs de France.* Ministère de la Qualité de la Vie-Environnement, Paris, France.

Zhmud, M.Y. 1992. Territorial relations and population structure of the Redshank *Tringa totanus* during the nesting period in the south of the Ukraine. W.S.G. Bull. 64: 45-46.

Zink, R.M., S. Rohwer, A.V. Andreev & D.L. Dittmann. 1995. Trans-Beringia comparisons of mitochondrial DNA differentiation in birds. Condor 97: 639-649.

Zwarts, L., B.J. Ens, M. Kersten & T. Piersma. 1990c. Moult, mass and flight range of waders ready to take off for long-distance migrations. Ardea 78: 339-364.

Annex I

Summary of morphometric data

The mean measurements per species are summarized her. They are presented in more detail in the species' chapters. The means are given per sex and geographical group as resulting from the species' chapters. Presented are the means, standard deviations and total *n*.

Species	Subspecies/ population	Wing length			Culmen length			Tarsus length			Secondary length			Tail length			Middle toe length		
		x	SD	*n*	*x*	SD	*n*	*x*	SD	*n*	*x*	SD	*n*	*x*	SD	*n*	*x*	SD	*n*
♂♂																			
Ringed Plover	*psammodroma*	134.1	3.4	90	14.3	0.8	89	25.2	1.0	90	56.5	1.9	53	55.1	2.9	53	21.6	1.1	29
	hiaticula	136.6	3.7	38	15.2	0.7	38	26.1	0.8	38	58.9	2.7	18	57.1	2.7	18	22.6	1.0	15
	tundrae	131.3	3.2	176	14.0	0.7	176	25.5	1.1	173	55.4	1.7	123	56.6	4.2	105	22.2	0.9	108
Semipalmated Plover		124.8	2.9	91	12.7	0.7	91	23.9	1.1	91	54.2	1.5	91	52.0	2.6	91	22.1	0.9	39
Dotterel		150.3	2.8	6	16.0	0.9	6	35.5	1.3	6	63.5	1.0	6	68.7	3.6	6	25.2	1.2	6
Eurasian Golden Plover		190.7	4.4	65	22.9	1.1	64	41.0	1.6	64	79.4	1.7	52	67.8	3.3	52	33.6	1.4	48
Grey Plover	*squatarola* - W Palearctic	204.7	4.1	34	29.1	1.5	34	47.5	1.9	34	83.2	2.9	24	72.8	3.5	24	34.8	1.6	19
	squatarola - E Palearctic	204.8	5.7	38	30.7	1.4	38	48.7	2.2	37	84.1	2.2	20	73.6	4.0	20	36.1	1.3	19
	tomkovichi	208.6	4.2	20	29.4	1.5	20	46.4	1.4	20	84.7	1.5	20	72.7	3.0	20	34.8	1.5	18
	Alaska	204.2	5.5	34	30.8	1.8	34	49.0	1.9	34	83.2	2.8	30	73.7	2.6	30	34.8	1.3	26
	cynosurae	196.8	4.5	53	30.1	1.5	53	45.5	1.6	52	80.4	2.0	39	69.4	3.6	39	34.2	1.1	34
Red Knot	*islandica*	169.5	4.6	78	31.7	1.3	78	30.6	1.4	51	71.2	2.1	38	58.4	2.6	38	26.7	1.1	35
	canutus - Taymyr	168.5	2.9	11	31.1	1.8	11	30.5	1.3	11	70.4	2.4	9	57.7	1.9	9	27.2	1.3	9
	rogersi	161.6	4.4	8	30.0	0.9	8	29.5	1.2	8	68.5	1.9	8	56.3	1.4	8	26.9	1.2	8
	roselaari	173.4	2.8	17	35.1	1.4	17	32.6	1.1	17	74.0	1.6	15	57.4	2.2	15	27.0	1.5	15
	Alaska	170.8	3.0	21	34.9	1.5	21	32.1	1.4	21	71.9	1.5	21	59.3	2.0	21	27.1	0.8	20
	rufa	169.9	3.0	15	34.8	0.9	15	32.1	1.2	15	70.2	1.9	13	58.3	1.4	13	27.6	1.4	10
Sanderling	*alba*	124.7	2.6	82	23.8	1.2	81	24.3	0.9	50	54.5	1.5	30	48.0	2.2	31	19.7	0.8	28
	rubidus	125.5	2.7	62	24.7	0.9	62	24.7	1.0	59	56.2	1.6	53	48.6	2.5	53	20.0	0.8	47
Little Stint		97.5	2.0	11	18.5	1.1	11	20.3	1.2	11	44.2	1.3	11	41.0	1.2	7	19.1	0.6	7
Temminck's Stint		98.8	2.4	21	17.1	0.8	21	17.7	0.6	21	47.0	1.3	21	43.2	2.7	21	19.1	0.9	21
Curlew Sandpiper		132.1	2.9	42	36.4	1.8	42	29.3	1.3	42	57.0	1.6	42	46.7	3.4	42	24.2	0.8	38
Rock Sandpiper		126.0	3.4	16	28.4	1.6	16	23.1	0.7	16	61.1	2.9	16	52.0	2.3	15	27.7	1.0	13
Purple Sandpiper	*maritima* - N Canada	131.8	2.3	42	30.7	1.8	42	22.8	0.8	42	64.0	1.3	37	55.3	2.5	37	27.5	0.9	36
	belcheri	125.8	3.6	12	29.2	1.3	12	21.4	0.6	11	60.8	1.6	12	53.6	2.0	12	26.5	0.9	9
	maritima - N Atlantic	128.5	2.5	111	28.1	1.7	110	22.3	0.8	101	61.3	1.4	39	56.2	4.0	45	26.4	1.2	55
	littoralis	135.9	2.6	27	30.2	1.1	27	23.1	0.8	26	65.0	1.7	21	59.0	2.0	25	27.4	0.9	21
Dunlin	*arctica*	113.2	2.2	62	26.8	1.0	60	22.5	0.6	26	51.0	1.5	12	44.4	2.2	15	21.1	0.7	12
	schinzii	113.5	2.5	151	28.7	1.5	150	23.2	0.9	138	52.0	1.5	97	43.7	2.4	94	22.5	1.0	89
	alpina	117.4	2.4	50	31.5	1.7	50	24.1	1.0	49	54.0	2.1	25	46.2	2.0	25	22.9	1.1	25
	centralis	117.5	1.7	32	31.7	1.3	33	24.7	1.0	32	53.9	1.1	12	45.2	2.8	13	22.5	1.4	13
	actites	109.0	1.0	3	28.3	0.4	3	23.7	1.0	3	53.3	2.1	3	48.3	5.1	3	22.9	0.8	3
	kistchinski	121.3	2.6	42	33.8	1.8	42	25.5	0.9	42	55.1	1.7	35	47.1	3.4	35	24.3	0.9	35

Species	Subspecies/ population	Wing length			Culmen length			Tarsus length			Secondary length			Tail length			Middle toe length		
		x	SD	*n*	*x*	SD	*n*	*x*	SD	*n*	*x*	SD	*n*	*x*	SD	*n*	*x*	SD	*n*
♂♂ - continued																				
Dunlin	*sakhalina*	122.4	2.7	53	33.7	2.2	54	25.7	0.9	53	57.1	2.6	23	48.7	3.9	27	24.4	1.1	23
	arcticola	124.1	2.6	49	34.3	2.3	50	25.5	1.2	50	56.7	1.7	46	49.7	3.4	49	24.0	1.0	39
	pacifica	122.9	2.5	38	37.4	1.4	38	25.8	0.8	38	57.4	1.9	38	49.8	2.4	38	24.4	1.0	36
	hudsonia	121.5	2.6	61	36.8	1.3	61	26.3	0.9	61	58.3	1.3	59	47.1	2.3	61	25.1	0.9	52
Broad-billed Sandpiper		105.7	2.5	10	29.4	1.4	10	21.4	0.9	10	46.2	1.5	10	42.0	4.2	7	20.7	0.9	10
Spoonbill Sandpiper		102.9	2.5	23	21.9	0.7	23	20.8	0.8	23	45.8	1.3	23	36.5	1.9	17	20.9	0.7	14
Western Black-tailed Godwit	*islandica*	210.3	4.9	22	79.9	3.6	21	65.9	3.4	22			0			0			0
	limosa - W Europe	213.6	5.5	63	91.1	5.5	62	72.6	5.2	61	97.5	0.7	2	74.5	3.2	8	46.1	1.8	8
	limosa - E Europe	219.6	6.1	7	102.2	4.5	7	78.2	1.6	7	96.0		1	74.6	3.2	7	47.5	1.9	5
Eastern Black-tailed Godwit		199.4	7.3	15	76.8	5.0	15	62.0	3.6	15	90.1	3.9	15	66.1	5.1	15	43.1	1.8	13
Bar-tailed Godwit	*lapponica*	217.5	6.2	35	81.3	4.1	35	51.3	2.1	33	90.5	2.5	20	70.5	4.5	21	35.8	1.9	20
	taymyrensis	214.9	4.4	42	78.8	3.9	42	50.3	2.0	42	90.0	2.8	29	67.2	2.8	29	35.8	1.3	29
	menzbieri	219.8	4.2	13	85.3	3.8	12	52.4	1.5	12	92.0	6.0	6	69.8	2.0	6	37.3	1.7	5
	anadyrensis	227.8	5.3	21	84.7	4.4	21	54.3	1.9	21	92.7	3.1	13	72.8	3.5	13	36.7	0.9	13
	baueri	231.3	5.9	55	86.0	8.6	55	53.5	2.7	54	91.4	2.5	50	73.1	3.7	50	37.3	1.5	46
Eurasian Whimbrel	*islandicus*	257.0	6.1	22	80.5	3.3	22	59.3	1.8	22	113.7	3.4	20	94.4	3.3	21	42.3	1.4	16
	phaeopus	250.0	5.7	31	77.1	3.3	30	58.0	2.9	31	109.2	3.1	29	92.4	1.5	7	40.1	1.9	26
	alboaxillaris	264.1	7.1	12	81.0	2.7	12	61.3	1.9	12	114.7	3.2	10	92.5	2.1	2	42.7	1.7	11
	variegatus	243.5	5.6	22	80.2	5.4	22	58.0	2.6	22	106.9	1.8	7	91.6	3.8	15	40.5	2.1	14
Hudsonian Whimbrel	*rufiventris*	247.4	9.2	22	83.2	4.6	22	55.9	2.4	22	106.9	3.1	20	87.2	4.2	22	41.6	1.7	16
	hudsonicus	254.6	5.8	12	85.1	3.8	12	58.9	1.4	11	109.6	3.3	11	89.7	3.6	11	45.1	1.5	7
Slender-billed Curlew		266.5	7.8	2	86.7	3.7	2	64.4	4.2	2			0	90.5	0.7	2	41.0		1
Eurasian Curlew	*arquata*	299.0	7.2	56	122.0	7.7	57	78.2	3.9	57	138.2	6.3	37	97.7	0.6	3	49.4	2.3	38
	suschkini	301.3	5.5	8	126.5	6.0	8	83.8	5.5	8	137.8	4.5	8	92.0		1	51.0	4.1	8
	orientalis	301.4	6.7	15	139.4	6.4	14	84.3	4.5	15	137.3	4.3	15	100.0	2.8	2	50.1	1.8	15
Spotted Redshank		168.6	3.9	49	57.4	2.6	49	56.2	2.0	49	81.3	2.4	39	62.2	2.6	41	38.9	1.5	34
Redshank	*robusta*	168.7	4.5	67	40.0	2.0	67	47.0	2.0	67	86.1	3.0	23	64.1	2.9	26	35.1	1.2	12
	britannica	160.0	3.6	74	42.6	1.8	72	48.8	2.1	71	80.6	2.0	44	59.7	2.7	44	34.2	1.5	40
	totanus	156.7	2.6	31	40.6	1.8	31	45.6	2.1	31	78.6	1.7	23	58.0	2.5	23	33.5	1.2	23
	ussuriensis	163.5	4.8	11	43.7	2.0	11	51.1	2.1	11	84.3	3.0	11	61.1	2.5	11	34.9	1.5	9
	Central Asia	161.3	4.5	23	44.0	1.9	23	48.6	2.0	23	83.4	2.7	23	59.1	3.4	23	33.4	1.8	21
	terrignotae	154.3	3.9	5	42.7	1.0	5	46.9	2.9	5	80.2	0.8	5	56.8	2.2	5	33.1	1.2	5
Greenshank		192.4	5.0	84	54.2	2.4	84	59.4	2.7	84	88.0	2.7	72	72.8	3.5	72	38.4	1.5	60
Ruddy Turnstone	*interpres*	154.9	4.2	209	22.7	1.2	208	25.6	1.0	161	66.8	2.2	120	57.9	2.7	111	27.0	1.2	102
	morinella	152.0	3.4	67	23.5	1.1	67	26.0	0.9	66	66.6	1.8	58	55.7	2.7	58	27.1	1.3	46
Black Turnstone		153.8	0.7	2	22.6	1.9	2	25.9	0.4	2	73.5	2.1	2	58.5	0.7	2	27.8	0.2	2

Species	Subspecies/ population	Wing length			Culmen length			Tarsus length			Secondary length			Tail length			Middle toe length		
		x	SD	*n*	*x*	SD	*n*	*x*	SD	*n*	*x*	SD	*n*	*x*	SD	*n*	*x*	SD	*n*
						♀♀													
Ringed Plover	*psammodroma*	133.8	3.5	48	14.2	0.7	48	24.9	0.8	48	56.3	1.5	24	55.5	2.3	24	21.6	0.8	15
	hiaticula	140.1	4.2	20	15.5	1.0	20	26.0	1.0	20	59.4	2.1	7	57.6	4.5	7	22.2	1.3	7
	tundrae	132.0	3.5	89	13.8	0.8	89	25.0	1.1	86	55.8	1.6	63	57.9	4.8	55	21.7	0.9	60
Semipalmated Plover		126.0	3.5	58	12.5	0.6	58	23.5	1.0	58	54.4	2.0	58	51.8	2.2	58	21.3	0.7	22
Dotterel		157.6	3.2	7	16.5	0.5	7	36.4	1.6	7	66.0	1.2	7	69.1	3.5	7	24.9	1.3	7
Eurasian Golden Plover		193.9	4.8	35	23.1	0.9	35	40.7	1.9	35	80.3	4.3	30	68.1	3.5	30	33.2	1.0	29
Grey Plover	*squatarola* - W Palearctic	204.4	4.1	21	29.0	1.3	19	47.2	1.8	21	83.4	2.2	13	69.8	2.3	13	34.9	1.1	13
	squatarola - E Palearctic	206.7	4.4	15	30.6	1.5	15	47.0	2.1	15	83.9	2.0	9	72.5	2.2	10	34.5	1.3	8
	tomkovichi	213.7	5.8	9	28.9	2.2	9	47.1	1.7	9	85.0	2.2	7	74.7	2.0	7	34.5	1.6	7
	Alaska	204.0	5.0	22	29.9	1.0	22	48.3	2.1	22	83.7	2.7	21	71.5	2.7	21	35.0	1.6	15
	cynosurae	196.6	3.9	28	30.0	1.5	29	45.1	1.7	29	79.1	1.6	18	67.4	3.7	18	33.6	1.3	15
Red Knot	*islandica*	173.7	4.4	55	34.0	1.3	55	31.2	1.4	37	73.6	2.0	28	59.9	2.7	28	27.0	1.0	20
	canutus - Taymyr	173.9	4.3	13	34.7	1.5	13	31.8	1.1	13	72.8	2.1	8	59.4	2.0	8	27.5	0.7	8
	rogersi	169.0	3.2	5	33.3	0.6	5	31.1	1.8	5	73.4	2.9	5	59.6	2.1	5	27.6	0.9	5
	roselaari	178.5	4.2	13	36.7	1.6	13	32.8	1.5	13	76.6	1.9	12	61.3	2.9	12	27.3	1.0	12
	Alaska	172.7	3.8	5	37.3	1.9	5	33.5	1.3	5	72.4	2.1	5	60.0	1.9	5	27.5	1.3	5
	rufa	174.1	4.0	23	36.7	1.4	23	32.2	1.1	23	72.1	2.0	17	59.2	2.8	17	27.8	1.1	14
Sanderling	*alba*	128.1	2.8	54	25.6	1.1	53	25.1	0.7	29	56.6	2.1	14	49.0	2.4	18	20.0	0.8	13
	rubidus	128.7	2.8	50	26.7	1.1	50	25.6	0.9	49	57.1	1.5	48	49.5	2.1	48	20.1	0.9	45
Little Stint		102.0	1.0	3	19.2	0.6	3	20.8	0.6	3	46.0	0.0	3	41.5	6.4	2	19.2	2.4	2
Temminck's Stint		100.6	4.4	9	17.8	0.6	9	17.5	0.7	9	47.4	1.7	9	42.8	2.5	9	19.5	3.5	9
Curlew Sandpiper		134.6	3.9	26	40.4	1.5	26	30.9	1.3	26	57.8	1.4	26	46.0	3.2	26	25.2	1.1	25
Rock Sandpiper		132.7	3.5	12	32.7	2.2	13	23.8	1.1	13	64.9	2.4	12	54.2	4.8	13	28.7	1.2	12
Purple Sandpiper	*maritima* - N Atlantic	134.0	2.5	90	32.7	1.8	87	23.0	0.7	83	64.9	1.6	27	57.7	4.2	28	27.2	1.1	41
	belcheri	129.9	3.0	11	33.5	1.3	11	22.8	0.7	11	63.7	2.0	11	54.5	2.2	11	27.4	0.9	11
	maritima - N Canada	135.7	3.3	24	35.2	1.4	24	23.5	0.6	24	66.0	1.2	23	56.4	3.3	23	28.1	0.9	20
	littoralis	140.8	4.5	21	34.2	2.0	19	24.1	1.0	21	67.9	2.3	15	60.2	3.0	19	28.5	1.5	18
Dunlin	*arctica*	117.6	2.5	42	30.0	1.1	41	23.5	0.8	23	53.9	2.7	7	45.1	3.2	10	21.7	0.6	10
	schinzii	117.6	2.6	110	32.9	1.5	108	24.6	1.0	101	53.8	1.6	72	44.6	2.6	70	23.4	1.1	61
	alpina	120.2	2.4	47	34.5	1.8	46	25.1	0.9	47	55.5	1.3	15	48.0	2.0	14	23.5	0.9	15
	centralis	121.3	1.7	16	36.5	1.8	16	26.3	1.1	16	55.6	1.9	5	46.0	1.9	5	23.9	1.2	5
	actites	113.4	0.6	3	31.0	1.3	3	24.5	0.2	3	55.7	1.2	3	47.0	3.6	3	23.9	1.0	3
	kistchinski	124.3	3.0	23	38.1	1.9	23	26.7	1.1	23	56.8	1.9	21	48.0	3.2	21	25.1	0.8	21
	sakhalina	124.5	2.9	37	38.0	2.8	37	26.7	0.9	36	58.0	1.8	21	49.2	2.8	21	24.5	1.0	20
	arcticola	125.7	2.8	31	37.7	2.0	33	26.4	0.9	33	57.9	2.1	30	49.0	2.8	31	25.0	0.8	26
	pacifica	127.2	2.7	32	40.6	1.4	32	26.8	0.8	32	59.5	1.7	30	50.5	2.6	30	25.2	1.0	30
	hudsonia	122.7	3.0	44	39.4	1.7	44	26.9	0.9	43	59.2	1.4	43	46.9	2.8	44	25.2	1.0	38
Broad-billed Sandpiper		108.5	2.6	6	31.0	3.1	6	22.1	0.4	6	46.8	1.3	6	38.5	2.1	2	20.3	1.0	4

Species	Subspecies/ population	Wing length			Culmen length			Tarsus length			Secondary length			Tail length			Middle toe length		
		x	SD	*n*	*x*	SD	*n*	*x*	SD	*n*	*x*	SD	*n*	*x*	SD	*n*	*x*	SD	*n*
♀♀ - continued																			
Spoonbill Sandpiper		106.4	4.2	9	22.7	0.8	9	21.0	0.4	8	46.9	1.8	9	37.0	0.8	4	21.3	0.5	4
Western Black-tailed Godwit	*islandica*	223.6	5.1	9	96.6	4.4	7	74.0	3.3	9			0			0			0
	limosa - W Europe	225.1	5.3	42	108.0	4.6	41	80.3	4.4	39	109.0		1	79.0	5.0	3	48.0	3.5	2
	limosa - E Europe	229.9	5.4	10	113.7	9.7	10	83.0	6.7	10			0	77.9	3.7	8	48.9	2.9	7
Eastern Black-tailed Godwit		203.0	5.5	8	88.5	11.4	8	68.3	4.8	8	94.0	3.3	8	70.0	3.3	8	44.2	2.2	8
Bar-tailed Godwit	*lapponica*	230.8	6.1	23	101.6	5.3	23	55.9	2.8	20	98.2	3.8	14	71.9	3.7	14	38.2	1.9	15
	taymyrensis	226.6	4.5	33	95.5	4.5	33	54.1	2.3	33	95.6	2.3	26	70.3	2.9	28	37.9	1.6	27
	menzbieri	231.2	5.3	16	108.6	4.2	16	55.6	1.6	16	97.0	0.7	5	74.3	1.9	6	38.6	1.4	6
	anadyrensis	240.1	8.0	13	106.5	8.7	13	58.3	2.7	13	99.4	3.3	7	74.9	3.4	7	39.8	0.6	7
	baueri	245.9	7.0	46	108.2	7.2	45	59.1	2.3	46	96.9	2.6	36	75.8	3.1	42	40.5	1.4	35
Eurasian Whimbrel	*islandicus*	266.4	7.1	18	86.9	5.4	18	61.5	3.6	18	118.1	1.8	15	96.5	1.5	13	43.7	1.9	14
	phaeopus	259.5	5.6	22	86.1	4.7	22	60.4	2.1	22	113.8	3.7	20	95.9	2.4	7	42.4	1.7	18
	alboaxillaris	268.1	6.3	8	86.3	3.5	8	64.4	1.9	8	115.2	1.9	6	94.0		1	42.6	2.1	6
	variegatus	252.2	5.8	17	88.3	5.4	17	61.4	2.8	17	112.0		1	94.9	3.1	8	41.2	1.5	8
Hudsonian Whimbrel	*rufiventris*	260.9	9.1	16	92.8	4.1	16	60.0	2.4	16	110.8	4.6	14	89.7	3.3	15	43.2	1.9	12
	hudsonicus	263.9	6.5	11	94.2	4.7	11	60.6	2.6	11	114.5	3.4	11	92.8	6.4	9	44.5	1.9	10
Eurasian Curlew	*arquata*	317.9	9.5	33	148.0	8.5	33	82.4	4.1	33	145.3	5.9	23	96.0		1	50.7	2.4	20
	suschkini	310.0	6.8	6	137.6	5.7	6	83.5	1.3	6	136.3	9.6	6	95.0	1.4	2	49.6	1.9	5
	orientalis	319.4	5.1	11	175.6	8.2	11	86.8	2.4	11	147.2	6.4	10	116.0		1	52.8	2.5	8
Spotted Redshank		172.6	4.7	21	58.9	3.7	21	56.8	2.7	21	83.6	3.0	19	62.2	4.4	19	39.2	1.8	17
Redshank	*robusta*	172.0	4.2	53	42.5	1.4	53	48.1	1.9	53	88.3	2.2	19	63.7	2.8	21	36.6	1.2	13
	britannica	163.7	4.2	47	44.7	1.7	47	49.6	3.0	47	82.1	1.4	23	59.6	2.2	23	35.0	1.8	20
	totanus	161.4	2.8	29	41.8	1.7	29	46.9	2.2	29	80.6	2.2	19	58.8	3.8	19	33.7	0.9	19
	ussuriensis	170.7	5.2	10	45.9	2.4	10	51.5	2.3	10	86.2	3.9	10	61.9	2.2	10	35.2	0.9	8
	Central Asia	164.3	4.9	18	44.1	1.9	18	48.1	2.5	18	85.5	3.2	18	57.7	4.3	18	33.5	1.6	17
	terrignotae	158.1	6.9	9	42.3	2.7	9	45.0	1.8	9	81.6	2.9	9	58.2	3.2	9	32.8	1.6	8
Greenshank		196.0	4.7	60	54.4	2.7	60	59.9	3.1	60	89.9	3.2	49	74.6	3.6	49	38.7	1.6	47
Ruddy Turnstone	*interpres*	157.5	4.3	124	23.4	1.0	123	25.9	1.0	90	67.8	1.9	63	59.0	2.8	60	27.0	1.1	54
	morinella	154.1	4.2	40	24.2	1.1	40	26.3	0.8	40	67.6	1.7	35	57.1	2.4	35	27.3	0.9	31
Black Turnstone		155.9	1.4	5	23.2	1.4	5	26.0	0.9	5	73.8	1.1	5	57.2	2.8	5	28.4	1.7	4

Annex 2

Lengths of primaries in relation to maximum wing length

The lengths of the primaries of the various species and their populations are related here to the total wing lengths. These percentages can be useful for estimating total wing length. For instance, when birds are moulting their outermost primaries, total wing lengths cannot be measured. In most situations primary 7, primary 1or secondary 1 can then be measured.Two tables with relative feather lengths are presented: (1) mean values (SD and *n*) per species, and (2) mean values (SD and *n*) per subspecies/ population.

Species	Primary 10			Primary 9			Primary 8			Primary 7			Primary 1			Secondary 1		
	x	SD	*n*	*x*	SD	*n*	*x*	SD	*n*	*x*	SD	*n*	*x*	SD	*n*	*x*	SD	*n*
								♂♂										
Ringed Plover	68.9	1.4	144	67.1	1.4	107	61.7	1.4	108	54.7	1.7	109	44.5	1.2	142	42.4	1.4	217
Semipalmted Plover	70.4	1.8	78	68.9	1.9	78	64.0	2.0	78	56.5	2.2	78	45.6	1.3	77	43.5	1.3	91
Dotterel	-	-	0	-	-	0	-	-	0	-	-	0	-	-	0	42.3	0.7	6
Eurasian Golden lover	67.3	1.2	14	65.0	1.2	14	59.7	1.5	14	53.8	1.1	11	44.8	1.6	12	41.7	1.0	53
Grey Plover	67.6	1.1	67	65.2	1.2	67	60.2	1.6	67	54.3	1.3	67	43.9	1.4	58	40.7	1.0	133
Red Knot	67.9	1.3	80	65.9	1.5	80	60.9	1.5	80	54.2	1.5	80	44.3	1.1	85	42.1	1.2	108
Sanderling	68.9	1.0	48	67.4	1.3	48	62.6	1.5	48	55.3	2.1	48	46.3	1.1	48	44.5	1.2	83
Little Stint	-	-	0	-	-	0	-	-	0	-	-	0	-	-	0	45.3	1.2	11
Temminck's Stint	-	-	0	-	-	0	-	-	0	-	-	0	-	-	0	47.6	1.1	21
Curlew Sandpiper	68.7	1.1	14	67.5	1.0	14	62.7	1.0	14	56.0	1.1	14	46.1	1.3	14	43.1	1.2	42
Purple Sandpiper	68.8	1.3	47	67.3	1.4	47	62.7	1.4	47	56.6	1.6	47	50.0	1.2	47	48.1	1.1	109
Rock Sandpiper	-	-	0	-	-	0	-	-	0	-	-	0	-	-	0	48.5	1.6	16
Dunlin	67.6	1.5	158	66.5	1.6	154	61.9	1.5	151	55.6	2.0	151	48.7	1.7	162	46.2	1.7	363
Broad-billed Sandpiper	-	-	0	-	-	0	-	-	0	-	-	0	-	-	0	43.7	1.1	10
Spoonbill Sandpiper	68.3	1.7	17	67.5	1.7	17	63.2	1.4	17	56.0	1.6	17	46.8	1.7	17	44.6	1.5	23
Western Back-tailed Godwit	65.7	1.6	10	64.0	1.9	10	60.1	2.1	10	53.8	2.6	10	49.7	2.5	9	46.3	1.5	3
Eastern Black-tailed Godwit	66.8	2.7	7	65.7	3.0	7	61.5	3.6	7	56.0	4.3	7	48.7	1.7	7	45.2	1.4	15
Bar-tailed Godwit	67.0	1.2	60	64.8	1.4	60	60.2	1.7	60	54.0	1.9	60	44.0	1.2	87	40.7	1.6	120
Eurasian Whimbrel	66.5	1.4	20	64.9	1.6	20	61.0	2.0	20	55.9	2.2	20	47.2	1.1	20	43.9	1.2	66
Hudsonian Whimbrel	66.3	2.4	21	64.5	2.0	21	60.5	2.4	21	54.5	2.7	21	46.3	1.1	21	43.1	1.3	31
Slender-billed Curlew	-	-	0	-	-	0	-	-	0	-	-	0	-	-	0	-	-	0
Eurasian Curlew	64.5	2.6	18	63.2	1.5	18	59.3	1.9	18	53.9	1.6	18	48.7	0.9	18	46.1	2.0	59
Spotted Redshank	65.5	1.7	10	63.9	1.6	10	60.6	1.9	10	55.1	1.9	10	51.3	0.9	10	48.2	1.0	39
Redshank	66.8	1.5	51	66.1	1.5	50	63.6	1.9	51	58.1	1.8	51	53.7	1.5	51	50.8	1.4	148
Greenshank	67.0	1.7	17	64.8	1.7	17	60.2	2.4	17	54.1	2.2	17	48.9	1.5	17	45.8	1.4	72
Ruddy Turnstone	69.4	2.0	142	67.6	1.4	142	63.1	1.5	143	56.7	1.5	142	46.2	1.3	143	43.4	1.4	178
Black Turnstone	-	-	0	-	-	0	-	-	0	-	-	0	-	-	0	47.8	1.2	2

Species	Subspecies/ population	Primary 10			Primary 9			Primary 8			Primary 7			Primary 1			Secondary 1		
		x	SD	*n*	*x*	SD	*n*	*x*	SD	*n*	*x*	SD	*n*	*x*	SD	*n*	*x*	SD	*n*
♂♂ - continued																			
Eurasian Golden Plover		67.2	1.2	13	64.9	1.2	13	59.7	1.5	13	53.8	1.1	10	45.1	1.1	11	41.7	1.0	52
Grey Plover	*squatarola* - W Palearctic	67.9	0.8	2	64.2	0.5	2	58.5	0.8	2	52.6	0.2	2	42.2	0.5	2	40.6	1.4	24
Grey Plover	*squatarola* - E Palearctic	66.7	0.2	2	64.3	0.3	2	59.0	0.4	2	52.8	1.1	2	43.2	0.7	2	40.9	1.1	20
	tomkovichi	-	-	0	-	-	0	-	-	0	-	-	0	-	-	0	40.6	0.8	20
	Alaska	67.8	1.2	25	65.6	1.3	25	60.8	1.5	25	54.6	1.3	25	44.2	1.2	20	40.8	1.1	30
	cynosurae	67.4	1.1	38	65.1	1.2	38	60.1	1.5	38	54.2	1.3	38	43.8	1.5	34	40.7	0.9	39
Red Knot	*rufa*	68.7	1.0	13	66.4	0.9	13	61.3	0.9	13	54.3	1.1	13	43.5	1.4	13	41.4	1.4	13
	islandica	68.3	1.4	33	66.7	1.4	33	61.7	1.4	33	54.9	1.2	33	44.2	0.9	38	42.1	1.1	38
	canutus	67.8	0.9	9	65.7	1.1	9	60.6	1.3	9	53.3	1.8	9	44.1	1.2	9	42.0	1.4	9
	rogersi	67.7	0.7	8	65.2	1.0	8	59.7	0.9	8	53.1	1.3	8	44.6	1.1	8	42.4	1.3	8
	roselaari	66.7	0.9	15	64.1	0.9	15	59.5	1.1	15	53.4	1.4	15	44.9	1.0	15	42.7	1.2	15
	Alaska	67.7	0.3	2	65.9	0.5	2	61.1	0.4	2	55.2	1.2	2	45.4	0.3	2	42.1	1.0	21
Sanderling	*rubidus*	68.9	1.0	42	67.4	1.2	42	62.7	1.2	42	55.5	1.7	42	46.3	1.1	42	44.8	1.2	53
	alba	69.4	0.8	6	67.4	2.1	6	61.9	2.7	6	53.7	3.5	6	45.9	1.3	6	44.0	1.1	30
Little Stint		-	-	0	-	-	0	-	-	0	-	-	0	-	-	0	45.3	1.2	11
Temminck's Stint		-	-	0	-	-	0	-	-	0	-	-	0	-	-	0	47.6	1.1	21
Curlew Sandpiper		68.7	1.1	14	67.5	1.0	14	62.7	1.0	14	56.0	1.1	14	46.1	1.3	14	43.1	1.2	42
Rock Sandpiper		-	-	0	-	-	0	-	-	0	-	-	0	-	-	0	48.5	1.6	16
Purple Sandpiper	*maritima* - N Atlantic	68.2	0.3	3	66.7	0.9	3	62.0	0.7	3	55.0	1.0	3	49.9	0.6	3	47.8	1.0	39
	maritima - N Canada	68.8	1.4	37	67.4	1.4	37	62.8	1.4	37	56.8	1.6	37	50.1	1.3	37	48.6	1.1	37
	belcheri	68.7	1.3	7	66.9	1.4	7	62.5	1.7	7	56.4	1.7	7	49.8	0.9	7	48.4	0.8	12
	littoralis	-	-	0	-	-	0	-	-	0	-	-	0	-	-	0	47.8	1.0	21
Dunlin	*arctica*	69.0	0.8	2	67.3	0.8	2	62.0	0.9	2	55.9	0.4	2	47.2	2.8	2	45.1	1.2	12
	schinzii	67.4	1.5	32	66.2	1.7	32	62.2	1.6	32	55.8	1.7	32	48.0	1.2	32	45.7	1.2	97
	alpina	66.9	1.3	11	65.9	0.9	11	61.0	1.2	11	54.0	1.1	11	48.1	1.6	11	46.0	1.7	25
	centralis	67.8	-	1	68.6	-	1	-	-	0	-	-	0	48.3	-	1	45.9	1.1	12
	kistchinski	67.6	1.1	7	65.6	1.5	7	61.1	1.4	7	54.2	2.0	7	48.6	2.6	13	45.3	1.6	35
	actites	-	-	0	-	-	0	-	-	0	-	-	0	-	-	0	48.9	1.5	3
	sakhalina	68.5	0.6	7	67.2	1.4	4	60.9	2.0	2	53.7	5.8	4	48.4	1.9	18	46.6	2.3	23
	arcticola	68.3	1.2	25	67.2	1.3	25	62.0	1.4	25	55.5	1.9	23	47.8	1.6	22	45.7	1.4	45
	pacifica	67.4	0.1	4	66.2	0.4	4	62.8	0.8	4	56.1	0.4	4	48.8	1.4	4	46.7	1.5	38
	hudsonia	67.5	1.7	60	66.7	1.7	59	62.2	1.3	59	56.4	1.5	59	50.1	0.9	50	48.0	1.3	59
Broad-billed Sandpiper		-	-	0	-	-	0	-	-	0	-	-	0	-	-	0	43.7	1.1	10
Spoonbill Sandpiper		68.3	1.7	17	67.5	1.7	17	63.2	1.4	17	56.0	1.6	17	46.8	1.7	17	44.6	1.5	23
Western Black-tailed Godwit	*islandica*	-	-	0	-	-	0	-	-	0	-	-	0	-	-	0	-	-	0
	limosa - W Europe	65.8	1.7	7	63.8	2.0	7	59.9	2.5	7	53.4	3.0	7	50.4	2.1	6	47.0	1.1	2

Species	Primary 10			Primary 9			Primary 8			Primary 7			Primary 1			Secondary 1		
	x	SD	*n*	*x*	SD	*n*	*x*	SD	*n*	*x*	SD	*n*	*x*	SD	*n*	*x*	SD	*n*
								♀♀										
Ringed Plover	69.0	1.3	66	67.4	1.4	44	61.9	1.6	45	55.6	2.4	45	44.7	2.1	66	42.5	1.4	110
Semipalmted Plover	70.3	1.5	55	68.6	1.6	55	63.4	2.0	54	56.2	2.1	55	45.0	1.3	55	43.2	1.5	58
Dotterel	-	-	0	-	-	0	-	-	0	-	-	0	-	-	0	41.9	0.8	7
Eurasian Golden lover	67.4	1.0	9	64.6	0.9	9	59.7	1.1	9	52.7	1.2	8	45.7	2.3	7	41.5	2.1	31
Grey Plover	67.9	1.2	37	65.4	1.6	37	60.5	1.5	37	54.6	1.7	37	44.2	1.5	34	40.6	1.2	68
Red Knot	67.5	1.2	71	65.4	1.3	71	60.4	1.5	71	53.6	1.4	71	44.2	1.0	71	42.2	1.2	76
Sanderling	69.0	1.3	42	67.4	1.4	42	62.4	1.7	42	55.4	1.7	42	46.1	1.3	41	44.3	1.2	62
Little Stint	-	-	0	-	-	0	-	-	0	-	-	0	-	-	0	45.1	0.4	3
Temminck's Stint	-	-	0	-	-	0	-	-	0	-	-	0	-	-	0	47.2	1.5	9
Curlew Sandpiper	68.8	2.5	7	66.6	2.5	7	61.1	3.2	7	54.8	3.4	7	46.3	1.6	7	43.0	1.2	26
Purple Sandpiper	68.9	1.0	34	67.7	1.2	34	63.3	1.4	34	56.8	1.4	34	50.5	1.0	34	48.6	1.2	76
Rock Sandpiper	-	-	0	-	-	0	-	-	0	-	-	0	-	-	0	48.9	1.6	12
Dunlin	67.6	2.0	100	66.5	2.0	99	62.1	2.0	99	55.7	2.2	96	49.1	1.8	103	46.5	1.6	248
Broad-billed Sandpiper	-	-	0	-	-	0	-	-	0	-	-	0	-	-	0	43.2	1.2	6
Spoonbill Sandpiper	65.9	4.6	4	64.7	3.3	4	61.0	3.2	4	54.6	2.8	4	45.6	2.1	4	44.1	1.1	9
Western Back-tailed Godwit	65.2	1.5	8	64.0	1.7	8	59.9	1.7	8	54.4	3.0	8	50.9	1.3	8	48.4	-	1
Eastern Black-tailed Godwit	65.4	0.9	3	64.6	0.3	3	60.6	0.6	3	55.6	1.3	3	49.6	1.9	3	46.3	1.0	8
Bar-tailed Godwit	66.2	1.3	45	64.1	1.3	45	59.4	1.6	45	53.2	1.8	45	44.3	1.7	69	41.0	1.7	89
Eurasian Whimbrel	66.7	1.6	12	65.4	1.9	12	61.8	2.1	12	56.2	2.0	12	47.9	1.2	12	43.8	1.3	42
Hudsonian Whimbrel	66.2	0.9	20	64.0	0.9	20	60.4	1.0	20	54.7	1.0	20	45.5	1.2	20	42.9	1.1	25
Eurasian Curlew	64.4	1.8	14	63.1	1.8	14	59.8	1.8	14	54.3	1.5	14	49.7	2.1	14	45.7	2.2	40
Spotted Redshank	65.4	1.6	8	63.3	1.5	8	60.1	0.9	8	54.9	0.8	8	50.9	1.0	8	48.3	1.2	19
Redshank	66.4	2.3	36	65.5	2.4	36	62.5	2.2	36	57.1	2.5	36	53.7	1.8	36	50.8	1.5	103
Greenshank	66.5	1.0	20	64.7	1.6	20	60.3	1.8	20	54.7	2.4	20	49.3	1.4	20	46.0	1.4	49
Ruddy Turnstone	69.3	1.5	76	67.9	1.5	76	63.5	1.6	76	57.1	1.5	76	46.4	1.4	76	43.5	1.4	98
Black Turnstone	-	-	0	-	-	0	-	-	0	-	-	0	-	-	0	47.3	0.8	5

Species	Subspecies/ population	Primary 10			Primary 9			Primary 8			Primary 7			Primary 1			Secondary 1		
		x	SD	*n*	*x*	SD	*n*	*x*	SD	*n*	*x*	SD	*n*	*x*	SD	*n*	*x*	SD	*n*
									♂♂										
Ringed Plover	*psammodroma*	69.3	1.5	50	67.2	1.3	49	61.6	1.5	50	54.8	1.4	50	44.5	1.2	48	42.2	1.2	53
	hiaticula	67.8	1.4	8	66.3	1.4	8	61.1	1.2	8	54.3	1.4	8	45.5	1.3	8	43.4	1.7	18
	tundrae	68.8	1.3	85	67.2	1.4	49	61.7	1.4	49	54.5	2.1	50	44.5	1.2	85	42.2	1.3	123
Semipalmated Plover		70.4	1.8	78	68.9	1.9	78	64.0	2.0	78	56.5	2.2	78	45.6	1.3	77	43.5	1.3	91
Dotterel		-	-	0	-	-	0	-	-	0	-	-	0	-	-	0	42.3	0.7	6

Species	Subspecies/ population	Primary 10			Primary 9			Primary 8			Primary 7			Primary 1			Secondary 1		
		x	SD	*n*	*x*	SD	*n*	*x*	SD	*n*	*x*	SD	*n*	*x*	SD	*n*	*x*	SD	*n*
♀♀ - continued																			
	squatarola - E Palearctic	69.2	-	1	66.8	-	1	61.1	-	1	55.3	-	1	44.2	-	1	40.8	0.9	9
	tomkovichi	-	-	0	-	-	0	-	-	0	-	-	0	-	-	0	39.9	1.2	7
Grey Plover	Alaska	67.9	1.2	18	65.5	1.6	18	60.8	1.5	18	54.7	1.7	18	44.7	1.2	17	41.0	1.2	21
	cynosurae	67.7	0.9	17	65.0	1.3	17	60.1	1.2	17	54.3	1.6	17	43.6	1.6	15	40.1	1.1	18
Red Knot	*rufa*	67.6	0.7	17	65.2	1.1	17	59.9	1.1	17	53.5	1.0	17	43.9	1.0	17	41.4	1.1	17
	islandica	68.0	1.2	27	66.2	1.2	27	61.0	1.2	27	54.2	1.3	27	44.2	1.0	27	42.4	0.9	28
	canutus	67.1	1.4	8	64.5	1.5	8	59.9	3.1	8	52.5	1.7	8	43.8	0.7	8	41.9	0.9	8
	rogersi	67.3	1.6	5	65.9	1.3	5	60.2	0.7	5	53.0	1.6	5	44.3	0.9	5	43.5	2.1	5
	roselaari	66.7	0.7	12	64.3	0.7	12	59.9	1.1	12	53.2	1.3	12	45.0	1.0	12	42.9	0.7	12
	Alaska	67.8	0.1	2	66.0	0.2	2	62.1	0.6	2	55.3	1.2	2	45.0	0.9	2	41.9	0.5	5
Sanderling	*rubidus*	69.1	1.3	39	67.5	1.3	39	62.6	1.6	39	55.6	1.6	39	46.2	1.3	38	44.4	1.3	48
	alba	68.1	1.1	3	66.0	1.6	3	60.5	0.9	3	52.9	1.1	3	45.3	0.6	3	44.0	1.0	14
Little Stint		-	-	0	-	-	0	-	-	0	-	-	0	-	-	0	45.1	0.4	3
Temminck's Stint		-	-	0	-	-	0	-	-	0	-	-	0	-	-	0	47.2	1.5	9
Curlew Sandpiper		68.8	2.5	7	66.6	2.5	7	61.1	3.2	7	54.8	3.4	7	46.3	1.6	7	43.0	1.2	26
Rock Sandpiper		-	-	0	-	-	0	-	-	0	-	-	0	-	-	0	48.9	1.6	12
Purple Sandpiper	*maritima* - N Atlantic	69.8	1.1	7	68.8	1.3	7	64.2	2.2	7	58.0	1.7	7	51.0	1.5	7	48.5	1.2	27
	maritima - N Canada	68.4	0.7	22	67.1	0.8	22	62.7	1.0	22	56.1	1.0	22	50.3	0.9	22	48.7	1.0	23
	belcheri	69.6	0.6	5	68.5	0.7	5	64.2	0.6	5	57.7	0.7	5	50.5	0.9	5	49.1	1.3	11
	littoralis	-	-	0	-	-	0	-	-	0	-	-	0	-	-	0	48.1	1.4	15
Dunlin	*arctica*	69.2	-	1	68.3	-	1	63.3	-	1	55.0	-	1	48.3	-	1	46.0	1.5	7
	schinzii	67.0	1.7	21	65.9	2.0	21	61.6	2.6	21	55.4	2.7	21	48.4	1.4	20	45.7	1.1	72
	alpina	66.7	0.8	2	65.4	2.5	2	60.5	2.6	2	54.3	2.0	2	47.7	2.1	2	46.2	1.0	15
	centralis	-	-	0	-	-	0	-	-	0	-	-	0	-	-	0	45.6	1.2	5
	kistchinski	67.8	1.4	8	66.3	1.2	8	61.7	1.0	8	54.7	1.3	8	48.1	2.1	12	45.7	1.4	21
	actites	-	-	0	-	-	0	-	-	0	-	-	0	-	-	0	49.1	1.2	3
	sakhalina	67.2	2.3	9	65.8	1.5	8	61.7	2.1	8	54.4	0.9	6	48.3	0.8	14	46.5	1.0	21
	arcticola	67.4	1.3	12	66.3	1.2	12	62.4	1.8	12	56.0	2.0	12	47.7	1.3	10	45.9	1.6	28
	pacifica	66.4	-	1	65.6	-	1	60.8	-	1	53.6	-	1	49.6	-	1	46.9	1.4	30
	hudsonia	67.9	2.3	44	67.0	2.3	44	62.4	1.7	44	56.3	2.1	43	50.5	1.4	41	48.2	1.1	43
Broad-billed Sandpiper		-	-	0	-	-	0	-	-	0	-	-	0	-	-	0	43.2	1.2	6
Spoonbill Sandpiper		65.9	4.6	4	64.7	3.3	4	61.0	3.2	4	54.6	2.8	4	45.6	2.1	4	44.1	1.1	9
Western Black-tailed Godwit	*islandica*	-	-	0	-	-	0	-	-	0	-	-	0	-	-	0	-	-	0
	limosa - W Europe	67.2	0.7	2	65.8	0.7	2	61.6	1.7	2	53.7	7.1	2	51.0	1.1	2	48.4	-	1
	limosa - E Europe	64.6	0.9	6	63.4	1.5	6	59.3	1.3	6	54.6	1.4	6	50.8	1.4	6	-	-	0
Eastern Black-tailed Godwit		65.4	0.9	3	64.6	0.3	3	60.6	0.6	3	55.6	1.3	3	49.6	1.9	3	46.3	1.0	8

Species	Subspecies/ population	Primary 10			Primary 9			Primary 8			Primary 7			Primary 1			Secondary 1		
		x	SD	*n*	*x*	SD	*n*	*x*	SD	*n*	*x*	SD	*n*	*x*	SD	*n*	*x*	SD	*n*
♂♂ - continued																			
	limosa - E Europe	65.3	1.6	3	64.6	1.9	3	60.5	1.5	3	54.7	1.2	3	48.4	3.2	3	44.9	-	1
Eastern Black-tailed Godwit		66.8	2.7	7	65.7	3.0	7	61.5	3.6	7	56.0	4.3	7	48.7	1.7	7	45.2	1.4	15
Bar-tailed Godwit	*lapponica*	66.2	1.7	10	64.1	1.4	10	59.9	2.2	10	54.0	2.4	10	45.3	0.8	11	41.9	1.0	20
	taymyrensis	66.8	0.8	14	64.5	1.2	14	60.5	1.5	14	54.1	2.4	14	44.4	1.2	29	41.8	1.5	29
	menzbieri	67.1	-	1	64.4	-	1	60.4	-	1	54.5	-	1	44.1	1.4	5	41.9	3.0	6
	anadyrensis	67.3	1.0	8	65.3	1.2	8	60.5	1.3	8	53.9	1.3	8	43.7	1.0	13	40.5	0.9	13
	baueri	67.2	1.1	27	65.1	1.5	27	60.1	1.9	27	53.9	1.8	27	43.2	1.0	27	39.6	1.0	50
Eurasian Whimbrel	*phaeopus*	-	-	0	-	-	0	-	-	0	-	-	0	-	-	0	43.7	1.2	29
	islandicus	66.2	1.5	14	64.7	1.8	14	60.7	2.4	14	55.9	2.6	14	47.2	1.2	14	44.3	1.1	20
	variegatus	67.3	0.2	4	65.3	1.0	4	61.6	0.9	4	56.0	0.9	4	47.1	1.2	4	43.7	0.7	7
	alboaxillaris	67.0	0.3	2	65.5	0.4	2	61.3	0.1	2	56.0	0.9	2	47.8	1.1	2	43.5	1.1	10
Hudsonian Whimbrel	*rufiventris*	66.8	0.3	4	64.4	0.8	4	60.7	0.9	4	54.9	0.9	4	46.1	0.8	4	43.1	1.4	11
	hudsonicus	66.2	2.7	17	64.5	2.3	17	60.5	2.7	17	54.4	3.0	17	46.3	1.2	17	43.1	1.3	20
Slender-billed Curlew		-	-	0	-	-	0	-	-	0	-	-	0	-	-	0	-	-	0
Eurasian Curlew	*arquata*	64.4	2.8	16	63.2	1.6	16	59.2	2.0	16	53.9	1.7	16	48.6	0.9	17	46.4	2.2	36
	suschkini	-	-	0	-	-	0	-	-	0	-	-	0	-	-	0	45.7	1.4	8
	orientalis	65.0	0.3	2	63.2	0.5	2	59.5	0.3	2	54.4	0.8	2	49.3	-	1	45.6	1.4	15
Spotted Redshank		65.5	1.7	10	63.9	1.6	10	60.6	1.9	10	55.1	1.9	10	51.3	0.9	10	48.2	1.0	39
Redshank	*robusta*	66.0	1.5	18	65.3	1.2	17	62.4	1.9	18	57.0	1.4	18	53.6	1.4	18	51.0	1.3	23
	britannica	67.4	1.0	6	66.3	1.4	6	63.7	1.4	6	58.0	1.2	6	53.4	1.0	6	50.4	1.2	44
	totanus	66.7	0.9	12	65.9	0.9	12	63.4	0.9	12	57.6	0.8	12	52.8	1.1	12	50.2	1.3	23
	ussuriensis	67.6	1.8	5	66.9	2.0	5	64.8	2.1	5	60.1	2.4	5	55.7	1.5	5	51.6	1.5	11
	Central Asia	67.1	1.5	4	66.8	1.6	4	65.3	1.3	4	60.4	0.6	4	54.9	1.2	4	51.7	1.1	23
	terrignotae	66.8	1.1	2	66.8	1.1	2	64.5	1.4	2	59.7	1.7	2	54.6	2.5	2	52.0	0.9	5
Greenshank		67.0	1.7	17	64.8	1.7	17	60.2	2.4	17	54.1	2.2	17	48.9	1.5	17	45.8	1.4	72
Ruddy Turnstone	*interpres*	69.5	2.4	87	67.7	1.3	87	63.3	1.5	87	56.9	1.5	86	46.1	1.3	87	43.1	1.4	120
	morinella	69.2	1.2	55	67.5	1.5	55	62.9	1.4	56	56.5	1.5	56	46.4	1.3	56	43.9	1.2	58
Black Turnstone		-	-	0	-	-	0	-	-	0	-	-	0	-	-	0	47.8	1.2	2
♀♀																			
Ringed Plover	*psammodroma*	69.7	1.2	20	67.7	1.4	20	62.1	1.5	21	55.4	1.4	21	44.6	1.2	20	42.4	1.3	24
	hiaticula	-	-	0	-	-	0	-	-	0	-	-	0	-	-	0	42.8	1.5	7
	tundrae	68.7	1.2	46	67.1	1.4	24	61.7	1.6	24	55.7	3.1	24	44.8	2.4	46	42.4	1.3	63
Semipalmated Plover		70.3	1.5	55	68.6	1.6	55	63.4	2.0	54	56.2	2.1	55	45.0	1.3	55	43.2	1.5	58
Dotterel		-	-	0	-	-	0	-	-	0	-	-	0	-	-	0	41.9	0.8	7
Eurasian Golden Plover		67.5	1.0	8	64.7	0.9	8	59.7	1.2	8	52.8	1.3	7	45.9	2.4	6	41.5	2.2	30
Grey Plover	*squatarola* - W Palearctic	71.4	-	1	68.8	-	1	63.3	-	1	57.8	-	1	46.2	-	1	40.9	1.1	13

Species	Subspecies/ population	Primary 10			Primary 9			Primary 8			Primary 7			Primary 1			Secondary 1		
		x	SD	*n*	*x*	SD	*n*	*x*	SD	*n*	*x*	SD	*n*	*x*	SD	*n*	*x*	SD	*n*
♀♀ - continued																			
Bar-tailed Godwit	*lapponica*	66.7	2.4	5	64.0	2.4	5	59.5	2.9	5	53.0	2.7	5	45.7	1.5	7	42.6	1.6	14
	taymyrensis	66.2	1.2	17	64.3	0.9	17	59.8	1.3	17	54.0	1.8	17	45.2	1.1	29	42.2	1.1	26
	menzbieri	66.5	-	1	63.5	-	1	58.8	-	1	53.7	-	1	44.4	1.3	6	41.6	0.8	5
	anadyrensis	66.1	0.9	3	64.6	0.9	3	59.5	0.9	3	52.8	1.0	3	43.3	1.4	7	40.7	0.9	7
	baueri	66.1	1.3	19	63.9	1.4	19	59.1	1.6	19	52.6	1.3	19	42.7	1.2	19	39.6	0.9	36
Eurasian Whimbrel	*phaeopus*	-	-	0	-	-	0	-	-	0	-	-	0	-	-	0	43.8	1.3	20
	islandicus	66.6	1.7	11	65.4	2.0	11	61.8	2.2	11	56.1	2.1	11	47.9	1.2	11	44.2	1.3	15
	variegatus	66.9	-	1	65.8	-	1	61.8	-	1	56.7	-	1	48.0	-	1	44.1	-	1
	alboaxillaris	-	-	0	-	-	0	-	-	0	-	-	0	-	-	0	42.6	1.0	6
Hudsonian Whimbrel	*rufiventris*	65.7	0.6	6	63.6	1.1	6	60.3	1.4	6	54.7	1.2	6	46.4	1.3	6	43.4	1.3	11
	hudsonicus	66.4	0.9	14	64.1	0.9	14	60.4	0.9	14	54.7	1.0	14	45.2	1.1	14	42.5	0.9	14
Eurasian Curlew	*arquata*	64.0	1.6	12	62.7	1.7	12	59.5	1.7	12	54.0	1.5	12	49.7	2.3	12	45.9	1.8	23
	suschkini	-	-	0	-	-	0	-	-	0	-	-	0	-	-	0	44.0	3.7	6
	orientalis	66.6	0.7	2	65.2	0.4	2	61.6	0.7	2	56.0	0.3	2	49.9	0.3	2	46.1	2.0	10
Spotted Redshank		65.4	1.6	8	63.3	1.5	8	60.1	0.9	8	54.9	0.8	8	50.9	1.0	8	48.3	1.2	19
Redshank	*robusta*	65.2	2.2	17	64.1	2.4	17	61.2	2.0	17	55.6	2.4	17	53.2	1.3	17	51.0	1.3	19
	britannica	67.7	1.1	4	66.9	1.3	4	63.5	1.5	4	58.1	1.3	4	53.4	2.3	4	50.2	1.1	23
	totanus	67.1	1.5	5	66.6	1.6	5	63.7	1.5	5	57.4	1.5	5	52.6	1.6	5	50.0	1.1	19
	ussuriensis	67.7	1.4	3	66.9	1.7	3	63.7	0.4	3	58.6	0.8	3	54.8	2.5	3	50.5	1.4	10
	Central Asia	68.9	0.9	4	67.6	1.1	4	65.4	1.3	4	60.7	1.5	4	55.7	0.5	4	52.1	1.7	18
	terrignotae	65.3	3.8	2	64.3	3.4	2	61.0	2.5	2	56.0	1.3	2	56.0	0.3	2	51.6	1.1	9
Greenshank		66.5	1.0	20	64.7	1.6	20	60.3	1.8	20	54.7	2.4	20	49.3	1.4	20	46.0	1.4	49
Ruddy Turnstone	*interpres*	69.1	1.5	42	67.8	1.5	42	63.5	1.4	42	57.0	1.2	42	46.2	1.5	42	43.2	1.4	63
	morinella	69.6	1.5	34	68.0	1.6	34	63.5	1.8	34	57.2	1.7	34	46.6	1.3	34	43.9	1.3	35
Black Turnstone		-	-	0	-	-	0	-	-	0	-	-	0	-	-	0	47.3	0.8	5

Annex 3

POSCON and the prediction of a bird's breeding origin or sex

Part I of this annex is a reprint of the user manual produced by D.M. van der Sluis, W. Schaafsma and A.W. Ambergen in February 1991. The programs were written by Van der Sluis and the theory was developed by Schaafsma and Ambergen. The manual is reprinted here with their permission. When using this text, please quote as follows:

Van der Sluis, D.M., W. Schaafsma & A.W. Ambergen. 1991. POSCON User Manual. A decision support system in diagnosis and prognosis. University of Groningen, Groningen.

Introductory notes

This appendix is aimed as an introduction to the use of POSCON in combination with the database of wader-measurements. Both the POSCON-programs and the set of sample files are included on the inserted floppy disk. This appendix thus consists of two parts:

I. The user manual of POSCON.

II. The *samples of measurements of breeding waders*. These samples are species-specific with varying sample sizes, populations and measured variables.

The inserted disk contains a set of DOS-based programs and the wader-database. We have not had the pretention to present a smoothly running computer program. The only aim was to facilitate the use of POSCON for predicting a bird's breeding origin or sex. The sample files can also be used in combination with other computer programs. For instance, we recently used the samples in a series of analyses to predict the population composition of waders in the Dutch part of the Wadden Sea. For this purpose, it was necessary to use a procedure from the NAG Fortran Library (NAG, 1991. *The NAG Fortran Library Introductory Guide*, Mark 15. NAG, Oxford, UK). In order to serve such a pupose, a Dbase IV file was added with all the collected data of the measurements of sexed waders.

PART I - The POSCON user manual
A decision support system in DIAGNOSIS and PROGNOSIS

D.M. van der Sluis, W. Schaafsma & A.W. Ambergen

Contents

1. INTRODUCTION

There are many instances where one likes to discuss the probability that a specific individual, given some of its characteristics, has a certain property, i.e. belongs to a certain population. Such individual-dependent probabilities can often be defined as posterior probabilities. Basic assumptions are that k mutually exclusive populations are involved and that the individual belongs to one of these k populations of more precisely that it is a random drawing from one of these populations. The posterior probabilities can then be defined by means of Bayes' theorem. However, to obtain numerical values for the thus defined individual-dependent probabilities we not only need the scores for the characteristics of the individual but also numerical values for the k prior probabilities and for the values of the k densities in the scorevector x. These values ares usually based on some estimation procedure. Accordingly there are uncertainties involved. The aim of POSCON is to express these uncertainties, especially those involved in the ultimately obtained estimates for the posterior probabilities. A nice way to express uncertainties is by means of confidence intervals. The term 'POSCON' is a contraction of '**pos**terior probabilities' and '**con**fidence intervals'. However, POSCON actually provides only standard errors of and correlations between estimates.

Sometimes, especially when dealing with normal distributions, one can define typicality probabilities characterizing the degree of belongingness of the individual to each of the k populations. Our definitions of the typicality probabilities are such that they appear as unknown parameters. Computing confidence intervals for these typicality probabilities is an important secondary aim of POSCON.

2. THEORY

2.1. General

The formula of Bayes gives the posterior probabilities as function of the prior probabilities and the values of k probability density functions for the vector of scores for the individual in study. The prior probabilities have to be given or estimated. In the case that they are estimated, the corresponding uncertainties are given by certain standard deviations. With the probability density functions usually being unknown, we need training samples from each of the k populations. In practice this often means that the cases, belonging to population h are regarded as an independent random sample. This assumption can very often be criticized. However, without this assumption nothing is possible. Training samples, together with a mathematical model for the corresponding probability density functions and statistical estimation theory, determine the estimates of the posterior probabilities. Additional theory gives the asymptotic distributions of the underlying estimators. These asymptotic results are used to propose approximations for the standard deviations of and correlation coefficients between the employed estimators. The

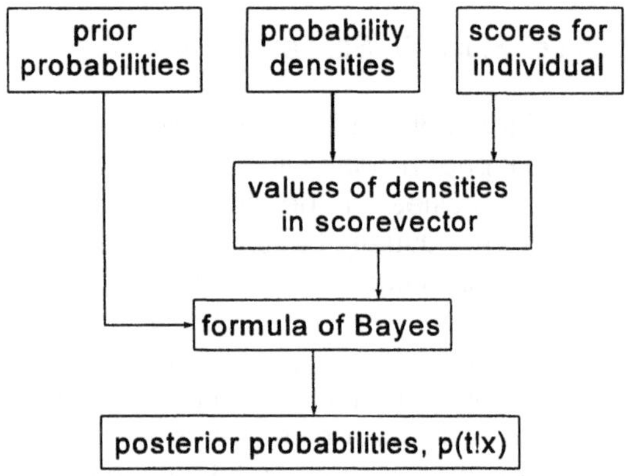

Fig. 1.

typicality probabilities are defined as functions of the Mahalanobis distance between the vector of scores and the vector of expectations. The program estimates the Mahalanobis distances and produces corresponding standard deviations. These are transformed into approximate confidence intervals for the typicality probabilities. More theory can be found in the references.

2.2. POSCON's tasks

It was remarked in section 2.1 that Bayes' theorem as represented by Fig. 1 can be used to define the posterior probability but not to compute them, because in Fig. 1 several unknown parameters appear, e.g. the prior probabilities and the probability densities. Hence Bayes' theorem has to be adapted if one wants to do more than simply replace unknown parameters by their estimates. This is the main aim of POSCON as outlined in Fig. 2. The ingredients to be used by POSCON and expressed in Fig. 2 are successively discussed.

1. The **prior probabilities**, possibly with involved uncertainties. POSCON uses fixed prior probabilities, expressed without uncertainties because they are more or less postulated, or estimated prior probabilities in which case standard deviations of estimates can be introduced. In the last case the expressed uncertainties are implemented by means of a simple fictitious experiment (in which case the user has to agree with some modification of his input standard deviations) or by means of a more complicated fictitious experiment (which may lead to a degeneracy).

2. **Expert opinions**, expressed with uncertainties. In clinical medicine it may happen that an independent expert is consulted. POSCON offers two ways to incorporate expert opinions. Way 1 is similar to the way prior probabilities are introduced together with uncertainties. Way 2 is such that the expert opinion is expressed in the form of estimated likelihoods together with corresponding standard errors.3.**Training samples** for the k populations. Handling the training samples is the task of two special parts of POSCON called CREATE and CHANGE and is discussed more extensively in § 3.4 and § 3.5. An important requirement is that the training samples should be complete. This restriction of the domain of application is in line with

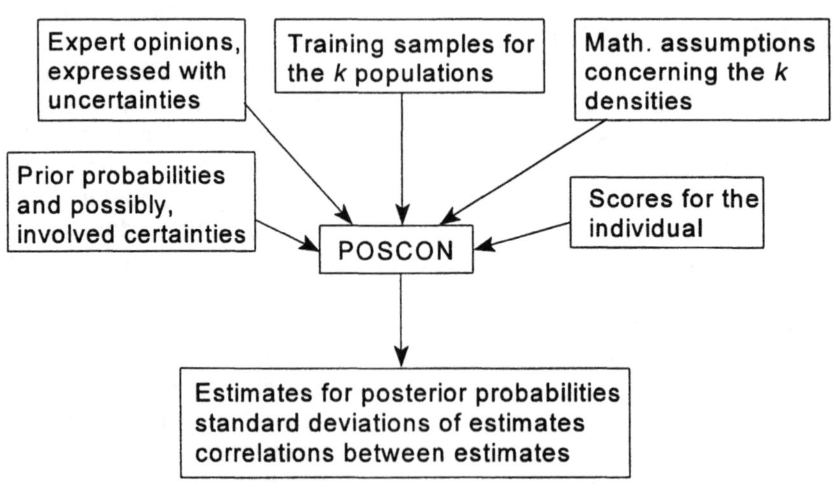

Fig. 2.

POSCON's aim to express the uncertainties in probabilities. This aim is only of interest in situations where populations are so closely related that it makes sense to require that all variables are observed for all individuals in the training samples.

4. **Mathematical assumptions** concerning the k densities. An important flexibility of POSCON is that the set of variables can be partitioned into a number of subsets. These subsets will then be regarded as stochastically independent for each of the k populations. For the separate subsets a probabilistic model has to be chosen. This can be:

DIS indicating discrete data;
NEC normal data with equal variance/ covariance matrices;
NOR normal data with unequal variance/ covariance matrices;
NON nonparametric data;
MIX mixed data.

At the moment all options have been implemented except for NON. Theory for NON is available but it is expected that NON is usually not useful because of low sample size. That is why NON has nog been and will probably never be implemented. The separate subsets are successively dealt with according to the flow diagram presented in Fig 3.

5. **Scores** of the individual. Contrary to the training samples, it is allowed that some scores are missing. The available scores are read in one run. Partitioning into independent subclasses is of later concern. Users often worry about measurement errors in the scores. If measurement errors are caused by the same random mechanism for the training samples and for the individual under study, then they are actually working with the posterior probabilities with respect to the variables as they are measured, instead of with respect to the true values of these variables. Hence they should not worry too much.

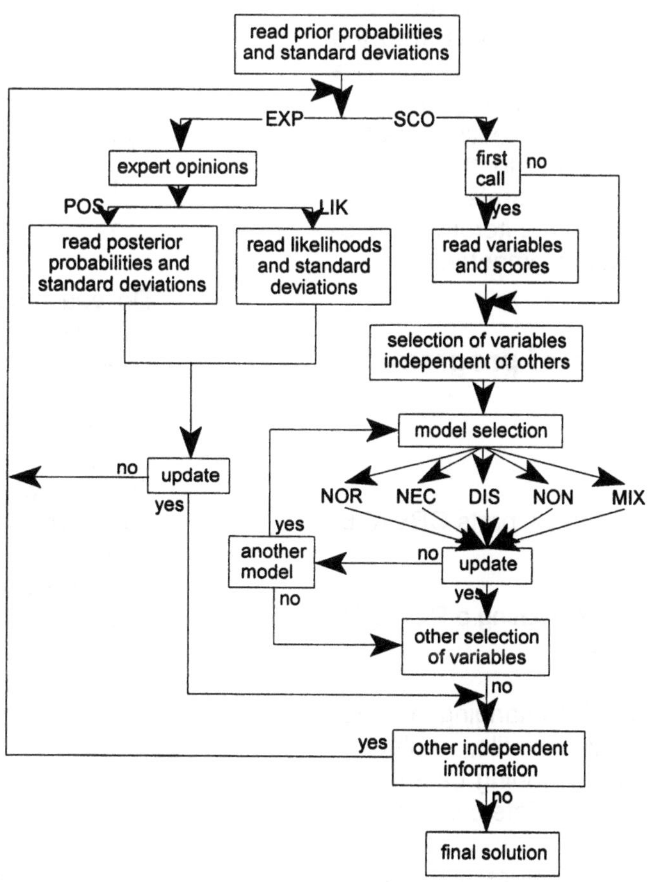

Fig. 3.

POSCON works as indicated in the flow diagram of Fig 3. The prior probabilities and, possibly, the involved uncertainties are used to make the neccessary preparations for the basic loop. Completing a loop corresponds with incorporating a piece of information which is supposed to be independent of the other pieces of information. It can consist of the optinion of an independent expert or of the scores for one of the independent subsets of the set of variables. After completing the loop, the program displays the posterior probabilities and standard errors as they were just before the loop was started and as the become after completing the loop. The user is allowed to choose between the two options: delete the information employed in this loop in which case the program continues with the posterior probabilities as they were before the loop was started, and incorparate the information which means that the posterior probabilities etc are appropriately updated.

F-tests, Mahalanobis distances, lower and upper bound of typicality probabilities, etc. are only presented with respect to the sub-loops indicated by NOR and NEC. These results refer to the selected set of variables only. The F-tests refer to testing the null-hypothesis that this individual with its vector of scores is from the same population as training sample h. The Mahalanobis distance is estimated by using the square root of the best unbiased estimator of its square. The confidence bounds for the typicality probabilities are rather inaccurate if sample sizes are small. More accurate iterative procedures have not yet been implemented.

3. Action and use of program

3.1. Survey

The program consists of three parts CREATE, CHANGE and RUN.

CREATE is used for creating a POSCON system file.

CHANGE is used for changing a POSCON system file, replacing, deleting or adding cases and/or even complete samples.

PRRUN is used for interactive evaluation of an individual's vector of scores as indicated in section 2.

POSCON is an interactive program, written in FORTRAN77 and has a dynamic memory allocation (system dependent), but can also be used with fixed memory allocation. Also saving and getting the system files is system dependent. POSCON is loaded segmentally only for a quick access and execution.

At each time the program can be finished with the keyword STOP, beginning in the first position of a line.

3.1.1. Input mode

POSCON has a short (FAST) and an extensive (SLOW) input mode. In the fast mode more than one answer can be given on the same input line and the program does not generate the corresponding requests. For example:

- the **SLOW** mode (for the unexperienced user):
  Do you want standard prior probabilities? Type YES or NO
  ? **no**
  Type prior probability sam1
  ? **.6**
  Type prior probability sam2
  ? **.4**

NO	NAME	PRIOR
1	sam1	0.6
2	sam2	0.4

  Do you have the right prior probabilities? Type YES or NO
  ? **yes**

- the **FAST** mode (for the experienced user):
  Do you want standard prior probabilities? Type YES or NO
  ? **no**
  Type the 2 prior probabilities
  ? **.6 .4**

NO	NAME	PRIOR
1	sam1	0.6
2	sam2	0.4

  Do you have the right prior probabilities? Type YES or NO
  ? **yes**

- the **FAST** mode with the **TYPE AHEAD** facility (for the most experienced user):
  Do you want standard prior probabilities?
  Type YES or NO
  ? **no .6 .4 yes**

NO	NAME	PRIOR
1	sam1	0.6
2	sam2	0.4

In the slow mode each set of data is asked for by an explicit request. If the user makes some mistake the request is repeated. If more than one response is needed by the request then the fast mode is automatically replaced by the slow mode when the user makes some mistake. All input has to be checked. Before computation the program asks for correctness of the input. If the user's answer is negative then he is asked to provide the right input.

HELP facility
POSCON has a HELP facility. At each moment a user can type a ? (question mark) afther which the program gives a short description of the expected input.

Data and or instructions can be read from file or from terminal.

3.1.2. Menu selection

For almost all input data, menu selection is applied. This means that the request will be followed by the list of all possible answers. The first time with a brief explanation, the other times without.

3.1.3. POSCON system file

Before computations are carried out, a POSCON system file has to be created. This consists of samples of individuals, each individual characterized by the scores for a given number of variables. For each variable upper and lower bounds are required to carry out certain checks. Except for the facility to introduce training samples, there is also a facility to introduce, for one particular population and instead of a vector of means and a matrix of sample covariances. This facility is of interest in some particular cases, where the above-mentioned particular population corresponds with the population of normal individuals. See § 3.3 for creating and § 3.4 for changing a POSCON system file.

3.1.4. Short input

Names and numbers of variables and samples can be used interchangeably. A sequence of names or numbers can be mentioned in a short way by giving the first and the last name, or number, linked by a dash (-). A dash can only link names of numbers. For example, the next 5 lines for input of variables have the same meaning:

var1,var2,var3,var4
var1 - var4
1, 2, 3, 4
1 - 4
1, var2 - var4

Only the first three characters of the reserved words TYPE, PAGE, SEARCH, RECODE, FILE, MEAN, DISPERSION, DELETE, VIEW, FINISH, ELSE, CREATE, CHANGE and RUN are needed.

3.1.5. Scratch files

POSCON uses a number of scratch files. Output, not essential for the computation but containing extra information will not be written on the terminal but on a scratch file. The contents of this scratch file can be made visible at any moment if the command 'Type the output parameter PAGE or TYPE or SEARCH or SAVE or END' appears, with:

TYPE - all output is written on terminal

PAGE - all output can be written on terminal in blocks of 20 lines. The content of one block is written on terminal together with the command 'Type NEXT (+M or -M) for another PAGE or END for termination'.
With NEXT the following 20 lines, if possible, are written on terminal and with END the routine PAGE will be ended.
With NEXT m a number of m-1 blocks of 20 lines will be skipped (m positive means looking forward, m negative looking backward). Default m=1.

SEARCH - the command 'Type lines a-b,c (b a) or text / xyz / or END' appears, with:

a-b,c — the lines a to b and the line c will be written on terminal
/xyz/ — all lines with the text xyz will be written on terminal
END — the routine SEARCH will be ended.

SAVE - all output will be written on file CSFIL. Later it can for example be printed on a line printer after finishing POSCON.

END - no more output from scratch file is needed. The scratch file will be destroyed unless SAVE has been specified before.

3.1.6. BCKUP file

POSCON writes all input to scratch file BCKUP and this file can be used as input file for a next run (see § 3.2). POSCON can read all the instructions from input and from file. Also in the case of creating (see § 3.3) and changing (see § 3.4), the input can be read from file.

3.1.7. Overview

An overview of all update steps, which have been done, can be written to file CSFIL and, after finishing POSCON, to a line printer.

3.2. General

POSCON begins with printing the time and the day and asks for the input mode (FAST of SLOW). Further it asks for the name of the POSCON system file and a user identification (system dependent). With CREATE a new system file will be made and with CHANGE the system file will be altered. The FILE=xxx indicates that the next instructions have to be read from file xxx. The file xxx must be 'local'. A line with text is printed from the system file to identify the file with training samples in use.

3.3. PRRUN

After a check whether the right file has been specified, POSCON asks for the number K of samples which will be used in the computations. Next the prior probabilities can be read possibly together with the standard deviations (uncertainties). If no information is given, then POSCON computes standard prior probabilities 1/K. If the uncertainties are not equal to zero, POSCON proposes to modify the standard deviations such that they are in agreement with those of a fictituous experiment. If the user rejects the proposed modification, then another much more complicated fictituous experiment is tried but this is not recommended because the computation becomes difficult and may degenereate. Next the evaluation starts, either with the introduction of independent expert opinions (EXP) or by reading a vector of scores (SCO).

EXP POSCON has two ways for introducing expert opinions, namely: estimates of likelihoods (**LIK**) with corresponding uncertainties, and estimates of posterior probabilities (**POS**) with corresponding uncertainties.

SCO POSCON reads the scores of the available variables. Unlike with CREATE and CHANGE, some missing values are allowed. An overview of the available scores with the corresponding sample means will be given and, if wished, the pooled and also the separate covariance matrices are printed. For the evaluation of a subset of the set of variables the following probability models can be chosen:

DIS for discrete data. POSCON counts the number of times the combination of scores is present in the different samples. This provides estimates of the probabilities and for the covariance matrix of the logarithms of the corresponding estimates.

NOR POSCON computes, for normally distributed data possibly with unequal covariance matrices, the estimates of the probability density functions and of the cavariance matrix of the estimated logarithms of the estimates of the probabiliy density functions.

NEC POSCON computes, for normally distributed data with equal covariance matrix, similar estimates as under NOR.

NON POSCON computes, for nonparametric data, similar estimates as under NOR (not yet implemented).

MIX POSCON computes, for mixed data, similar estimates as under NOR. If the discrete variable(s) and the continuous variable(s) are not dependent, the option MIX should be used. If there is insufficient evidence for dependence between the discrete and the continuous variables then one should avoid MIX by using a finer partitioning into independent subsets. The option MIX is very complicated. MIX works as follows. For each combination of the outcomes of the discrete variables, a subsample is made from the original training samples by collecting all individuals with the corresponding scores. With K training samples and D possible outcomes of the vector of discrete variable(s), a number of K*D training subsamples is obtained.

	1	2			D
sample 1					
sample K					

It happens very often that many subsamples contain no or too few cases. In that case the user can pool various outcome combinations of the discrete variable(S) with the option RECODE. A possible form of the RECODE instruction is:

v1,v2 - v3 (c1 = a) (c2 - c3 = b) (ELSE = c) v4 (c4 = d) END

with v1, v2-v3, v4 discrete variables, c1, c2-c3, c4 categories of the discrete variables before recoding and a,b,c,d after recoding. For variable v1 and the variables v2 until v3 the category c1 will be changed into category a, the categories c2 until c3 into category b and the other categories will be replaced by the new category c.

In this case the variable v1 and the variables v2 until v3 have -after recoding- 3 possible outcomes. The result will be:

	1	2		E
sample 1				
sample K				

with E << D

For example:
A sample s contains 2 continuous variables and 2 discrete variables d1 and d2, each with three categories and 15 cases or individuals (Fig. 4a). This means that 9 subsamples will be created (Fig. 4b).
For an individual with scores 1.3, 2.2, 1 and 3 the program uses subsample s3 for computing several parameters as the mean value and the dispersion matrix. But subsample s3 consists of only one case which is not enough for computation. With the RECODE instruction we have a way to combine categories.

RECODE d1(1,2=1)(3=2) d2 (2,3=2) END

The purpose of this recoding is to find a subset with a sufficiently large number of cases for computation. The vector of scores actually will not be changed. The result is 1.3, 2.2, 1, 2 hence the program needs subsample s22 for the computation(fig 4c, note that the individuals have been rearranged by pooling certain subsamples for fig 4b). After recoding, the program gives a survey of the transformations and of the new numbers of cases in the subsamples. Now the user can decide to go on with the obtained recoded variables or to make additional changes.

Before the computation a choice has to be made from the suboptions NOC, CLC, RWC and MXC:

NOC NO Constraints; no restriction for the covariance matrices.
CLC CoLumn Constraints; equal covariance matrices in the different columns.
RWC RoW Constraints; equal covariance matrices in the different rows.
MXC MaX Constraints; all covariance matrices are equal.

The flow diagram is presented in Fig. 5. After evaluating an expert opinion or a subset of the set of scores, POSCON displays the probabilities prior to and posterior after the possible incorporation of this piece of information. The user can decide to reject the incorporation of this

s ↓			d1	d2
	1.2	2.1	1	2
	1.3	2.1	1	1
	1.4	2.2	1	2
	1.1	2	3	1
	1.4	2	1	3
	1.2	2.2	2	1
	1.3	2.1	3	2
	1.4	1.9	2	1
	1.4	2.2	2	2
	1.2	2.1	1	2
	1.3	2.2	2	3
	1.2	2	2	3
	1.2	1.9	3	1
	1.1	2.2	3	2
	1.4	2.5	3	3

Fig. 4a

		d1	d2	
1.3	2.1	1	1	s1
1.2	2.1	1	2	
1.4	2.2	1	2	s2
1.2	2.1	1	2	
1.4	2	1	3	s3
1.2	2.2	2	1	s4
1.4	1.9	2	1	
1.4	2.2	2	2	s5
1.3	2.2	2	3	s6
1.2	2	2	3	
1.1	2	3	1	s7
1.2	1.9	3	1	
1.3	2.1	3	2	s8
1.1	2.2	3	2	
1.4	2.5	3	3	s9

Fig. 4b

				temporary		
		d1	d2	d1	d2	
1.3	2.1	1	1	1	1	s11
1.2	2.2	2	1	1	1	
1.4	1.9	2	1	1	1	
1.2	2.1	1	2	1	2	s22
1.4	2.2	1	2	1	2	
1.2	2.1	1	2	1	2	
1.4	2	1	3	1	2	
1.4	2.2	2	2	1	2	
1.3	2.2	2	3	1	2	
1.2	2	2	3	1	2	
1.1	2	3	1	2	1	s33
1.2	1.9	3	1	2	1	
1.3	2.1	3	2	2	2	s44
1.1	2.2	3	2	2	2	
1.4	2.5	3	3	2	2	

Fig. 4c

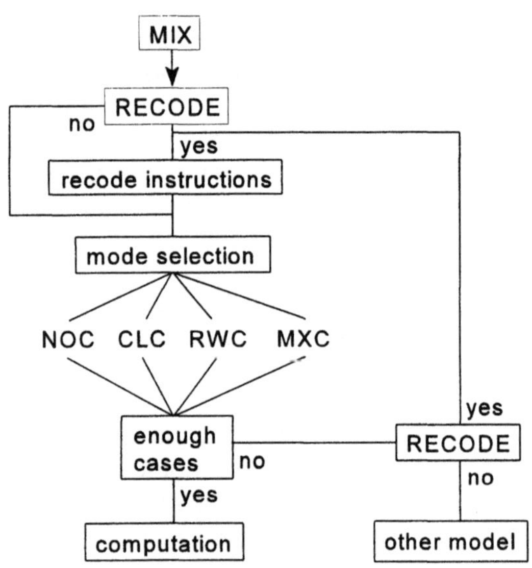

Fig. 5

information. In that case another model or another selection of variables can be used. If he decides to incorporate the information, then he can proceed by introducing new independent information. POSCON verifies that the variables to be introduced have not been introduced before. If the user decides not to introduce further information, then POSCON gives the ultimately obtained posterior probability estimates together with the corresponding standard-deviations, the correlation matrix and, if desired, an overview of all steps of the updating process.

3.3.1. Update

Independent information in the form of a set of variables or expert opinions can be 'updated' which means the information can be incorporated into the posterior probabilities and 95% confidence intervals. The result can be:

		ORIGINAL		UPDATED	
		PRIOR	STDEV	POST	STDEV
1	sam1	0.1	0.1	0.2	0.15
2	sam2	0.9	0.1	0.8	0.2

```
0-------------------------------1
 ---*------                      (.2000) sam1
               -------*-------   (.8000) sam2
```

3.3.2. Data from file

The data of the individual (case) whose probabilities have to be evaluated can be read from input (terminal) or from a file. If diagnoses have to be made or more than one individual then the question 'Do you want another attempt? Type YES or NO' is positively answered. The program asks then: 'With the same instruction set? Type YES or NO'. If the answer is yes, then:

- in the case of data on file, POSCON computes results for all the cases (individuals) on file with the same instruction set and no more questions will be asked.

- in the case of data from input, POSCON asks for the scores of the variables of one new individual. After computation POSCON asks for a new attempt.

Remark: Input of the scores of each case have to start on a new line. So the scores of the variables of two or more cases may not be written on the same line.

3.4. CREATE

The create part of POSCON generates a new POSCON system file. A line with text will be read and is printed above each output to identify the job. Next the number of samples (k), a password and the number of variables (p) will be read. The password forms the protection of the system file. Without the right password the system file cannot be changed by the program. Now POSCON reads p variable names with lower and upper bounds of the variables. The scores on the variables must have values between these bounds.

POSCON can only read samples consisting of cases with complete score vectors, missing values are not allowed. For each sample, POSCON reads the name of the sample and the scores of the corresponding cases. The scores will be separated by a comma and each case starts on a new line. The scores can also be read from a file given by FILE=<xxx>. The scores and also the sum of the variables and the matrix of crossproducts of each sample will be written on the system file. POSCON has an option to read own mean values (MEAN) and own covariance matrix (DISPERSION) per sample. The system file will automaticlly be made permanent which means that the file remains when the terminal session has been ended (system dependent).

3.5. CHANGE

The changing part of POSCON changes an existing POSCON system file. The file can only be changed by giving the right password. For the file organization it is needed that at first the number of samples to add and/or to delete will be given. The samples can be changed in the order they are placed on the POSCON system file which means that for example changing the fourth sample implies that the preceding three samples cannot be changed. New samples will be written at the end of the file. The instructions and the scores can be read from a file with FILE=<xxx>. The general form of the instruction is:

<sample name>,ADD or DELETE or VIEW,<n>
and/or FILE=<xxx>

with *n* is the number of cases, that will be added or deleted in the corresponding sample. If the sample name is not available on the system file, that sample will be interpreted as new sample which can be verified by the user. In this case only the add parameter can be used. With VIEW the program writes the scores of the corresponding sample to a scratch file which can be made visible with 'TYPE THE OUTPUT PARAMETER etc.' and can also be written to file 'csfil' for printing or saving. The keyword FINISH terminates the program.

The original system file will not be destroyed. POSCON makes a new system file which will automatically be made permanent. The original system file cannot be destroyed by misuse or faults. The user, who is responsible for management of POSCON system file, must destroy the old system files by him/herself.

4. EXAMPLES

4.1. CREATE

POSCON , Version 04-01-1990
    Date - 1990-10-10; Time - 11:27:18
    POSCON finishes with keyword STOP in first position of a line
    With a ? always more explanation will be given
Type conversation mode FAST or SLOW
? **FAST**
Type <file name>,RUN or CREATE or CHANGE and/or FILE=<...>
? **Demon,CREATE**
Type <project text>
? **Demonstration of creation of a POSCON system file**
Type <k>,<password>,<p>
? **2, password, 2**
Type 2 * (<variable name>,<min value>,<max value>
? **var1,1,10**
Type 1 * (<variable name>,<min value>,<max value>
? **var2,1,20**

Type <sample name>,<n>, and/or MEAN and/or DISPERSION, and/or FILE=<name file with data>
? **sam1,2**
Type 2 scores of case 1
? **1,1**
Type 2 scores of case 2
? **1,2**
Sample SAM1 contains 2 cases
Type <sample name>,<n>, and/or MEAN, and/ or DISPERSION, and/ or FILE= <name file with data>
? **sam2,3,FILE=dat2**
Sample SAM2 contains 3 cases

System file saved on file Demon

Type the output parameter PAGE or TYPE or SEARCH or SAVE or END
? **TYPE**

*****
***** 17 lines output
*****

Type the output parameter PAGE or TYPE or SEARCH or SAVE or END
? **END**

After computation the file 'BCKUP' contains:
FAST
Demon,CREATE
Demonstration of creation of a POSCON system file
2,password.2
var1,1,10
var2,1,20
sam1,2
1,1
1,2
sam2,3,FILE=dat2
1,1
1,2
1,2
TYPE
END

The ASCII file 'dat2' contains the next three cases:
1,1
1,2
1,2

4.2. CHANGE

POSCON , Version 04-01-1990

Date - 1990-10-10; Time - 11:27:20
POSCON finishes with keyword STOP in first position of a line
With a ? always more explanation will be given
Type conversation mode FAST or SLOW
? **FAST**
Type <file name>,RUN or CREATE or CHANGE and/or FILE=<...>
? **Demon,CHANGE**
Type <password>
? **password**
Type <new system file name>
? **Demon1**
Do you want to add new samples? Type YES or NO
? **YES**
Type <number of new samples> to add
? **1**

File Demon contains 2 variables
2 samples

For more information about system file,
Type the output parameter PAGE of TYPE or SEARCH of SAVE or END
? **END**
Do you want to delete samples? Type YES or NO
? **YES**
Type <number of samples> to delete
? **1**
Type 1 samples (names or numbers) to delete
? **sam2**

Selected samples to delete:
2 sam2

Do you delete the right samples? Type YES or NO
? **YES**

Number of new samples in system file Demon1 become
2(old) + 1(add) - 1(delete) = 2

Do you go on with these changes? Type YES or NO
? **YES**
Do you want a new project name? Type YES or NO
? **YES**
Type <project text>
? **Demonstration of changing a POSCON system file**
Type <sample name>, ADD or DELETE or VIEW, <n> and/or MEAN and/or DISPER-

DISPERSION and/ or FILE= <name file> or FINISH
? **sam1,ADD,1**
Type 2 scores of case 1
? **2,2**

Total number of cases in sample sam1 = 3

Type <sample name>, ADD or DELETE or
VIEW, <n> and/or MEAN and/or DISPERSION and/ or FILE= <name file> or FINISH
? **sam1,DELETE,1**
Type 1 case numbers to delete
? **2**

Selected case numbers to delete:
2

Do you delete the right cases? Type YES or NO
? **YES**
Type <sample name>, ADD or DELETE or VIEW, <n> and/or MEAN and/or DISPERSION and/ or FILE= <name file> or FINISH
? **sam22,ADD.3,FILE=dat22**

Create new sample sam22

Will the right sample be made? Type YES or NO
? **YES**

Total number of cases in sample sam22 = 3

Type <sample name>, ADD or DELETE or
VIEW, <n> and/or MEAN and/or DISPERSION and/ or FILE= <name file> or FINISH
? **FINISH**

System file saved on file Demon1

Type the output parameter PAGE or TYPE or SEARCH or SAVE or END
? **TYPE**

Cases in sample sam22
1.00 1.00
1.00 2.00
2.00 3.00
Total number of cases in sample sam22 = 3
--- Number of lines is 7 ---

Type the output parameter PAGE or TYPE or SEARCH or SAVE or END
? **SAVE**
--- Number of lines is 7 ---
Type the output parameter PAGE or TYPE or SEARCH or SAVE or END
? **END**

POSCON has made the file CSFIL
POSCON has made the file BCKUP with the input instructions
End of POSCON

After computation file BCKUP contains:
FAST
Demon,CHANGE
password
Demon1
YES
1
END
YES
1
sam2
YES
YES
YES
Demonstration of changing a POSCON system file
sam1,ADD,1
2,2
sam1,DELETE,1
2
YES
sam22,ADD,3,FILE=dat22
YES
1,1
1,2
2,3
FINISH
TYPE
SAVE
END

The ASCII file dat22 contains the next three cases:
1,1
1,2
2,3

4.3. PRRUN

POSCON , Version 04-01-1990
Date - 1990-10-10; Time - 11:27:22
POSCON finishes with keyword STOP in first position of a line
With a ? always more explanation will be given
Type conversation mode FAST or SLOW
? **FAST**
Type <file name>,RUN or CREATE or CHANGE

```
and/or FILE=<...>
? ?
<file name> - file name of POSCON system file
RUN         - RUN with a POSCON system file
CREATE      - CREATE a POSCON system file
CHANGE      - CHANGE a POSCON sytem file
FILE=<...>  - read the instructions from file ...
Type <file name>,RUN or CREATE or CHANGE
and/or FILE=<...>
? Hor,RUN

Project name
Urinary research

File Hor contains   2 variables
                    2 samples

For more information about system file,
Type the output parameter PAGE or TYPE or
SEARCH or SAVE or END
? Type END

Project name
Urinary research

File Hor contains   2 variables
                    2 samples

Nr   Variable  Min value   Max value
1    ANDRO     1.00        10.00
2    ETIO      1.00        10.00

Nr Sample  N of cases  Nr Sample N of cases
1   Hat     11          2  Hor    15
--- Number of lines is 13 ---

Do you have the right file? Type YES or NO
? YES
Do you need all 2 samples? Type YES or NO
? YES

Selected samples
1 Hat (11)  2 Hom (15)

Do you want standard prior probabilities? Type
YES or NO
? YES

               Hat    Hor
Probabilities  .5000  .5000

Type the 2 standard deviations
? 0,0

               Hat    Hor
Stand. dev.    0.0000 0.0000

Do you have the right standard deviations? Type
YES or NO
? YES
Evaluate expert opinion or scores; Type EXP or
SCO
? SCO
Do you need all 2 variables? Type YES or NO
? YES

Selected variables
1  ANDRO  2  ETIO

Type the 2 scores or FILE=<name file with data>
? 2,2

Measurement  Score  Hat    Hor
1     ANDRO  2.00   3.52   2.50
2     ETIO   2.00   2.10   3.24

Do you have the right scores? Type YES or NO
? YES
Do you want to display pooled covariance ma-
trix? Type YES or NO
? YES
Type the output parameter PAGE or TYPE or
SEARCH or SAVE or END
? TYPE

Measurement  Score  Hat    Hor
1     ANDRO  2.00   3.52   2.50
2     ETIO   2.00   2.10   3.24

POOLED covariance matrix

Variable   STDEV
1 ANDRO    0.84   1.00
2 ETIO     1.05   0.70    1.00
--- Number of lines is 11 ---

Type the output parameter PAGE or TYPE or
SEARCH or SAVE or END
? END

Do you want to display separate covariance
matrices? Type YES or NO
? YES
Type the output parameter PAGE or TYPE or
SEARCH or SAVE or END
? TYPE

SAMPLE covariance matrix

Sample Hat     F =  11

Variable   STDEV
1 ANDRO    0.72   1.00
2 ETIO     0.69   0.90    1.00
```

Sample Hor F = 15

Variable STDEV
1 ANDRO 0 .92 1.00
2 ETIO 1.24 0.64 1.00
--- Number of lines is 15 ---

Type the output parameter PAGE or TYPE or SEARCH or SAVE or END
? **END**
Do you continue with these variables? Type YES or NO
? **YES**
Type number of variables for updating
? **2**

Name	F HOT	P	F-P+1	P-EXC	SQMAH	STDEV	P-under	P-upper
Hat	8.3874	2	9	0.01	14.051	9.3932	0	1
Hor	0.4395	2	13	0.6536	0.66	0.6137	0.394	1

Fig. 6

Selected variables

1 ANDRO 2 ETIO

Do you have the right variables? Type YES or NO
? **YES**
Type probabilistic model DIS or NOR or NEC or MIX or NON or END
?
DIS - model with discrete data
NOR - model with continuous data with unequal covariance matrices
NEC - model with continuous data with equal covariance matrices
MIX - model with discrete and continuous data
NON - nonparametric model; not implemented
END - no more models

Type probabilistic model DIS or NOR or NEC or MIX or NON or END
? **NOR**

Model with normal data and different covariance matrices (example of output in Fig. 6):

The meaning of this output is as follows. On the basis of the model with assumptions of normality but without assumptions of equality of covariance matrices, the Mahalanobis squared distance

$$\Delta^2_{h;x}=(\mu_h-x)^T\Sigma^{-1}(\mu_h-x)$$

between the individual, in our case that with scores x=(2,2), and the unknown vector of expectations of population h, is estimated. The obvious plug-in estimate

$$D^2_{h;x}=(x_{h.}-x)^T[(n_h-1)^{-1}S_h]^{-1}(x_{h.}-x)$$

can be used as the basis for the Hotelling test-statistic F HOT which, under the null-hypothesis that x is from the same normal population as sample h, has on F distribution with:

$$p \text{ and } (n_h - 1) - p + 1$$

degrees of freedom. The corresponding probability of exceedance shows that, even at level $\alpha = 0.01$, the null hypothesis is rejected that the individual is from the same population as the 'Hat' sample. The null-hypothesis is not rejected that it comes from the 'Hor' sample.

Interpretation of F HOT can be supplemented by constructing an interval estimate for the typicality probability P (chi^2 with p degrees of freedom ≥). This interval is simply obtained by transforming the corresponding interval for . The latter interval is obtained by using a crude method: the unbiased estimator for (=SQMAH) is determined and its standard deviation (=STDEV) is estimated. Using the symmetric interval [estimate ± 2 standard deviation], the present version of POSCON can easily produce degenerate results, especially if sample sizes are small. This degeneracy is displayed in the output for 'Hat'. The confidence interval [0,1] is not reasonable at all, because the Hotelling test produced a significant result. Better results can be based on more accurate approximations for the noncentral F distribution.

		Original		Updated	
		PRIOR prob	STDEV	POSTERIOR prob	STDEV
1	Hat	0.5	0	0.0003	0
2	Hor	0.5	0	0.9997	0

```
0--------------------------------------1
*-                              (.0003) Hat
                             -* (.9997) Hor
```

The interpretation of these results is in line with that based on F HOT. This individual belongs to the population of 'Hor'. This interpretation would not change much if other prior probabilities were used. The standard errors of the posterior probabilities are not of much value in this situation because there is almost no uncertainty.

Original values to be replaced by updated values? Type YES or NO
? **YES**
Do you have more independent information? Type YES or NO
? **YES**
Evaluate expert opinion or scores; Type EXP or SCO
? **EXP**
Information of likelihoods or posterior probabilities; Type LIK or POS
? **POS**
Type the 2 posterior probabilities
? **.4,.6**

	Hat	Hor
Probabilities	.4000	.6000

Do you have the right posterior probabilities? Type YES or NO
? **YES**
Type the 2 standard deviations
? **.2,.3**

Note that in this case the diagnosis of our individual is adapted to new information in the form of an expert opinion. An expert, e.q. a psychologis, comparing the individuals psychological scores with those of his training samples, provides his opinion in the same form as the output from POSCON.

	Hat	Hor
Stand. dev.	.2000	.3000

Do you have the right standard deviations? Type YES or NO
? **YES**

Model with POSTERIOR probabilities criterion

	Hat	Hor
RHO	.4000	.6000
STD	.2000	.3000
MOD	.2500	.2500

The interpretation of this result is quite obvious. The original deviations .2000 and .3000 were not consistent with each other because the uncertainty in P is exactly as large as that in 1-P. It is quite reasonable that these inconsistent values are replaced by the ones prescribed.

Modification allowed? Type YES or NO
? **YES**

		Original		Updated	
		PRIOR prob	STDEV	POSTERIOR prob	STDEV
1	Hat	0	0	0.0002	0
2	Hor	0.9997	0	0.9998	0

```
0--------------------------------------1
*-                              (.0002) Hat
                             -* (.9998) Hor
```

Original values to be replaced by updated values? Type YES or NO
? **YES**
Do you have more independent information? Type YES or NO
? **NO**

Final result

	POSTERIOR prob	STDEV
1 Hat	.0002	.0005
2 Hor	.9998	.0005

For further information, type the output parameter PAGE or TYPE or SEARCH or SAVE or END
? **SAVE**
--- Number of lines is 11 ---
Type the output parameter PAGE or TYPE or SEARCH or SAVE or END
? **TYPE**

Final result

	POSTERIOR prob	STDEV
1 Hat	.0002	.0005
2 Hor	.9998	.0005

```
Sample      STDEV    Correlation matrix
Hat         .0005     1.00
Hor         .0005    -1.00   1.00
                        1       2
--- Number of lines is  11 ---
```

Type the output parameter PAGE or TYPE or SEARCH or SAVE or END
? **END**
For a SURVEY of ALL the steps,
Type the output parameter PAGE or TYPE or SEARCH or SAVE or END
? **TYPE**

```
***********************************
*** Survey of all the steps ***
***********************************

Project name
Urinary research

***** Updated with NORmal with diff cov

and with the variables
      ANDRO  ETIO
       2.00   2.00

            POSTERIOR       STDEV
            prob
1  Hat          .0003        .0008
2  Hor          .9997        .0008

0----------------------------------------1
*-                                  (.0003) Hat
                                 -* (.9997) Hor

***** Updated with expert opinion POSTERIOR

and with the POSTERIORS
.4000          .6000
and with the DEVIATIONS
.2000          .3000

            POSTERIOR       STDEV
            prob
1  Hat          .0002        .0005
2  Hor          .9998        .0005

0----------------------------------------1
*-                                  (.0002) Hat
                                 -* (.9998) Hor

Final result
            POSTERIOR       STDEV
            prob
1  Hat          .0002        .0005
2  Hor          .9998        .0005

Sample      STDEV    Correlation matrix
Hat         .0005     1.00
Hor         .0005    -1.00   1.00
                        1       2
--- Number of lines is  49 ---
```

Type the output parameter PAGE or TYPE or SEARCH or SAVE or END
? **END**
Do you want another attempt? Type YES or NO
? **NO**

POSCON has made the file CSFIL
POSCON has made the file BCKUP with the input instructions
End of POSCON

After computation file BCKUP contains:

```
FAST
?
Hor,RUN
TYPE END
YES
YES
YES
0,0
YES
SCO
YES
2,2
YES
YES
TYPE
END
YES
TYPE
END
YES
2
YES
?
NOR
YES
YES
EXP
POS
.4,.6
YES
.2,.3
YES
YES
YES
NO
SAVE
```

TYPE
END
TYPE
END
NO

5. References

- Ambergen, A.W. & W. Schaafsma. 1982. Interval estimates for posterior probabilities, (Case k≥2,p≥1; normality without equality of covariance matrices; nonparametric density estimation models). Report Mathematical Center, Amsterdam.
- Ambergen, A.W. & W. Schaafsma. 1984. Estimates for posterior probabilities, applications to Border Cave, p. 115-134 *In*: G.N. van Vark and W.W. Howells (eds.) *Multivariate Statistical Methods in Physical Anthropology*. Reidel Publ. Dordrecht.
- Rigby, R.A. 1982. A credibility interval for the probability that a new observation belongs to one of two multivariate populations. J. Roy. Statist. Soc. B 44 (2): 212-220.
- Schaafsma, W. 1985. Standard errors of posterior probabilities and how to use them in multivariate analysis. *In*: P.R. Krishnaiah (ed.). *Proceedings of the 6th international symposium on multivariate analysis*. Noord-Holland, Netherlands.
- Schaafsma, W. & G.N. van Vark. 1977. Classification and discrimination problems with applications (part 1, case k= 2, p= 1). Statistica Neerlandica 31: 25-45.
- Schaafsma, W. & G.N. van Vark. 1979. Classification and discrimination problems with applications (part 2a, case k= 2, p≥ 1, normality and equality of covariance matrices). Statistica Neerlandica 33: 91-126.
- Tatsuoka, M.M.. 1982. *Multivariate Analysis: Techniques for educational and psychological research*. John Wiley.

PART II - The samples of breeding populations in waders

Generally large sample sizes coincide with few measured variables per skin and small sample sizes with more measured variables per skin. This is caused by (1) variation in number of variables measured over the project period, and (2) not every variable can always be measured due to e.g. a broken bill or crooked middle toes. Statistical analyses just require the opposite: the measuring and analysis of more variables needs larger sample sizes. We thus decided to present principally 3 types of sample files per species (Table A):

- sample files with wing- and culmen lengths.
- sample files with wing-, culmen- and tarsus lengths.
- sample files with more than 3 measurements included.

However, when sample sizes in files with 2 measurements were (nearly) just as large as in files with 3 measurements, it was not needed to construct also a sample file including only 2 measurements, since within the POSCON-program variables can also be selected. The same applies to the situation when sample sizes in files with 3 measurements were (nearly) the same as in the corresponding files with > 3 measurements.

Table A. *Presentation of the sample files to be used in POSCON analysis. The total sample sizes are given between brackets. When no other measurements than wing-, culmen- and tarsus lengths were measured within a species, or when this was done with too few birds, this is indicated with the entry 'not possible' in the last column.*

Variables / Species	Wing length, culmen length		Wing length, culmen length, tarsus length		> 3 measurements	
	File name	*N*	File name	*N*	File name	*N*
Ringed- and Semipalmated Plover			rsplov3.pos	648	rsploval.pos	258
Dotterel					dotterel.pos	13
Eurasian Golden Plover			goplov3.pos	80	goploval.pos	53
Grey Plover	grplov2.pos	272	grplov3.pos	254	grploval.pos	141
Red Knot	rknot2.pos	290	rknot3.pos	230	rknotal.pos	183
Sanderling	sander2.pos	276	sander3.pos	200	sanderal.pos	143
Little Stint	lstint2.pos	32	lstint3.pos	22	lstintal.pos	14
Temminck's Stint			tstint3.pos	30	*not possible*	
Curlew Sandpiper			csand3.pos	89	csandal.pos	63
Purple- & Rock Sandpiper	prsand2.pos	372	prsand3.pos	336	prsandal.pos	171
Dunlin	dunlin2.pos	1031	dunlin3.pos	881	dunlinal.pos	528
Broad-billed Sandpiper			bsand3.pos	16	*not possible*	
Spoon-billed Sandpiper			ssand3.pos	31	*not possible*	
Black-tailed Godwit			blgodw3.pos	166	*not possible*	
Bar-tailed Godwit	bargw2.pos	302	bargw3.pos	295	bargwal.pos	191
Whimbrel			whimb3.pos	210	whimbal.pos	81
Eurasian Curlew			curlew3.pos	129	curlewal.pos	87
Spotted Redshank			spottr3.pos	70	spottral.pos	58
Redshank			redsh3.pos	373	redshal.pos	195
Greenshank			green3.pos	146	greenal.pos	123
Ruddy Turnstone	turnst2.pos	528	turnst3.pos	385	turnstal.pos	217

Table B. *Selected variables in the sample files with more than 3 measurements (last column Table A).*

File name	Species	Wing-length	Culmen-length	Tarsus-length	Secondary-length	Tail-length	Middle toe-length	Culmen-height
rsploval.pos	Ringed- & Semipalmated Plover	+	+	+	+	+	+	+
dotterel.pos	Dotterel	+	+	+	+	-	-	-
goploval.pos	Eurasian Golden Plover	+	+	+	+	-	+	+
grploval.pos	Grey Plover	+	+	+	+	+	+	+
rknotal.pos	Red Knot	+	+	+	+	+	-	-
sanderal.pos	Sanderling	+	+	+	+	+	-	-
lstintal.pos	Little Stint	+	+	+	+	-	-	-
csandal.pos	Curlew Sandpiper	+	+	+	-	-	+	-
prsandal.pos	Purple- & Rock Sandpiper	+	+	+	+	+	+	-
dunlinal.pos	Dunlin	+	+	+	+	+	+	-
bargwal.pos	Bar-tailed Godwit	+	+	+	+	+	+	-
whimbal.pos	Whimbrel	+	+	+	+	+	+	-
curlewal.pos	Eurasian Curlew	+	+	+	+	-	+	-
spottral.pos	Spotted Redshank	+	+	+	+	-	-	-
redshal.pos	Redshank	+	+	+	+	+	-	-
greenal.pos	Greenshank	+	+	+	+	+	+	-
turnstal.pos	Ruddy Turnstone	+	+	+	+	+	+	-

In most species, it also appeared to be important to include other variables than wing-, culmen- and tarsus lengths: first secondary length, tail length, middle toe length, culmen height, axillary score or upper-tail covert score (see the species chapters). Then, a third set of sample files (**AL.POS files) was constructed including the additional relevant variables. These files are presented in the last column of Table A. There were important differences between species in the relevance of the various measurements. The selected measurements per species are presented in Table B. Their selection is mainly based upon the outcome of the morphometric analyses in the various species-chapters. There is an important advantage in analysing as few measuremnts as possible: generally the sample sizes will be larger.

The distinguished samples per species(-group) are shown in Table C. Mostly, evidently different populations were distinguished, but such was not always possible when data were scarce. In case of doubt, groups were differentiated. In some cases, groups with few cases were distinguished because they were known to be different (e.g. Dunlin on Sakhalin Island) or were suspected to be different (e.g. Purple Sandpiper on Greenland). It is easy to combine the combine the calculations for different groups. The resemblance to a combined group equals the highest calculated typicality probability: when the typicality probability for population A is 0.8533 and for population B 0.4532, the typicality probability for the combination of both populations must be regarded as 0.8533.

Many sample groups include few skins. Whether this is enough to perform the calculations depends upon the number of sample groups, the number of variables included and the chosen models. Generally, there must be always 1 bird more in a sample than the sum of sample groups and variables together. Sometimes it is necessary to assume equal covariance matrices (model '**NOR**') since then calculations can be performed, whilst such would be impossible with the assumption of unequal covariance matrices (model '**NEC**'). It was decided not to include nonparametric data in the sample files because of the required large sample sizes.

Table C. *Group distinctions and sample sizes in the presented sample files (*.POS). In the first column the numbers of the samples within the sample files are given. In the second column the names of these samples are presented. A short description concerning the content of these samples follows in the third column. The last 3 columns present the sizes of the samples in the various sample-files: (1) the sample files with wing- and culmen lengths, (2) the ones with wing-, culmen- and tarsus lengths, and (3) the ones with more than 3 measurements (as detailed in Table B). The group names are mostly a combination of breeding location (first 3 characters) and sex, and since only 7 characters were allowed as a group name, 'male' stands for ♂♂, and 'fema' stands for ♀♀. In some species only ♂♂ and ♀♀ were distinguished without considering their breeding locations. In the Grey Plover, measurement-data of both sexes were combined (see species chapter).*

Nr.	SAMPLE NAME	AREA/ SUBSPECIES/ SEX	*N* with 2 variables	*N* with 3 variables	*N* with >3 variables
		Ringed- and Semipalmated Plover			
1	CGYmale	NE Canada, Greenland, Iceland; ssp. *psammodroma* ♂♂	-	109	29
2	CGYfema	idem ♀♀	-	66	15
3	WEUmale	W Europe; ssp. *hiaticula* ♂♂	-	37	15
4	WEUfema	idem ♀♀	-	21	7
5	SIBmale	N Fennoscandia, N Russia, Siberia; ssp. *tundrae* ♂♂	-	176	86
6	SIBfema	idem ♀♀	-	90	49
7	SEMmale	Semipalmated Plover ♂♂	-	91	35
8	SEMfema	idem ♀♀	-	58	22
		Eurasian Dotterel			
1	male	♂♂	-	-	6
2	female	♀♀	-	-	7
		Eurasian Golden Plover			
1	male	♂♂	-	58	37
2	female	♀♀	-	22	16
		Grey Plover			
1	ARKHANG	Archangelsk east to Yamal; ssp. *squatarola*	8	8	6
2	WPALEAR	Yamal and Taymyr; ssp. *squatarola*	78	62	25
3	EPALEAR	Siberia east of the Lena & W Alaska; ssp. *squatarola*	76	75	36
4	WRANGL	Wrangel Island; ssp. *tomkovichi*	29	29	25
5	CANADA	N Canada; ssp. *cynosuræ*	81	80	49
		Red Knot			
1	RUFmale	low arctic Canada; ssp. *rufa* ♂♂	15	15	13
2	RUFfema	idem ♀♀	23	23	17
3	ISLmale	NE Canada, Greenland; ssp. *canutus* ♂♂	79	52	37
4	ISLfema	idem ♀♀	57	39	28
5	CANmale	Taymyr; ssp. *canutus* ♂♂	17	11	9
6	CANfema	idem ♀♀	22	13	8
7	ROGmale	New Siberian Islands; ssp. *rogersi* ♂♂	8	8	8
8	ROGfema	idem ♀♀	5	5	5
9	YAKmale	Yakutia & Chukotsk ♂♂	6	6	4
10	YAKfema	idem ♀♀	2	2	1
11	ROSmale	Wrangel Island; ssp. *roselaari* ♂♂	17	17	15
12	ROSfema	idem ♀♀	13	13	12
13	ALAmale	Alaska ♂♂	21	21	21
14	ALAfema	idem ♀♀	5	5	5

Nr.	SAMPLE NAME	AREA/ SUBSPECIES/ SEX	*N* with 2 variables	*N* with 3 variables	*N* with >3 variables
		Sanderling			
1	RUBfema	Alaska & N Canada; ssp. *rubidus* ♀♀	50	49	48
2	ALBfema	Greenland & Siberia; ssp. *alba* ♀♀	64	34	14
3	RUBmale	Alaska & N Canada; ssp. *rubidus* ♂♂	62	59	53
4	ALBmale	Greenland & Siberia; ssp. *alba* ♂♂	100	58	28
		Little Stint			
1	male	♂♂	20	12	11
2	female	♀♀	19	10	3
		Temminck's Stint			
1	male	♂♂	-	21	-
2	female	♀♀	-	9	-
		Curlew Sandpiper			
1	male	♂♂	-	42	25
2	female	♀♀	-	47	38
		Purple- and Rock Sandpiper			
1	NCAmale	N Canada; ssp. *maritima* ♂♂	42	42	36
2	NCAfema	idem ♀♀	24	24	20
3	HUDmale	Hudson Bay; ssp. *belcheri* ♂♂	12	11	8
4	HUDfema	idem ♀♀	11	11	11
5	GREmale	Greenland; ssp. *maritima* ♂♂	11	11	4
6	GREfema	idem ♀♀	5	5	4
7	ICEmale	Iceland, ssp. *littoralis* ♂♂	27	26	16
8	ICEfema	idem ♀♀	19	19	14
9	ATLmale	N Atlantic, ssp. *maritima* ♂♂	98	88	24
10	ATLfema	idem ♀♀	79	71	10
11	SNOmale	S. Norway, ssp. *maritima* ♂♂	12	-	-
12	SNOfema	idem ♀♀	4	-	-
13	ROCmale	Rock Sandpiper; ssp. *tschuktschorum* ♂♂	16	16	13
14	ROCfema	idem ♀♀	12	12	11
		Dunlin			
1	ARImale	NE Greenland; ssp. *arctica* ♂♂	87	38	9
2	ARIfema	idem ♀♀	59	32	6
3	SCHmale	Iceland, British Isles, Baltic; ssp. *schinzii* ♂♂	178	144	84
4	SCHfema	idem ♀♀	138	109	61
5	ALPmale	N Fennoscandia, N Russia; ssp. *alpina* ♂♂	50	49	25
6	ALPfema	idem ♀♀	46	46	14
7	CENmale	Taymyr east to Indigirka; ssp. *centralis* ♂♂	36	31	12
8	CENfema	idem ♀♀	21	19	5
9	KISmale	SW Koryak, N & E coast Okhotsk Sea, Kamchatka, N Kurilen; ssp. *sakhalina* ♂♂	42	42	35
10	KISfema	idem ♀♀	23	23	21
11	ACTmale	Sakhalin Island; ssp. *actites* ♂♂	3	3	3
12	ACTfema	idem ♀♀	3	3	3
13	SAKmale	Chukotsk west to Chaunsk Bay, Wrangel; ssp. *sakhalina* ♂♂	53	52	22
14	SAKfema	idem ♀♀	36	35	14
15	AROmale	N Alaska; ssp. *sakhalina* ♂♂	49	49	37
16	AROfema	idem ♀♀	31	31	23

Nr.	SAMPLE NAME	AREA/ SUBSPECIES/ SEX	*N* with 2 variables	*N* with 3 variables	*N* with >3 variables
17	PACmale	W Alaska; ssp. *pacifica* ♂♂	38	38	36
18	PACfema	idem ♀♀	33	33	31
19	HUDmale	N Canada; ssp. *hudsonia* ♂♂	61	61	50
20	HUDfema	idem ♀♀	44	43	37
		Broad-billed Sandpiper			
1	male	♂♂	-	10	-
2	female	♀♀	-	6	-
		Spoon-billed Sandpiper			
1	male	♂♂	-	23	-
2	female	♀♀	-	8	-
		Black-tailed Godwit			
1	ISLfema	Iceland; ssp. *islandica* ♀♀	-	7	-
2	WEUfema	W Europe; ssp. *limosa* ♀♀	-	38	-
3	EEUfema	E Europe; ssp. *limosa* ♀♀	-	10	-
4	MELfema	E Asia; ssp. *melanuroides* ♀♀	-	8	-
5	ISLmale	Iceland; ssp. *islandica* ♂♂	-	21	-
6	WEUmale	W Europe; ssp. *limosa* ♂♂	-	60	-
7	EEUmale	E Europe; ssp. *limosa* ♂♂	-	7	-
8	MELmale	E Asia; ssp. *melanuroides* ♂♂	-	15	-
		Bar-tailed Godwit			
1	LAPmale	N Fennoscandia, Murmansk, Kanin; ssp. *lapponica* ♂♂	35	33	19
2	LAPfema	idem ♀♀	23	20	13
3	TAYmale	Yamal east to Anabar; ssp. *taymyrensis* ♂♂	46	46	29
4	TAYfema	idem ♀♀	36	36	23
5	MENmale	Lena east to Chaunsk Bay; ssp. *menzbieri* ♂♂	12	11	5
6	MENfema	idem ♀♀	16	16	5
7	ANAmale	Anadyr, south part Chukotsk; ssp. *anadyrensis* ♂♂	21	21	13
8	ANAfema	idem ♀♀	13	13	7
9	BAUmale	Alaska; ssp. *baueri* ♂♂	55	54	46
10	BAUfema	idem ♀♀	45	45	31
		Whimbrel			
1	ALAmale	Alaska; ssp. *rufiventris* ♂♂	-	21	13
2	ALAfema	idem ♀♀	-	16	12
3	HUDmale	Hudson Bay; ssp. *hudsonicus* ♂♂	-	13	9
4	HUDfema	idem ♀♀	-	12	7
5	ICEmale	Iceland; ssp. *islandicus* ♂♂	-	16	11
6	ICEfema	idem ♀♀	-	15	8
7	FENmale	Fennoscandia; ssp. *phaeopus* ♂♂	-	29	5
8	FENfema	idem ♀♀	-	20	4
9	VARmale	NE Siberia west to Yana basin; ssp. *variegatus* ♂♂	-	21	6
10	VARfema	idem ♀♀	-	17	1
11	BRImale	British isles; ssp. *islandicus* ♂♂	-	5	2
12	BRIfema	idem ♀♀	-	3	-
13	ALBmale	Kazakhstan, lower Volga, S Ural; ssp. *alboaxillaris* ♂♂	-	13	2
14	ALBfema	idem ♀♀	-	9	1
		Eurasian Curlew			
1	WEUmale	W Europe; ssp. *arquata* ♂♂	-	23	5

Nr.	SAMPLE NAME	AREA/ SUBSPECIES/ SEX	*N* with 2 variables	*N* with 3 variables	*N* with >3 variables
2	WEUfema	idem ♀♀	-	9	1
3	SCAmale	Fennoscandia; ssp. *arquata* ♂♂	-	33	28
4	SCAfema	idem ♀♀	-	24	18
5	SRUmale	S Ural, Kazakhstan; ssp. *suschkini* ♂♂	-	8	8
6	SRUfema	idem ♀♀	-	6	5
7	ORImale	Central Asia; ssp. *orientalis* ♂♂	-	14	14
8	ORIfema	idem ♀♀	-	11	7
		Spotted Redshank			
1	WPAmale	W Palearctic ♂♂	-	38	28
2	WPAfema	idem ♀♀	-	16	14
3	EPAmale	E Palearctic ♂♂	-	11	11
4	EPAfema	idem ♀♀	-	5	5
		Redshank			
1	ROBmale	Iceland, Faeroes; ssp. *robusta* ♂♂	-	67	12
2	ROBfema	idem ♀♀	-	53	13
3	BRImale	S & SE England; Netherlands; W Germany; ssp. *britannica* ♂♂	-	68	39
4	BRIfema	idem ♀♀	-	45	20
5	TOTmale	N Fennoscandia; ssp. *totanus* ♂♂	-	32	23
6	TOTfema	idem ♀♀	-	29	19
7	SRUmale	S Russia, Kazakhstan ♂♂	-	7	5
8	SRUfema	idem ♀♀	-	10	8
		Greenshank			
1	WPAmale	W Palearctic ♂♂	-	59	49
2	WPAfema	idem ♀♀	-	42	34
3	EPAmale	E Palearctic ♂♂	-	16	16
4	EPAfema	idem ♀♀	-	18	18
		Ruddy Turnstone			
1	INTmale	NE Canada, Palearctic, Alaska; ssp. *interpres* ♂♂	246	166	89
2	INTfema	idem ♀♀	163	101	45
3	MORmale	low arctic Canada; ssp. *morinella* ♂♂	70	69	46
4	MORfema	idem ♀♀	42	42	31

Table D. *Minimum- and maximum values of the measurements as used in the POSCON programs. The values are presented per species.*

EURING code	Species	Wing length	Culmen length	Tarsus length	Secondary length	Tail length	Middle toe length	Culmen height
4700	Ringed Plover	100-160	10-20	15-35	40-70	30-80	15-30	2-5.5
4820	Eurasian Dotterel	140-170	10-25	25-50	50-80	-	-	-
4850	Eurasian Golden Plover	165-210	15-30	30-50	65-90	-	25-45	3-6
4860	Grey Plover	175-230	20-40	35-60	65-100	50-95	25-45	4-9.5
4960	Red Knot	145-195	20-50	20-45	55-90	40-80	-	-
4970	Sanderling	110-145	15-35	15-35	45-70	30-65	-	-
5010	Little Stint	80-120	10-30	10-30	30-60	-	-	-
5020	Temminck's Stint	85-120	10-30	10-30	-	-	-	-
5090	Curlew Sandpiper	115-155	25-50	20-40	-	-	15-35	-
5100	Purple Sandpiper	110-160	15-50	10-40	50-80	35-80	15-40	-
5120	Dunlin	100-150	20-50	15-35	40-70	30-70	10-35	-
5140	Broad-billed Sandpiper	90-130	20-45	10-30	-	-	-	-
5150	Spoon-billed Sandpiper	85-130	15-30	10-30	-	-	-	-
5320	Black-tailed Godwit	175-250	60-140	50-110	-	-	-	-
5340	Bar-tailed Godwit	200-265	60-140	40-70	75-115	50-95	25-50	-
5380	Whimbrel	215-295	60-120	40-80	90-130	70-110	25-55	-
5410	Eurasian Curlew	200-350	50-250	60-110	100-180	-	35-70	-
5450	Spotted Redshank	150-190	40-70	40-70	70-100	-	-	-
5460	Redshank	140-190	30-55	30-65	65-105	40-80	25-45	-
5480	Greenshank	165-220	40-70	40-80	70-110	50-100	-	-
5610	Ruddy Turnstone	130-180	10-35	15-35	50-85	40-75	15-40	-

MIX
Papier aus verantwortungsvollen Quellen
Paper from responsible sources
FSC® C105338

If you have any concerns about our products,
you can contact us on
ProductSafety@springernature.com

In case Publisher is established outside the EU,
the EU authorized representative is:
Springer Nature Customer Service Center GmbH
Europaplatz 3, 69115 Heidelberg, Germany

Printed by Libri Plureos GmbH
in Hamburg, Germany